Wild Fruits: Composition, Nutritional Value and Products

Abdalbasit Adam Mariod
Editor

Wild Fruits: Composition, Nutritional Value and Products

Springer

Editor
Abdalbasit Adam Mariod
Ghibaish College of Science & Technology
Indigenous Knowledge and Heritage Center
Ghibaish, Sudan

ISBN 978-3-030-31884-0 ISBN 978-3-030-31885-7 (eBook)
https://doi.org/10.1007/978-3-030-31885-7

This Springer imprint is published by the registered company Springer Nature Switzerland AG
The registered company address is: Gewerbestrasse 11, 6330 Cham, Switzerland

This book is dedicated to the soul of my beloved wife Amal.

Preface

Although there have been articles in journals and chapters in books which have described some chemistry, composition, and health issues of wild fruits, a book devoted entirely to subjects such as composition, nutritional value, products, and medicinal uses has not previously been available.

This book presents 41 chapters on wild fruits. The first nine chapters (except Chap. 6) deal with economic contribution of wild fruits in rural people's income, sustainable rural development, and implications of wild fruits on food security and poverty alleviation. Chapter 6 describes wild fruits domestication using participatory methodology. Chapter 10 deals with wild fruits medicinal use and health benefits, as wild fruits are the potential supply of human health due to its energetic compounds that are chargeable for its various antioxidant, antidiabetic, antibacterial, antimalaria, and anticancer activities. Chapters 11–41 cover different contents from Africa to Asia to South America, investigating the phytochemical constituents, bioactive compounds, and traditional and medicinal uses of different selected wild fruits.

This book has been written to ensure that it will be of benefit to the industry, medicine, and food scientists. In addition, the book should appeal to academic scientists who require a good source of applications and a good set of references. Since wild fruits have diverse uses, it is hoped that the use of the book will not be limited to the food and medicine industry but will extend to related industries such as the pharmaceutical industry.

Ghibaish, Sudan — Abdalbasit Adam Mariod

Contents

Chapter 1
The Role of Wild Fruits in Mitigating Hunger

Izzeldien Abdelhamid Al Faki Adam

1.1 Introduction

During the last decades, many rural communities in the African countries have faced successive short rainfall and drought hits that weakened their productive capacity, and hence resulted in profound food deficit and social problems. Moreover, these environmental problems are intensified by the effects of the economic crisis and the impact of the local war on the community of the war affected regions. For Sudan, environmental disasters and afflictions such as short rainfall, locust and agricultural pests…etc., which have been held responsible for its frequent traditional economy collapse were normal features of rural life, and until recent time people were managing to cope and recover without acute suffering. The implementation of the trade liberalization policy (i.e. devaluation, reduction of government expenditure, removal of subsidization) and deterioration of term of trade against primary products have intensified the effect of harvest failure particularly at the local level. The accumulation of these factors compounded by successive epochs of drought has weakened the traditional resilience systems and the capacity of the rural people to cope with and recover from the effect of the successive harvest failure. The situation culminated in the 1985 famine and the successive food shortage periods over many parts of the country during the last decades.

Like the other traditional African societies, people in rural Sudan have been subject to profound economic and social problems and crises due to drought and local resources decline. The people reacted to these crises by introducing new coping strategies. Most of these coping/survival strategies, including resort to wild food are already known in traditional areas, and usually resort to or being adopted during the frequent short-rainfall and food deficit periods.

I. A. Al Faki Adam (✉)
University of Khartoum, Khartoum, Sudan

A. A. Mariod (ed.), *Wild Fruits: Composition, Nutritional Value and Products*,
https://doi.org/10.1007/978-3-030-31885-7_1

Collection of wild fruits is a traditional activity practiced by traditional people in rural Sudan as supplementary food and for delicious consumption; further some types of fruits became significant during acute food deficit and also a trade activity for the national and international markets. Beside that majorities of the rural Sudan rely on some wild fruits to ensure their primary health care, while in some urban Sudan access and use of some certain wild fruits is mainly for medicine or delicious consumption. Currently collection and use of wild fruit become more widespread and even more marketed in the rural markets of Sudan.

1.2 Wild Fruit Types and Uses in the Sudan

Throughout the diverse ecological zones of the Sudan a remarkable variety of wild fruit trees are grown in their own almost everywhere in the country, and most its fruits is utilized for subsistence. Even in some cases, some wild fruits trees and shrubs products trade have been well established activities not only for the local markets, but also found its way to the international markets for use in several industries. However collection and enterprise of wild fruit become more widespread and marked not only in the Sudan but in many countries in the tropical regions (FAO 2013).

However the use of particular wild fruit in the Sudan is often restricted to specific regions in which they are grown, but there is a wide range of wild fruits are used widely by both the rural and urban population all over the Sudan. The most important trees that are widely utilized by local communities for domestic consumption and sale include: *Balanites aegyptiaca* "Higlig" or "Lalob", *Ziziphes spina-Christi* (Sidir tree) "Nabag", *Hyphaene thebeica* "Dom", *Adansonia digitata* "Tabaldi", *Tamarindus indica* "Aradieib" and *Grewia tenax* "Godeim".

1.2.1 Balanites aegyptiaca (Heglig)

"Heglig" is a small tree found in Africa in more humid areas, is an important multi-use tree in the traditional sector in the Sudan where it is known as soap berry and source of traditional medications, pesticides, edible oil and livestock forage. It is widespread throughout the Sudan, spreads all over the country in a diverse ecological distribution ranging from sandy soil, heavy clay soil, to the river soils along the water watercourses. The fruit of *balanites aegyptiaca* locally known as "Lalob", Sudan is the main producer of it and consider as the most wild fruit trees that is grows in its own in almost all urban centers, including Khartoum, where is not uncommon to observe some *Balanites aegyptiaca* trees along street within the old residential areas of Khartoum and other urban centers of the Sudan. The fleshy pulp of the ripe and immature fruit is eaten dry or fresh for delicious consumption. In many parts of Sudan the *balanites aegyptica* lasted throughout the dry season, many rural people relied on revenue generated from its fruit heavily until the rain returned. The fruit usually collected directly from the trees or from the ground, in western and central Sudan the fruits are collected by females and children and sold in the local market to provide supplementary income for the family especially in the short rainfall season and harvest failure. It considered as a laxative and purgative for stomach beside other traditional medical uses for treatment of several diseases include stomach pains, diabetics, healing of wounds, jaundice. Its oil is used locally for treating headaches; traditionally it is released from the kernels by boiling nuts in several water changes at least for four several times, to remove the unpleasant taste of the kernel. Soaked kernel is then used as snacks or mixed with honey and used to increase the male sexual motivation (Elfeel and Warrag 2011).

Generally speaking, utilized fruit plays a significant role in supplementing household income in many villages of the traditional sector in the Sudan especially during the drought season. In most rural Sudan "Heglige" is a very important tree, produces edible fruits and oil, potentially a source of traditional medications, pesticides, livestock fodder as well as income generating source (Fig 1.1).

1.2.2 Ziziphus spina-christi *(Nabig Sidr)*

Sidr *Ziziphusspina-christi* is a tropical evergreen tree of Sudanese origin belongs to the Rhamnaceae family; it is an edible and medicinal forest fruits distributed in warm areas and tropical regions. The common Arabic name is Sider as it mentioned

Fig. 1.1 *Balanites aegyptiaca* "Heglig" Fruit

in the Quran, accordingly this species are highly respected by the Muslims through the Middle East (Al-Khalifa and Al-Arify 1999). In the Sudan known as "Nabig" tree, grown throughout the country regions; desert and semi desert arid parts, found abundantly in Kordofan, Darfur and central Sudan states. It is one of the most common known indigenous fruits, bally in figure; the shiny skin colour is brownish to red.

"Nabig" is one of the most valuable fruits in the rural areas; usually small boys spend most of their time in the slack season, collecting the "Nabig" usually to generate income to supplement their own spending or household income. Some adults allocate much time for "Nabig" collection as a major source of income during the season. However, due to the increased demand for "Nabig" in the urban centers and the remunerative income from it, some communities in the productive regions have been incorporating their whole household labour supply in the process of the fruit collection.

The fruit is eaten fresh or dry, but sometimes processed as sheet for sale in the local markets, and also crushed to make liquid like porridge called "Nasha" or "Madeda". Moreover, preservation of "Nabig" by jam making is also carried out by some housewives in the traditional sector for own consumption. Traditionally raw "Nabig" usually eaten by young people and school students in the urban center, currently the processed or added value products of this accessible fruit have gained considerable important among adults urban as good nutritional value and heath fitness fruit. Also in some part of the Sudan i.e. South Kordofan children crush the hard seed and eat the internal kernel of the "Nabig". In most rural Sudan "Nabig" is used as delicious consumption rather than household supplementary consumption. But in some rural of western Sudan it, supplement households consumption and constitutes a basic source of income for many poor people in times of harvest deficit or acute lack of income (Fig 1.2).

1.2.3 Tabaldi Adansonia digitata

In Sudan, the common Arabic lingua franca name of the *Adansonia digitata* is "Tabaldi" and its fruit is known as "Gungolez" or "Tabaldi" is predominant in the hot, dry regions of the Sudan, most frequently found on sandy soils and seasonal watercourses in the traditional sector of western Sudan and Blue Nile in the low grass and Savannas. In the Sudan, people have rights to collect wild food and fruits and seeds from the forests, from which they get most of their snacks or others food, no civil or customary law restrict their rights in utilizing their surrounding natural resources, but in some parts of Kordofan the "Tabaldi"Adansoniadigitata are commonly owned and protected by households as water reservouir. However, in rural Kordofan people naturally settled close to the "Tabaldi" tree indicated that it is integral parts of human settlements in that area while, other people planted it as an avenue tree for a public gathering of the village.

Fig. 1.2 *Ziziphusspina-christi* (Sidr) "Nabig" Fruit

Similar to other wild fruits in rural Sudan "Gungulaize" is usually collected/harvested as off-farm activities, e.g. after crop harvest, fruits and seeds are visually collected starting from January onwards. The fruits are usually collected by picking the capsules attached to the branches using simple implements or by hand then are smashed with a stone or other solid tools. The capsule contains very delicious white-coloured pulp. The fruit is an edible powdery pulp, it is one of the most favourite wild fruits in the county, is eaten raw commonly chewed and sucked, unripe fruits are boiled and used as salad. The fruit pulp is also processed as a refreshing drink, it dissolved in water or milk or made into "Nasha" liquid like porridge (It used for juice usually presented in the tourist sites and hotels). Porridge is also prepared by mixing the acid pulp with milk. In recent years, processing and use of "Gungulaize" in the Sudan has increased and more and more it has been processed at a commercial scale in form of packed powder or pulp in the urban supper markets (Abdel Galil 1996). Indigenous people also use the early matured leaves at the start of rainfall for cure of some diseases. The fruits are usually stored for long time under dry conditions (Fig 1.3).

1.2.4 Hyphaene thebaica *(Doum)*

Hyphaene thebaica is a desert palm tree with edible oval shape fruit, locally in the Sudan known as "Doum", originally native to the Nile valley, but it founds across the Sahel countries. The tree usually grows in groundwater soil, but also found in

Fig. 1.3 *Adansonia digitata* "Tabaldi" Fruits

oases and wades besides that it occurs on salty soils on river and watercourses banks and on the mountains slopes throughout Sudan (Kees 1995). In some parts of Eastern Sudan particularly in Dordaib the Doum Hyphae is an Indigenous tree fulfills many subsistence and economic needs in the district, it provides a range of products which contribute to livelihoods of community, particular importance is its income-generating capacity through the sale of the processed fruits and the use of palm leaves for basket production.

The fruit has a brown external fleshy cover which is normally chewed and spewed out; in urban Khartoum usually, people soaked it in water until the pulp becomes soft and syrupy is much enjoyed by children and adults. The internal unripe kernel can be processed and eaten, but it gets hard and inedible when it is ripped. The external coat of the fruit usually cut off in slices or processed to form powder for further processing and uses i.e. a flavouring additive. The fruit is used in different forms and have a wide use in folk medicine in everyday life of local population (El Gazali et al. 1987) (Fig 1.4).

1.2.5 Borassus aethiopum *(Daleib)*

The"Daleib", *Borassus aethiopum* is a strong to massive tree with a very single trunk of about 50 cm diameters commonly known as African fan palm, it grows up to 30 m high. This palm grows in riverine forests and savannas particularly on sandy or alluvial soils (Bayton 2007). The fruit of the tree is massive with ovoid shape and

Fig. 1.4 Hyphaene thebaica Fruits "Doum"

orange colour at maturity. In the Sudan found in South Kordofan, is known as "Dalieb" and consumed in both rural collection districts and large urban centers, however the consumption pattern of the fruit is more diverse in the rural area. The fresh fruit contains yellow juicy fibrous flesh and have a nice smell, in rural of South Kordofan and Darfur circumstances people put in the ground the fruit seeds in large pits for more than 5 weeks to germinate. The germination shoot locally known as "Halook", it boiled and eaten as delicious snacks which is most favoured and widespread in South Kordofan (Fig 1.5).

1.2.6 Grewia tenax *(Gudaim)*

Grewia tenax "Guddaim", is one of most valuable wild shrub in the Sudan, it is largely spread in arid and semi-arid areas in the country in rocky and semi-dessert areas, at the Northern and Middle of Sudan. "Guddaim" almost is a 2 m tall shrub, its fruit is red to orange colored with 2-3-4 spheroid lobes (FAO 1988). The fruit collected from the shrub usually by women and children, especially in the traditional sector of the Sudan. In the past the fruit was collected in the production area mainly for luxury consumption, currently it gained good reputation among the Sudanese as favour iron transfer, consequently the demand for it has increased while its price has soared up very significantly. In some parts of Sudan especially in Kordofan, the traditional "Guddaim" juice is made by soaking the fruits for long time, pressing and purifying, then it either sweetening and gets ready for use or

Fig. 1.5 *Borassus aethiopum* "Daleib" Fruits

processed as Nesha by the addition of sugar and other additive. Traditionally the Nesha usually prepared for lactating mothers to improve their health and lactation. Guddaim is most favorite wild fruit among the Sudanese may eaten ripe as luxury consumption or kept for later usag, however currently it has been extensively consumed by both rural and urban population for its nutritional value (Fig 1.6).

1.2.7 Tamarindus indica

Tamarinds indica, "Aradeib" tree grows in the tropical region across African countries, in Sudan, it is located in the central Sudan and the traditional sector extends to the South, locally known as "Tamor-Hindi", meaning date of India. The ripe fruit is edible, hard green pulp of early matured fruit somewhat sour but palatable eaten fresh or processed. This type of wild fruit is typically sweet and sour in taste and even get sweeter as it ripen and stored for long time. However the fruit is widely used in the cafeterias and hotels around the country, while the international hotels and cafeterias in Khartoum used to present fresh tamarinds, the local people usually prefer the juice of long stored fruits which is commonly presented in the local cafeterias and restaurants (Hassan 2006).

However, "Tamor-Hind" or "Aradeib" is a fruit that is popular in Sudan used to make juice which is used almost in all Sudanese households especially during Ramadan the fasting month. There are other several local uses of the "Tamor-Hind" such as seasoning or spice, in desserts as a jam and as traditional medicine specially treatment against malaria in the rural areas. "Tamarind" fruit is an important source

Fig. 1.6 *Grewia tenax* (Godeim)

of income for many people in both rural and urban areas in the country, is sold on local markets in the Sudan as well as on international markets (Fig 1.7).

1.3 Impact of Wild Food on Local Household Food

In the arid and semiarid land of the Sahel drought is a frequent natural disaster, which often causes frequent large food and resources deficit, famines and other economic and social problems. Food insecurity and malnutrition affect much of the rural African communities' population, especially in the sub- Saharan Africa (FAO 2011). The degree of vulnerability of the Sahel communities to the negative impacts of drought varies from country to another according to climate and prevailing environment and national policy in the country concerned. Moreover, the livelihood systems of the Sahel communities rely mainly on the rain fed cultivation and livestock, which are seriously affected by drought and short rainfall. Thus, drought

Fig. 1.7 *Tamarindus indica* Fruits

usually causes severe economic problems and affects the ability of communities to maintain their normal livelihoods. Rural communities that lack resilience or capacity to offset the resources deficit, usually tend to rely on a range of coping strategies to mitigate food shortage. These coping mechanisms usually include set of mechanisms and environmental adaptation aim at creation and improvement of the use of existing resources to maintain lowest level of sustainable existence. Wild fruits are known to make important contributions to livelihoods and hunger mitigation during acute food deficit periods in most sub-Saharan and Sahel countries. One way of diversification of food sources during drought and acute food shortage among these communities is the resort to wild fruits. There is, however many evidences emphasized the importance of wild fruits in contributing to rural livelihoods, mitigation of hunger and supplement people's income during periods of food shortage (Christine and Hannington 2016; Kabiru et al. 2015; Jatau 2008). Wild fruit trees play very significant role in most dry regions where harvests failure often cause food shortage and malnutrition of the local population (Maxwell 1991). Wild fruits provide a livelihood support system for many rural communities in terms of food, medicines, income and employment.

Collection of wild fruits and food in rural Sudan play significant role in mitigating hunger as substitute for some food items or supplementary consumption, moreover it constitutes the most important income generation activity for rural people in the traditional sector during the slack period. The process of wild fruits collection usually fills the slack period gap where farmers badly seek supplementary income generating activities. Furthermore, fruits gathering job in the dry seasons provides alternative opportunities for household labour supply to earn money.

In fact, due to successive epoch of drought caused by climate change and civil conflicts in Kordofan and Darfur in general and other parts of the Sudan many rural people in these regions have been diversifying their livelihood systems so that they have alternative resources or enough income to maintain subsistence. This situation is particularly severed during the short rain periods and acute food shortage that people move to where they can collect wild fruits to maintain subsistence or sell for additional income. Yet these coping strategies constitute the main buffering mechanism through which most households offset food deficit and maintain the subsistence during acute food shortage periods in the traditional sector of Western Sudan. The wild food and fruit consumption varies from one area to another according to several factors including traditions and social stigma, food deficit or famine intensity and surrounding environment. However, there are many evidences indicated that wild food consumption is a function of income level in the drought times that is no one of high income group reported consumption of wild fruits or food hunger in famines and acute food deficit times (Al Faki 2006).

Most of the wild fruits are collected from natural forests and nearby resources, some of the fruits such as Higlig/Lalob from Balanitesaegyptiaca or Tabaldi from Adansoniadigitata are also collected from farm localized trees or natural garden which are owned by producers.

Collection of wild fruits usually take place in the slack period, some wild fruits, e.g. Tabaldi, Gudaim, Higlig and Daleib supplement people's income while certain wild fruits e.g. Kurrssan supplement people's consumption as substitute for stable food at times of food scarcity. People have rights to collect the wild fruits from the forest and nearby bushes; no civil law or customary regulations restrict them in practicing these rights as far as its communal ownership.

Some wild fruit are typically seasonal and usually in rural Sudan be utilized in the situation of acute food shortage only that to serve as main substitute for the stable food, especially in states of drought, harvest failure or other natural afflictions causing food deficit. The list of these traditional wild food and fruits is fairly lengthy, but the most important one, which maintained many poor people during the famines of 1985 and 1991, is "Mukhiet" fruits, which sometimes called "Kurssan", the seeds of genagalensis shrub. "Kurssan" is already known in Sudan, but since 1984–85 famine have been playing significant role in mitigating food deficit among poor families during food crisis. According to Dewaal1 the wild food mainly Mukhiet fruits impact in mitigating the initial effect of harvest failure in 1984, 85 was greater than food relief (De waal 1989). Processing of "Mukhiet" fruit for consumption takes long time, soaked in several water exchanges over long period before being edible. It is either mixed with grain or use in its pure form. Most of the people who used "Mukhiet" during drought reported dilution of grain with "Mukhiet" cereal. The "Mukhiet" seed are very small and its shrub spread over wide area, therefore, its collection requires long marching for long periods. Its collection was not confined to own or immediate consumption, stocking and sale of "Mukhiet" is reported post 1985 famine in different parts of North Darfur and Kordofan (De waal 1989).

However, with the successive drought hits, and climatic changes, in the Sahel zone some writers have assumed greater role for wild food and fruits particularly "Mukhiet" within Darfur economic and environmental crisis Elbashir Hamed has pointed out that:

> …and God knows if it will one day be calculated as part of the regional Gross Domestic Production by economists. (Elbashir 1993)

It seems that such analysis stems from the investigation of the traditional coping strategies in their initial process of formation rather than deep investigation of the potential capacity and prospect of human interaction with the environment. Nowadays collection of wild food and fruits in Northern Darfur has been confined to delicious and supplementary consumption and sale in the urban centers. The people of Darfur have been able to solve the problem of frequent food shortage in the area in more dignified way, mainly expansion into Wades and clay soil cultivation.

One way of help to full the hunger gap of food during drought and acute food shortage in most parts of rural Sudan particularly in the traditional sector of Kordofan is the resort to wild food and fruits. (Elmola et al. 2015) reported that wild fruits in West Kordofan state provide significant source of income and secure alternative food source for many communities in the state moreover, it contains vital food ingredients which are essential for health. Beside that some fruits have flourished some local markets in state very significantly.

In Southern Kordofan state collection of wild fruits is an old established tradition, the rich forests of the region support collection of a wide variety of forest fruits. The easy and economic access of wild fruits counterbalances the low land productivity and prevalence of poverty intensified by the civil war. Most of the wild fruits of all types are normally eaten fresh for delicious consumption, but some are most preferable over the other edibles. The Nuba of SouthKordofan were found to appreciate some edibles over the others (such as A. digitata, B. aegyptiaca and Z. spina-christi) in their utilization as both supplementary food and cash source. The hypocotyl of the newly germinated seeds of Borassusaethiopum, locally named "Halook" is highly popular in South Kordofn as traditional food usually eaten with roasted ground nut or boiled. As to economic important (Salih and Ali 2014) reported that most of the rural families in South Kordofan are wild fruits collectors to supplement household consumption or sale for cash as well. Wild fruits role in poverty reduction among some Nuba mountain population is very significant, the income derived from the wild fruit is of particular importance to the poor household. In some parts of the state the contribution of wild fruit in the household budget is very significant amount to more than 75 percent of the total annual household income (Adam 2011).

Some wild fruits are highly marketable especially the ones that have high edible preferences such A. digitata, Z. spina-christi and B. aegyptiaca which contribute very significantly to households trade and income in the area. However, under the current situation, of the prevailing armed conflict and displaced people and closed districts in the Nuba mountain it is most likely that the wild edibles is increasingly playing vital role in securing food for the blocked population.

In some parts of the traditional sector of Western Sudan, mainly the far south of South Kordofan people usually practice subsistence economy with limited cash crops production and livelihoods options or employment opportunities. The ecological zone of the area is rich with diverse wild fruits trees, hence the fruits are collected for households' consumption needs and incomes generation. (Ibrahim et al. 2014) reported Wild fruits collection is the sole source for income generating in Abyii, Elsalam, Elsanout, Lagawa, Elmerum and Keilack localities during dry seasons.

> ... wild fruits can be used as substitutes of seed grains to reduce hunger and alleviate poverty... It was also shown that fruits returns contribute to family daily expenses and education fees. Moreover, it supported rural people food security, enterprises and met environmental objectives. (Ibrahim et al. 2014)

The role of wild fruits in hunger mitigation in the central Sudan and the other irrigated livelihood systems in the country is very limited due to the round year cultivation and existing of other sources of income. However, wild fruits in these areas are extensively used as delicious consumption includes "Gambil" *Cordiaafricana*, "Homeid" *Sclerocaryabirrea*, "Medaika" *Ximeniaamericana*, and "Joghan" *Diospyrosmespiliformis*. Some fruits are very popular, usually harvested and eaten by people as 'snacks' like "Joghan" *Diospyrosmespiliformis* and "Tamarhindi" *Pithecellobiumdulce* (Jens et al. 2002). The list of survival strategies adopted by food insecure households in the rural areas of White Nile State Central Sudan are vary from least severe to most severe including substitute food items, economization on food use and meals frequency and resort to food or cash borrowing. In the dry season indicated that rural people with lowest income levels in the White Nile of Central Sudan depend heavily on most severe coping strategies wild food and fruits to mitigate food shortage (Ahmed et al. 2011).

1.4 Conclusion

Sudan has numerous natural resource rich in edible food with many varieties of wild fruits and seeds as well as medicinal herbs, while remarkable sectors of the population, particularly in the traditional sector are involved in the collection and trade of wild fruits. The physical and economic accessibilities of such wild fruits make it a part of the normal diet of many people and of great importance during agricultural crops off seasons as well as coping with the adverse food conditions. Rural people in the Suda used to adopt various coping strategies to mitigate hunger during famine and food deficit periods based mainly on resorting to use of wild fruits as a substitute for some stable food as we have seen in Darfur people were tending to substitute millet cereal by "Mukhiet" cereal. However, wild fruits use and consumption are well recognized even during normal times when cultivated food is bandanas especially, as a free source of delicious consumption for children. The use of wild fruits as delicious consumption, i.e. snacks, juice, flavour has been becoming more familiar to the urban population in the Sudan.

Wild fruits in the traditional sector of Sudan constitute a substantial economic opportunity to the poor people, particularly women, in general sale of wild fruits plays significant role in supplementing their income to meet basic needs especially when other incomes, particularly agriculture, are out of reach.

Furthermore, traders and vendor chain of transaction of these products has been increasing as both rural and urban markets grow due to the increasing demand locally and abroad for wild fruits for consumption and medicinal purposes.

References

Abdel Galil A (1996) Evaluation of Baobab (Gonglase) solution for home management of diarrhea in Sudan children. Dissertation, University of Khartoum

Adam Y (2011) Contribution of local-level trade in non-timber forest products to rural development in Rashad locality of Nuba Mountains, Sudan. Dissertation, Technical University of Dresden

Ahmed O, Hamazi T, Hassan K, Siddig A (2011) Coping strategies as a food insecurity indicator in Sudan: the case of White Nile Stat. On line research paper, Research gate. Available https://www.researchgate.net/publication/235663027

Al Faki I (2006) The response to drought and coping strategies in semiarid areas: the case of North Darfur rural population in western Sudan. Dissertation, University of Khartoum

Al-Khalifa A, Al-Arify IA (1999) Physiochemical characteristic and poll spectrum of some Saudi honeys. On line research paper, Scientific.com/Books&Journal. Available www.sciencedirect.com/science/article/pii/S0308814699000965,

Bayton BP (2007) A revision of Borassus L (Arecaceae). Kew Bull 62(4):561–585

Christine O, Hannington O (2016). Analysis of the fresh pulps of Borassus aethiopum Fruits of Gulu District, Uganda. On line Journal, Sciepub.com/AJFN. Available http://pubs.sciepub.com/ajfn/4/6/6/index.html

De waal A (1989) Famine that kills, Darfur, Sudan. Clarendon press, Oxford, pp 1984–1988

El Gazali G, Bari EA, Bashir AK, Salih AM (1987) Medicinal plants of the Sudan. Part II: The Eastern Nuba mountains. Khartoum University Press, Khartoum

Elbashir H (1993) Coping with poverty or famine? Some dynamics in the pursuit of survival in rural (1985–1992). SCF, Khartoum

Elfeel A, Warrag E (2011) Uses and conservation status of Balanites aegyptiaca (L.) Del. (Hegleig Tree) in Sudan: local people perspective. Asian J Agric Sci 3(4):286–290

Elmola S, Ibrahim H (2015) Contribution of wild fruits to household income and food security among small scale farmers in West Kordofan State – Central-West of Sudan. On line research paper. Available http://sustech.edu/staff_publications/20151102082829176.pdf

FAO (1988) Traditional food plants: a resource book for promoting the exploitation consumption of food plant in arid, semi-arid and sub-humid lands of Eastern Africa. FAO, Rome

FAO (2011) How does international price volatility affect domestic economies and food security? On line report, fao.org/The state of food insecurity in the world. Available URL: http://www.fao.org/docrep/014/i2330e/i2330e00

FAO (2013) International Conference on the Role of Forests in the Food Security and Nutrition. [FAO document], Rome

Hassan S (2006) Production and quality evaluation of vinegar from tamarind (Tamarindus indica L.) fruit pulp. Dissertation, University of Gezira

Ibrahim S, Maruod M, Mohmoud, T, Makeen M (2014) Contribution of Wild fruits to household income, expenditure and livelihood in western sector localities of South Kordofan state, Sudan. Online research paper, ijern.com. Available http://www.ijern.com2014/October-2014/05.pdf/journal/

Jatau DF (2008) Profitability assessment of Borassus aethiopum (Mart) marketing in Adamawa State. Nigeria J Agri Soc Sci 4:159–164

Jens G, El-Siddigb K, Eberta G (2002) The potential of under-utilized fruit trees in Central Sudan. On line conference paper, Tropentage, de/Deutscher Tropentag 2002. Available http://www.tropentag.de/2002/abstracts/full/42.pdf

Kabiru U, Bello A, Badaru M, Sirajo M, Lawal H, Nasiru S (2015) Nutritional and antinutritional profile of Borassus aethiopum Mart (African palmyra palm) shoots. IJSBAR 24(3):39–49

Kees V (1995) A field guide to the identification, propagation and uses of common trees and shrubs of dry land Sudan. SOS Sahel International UK

Maxwell S (1991) To cure all hunger: food policy and food security in Sudan. Short Run Press, London

Salih N, Ali A (2014) Wild food trees in Eastern Nuba Mountains, Sudan: use, diversity, and threatening factors. J Agric Rural Dev Trop Subtrop 115(1):1–7

Chapter 2
Wild Fruits and Sustainable Rural Development

Eatimad Babikir and Ahmed Abdel Aziz Ahmed

Abbreviations

NTFPs	Non Timber Food Products
RD	Rural Development
RESAP	Red Sea Area Program
RSS	Red Sea State
SARDF	South African Rural Development Framework
SRD	Sustainable Rural Development

2.1 Introduction

The term "wild" is linked with plant species that grow spontaneously in localized natural or semi-natural environments and can exist independent of direct human action (Vernon 1999). The term is sometimes used for exotic, underutilized, or neglected and less known plants. Wild natural products contribute to energy by supplementing more calories to diets and offering quality attributes to enhance as well as providing micronutrients to rural population (Bell 1995; FAO 2005).

Human being, in their development, started by collecting what is available in nature from food items. Through the course of evolution, customary habits of utilization were introduced. Modern agriculture and urbanization have eroded a great part of local knowledge (Dweba and Mearns 2011).

Local populations have adapted their livelihood in order to maintain their food security and nourishment habits, in this context a sizable number of wild plants received attention through systematic conservation, propagation, processing and marketing. Their dispersion, preservation, mode of collect by local people and ideal utilization requires regional efforts.

E. Babikir (✉)
Ministry of Health, Food Safety Laboratory, Port Sudan, Sudan

A. A. A. Ahmed
Red Sea University, Port Sudan, Sudan

A. A. Mariod (ed.), *Wild Fruits: Composition, Nutritional Value and Products*,
https://doi.org/10.1007/978-3-030-31885-7_2

The nutritional qualities of some of these wild plants which were rich in their carbohydrate sugars, proteins, fats and some other qualities related to aroma and taste were used during food shortage (Saka and Msonthi 1994; Saka et al. 2002; Kwesiga et al. 1998; Kwesiga et al. 2000).

2.2 Wild Fruits Security Contribution in Food and Nutritional

The limited access of poor families to wild fruits, remain as key factor for under nourishment in developing countries (Andersen et al. 2003; Adeboye and Phllips 2006). According to Belcher et al. (2005); Narendran et al. (2001); Scherr et al. (2004), the survival and coping mechanisms of local inhabitants emanated from the significant added value of these wild plants as part of the food chain for rural areas (McSweeny 2004; Takasaki et al. 2004; Mojeremane and Tshwenyane 2004; Getachew et al. 2005; Redzic 2007).

Household subsistence and economy can be enhanced through production, utilization and sales of wild fruits which act as cushioning effect for socio-economic parameters (Mahapatra et al. 2005; Bharuch and Pretty 2010).

Other factors impacting the utilization and significance of Non Timber Food Products (NTFPs) (wild natural products) to households in their tribal groups: Ethnic groups vary by their social strata, territorial provenance and history as well as their livelihood generating activities. Besides, they appear to utilize certain plant parts with respect to alimentation, vitality supply and pharmaceuticals (Bussmann et al. 2006).

The South African Rural Development Framework (SARDF) the concept of Rural Development (RD) refers to processes of improving living conditions and infrastructure of rural areas. On the other hand, sustainable rural development (SRD) attempts to maximize the influence of rural communities with regard to the planning and execution of their development goals with the ultimate creation of awareness and confidence to manage their own affairs and development goals (Krzywinski and Holton 2001). RD concentrates on the diversification of activities related to community objectives which enhance competitiveness of their living and non-living resources of agriculture and forestry, land management and, at the same time taking into account the uniqueness of its, environment, quality of life and continuation of nature's diversity and culture for the continuation of generation (Harris et al. 1982). RD, also includes, the translation of the community priorities through assessment and analysis of the economic, social and environmental strategy within national priorities.

In rural areas, factors affecting the fate of a plant species survival and continuity require an understanding of the intimate relationship between the land, its vegetation, and the populations surviving on its resources and the cultures surrounding them. Rural populations, often need to adjust and cope with the changing nature and exploit the advantages for their continued presence as part of the overall adaptations for life.

Tree population's survival is correlated with a complex of edaphic and environmental factors of which rain fall pattern and amount is the most important. The tree cover may take contrasting effect if factors such as elevation and high land formation, exists within an ecological set which dictate different plant species availability and regeneration. In general, wild tree distribution depends on their establishment and maintenance and the availability of ample climatic resources. The vegetation responds to the limiting factors through adaptations which limits survival and reproduction of each individual plant and also dictates its zonation.

In nature, however, there is a definite vegetation zonation related to the cross section land undulations.

2.3 Red Sea State (RSS) Profile

The Red Sea State is located between latitudes 16° and 22° north and longitudes 35° and 37° east in the extreme northeastern part of Sudan. As an administrative unit, the RSS occupies an area of 218,887 km^2 (10% of the area of Sudan) with a total population of 1.7 million persons, representing 3.4% of Sudan population (reported from last Red Sea State census of 2008). The Average population density is about 7 persons per km^2, but it considerably varies between the localities of the state (Ahmed 2013). Physically, most studies classify the Red Sea region in to the broad physical units roughly parallel to the coast. The nature of each of these units as well as how humans interest with it, has its implications on the land use and the production systems. These units are:

(a) The Red Sea shore which extends as a demarked plain from the coast to the hill premises (20–30 km wide). Soils are composed silt and sand with gravel that gets coaster towards the fool of the hills. As narrow strip of recent coral reef deposits bound the plain from east, extending up to 2-km inland.
(b) The Red Sea Hills which has been classified as being developed by the extension of Great Rift Valley to East Africa. It extends for approximately 500 and 1200 meters above sea level. The base rock is predominantly pre-Cambrian Basement Complex and volcanic and as such is a very poor aquifer. This limited the amount of underground water and made recharge occurs only with exceptionally high rainfall, and *Khors* not helped with topography, it normally occurs at the bottom of the that intersect the mountain range, and
(c) The western plain is composed mainly of superficial deposits of alluvial and, mostly, Aeolian origin. The area is underlain by the basement complex rock and the surface dominated by scattered rock outcrops, gravel and sands which increase considerably toward the west.
(d) In the state, the status of cultivation and distribution of the various species of edible woody plants within the area or farming systems, are related to their various uses and roles in providing food. For instance, important fruits and vegetables are often located in Wadies crisscrossing the area near homesteads. According to (Knut) the vegetation cover which exist in highland areas of

Red Sea State (RSS) include Adansonia digitata, Annona squamosa, Grewia tenax (Guddaim), Ziziphus spina-christi, Balanites aegyptiaca, Detarium microcarpum, Nopal cactus (Opuntia ficus-indica), Antidesma montanum or Antidesma angustifolium, Calamus manillensis, Debregeasia longifolia, Azanza garckeana L, Hyphaene thebaica Doum (Fig. 2.1), Borassus flabellifer L Palmyra palm (Daleeb), Ceratonia siliqua (carob), and Vaccinium myrtoides. Away from highlands and particularly in Tokar Delta, Khor Langeb, and khors Arab and Derudeb, species such as Tamarix aphylla and Hyphane thebaica grow in deep soil with easily accessible ground water.

(e) These zonal variations dictate the types of wild fruits which can flourish and adapt to changing environmental conditions. In particular, trees figure prominently in several of our studies. They are not only the most visually prominent form of vegetation; they also set lower limits for the survival of pastoralism and thus become a reference point for important social values. There is clear evidence of recent deforestation in some areas.

These units are:

(a) The coastal plain extending from the coast to the foot of the Red Sea Hills (20–30 km wide) covered by slit, sand and gravel that gets coaster towards the fool of the hills. As narrow strip of recent coral reef deposits bound the plain from east, extending up to 2-km inland.

Fig. 2.1 *Hyphaene thebaica* (Doum)

(b) The Red Sea Hills which represent the western edge of the Great African Rift Valley. It extends for approximately 500 and 1200 m above sea level. The base rock is predominantly pre-Cambrian Basement Complex and volcanic and as such is a very poor aquifer. This limited the amount of underground water and made recharge occurs only with exceptionally high rainfall, and Khors not helped with topography, it normally occurs at the bottom of the that intersect the mountain range, and

The western plain is composed mainly of superficial deposits of alluvial and, mostly, Aeolian origin. The area is underlain by the basement complex rock and the surface dominated by scattered rock outcrops, gravel and sands which increase considerably toward the west.

These zonal variations dictate the types of wild fruits which can flourish and adapt to changing environmental conditions. In particular, trees figure prominently in several of our studies. They are not only the most visually prominent form of vegetation; they also set lower limits for the survival of pastoralism and thus become a reference point for important social values. There is clear evidence of recent deforestation in some areas.

2.4 Population Dynamics Ethnicity and Tribal Identity

The physical, practical and emotional capacity gained through this process has, over time, come to constitute an integral part of its cultures. Those cultures are, however, formed not only by external natural forces but also by the inner force of ideas embodied in myths, beliefs and traditions as well as by the transformations induced in them when natural conditions alter or people change their habitats. One of the predominant influences of local livelihoods is that of tribal identity. According to (knut) the tribal structure is segmented into lower kin-based units (diwabs), which control territories. Rights to land access, including also water sources are critical livelihood resources-land for agriculture, pasture, trees for firewood, charcoal, and water. Tribal norms dictate many aspects of life, including the management of resources.

2.5 Population and Demographic Features

The population growth and intensity measures are essential to strategic planning of growing nations and for food security estimations.

The estimated number of people in Red Sea State is found to be 1.7 million by midyear 2018 at a growth rate of 1.4% per annum. Whereas males constitute 57% of the total population females share the remainder (43%). Table (2) represent the locality, population and villages in Red Sea State. (UNDP 2014) human health measures on global bases indicated that Sudan showed a HDI (0.473) in year 2013 (ranking as number 166) which is a low indicator.

2.6 Livelihood

According to Roche (1974), the contribution of edible indigenous plants to the nutrition of people in Sudan in general and RSS in particular has not been calculated. There is also an imbalance between quantity and quality which contributes to ill health.

The livelihood opportunities for the local communities of RSS have been traditionally based on pastoralism supported by a wide range of other livelihood activities. These include: flooding or rain-fed crop cultivation in deltas and wadies, limited artisan production in some localized areas, fishing, wage labour migration, and petty trade. A conspicuous feature of the economy is its continuous change over time. The development of Suakin and Port Sudan ports has enhanced the pace for change. The urban markets created job opportunities for large indigenous and other groups especially at times when rural areas were suffering from outer-effects of drought and desertification. In the isolated areas of the Red Sea Hills, nature and culture always intermingle to create a new sentiment which is linked to these environments:

1. People who are linked to a particular environment for their livelihoods and are persistently available,
2. People who live outside this particular environment but they need to pass occasionally or periodically to the area and,
3. People who make use of the area but don't settle

Natural products, contribute significantly to the livelihoods in rural areas. Rural societies of RSS had indicated that they regularly utilized wild fruit as a coping mechanism for hunger. They have consumed both fresh as well as dry and preserved fruits. The fruits, contribute cash earnings as they are also sold in both rural and urban areas. Over 30 wild fruit tree species have been identified in the RSS, many of which bear local names and are widely used and marketed.

The reliance on nutritional food products in the poor areas of the world where HIV/AIDS still maintains a strong foothold can alleviate some of the persistent problems (World Bank 2000).

With such a valuable resource, wild fruits can contribute to the local needs for food and other related products. (Mithofer and Waibel 2003; Mithofer et al. 2006), linked the simulation of income flow of household using indigenous fruits indicates that the benefits from selling fruits with better opportunities.

The (World Bank 2004) estimated that the rate of growth in food crop production which is about 2% does not match the population growth rate of 2.5% per year.

2.7 Natural Vegetation and Plant Cover

The vegetation units in the RSS have generally been described in items of plant-geographical zones that are somewhat corresponding to the basic topographic and rainfall patterns. White (1983) classifies the area into four zones:

(i) Semi-desert grassland and shrub land,
(ii) Semi-evergreen bush land and thickest forests,
(iii) Red Sea coastal desert, and
(iv) Desert.

The first is the eastern fringe of the *Sahelian* transition zone. This covers most of the area south of latitude 19° N, except for the "semi-evergreen bush land thickest" in Erkawit. North of this, is the Red Sea coastal desert belt, also with some small mountainous areas and semi-evergreen vegetation.

It was clear from The Red Sea Area Program (RESAP) surveys that in all parts of the region, there are several vegetation assemblages which are primarily controlled by physical conditions such as water availability and depth and particle size of soil. However, these vegetation assemblages are by no means discrete types with distinct boundaries. Rather, borders of vegetation zones and changes from one "zone" to the other are extremely gradual, even in the areas where one should expect clear changes e.g. between salt marsh and the rest, between the summits and the plain. Many studies on the Sahel, also indicate that more important than the difference in precipitation is the gradual change in randomness of the rainfall pattern. This is also because the decrease in precipitation is positively correlated with the increase in variability in time and space. This means that the main ecological constraint, water, shows increased irregularity from relatively moist condition into the desert.

Hence, the main ecological influence in the Red Sea Hills is unpredictability of rain in time and space, rather than aridity per se. These environmental conditions are reflected in all other parameters correlated with water availability, such as vegetation patterns, livestock-raising and human settlement. Besides, the impact of the abrupt changes in rainfall and surface run-off, that is considered a problem to agriculture, also provides a basis for short-lived annual grasses, which is vital for sustaining pastoralism.

Thus, through the concentration of favorable factors in certain places, a viable microclimate is being produced that allows for vegetation and hence forms the basis for the human adaptation in the area. This is so because these environments make up areas of key resources to local population. *Beja* access to such resource and their ability to exploit them for humans and livestock, are basic issues that have to be understood in order to grasp how they survive.

2.8 Khors, Wadies and Surface Run-off

The surface run-off of the hilly areas of RSS is linked to internal factors related to the pattern and amount of rain fall as well as the hilly nature and topography of the area. The poor vegetation cover, the constitution of basement rocks of surroundings and the steep slope, all contribute to the high rates of run-off in the RSS. The hills represent a water divide for the numerous number of water courses (wadis and Khors) that intersect the hills east and west. The most important Khors, in terms of volume of water discharge and level of dependence on them, are Arba'at, Gowb, Khor Arab and Amur, Baraka and Diib. These Khors, together, have a total discharge of about $38M^3$. From the hydrological information available and in view of the regions topography, it can be said that:

A. there is a clear difference between places where rain falls and where the water could be utilized or harvested as a result of rocky surface, steep slope and lack of vegetation which enhances run-off.
B. on the western slopes due to the high temperature levels. Huge losses are experienced because of excessive evaporation; and.
C. the distribution of rainfall and run-off occurrence are both characterized by a high degree of variability both geographically and over time. Variability over time is caused mainly by the thunder storm origin of rainfall and over space is caused by the spatial distribution of rainfall and the nature of the slope water runs on. The intensity of rainfall variability in run-off frequency, for example, ranges between a minimum of zero to maximum of 35 times a year in the case of Khor Arba'at. Consequently, the volume of water flow range between zero and 164 million m^3 for the four largest Khors in the region.

A variability of water varies according to the variations in types of topography in different areas. Thus, in a plain like Odrus, though rainfall amounts are not very high, but because hills and mountains surround it, ground water level is fairly stable. But in summits like Erkowit, for example, which experiences high rainfall, because of the steep slope and mountain, ground water level is highly unstable. This is why permanent and seasonal wells that are dug in the *Khor* beds.

Water supply in most parts of the region is principally dependent on the water bearing capacity of the *Khors* and *wadis*. Except for Port Sudan, which is dependent on piped water from Arba'at, most of the State population depends on shallow surface wells for drinking, because of poor catchments in the area and the chemical constitution of underlying rock. Drinking water is generally saline at varying degree of brackishness depending on relative location.

2.9 Climate Change Impact

Studies by Aziz and Elsheikh investigated climate change parameters which are associated with Red Sea coast environment, both ambient and sea surface temperature, rainfall and marine inorganic carbon systems were analyzed for 40, 50 and 30 years comparatively. Data for coastal rainfall and ambient temperature were based on national meteorological data for 14 stations based at the Red Sea State. Data for sea surface temperature were based on six datasets. For fCO_2 values data were calculated by (CO2SYS) based on Lewis and Wallace (1998) model. Result indicated inter-annual variability in rainfall during the study periods which were associated with high fluctuations at seasonal levels. The mean annual rainfall varied from less than 20 mm in extreme north of the coast to 160 mm over hilly areas.

Wet years were reported for 1950, 1962, 1968 and 1986 while dry years were during 1954, 1973, 1981 and 1983. Ambient temperature fluctuation showed continuous mean increase during decades of 70′s, 80′s, and 90′s with high temperature fluctuations between summer and winter. Sea surface temperature increased from mean of 27 °C to 31 °C during 1977–2007, showing the highest values during 2000–2007. The rate of increase is about 0.14 °C $yr.^{-1}$ for SST and about 1.6 μatm $yr.^{-1}$ for the oceanic fCO_2. The atmospheric fCO_2 has increased by the same rate (about 1.5 μatm $yr.^{-1}$) as the oceanic fCO_2, which is brought by SST increase.

2.10 Wild Fruits Used in the Red Sea Area

Wild fruits are important source of vitamins and minerals. In RSS there are several types of wild fruits and rural people use them for traditional medicine as describe.

2.10.1 Balanites aegyptiaca (L.) Del

(Balanite: Heglig., Balantiaceae) is a semi-desert tree growing in the wild in south Kordofan, Darfur and Central Sudan it is belonging to the family Zygophyllaceae, and it is commonly known as desert date. The fruit is green during the first stages of growth and turns yellow to brownish at maturity with an average weight of 5.7 g and the kernel 0.7 g. The root contains steroidal sapo-genins, whereas the bulb contains sugars and saponins. The leaves and fruit contain diosgenin while the kernel with a high oil contents and valuable protein combination.

2.10.2 Traditional Uses

B. aegyptiaca is an evergreen tree distributed throughout the drier parts of RSS by the tribes of Beja. It is traditionally used in treatment of various ailments such as jaundice, intestinal worm infection, malaria, diarrhea. The pharmacological studies of *B. aegyptiaca* demonstrated insecticidal, antibacterial, antifungal, hepatoprotective, anticancerous, antihelminthic, antiparasitic, antidiabetic and anti-inflammatory activities of the plant.

Fruit and seed maceration is used as a laxative and anthelmintic. It is used as a precursor to production in the food industry, animal feed and the pharmaceutical industry. This plant has been detailed as an antihelmintic, laxative, vermifuge, febrifuge, emetic and can also cure other types of diseases such as skin bubbles, leucoderma, intestinal disease, wounds, colds, syphilis, disorders of the liver and spleen, and throbs (Hamid et al. 2006). The fruits are used in Sudanese folk medicine in the treatment of jaundice (Sarker et al. 2000).

2.10.3 *Tamrind*

Tamarindus is a mono-specific genus formerly in the Caesalpiniaceae family, which has now been developed in Fabaceae sensu lato. Tamarindus indica is a pantropic species commonly found worldwide (Morton 1987). In the central and southern parts of Sudan, Tamrindus grows near valleys and near-termite mounds in the savannah. In Beja language called (Wuhumar). It contains free and combined organic acids (tartaric, malic, citric) and 25–40% inverted sugars. Rural people in RSS use Tamarind for fever, abdominal pain, malaria, constipation, diarrhea and dysentery as a gentle laxative and acid refrigerant.

2.10.4 *Ziziphus mauritiana Lam*

The distribution of natural resources from Central Asia to Africa (Diallo 2002). In West Africa, it is ranked among the most preferred fruit tree species in all the countries of the Sahel farmers, but the fruits are very small (Fig. 2.2), farmers are interested in Indian and Thai varieties producing great and delicious fruit (Kalinganire et al. 2007). The main use of fresh or dry fruit pulp, which is also prepared in a juice. The Sahel's greatest potential is to sell juice, fresh fruit pulp and dry fruit paste.

Fig. 2.2 *Ziziphus mauritiana* tree

2.10.5 *Adansonia digitata L. (Baobab)*

It is one of the most characteristic species in the region due to its enormous size and importance in the lives of people. Baobab trees are a major source of food security and nutrition during the dry season. For medicines, fruit pulps, seeds, and roots are used. The juice, rich in vitamin C, is prepared from the pulp of fruit, the sun – dried pulp of fruit can also be eaten raw or added to the sauces, the leaves are the staple vegetable used in sauces consumed with cereal. Seeds are also used in soups or roasted and consumed as snacks it is one of the most important characteristic species in the local area since its gigantic estimate and its significance in the lives of the people. Mash, seed and roots are used for natural products for solutions. Juice is rich in vitamin C and ready for the pulp of the natural product, Moreover, mash of the sun – dried natural product can be eaten either raw or included in the sauces, the clears are the staple vegetables used in the sauces of cereal suppers. Seeds are too much used as a snack in soups or broiled (Sidibe and Williams 2002).

2.11 Adaptation Mechanisms

Krzywinski and Pierce (2001) predicted that the long spells of dry weather have led to deforestation in several areas of RSS rural areas. In a way, the issue is being looked from the point of view of values. So it might be expected that the heavy deforestation the lucrative business of charcoal which is demanded in growing urban areas Tribal conflicts over land and specially cutting trees receives severe punishment from tribal leaders, this always starts by convening a meeting *Majles* for the violator.

The major contextual variable that contribute to the recurrent situation of food insecurity RSS are climate, population, and the tribal identity-the impacts of which are emphasized repeatedly. With regard to climate, drought has been persistent and frequent over the last three decades, and the traditional pastoral livelihoods have suffered under this stress. Again, rains were inadequate to secure a grain harvest and to replenish the rangelands. Recurrent drought and what is perceived to be a changed climate regime have now resulted in a reduction of herd size, greater decentralization of the population, and a shift in livelihood strategies. Thus, the diminishing pattern of rainfall has both an immediate crisis impact on food security and a longer term effect on chronic and structural vulnerability (FAO 2008) Conflict has forced the movement of population away from native villages, making them "guests" on the lands of others. The major impact of such shifting is to disrupt traditional mechanisms of access to land (controlled by the diwab) and to make land locally scare. At the same time, indigenous knowledge of the homeland is reduced, as displaced peoples do not have the accumulated experience that is so important in agro-pastoral livelihoods. In addition, the readjustment of the demographic landscape has put pressure on local employment markets, where internally displaced people and refugees compete for wage-earning opportunities.

Finally, the importance of tribal identity and the close-knitted reciprocal obligations that it engenders cannot be underestimated. In effect, tribal membership has created sets of buffers in which the risks associated with an uncertain environment are shared among all households. At the same time, tribal affiliation has structured a unique town-country relationship in which tribal members shift frequently between the rural and the urban setting, using urban livelihood strategies to support the rural ones (Aziz 2015).

2.12 Health and Traditional Medicine

Traditionally, the usage of plant as sources of medicines are based on the experience and superstition come from generation to generation practically by the term of mouth (Sofowora 1993).

Fruits and vegetables, containing abundant dietary fiber, vitamins, and minerals, in particular large amounts of phytochemicals are recommended by nutritionists because of their health benefits, phytochemicals in these natural products are considered to be responsible for positive health outcomes. Particularly, it is widely noted that plants produce a great deal of antioxidants to combat the oxidative stress induced by oxygen and light in the natural environment oxidative stress performs an essential role in multiple chronic diseases. Therefore, antioxidants in fruit and vegetables have been extensively explored for their effects on several diseases. Epidemiological and nutritional studies suggested that the higher one's fruit and vegetable consumption, the lower the incidence of chronic diseases such as coronary heart problems, cancer, and Alzheimer's disease. Many wild fruits are safe to consume, and some have been developed as medicines. Due to different genotypes and environmental concerns, wild fruits contain rich phytochemicals such as anthocyanin and flavonoids (Krzywinski and Pierce 2001).

Therefore, wild fruits are often considered to be healthy foods. In recent years, wild fruits have attracted increasing attention, and accumulative investigations have been performed for their bioactive effects, such as antioxidant, antimicrobial anti-inflammatory, and anticancer effects. These studies pointed out that wild fruits could have the potential to prevent and treat some chronic diseases.

Wild plants are still a major source of medicine for the local people living in the Red Sea State of the Sudan. Modern health care services in this area are not adequate, and most people have limited economic means to buy Western medicine from the market. Phytochemicals in these natural products are considered to be responsible for positive health outcomes. Wild fruits are often considered to be healthy foods. In recent years, wild fruits have attracted increasing attention, and accumulative investigations have been performed for their bioactive effects, such as antioxidant, antimicrobial anti-inflammatory, and anticancer effects. These studies pointed out that wild fruits could have the potential to prevent and treat some chronic diseases (Tabuti 2008).

(Etkin (1986) clarified that the utilization of plants in conventional medication can be clarified by physiologically dynamic phytochemical compounds of a species conjointly by its credited meaning in a culture). Tabuti (2008), clarified that medicinal plants with a long history of safe and effective use are likely to have a pharmaceutical effect. In addition, the likeliness of a medicinal use being based on pharmaceutical properties rather than on a cultural context increases when this use is repeatedly found in different cultures.

2.13 Poverty

Although poverty is recognized as a multi-faceted phenomenon, encompassing issues of security and safety-nets, self-esteem and belonging, power and control, as well as income and wealth considerations (Poulton and Poole 2001), it is most frequently measured in income. Wild fruits in rural area in RSS play important role in people's.

In expansion, a few non-income prove is displayed of particular intrigued to the wellbeing and environment-related to illustrate that innate natural product tree exercises can contribute to destitution easing (within the sense of making destitution simpler to persevere) as well as to making a difference agriculturists effectively diminish their levels of destitution (Arnold 2002).

The vital role linked to natural product trees can make to destitution decrease has been perceived (Garrity 2004; Russell and Franzel 2004). Eradicating extreme poverty and hunger is the most important of the Millennium Development Goals (MDGs) that are currently the focus of the international development agenda. Other goals relate to improving education and health, empowering women and ensuring environmental sustainability. At a national level, the Poverty Reduction Strategy Papers (PRSPs), promoted by the World Bank and the International Monetary Fund (IMF), depict how governments may work with donors to attain the MDGs. However, even in forest-rich countries, the forestry sector gets little attention in PRSPs and the lack of examination of the links between poverty and the use of forest resources means that forest policy recommendations are rarely based on hard evidence (Bird and Dickson 2005).

However investigate into the destitution decreasing part of non-timber woodland items (NTFPs) has driven to the acknowledgment that numerous of these are now not collected from the wild but are developed on cultivate (Ruiz-Perez et al. 2004). Simons and Leakey (2004) suggested a new term 'Agroforestry Tree Product' (AFTP) to distinguish forest-collected NTFPs from the same products cultivated as crops on farms.

The important contribution that indigenous fruit trees make to many farmers' livelihoods (The vital commitment that innate natural product trees make to numerous farmers' vocations) (Poulton and Poole 2001) is often not acknowledged in national reporting. The proof should be communicated in terms and ideas with which the policy-makers are recognizable, and must moreover be connected to proposals that can be coordinates effortlessly into national-level arranging forms.

2.14 Economics

The contribution that domesticated indigenous fruit trees make to many farmers' livelihoods is often not acknowledged in either national- or international-level poverty reduction strategies.

Existing data on indigenous fruit are often not presented in the kinds of income-related terms used in the policy debate, nor are they linked to simple policy recommendations. The majority of the world's poor are concentrated in rural areas and, consequently, depend greatly on indigenous natural resources, which act as a buffer against poverty.

Rural economy may be broadly described as a mixed economy which embraces rural or rather medium or small manufacturing industry, service industry and agriculture, with their resultant socio-economic and cultural implications. It refers also to the economy and exchange system of people living outside designated urban centers SOS Sahel International UK (2013) reported that the rural areas of Red Sea State are one of the poorest parts of sub-Saharan Africa according to standard indicators of income and food security. The indigenous Beja people experience chronic hunger and malnutrition, and depend on low and unreliable rainfall to grow crops and support their livestock.

Some indigenous food trees occurring in the RSS forest zone, break through the rural economy level into the national economy of the country.

Garrity (2004), Russell and Franzel (2004) recognized the important contribution that indigenous fruit trees can make to poverty reduction. In spite of farmers' obvious interest in indigenous fruits for cash income and the range of food and medicinal products they provide, these species do not receive much attention from policy-makers, foresters or agriculturalists (Tchiegang-Meguenic et al. 2001). Where it is difficult to obtain information on the absolute income contributed by indigenous fruit, an estimate of the proportion of household income obtained from fruit may also be a useful measure.

The proof related to the utilization of wage is critical since it can clarify whether taming exercises simply give a safety-net when times are troublesome or are adequate to supply for a level of development that can in the long run offer assistance families or communities move out of destitution.

Educating a large number of people in rural areas is crucial for achieving sustainable development. Poverty reduction strategies are now placing emphasis on rural development that encompasses all those who live in rural areas. Such strategies need to address the provision of education for the many target groups: children, youth and adults, giving priority to gender imbalances. This complex and urgent challenge should be addressed systematically, through an intricate set of policy measures, at all levels of education systems (United Nation 2005).

In addition to their nutritional importance, the same plants serve many complementary purposes, including significance in traditional farming practices, soil erosion control and maintenance of soil fertility, mulch materials, stakes and structural materials, farming implements timber, firewood, chewing sticks, fibers, drugs, dyes, socio-cultural roles, and religious purposes. These multifarious uses indeed stress the economic importance of edible wild fruits.

(According to dealers within the Souq of Dordib, which offers the varieties of *Balanite* (Elli) and *Hyphaene thebaica* (Acat) and *Ziziphus mauritiana Lam* (Igaba). fruits where they are brought from sources and sold within the primary showcase for wholesalers and divisions)

Where the fruits are transported by camels and donkeys to Dordib and is sold dom in the range of 370–400 Sudanese pounds (SDG) per 50 kg and some of it is sold to the rest of the states. Lalob (balanite) and buckthorn are consumed locally for children as snacks and are sold by women and children and they sold by 250–300 SDG/50 Kg.

The economic impact of fruits and vegetables are however still very limited compared to the actual production potential. This is due to less attention paid to vegetables and fruits compared to cash crops like cotton, peanut, gum Arabic and staple food grains like sorghum. It is even difficult to obtain reliable data on the area and production of fruits and vegetables.

tourism—the general development strategy for rural areas—cannot be the sole saviour for SRD, but rather their locality characteristics and even their agricultural capacities can turn them into hot spots.

2.15 Nutrition

According to (Johns and Maundu 2006), the persistent availability of malnutrition in rural areas could be linked to the low development potential of rural environments, consistent poverty and the deterioration of basic ecological systems. The wide range of indigenous fruit trees available in many areas can enable farmers to meet their varied household needs for food, nutrition, medicines, etc. These species are often part of the traditional diet and culture and the subject of a body of indigenous knowledge regarding their management and use. Two well documented cases of such species are the marula tree *(Sclerocarya birrea)* in southern Africa (Shackleton and Shackleton 2005) and the shea tree *(Vitellaria paradoxa)* in the West African parklands (Baffo 1999). While these two species both have wide ranges, many indigenous fruit trees have very localized importance.

(Guinand and Dechassa 2000; Kebu and Fassil 2006) illustrated that several tropical nations, regularly gather vegetables, roots, tubers, natural products from wild because of its taste, social employments and nourishment supplements. This has assisted in combating nourishment. In rural countryside of many developing nations, wild fruits are often the only fruits consumed as people cannot afford cultivated commercial fruits as apple, grapes, pomegranate or orange. In RSS, the indigenous fruits collected from the wild, play significant role in the food and nutrient security of rural poor and tribal stability.

Table 2.1 physicochemical characteristics of Tamrind

Constituents	Amount per 100 g
Water	17.8–35.8
Ash	2.6–3.9 g
Fat/oil/lipid	0.50–3.10 g
Carbohydrates, total	56.70–82.60
Tartaric acid, total	8.00–18.00
Protein	2.00–4.00
Crude fibre	2.20–18.30
Thiamine	0.33
Vitamin C	44 mg

Source: Anon (1976)

There are many uses for tamarind (Gunasena and Hughes 2000), few are known or practiced in the Sahel. In the Sahel, the fruit pulp is used primarily for sauces, porridge and juice. Tamarind can be used as snacks, in sauces, confectionery, drinks, jam, ice cream, wine, and as a coffee substitute, a pectin, food stabilizer, dye, animal fodder, glue, edible oil and medicine (Maundu et al. 1999). Tamarind in RSS which called (Wuhumar) is used as snack, juice and make (Asida). The most valuable and commonly used part of the tamarind tree is the fruit. The pulp constitutes 30–50% of the ripe fruit (Purseglove 1987; Shankaracharya 1998). Table 2.1 shows the physicochemical characteristics of *Tamrind.*

2.16 Adaptation and Sustainability

Coping strategies of local communities used to exist through centuries of experience, the population growth, however, always outweighs food availability. The (World Bank 2004) has estimated that food crop production – about 2% does not match with the population increase rate of about 2.5% per year.

From the ancient, local tribes have discovered the importance of certain plant species to their subsistence needs. Unfortunately this knowledge is disappearing as subsequent generations knowledge is diminishing about wild fruits (Dweba and Mearns 2011).

The low food intake due to unavailability or difficult access especially in developing countries has seriously affected growth and development issues of many countries (Andersen et al. 2003; Adeboye and Phllips 2006).

have escaped recognition and scientific inquiry. Their distribution, conservation, mode of harvest by locals and optimal use require region-specific assessment in order to integrate them into developmental interventions.

A number of wild plants, used by rural and tribal populations have contributed significantly to rural development through providing alternative substitutes (Belcher et al. 2005; Fisher 2004; Narendran et al. 2001; Scherr et al. 2004). Uses, of wild fruits for direct intake or as local use have been an added value especially in developing countries (McSweeny 2004; Takasaki et al. 2004; Mojeremane and Tshwenyane 2004; Getachew et al. 2005; Redzic 2007).

(Bell 1995; FAO 2005), emphasized o the dietary contribution of wild fruits to the daily intake of villagers and also the quality attributes and flavour associated with these natural products.

Of late, further contribution of wild fruits to diversify family incomes has been an added value. Mahapatra et al. (2005) and Bharuch and Pretty (2010) investigated the effect of biodiversity of local fruits on the socio-economics of the end users of processed products.

2.17 Conclusion

In the RSS, local fruits contribution to food security is prominent. Many organizations and development centers need to coordinate their efforts in order to continue their efforts for confronting food access and shortage. Rural communities which wholly or partially depend on forests for their livelihood, are affected by multitude of factors which affect their decision making process. Land ownership issues as well as the rights for conservation and sustenance seem to be part of the local villagers.

Industries sustainability based and cultural heritage conservation in form of hand crafts is affected by ideas which can promote best alternatives developed in rural areas for their livelihoods so as to alleviate unemployment problems. The contribution of scientific research to maximize the potential economic, social and cultural potential can further enhance rural sustainability and community cushioning.

Among the multi-faceted uses of wild plants, products such as leaves, which could be cooked, or honey which can add to rural economy can be of additional value. Hand crafts based on timber quality, leaves, stem, and other non-timber products that sustain local mats industry. Strong marketing policies are also needed to augment local efforts.

Rural-based artisans are based mainly on women activities which tend to concentrate on their activities in their homes. In the rural RSS this has positive potential implications. The stability of rural population increases with generation of income at household level.

References

Africa. Dev South Afr 16:299–318

Aziz A (2015) Dynamism of livelihood systems in the Red Sea State. Red Sea State University, Port Sudan

Adeboye OC, Phllips OT (2006) Studies on seed characteristics and chemical composition of three morphotypes of *Mucuna urens* (L.). Medikus-Fabaceae. Food Chem 95(2006):658–663

Ahmed AAA (2013) Livelihood changes and opportunities in seven localities of the Red Sea State. Ocean docs.org

Andersen CF, Botter JI, Murray AS (2003) A mini X-raygenerator as an alternative to a ^{90}Sr beta source in luminescence dating. Radial. Meas., these proceedings

Anon (1976) The wealth of India (raw materials), vol 11. Council of Scientific and Industrial Research, New Delhi

Arnold M (2002) Clarifying the links between forests and poverty reduction. Int For Rev 4(3): 231–233

Baffo JM (1999) Agroforestry parklands in Sub Saharan Africa, Les parcs agroforesties en Afrique substarieme, Rome, Italia, pp 204–230

Belcher B, Sunderlin WD, Angelsen A, Burgers P, Nasi R, Santoso L, Wunder S (2005) Livelihoods, forests, and conservation in developing countries: an overview. Word Dev 33(9):1383–1402. https://doi.org/10.1016/J.Worlddev.2004.10.004

Bell AW (1995) Regulation of organic nutrient metabolism during transition from late pregnancy to early lactation. J Anim Sci 73(9):2804–2819

Bharuch Z, Pretty J (2010) The roles and values of wild foods in agricultural systems. Philos Trans Biol Sci 365(1554):2913–2926. Food security: feeding the world in 2050 (27 September 2010)

Bird N, Dickson C (2005) Poverty Reduction strategy papers: making the case for forestry. Odi Forestry Briefing Number 7, London

Bussmann RW, Sharon D (2006) Traditional medicinal plant use in Northern Peru:tracking two thousand years of healing culture. J Ethnobiol Ethnomed 2(1):47

Etkin NL (1986) Plants in indigenous medicine and diet: biobehavioral approaches. Redgrave Publishing Company, Bedford Hills, New York

Diallo I (2002) Etude de la biologie de la reproduction et de la variabilite genetique chez le jujubier (Ziziphu mauritiana lam). These de doctorat de troisieme cycle. Universite cheik Anta Diop, Dakar, Senegal

Dweba T, Mearns M (2011) Conserving indigenous knowledge as the key to the current and future use of traditional vegetables. Int J Inf Manage 31(6):564–571

FAO (2005) Impact of trawling and scallop dredging on benthic habitats and communities. FAO Fisheries Technical Paper 472. Rome

FAO (2008) Climate Change and food security: a framework document. Food and Agriculture Organization of the United Nation. Rome

Garrity DP (2004) Agroforestry and the achievement of the millennium development goals. Agroforestry Systems 61:5–17. Academic publishers, Netherland

Getachew Y, Hunde D, Tadesse M (2005) Survey of medicinal plants used to treat human diseases in Seka Chekorsa, Jimma zone, Ethiopia. Ethiop J Health Sci 15(2):89–107

Guinand Y, Dechassa L (2000) Indigenous food plants in Southern Ethiopia: reflections on the role of 'famine foods' at the time of drought. United Nations Emergencies Unit for Ethiopia (UNEUE), Addis Ababa

Gunasena HPM, Hughes A (2000) Tamarind: tamarindus indical. International Centre For Underutilized Crops:170

Hamid O, Wahab M, Hassan E (2006) *Balanites aegyptiaca* extract for treatment of HIV/AIDS and leukemia. International Publication Number WO 2001/49306 A1

Harris M, Kriebel CH, Raviv A (1982) Asymmetric information, incentives and intrafirm resource allocation. Manag Sci 28(6):604–620

Johns T, Maundu P (2006) Forest biodiversity, nutrition and population health in market-oriented food systems. Unasylva 224(57):34–40

Kalinganire A, Weber JC, Uwamariya A, Kone B (2007) Improving rural livelihoods through domestication of indigenous fruit trees in the park lands of the sahel

Kebu B, Fassil K (2006) Ethnobotanical study of wild edible plants in Derashe and Kucha Districts, South Ethiopia. J Ethnobiol Ethnomed 2:53

Krzywinski K, Pierce R (2001) Deserting the desert: A threatened cultural landscape between the Nile and the Sea. University of Bergen, Department of Botany Allegaten 41, N-5007 Bergen, NORWAY.

Krzywinski K, Holton RP (2001) Deserting the desert. a Threatened Cultural Landscape between the Nile and the Sea. Department of Botany, Department of Greek, Latin, and Egyptology University of Bergen All rights Reserved ISBN 82-Grafisk, Larvik, Norway

Kwesiga F, Maghembe JA, Simons AJ, Hagman J (1998) Plan of operation. project on domestication of indigenous fruit trees of the Miombo woodlands of southern Africa. October 1997–September 2000. ICRAF, Nairobi

Kwesiga FR, Akinnifesi FK, Ramadhani T, Kadzere I, Sako J (2000) Domestication of Indigenous Fruit trees of the miombo in southern Africa. In: Shumba EM, Lusepani E, Hangula R (eds) The domestication and commercialization of indigenous fruit trees in the SADC region. SADC Tree Seed Centre Network

Lewis E, Wallace D (1998) Co_2 SYS. Program development for the Co_2 system calculations. Carbon Dioxide Inf Anal Center Report OR NL/CDIAC

Mahapatra R, Nerur S, Mangalaraj G (2005) Challanges of migrating to agile methodologies. Communications of the ACM 48(5):72–78

Maundu MP, Ngugi WG, Kabuye HSC (1999) Traditional food plants of Kenya. National museum of Kenya

McSweeny K (2004) Forest product sales as natural insurance: the effect of household characteristics and the nature of shock in Eastern Honduras. Soc Nat Res 17:39–56

Mithofer D, Waibel H (2003) Income and labour productivity of collection of indigenous fruit tree products in Zimbabwe. Agroforestry Sys 59:295–305

Mithofer D, Waibel H. Akinnifesi FK (2006) The role of food from natural resources in reducing vulnerability to poverty: a case study from Zimbabwe. Paper accepted for the 26th Conference of the International Association of Agricultural Economists (IAAE), Queensland, Australia

Mojeremane W, Tshwenyane SO (2004) A valuable edible indigenous fruit tree of Botswana. Pak J Nutr 3(5):264–267

Morton ES (1987) Variation in mate guarding intensity by male purple martins. Behaviour 101: 211–224

Narendran K, Murthy IK, Suresh HS, Dattaraja HS, Ravindranath NH, Sukumar R (2001) Nontimber forest product extraction, utilization and valuation: a case study from the Nilgiri Biosphere reserve, southern India. Econ Bot 55(4):528–538

Poulton C, Poole N (2001) Poverty and Fruit Tree Research. Produced for DFID'S Forestry Research Programme. Agricultural Economics and Business Management Research group Imperial college at Wey, Kent, TN 25 5 AH, UK

Purseglove JW (1987) Tropical crops dicotyledons. Longman Singapore Publishers Ltd

Redzic S (2007) Syntaxonomic diversity as an indicator of ecological diversity-case study Varanica Mts in the central Bosnia. Biologia. Bratislava 62/2:173–184

Roche JF (1974) Synchronization of oestrus and fertility following artificial insemination in heifers given prostaglanding F_{2x}. J Report Fert 37:135

Russel D, Farnzel S (2004) Trees of prosperity: Agroforestry, Markets and the African smallholder. Agrofor Syst 61:345–355. Nairobi, Kenya

Ruiz-Perez M, Ezzine de Blas D, Nasi R, Sayer J, Sassen M, Angoue C, Gami N, Ndoye O, Ngono G, Nguinguiri JC, Nzala D, Toirambe B, Yalibanda Y (2004) Pilot study on the social, Environmental and Economic Sustainability of Industrial Concessions in the congo basin. International Tropical Timber Organisation, https://www.itto.int/. Display Handler pageId=161id=804

Saka JDK, Msonthi JD (1994) Nutritional value of edible fruits of indigenous wild trees in Malawi. For Ecol Manag 64:245–248

Saka JDK, Mwendo-Phiri E, Akinnifesi FK (2002) Community processing and nutritive value of some miombo indigenous fruits in central and southern Malawi. In: Kwesiga F, Ayuk E, Aggumya A (eds) Proceedings of the 14th Southern Africa Regional Review and Planning Workshops, 3–7 Sept., 2001, Harare, Zimbabwe

Sarker SD, Bartholomew B, Nash RJ (2000) Alkaloids from *Balanites aegyptiaca*. Fitoterapia 71:328–330

Shackleton SE, Shackleton CM (2005) Agroforestry tree products (AFTPs): targeting poverty reduction and enhanced livelihoods. Int J Agric Sustain 3(1)

Sidibe M, Williams JT (2002) Fruits for the future –Baobab (Adansonia digitata L.). International Centre for Underutilized Crops, Southampton. 96pp

Simon AJ, Leakey RRB (2004) Tree domestication in tropical agroforestry. Agrofor Syst 61:167–181. Kluwer Academic publishers. Netherland

Shankaracharya NB (1998) Tamarind-chemistry, technology and uses-a critical appraisal. J Food Sci Technol 35(35):193–201. 100

Sofowora A (1993) Medicinal plants and traditional medicine in Africa. Wiley, New York, p 150

Scherr S, White A, Khare A (2004) The current status and future potential of markets for the ecosystem services provided by tropical forests. International Tropical Timber Organization (ITTO). Technical Series No 2

Takasaki Y, Barham B, Coomes O (2004) Risk coping strategies in tropical forests: floods, illnesses, and resourceextraction. Environ Dev Econ 9:203–224. Cambridge University Press

Tabuti J (2008) Herbalmedicines used in the treatment of malaria in Budiope county, Uganda. J Ethnopharmacol 1161:33–42

Tchiegang-Meguenic C, Mapongmetsem PM, Akagou ZCH, Kapseu C (2001) An Ethnobotanical study of indigenous fruittree in North Cameroon. J Forests Trees Livelihoods 11(2):149–158

UNDP (2014) Human development report. Sustaining human progress: reducing vulnerabilities and building resilience. UNDP, New York

UN (2005) Market Brief in the European Union for Selected natural ingredients derived from native species Adansonia digitata L. (Baobab). http://www.unctad.org/biotrade/docs/biotradebriefbaobab.pdf. Site visited on 14th May 2009

Vernon HH (1999) Use and potential of wild plants in farm households. FAO Farm System Management Series 15

White F (1983) The vegetation of Africa. Nat Resour Res UNESCO 20:356. pp.ref.50

World Bank (2004) Educational in Rwanda:Rebalancing resources to accelerate post conflict development and poverty reduction

World Bank (2000) Intensifying action against HIV/AIDS in Africa: responding to a development crisis. Washington: The International Bank for Reconstruction and Development/The World Bank, Washington, DC, 63 p

Chapter 3
Wild Fruits and Its Implications on Food Security and Poverty Alleviation in Sudan

Abdalbasit Adam Mariod

3.1 Introduction

Wild fruits assume an important role in the household income and food security of numerous communities in the rural societies through the world. In addition, wild fruits contain fundamental supplements (Carbohydrates, protein, minerals) and basic nutrients, which are critical, particularly for the developing youngsters who are inclined to lack of healthy food and related infections. A portion of the fruits has progressed toward becoming articles of business in the local, national and universal markets, thus adding to household incomes and food security (FAO 2012). Wild fruits collecting empowers rural societies to broaden their income sources. In addition, wild fruit cultivation is most imperative cash income, used in the dry season and starvation time (Gebauer et al. 2002).

There is a ton of work have been done everywhere throughout the world in wild fruits of a few hundred wild eatable plants, which fulfill the yearning of the general population as well as have been demonstrated nutritious as well. So Wild fruits can be utilized replace cereals to lessen hunger and lighten poverty. The area exists in savannah and semi-arid regions, past examinations showed that these regions were vital source for the wild fruits accumulation and advertising (Eltahir and Gebauer 2004). At a macroscale level, wild foods can add to the alleviation of household food insecurity in numerous families, in South Africa yet this couldn't be the situation at microscale levels as watched. Wild foods were for the most part devoured by food insecure households with lower riches index, in this way, were for the most part going about as a 'safety net'. There is requirement for advancing the utilization and consumption of wild foods in communities, both at low-and high-income levels, with the goal that households might know about the advantages of these items in mitigating food uncertainty and malnutrition by expanding food access through

A. A. Mariod (✉)
Indigenous Knowledge and Heritage Center, Ghibaish College of Science & Technology, Ghibaish, Sudan

A. A. Mariod (ed.), *Wild Fruits: Composition, Nutritional Value and Products*, https://doi.org/10.1007/978-3-030-31885-7_3

lessening reliance on food obtaining. As food costs are proceeding to expand progressively and declining household food insecurity, approaches to move 'income conditions' of these households and furthermore to build awareness on possess generation of food at family level should be pushed. Concentrate ought to be set on capacity building, work creatively and providing assets for families to complete practice food production, which is one approach to improve food security of poor family units (Chakona and Shackleton 2017).

Wild eatable plants are characterized as plant species gathered in the wild to be expended as food or drink. Wild palatable plants have been a necessary piece of the human diet since forever and around the globe. Forest and farm trees make critical commitments to the food security of rural populaces, supply basic nutrient, particularly when food sources inaccessible. Nevertheless, commitment of wild fruits trade in a rural community and consumption isn't yet very much archived (Olawoye 1996). This chapter focus on wild fruits and how it can keep the food security and poverty alleviation in the world. Wild fruits are sources of cash income for many people in many places through the world.

3.2 Income Generation from Wild Fruits

Kell et al. (2015) created a technique for making an agenda of, and organizing China's harvest wild relatives, and uncovered that 871 local species are identified with yields that are of especially high financial significance in China—including rice, wheat, soybean, potato, sweet potato, millet and yam—crops which are additionally of quite high incentive for food and economic security in different pieces of the world. Inside this agenda they have recognized species that are specifically need of protection evaluation depending on their relative Red List status and potential for use in harvest improvement programs. Shumsky et al. (2014a) gathered wild fruits from Eastern Province, Kenya farming regions to government-oversaw peaks with changing degrees of progress and lawfulness. Their numerous contextual analysis examines the formal woods controls and land residency rights, just as local enforcement and comprehension of those rules, so as to get it their effect on the capacity of poor people to utilize wild edible fruits as an adapting system. These authors recommended that broad confusion, trust issues and a solid spotlight on the trading of wild foods are constraining the possible sharing of wild fruits to food security and expanded socio-ecological versatility. They distinguished various approach changes and expansion programs that could better help local communities depending on wild fruits for subsistence purposes to enhance their versatile capacity (Shumsky et al. 2014a).

Worldwide, there are several lesser-referred to indigenous fruits just as other food plants collected from the wild that add to food security and assume fundamental roles in the nutrition of the general population especially the rural people. These local crops and wild food plants that have gotten little research consideration or augmentation exercises and have been to a great extent dismissed by the standard of worldwide science incorporate a wide assortment of wild indigenous foods that enhance the eating regimen of the rural masses and could flourish with little consideration and without the utilization of exorbitant agrarian information sources, for example, fertilizers, herbicides and pesticides (Aworh 2015).

Leão et al. (2017) inspected 87 instances of wild collecting of vertebrates and plants in developing nations to comprehend the conditions impacting its maintainability. They utilized arbitrary forest and calculated regression to locate the most essential indicators among a scope of natural, demographic and monetary markers, and to detail the anticipated impacts on the sustainability of harvest. Species flexibility, GDP per capita and poverty ratio were negatively related. Species flexibility was emphatically identified with manageability of collect, though GDP per capita and poverty proportion were adversely related. Gathering species, varieties with low versatility is bound to be unsustainable when collected in a middle-income, high poverty proportion nation than in a low-income, low poverty proportion nation. Their investigation featured the dangers of wild populace gather under the difficult economic conditions inalienable in developing countries, and imagine where efforts are most expected to accomplish sustainable harvest. Beside the building of WEPs to household food security, they are additionally a vital way of dealing with coping mechanism amid terms of food instability expedited by drought, political distress and unsteady commodity markets, because of the availability of various species over the date-book year (Fentahun and Hager 2009) and generally high resistance to water stress (Addis et al. 2005). In spite of the fact that WEPs are a critical dietary asset in rural zones, their contribution to food security is regularly undervalued by policy-makers, prompting formal approaches on access, extraction and sale that can need comprehension of local conditions (Shackleton and Shackleton 2004).

Shumsky et al. (2014b) examined the formal forest guidelines and land residency rights in Eastern Province, Kenya, as well as local requirement and comprehension of those standards, so as to comprehend their effect on the capacity of vulnerable populaces to utilize WEPs as an adapting strategy. Their outcomes recommended that, broad confusion, trust issues and a solid spotlight on the commercialization of wild foods are constraining the conceivable commitment of WEPs to food security and expanded socio-environmental flexibility. They recognized various arrangement changes and extension programs that could all the more likely help local communities depending on WEPs for subsistence purposes to improve their adaptive capacity.

There is the need to advance the development and use of a large number of indigenous wild fruit trees, that are additionally of extensive natural centrality and shield them from uncontrolled wood felling exercises that outcome in deforestation. Research center examinations have appeared great quality fruit jams, juices and different beverages can be delivered from a few lesser known and under-used fruit foods utilizing basic methodology appropriate for little scale business creation, including osmotic drying out, jam-production process and mechanical juice extraction followed by heated water purification (Aworh 2015).

Precise correlations of human reliance on forests and environmental assets have been trying, because of heterogeneous philosophies. Specific Forestry Modules have been created, with the objective of filling present day records gaps concerning the economic importance of forest and wild fruits in own family unit welfare and rural livelihoods. Results from a pilot appraisal of the Forestry Modules in West Kalimantan, Indonesia, are displayed, demonstrating that the Forestry Modules carry out properly in extricating the ordinary facts:mean according to capita forest and wild product income shifts as indicated through the geographical "forest gradient". Altogether, inside the forests rich upstream village, imply woodland and wild item income and mean forest area-related pay and enterprise earnings surpasses present day mean agricultural income measurements for West Kalimantan and imply non-farming rural family earning within the most minimal phase. Usage of forest gadgets and significance as an adapting system changed into better within the maximum upstream village, in which advertising of forest items inside the midst of surprise was an increasing number of set aside in the most downstream village (in which forest adapting techniques have been likewise less essential). The Forestry Modules' specific and planned method can assist guarantee that commitments of forests and wild objects are not concept little of in country wide figures (Bakkegaard et al. 2017).

Evacuating the requirements to the advancement of little scale food enterprises which incorporate deficient power supply, utilization of improper innovation, insufficient working capital, high loan fees, and constrained access to banks and other money related establishments would encourage business creation of these value-added foods in rural communities in this way decreasing post-reap misfortunes, advancing food security, improving small farmers' income and adding to the feasible rural development (Aworh 2015).

Wild fruits are an especially normal and powerful adapting technique utilized to increase socio-biological strength in Sub-Saharan Africa, where farming frameworks are frequently sensitive to ecological annoyances and instability. Wild fruits are gathered over the scene, from agrarian areas to government-managed ridges with shifting degrees of achievement and legitimateness (Shumsky et al. 2014b).

3.3 Conclusion

To increase interest of food security and poverty allevation, poeple in a significant part of the world tried many ideas to harvest and domesticate wild fruits as a method for creating income and meeting their family needs.

References

Addis G, Urga K, Dikasso D (2005) Ethnobotanical study of edible wild plants in some selected districts of Ethiopia. Hum Ecol 33(1):83–118

Aworh OC (2015) Promoting food security and enhancing Nigeria's small farmers' income through value-added processing of lesser-known and under-utilized indigenous fruits and vegetables. Food Res Int 76:986–991

Bakkegaard RK, Hogarth NJ, Bong IW, Bosselmann AS, Wunder S (2017) Measuring forest and wild product contributions to household welfare: testing a scalable household survey instrument in Indonesia. For Pol Econ 84:20–28

Chakona G, Shackleton C (2017) Minimum dietary diversity scores for women indicate micronutrient adequacy and food insecurity status in South African Towns. Nutrients 9(8):87–94

Eltahir BA, Gebauer J (2004) Non-timber forest products: Opportunities and constrains for poverty reduction in Nuba Mountains, South Kordofan, Sudan. A paper presented in the conference on international Agricultural Research for Development, Deutscher Tropentag 2004, Berlin, October 5–7, 2004

FAO, WFP and IFAD (2012) The State of Food Insecurity in the World 2012. Economic growth is necessary but not sufficient to accelerate reduction of hunger and malnutrition. FAO, Rome

Fentahun M, Hager H (2009) Exploiting locally available resources for food and nutritional security enhancement: wild fruits diversity, potential and state of exploitation in the Amhara region of Ethiopia. Food Sec 1(2):207–219

Gebauer J, El-siddig K, Elbert G (2002) Baobab (Adansonia digitata L.): a Review on a multipurpose tree with promising future in the Sudan. Gatenbauwissenschaft 67(4):S155–S160. ISSN0016-478X

Kell S, Qin H, Chen B, Ford-Lloyd B, Wei W, Kang D, Maxted N (2015) China's crop wild relatives: diversity for agriculture and food security. Agric Ecosyst Environ 209:138–154

Leão TCC, Lobo D, Scotson L (2017) Economic and biological conditions influence the sustainability of harvest of wild animals and plants in developing countries. Ecol Econ Elsevier 140(C):14–21. https://doi.org/10.1016/j.ecolecon.2017.04.030. https://ideas.repec.org/a/eee/ecolec/v140y2017icp14-21.html; https://ideas.repec.org/s/eee/ecolec.html

Olawoye JE (1996) Sociological Issues in sustainable forest management. Department of Agricultural Extension services, University of Abadan. Ghana J Forest 3:13–18

Shackleton C, Shackleton S (2004) The importance of non-timber forest products in rural livelihood security and as safety nets: a review of evidence from South Africa. S Afr J Sci 100(11/12):654–658

Shumsky SA, Hickey GM, Pelletier B, Johns T (2014a) Understanding the contribution of wild edible plants to rural social ecological resilience in semi-arid Kenya. Ecol Soc 19(4):34. https://doi.org/10.5751/ES-06924-190434

Shumsky SA, Hickey GM, Johns T, Pelletier B, Galaty J (2014b) Institutional factors affecting wild edible plant (WEP) harvest and consumption in semi-arid Kenya. Land Use Policy 38:48–69

Chapter 4
Prospects for Utilization of Wild Fruits

Abdalbasit Adam Mariod

4.1 Introduction

Wild edible plants (WEPs) are the species those are neither developed nor domesticated, anyway are accessible in their wild common living space and utilized as sources of these plants have assumed an important role in the improvement and progress of the development and civilization of human history throughout the ages and the globe. These wild edible plants have assumed a huge role in providing nutrition and dietary prerequisites of poor group in numerous provincial areas of the world. These wild edibles can be advanced just when they are contrasted with their nutritional and medical advantages with major or generally utilized cultivated plants. The social, religious, and belief arrangements of the rural people are deficient without these wild edible plants. Planting of these wild edible plants can expand their utilization and their preservation too (Chakravarty et al. 2016).

Two hundred and three species of wild flowering plants of Ethiopia are consumed by the community. These represent 3% of the higher plant species in the country and comprise herbs (37%), shrubs (32%), and trees (31%). Species with edible fruits contributed to 61.6%, leaves 27.7%, stems 14.4%, roots 13.3%, and seeds 10.3%. About 15% are considered famine foods. Some species are wild and others slightly or strongly associated with humans, revealing a living analogue of the wild-semiwild-domesticated continuum. Full-scale ethnobotanical studies, implementing appropriate conservation measures and domesticating and integrating promising taxa into existing crops are favored for the purpose of diversification of food sources, ensuring food security, and maintenance of biodiversity and environmental integrity (Asfaw and Tadesse 2001).

A. A. Mariod (✉)
Indigenous Knowledge and Heritage Center, Ghibaish College of Science & Technology, Ghibaish, Sudan

A. A. Mariod (ed.), *Wild Fruits: Composition, Nutritional Value and Products*,
https://doi.org/10.1007/978-3-030-31885-7_4

4.2 Wild Fruits as Household Food Supply

Twenty four species of edible wild plants (EWPs) ranging from fruits and leafy vegetables to tubers were used as a source of food and beverage in Gweta Village, central Botswana. Most utilised EWPs include *Grewia flava*, *Grewia bicolor*, *Scloreocarya birrea*, *Amaranthus thunbergii*, *Cleome gynandra*, *Corchorus olitorius*. They are the main source of food and income, and are harvested by almost everyone in the Village. The EWPs are of significant importance to the rural economy, and the results revealed that EWPs also assist in dietary diversity, food security and income generation (Badimo et al. 2015).

An ethnobotanical study was conducted to investigate indigenous knowledge and socioeconomic of edible wild fruit trees and shrubs (EWFTSs) in Arsi Zone, Central Ethiopia. A total of 90 respondents were interviewed and consisted of men, women, and children who were selected by stratified random sampling. A total of 30 EWFTSs bearing species that belongs to 20 families were identified. Age and gender distribution of respondents on the choice of four of the top five species were homogeneous indicating that promotion can be planned indifferently to all households. A considerable proportion of the community acknowledged food values of EWFTSs, and more than half preferred EWFTSs over cultivated commercial fruit, suggesting that efforts towards their integration into the current farming system are appealing. Realizing the resource depletion, about 54% of the respondents planted EWFTSs while 87% showed interest to participate in domestication programs. The study explored a great potential of promoting EWFTSs in Arsi Zone and assisting the country's efforts of ensuring food security environ (Seyoum et al. 2015).

There is a ton of work have been done everywhere throughout the world in wild fruits of a few hundred wild eatable plants, which fulfill the yearning of the general population as well as have been demonstrated nutritious as well. So Wild fruits can be utilized as substitutes of seed grains to lessen hunger and lighten poverty. The area exists in savannah and semi-arid regions, past examinations showed that these regions were vital source for the wild fruits accumulation and advertising (Eltahir and Gebauer 2004). At a macro-scale level, wild foods can add to the alleviation of household food insecurity in numerous families, in South Africa yet this couldn't be the situation at micro-scale levels as watched. Wild foods were for the most part devoured by food insecure households with lower riches index, in this way, were for the most part going about as a 'safety net'. There is required for advancing the utilization and consumption of wild foods in communities, both at low-and high-income levels, with the goal that households might know about the advantages of these items in mitigating food uncertainty and malnutrition by expanding food access through lessening reliance on food obtaining (Chakona and Shackleton 2017).

4.3 Traditional Recipes Prepared Using Wild Edible Fruits

Deb et al. (2013) studied the status of wild edible plants and their traditional utilization as diet recipes by different ethnic communities of Tripura, Ethiopia. Total 41 species of 36 genera and 22 families of wild edible plants were documented through semi-structured interviews and preference ranking methods among three ethnic groups viz. Tripuri, Molsom and Rupini of Tripura. These authors identified eight different types of traditional recipes which, mostly prepared by 41 wild edible plants. Among the traditional recipes, Gudak and Chakhwi were found to be most preferred diet compliments among the communities. These wild edible plants also traditionally used as major food supplements by those ethnic communities. Traditional recipes are mostly prepared from wild plant and play vital role in community nutrition. Diverse plants are available in the nearby forest, so the communities use plant resources in a simplest way (Deb et al. 2013).

Considering the potential and the need for new resources, a study was undertaken to explore the diversity, plant parts used, the process of preparation, scope and potential of the wild edible plant species from Maharashtra State, India. The results, revealed that, about 172 wild species are used as food in Maharashtra. Out of these 63 species are commonly used by tribal and rural peoples, 35 species are occasionally used while the remaining species are rarely consumed, especially during the scarcity of food due to drought or other reasons. Some of these have high scope and potential for their improvement and exploitation on a commercial scale (Rahangdale and Rahangdale (2014).

4.4 Different Uses of Wild Edible Fruits

The wild edible fruits (WEFs) have been a source of 'hidden harvest' which had supplemented the community with food and income. The traditional knowledge system and economic demand of a community influences the exploitation of WEFs in a particular area (Sawian et al. 2007). The consumption of wild edible fruits, indicated user's evaluation of the fruits in terms of its availability, tastes and preferences along with the duration of the settlement near to a forest from where the community is procuring these fruits. The world's poorest of the poor, those who are bereft of a dignified life and are most vulnerable dominates South Asia, Latin American, Northern Africa, Sub-Saharan Africa and the Caribbean (FAO, WFP and IFAD 2012). There is a prevalence of undernourishment in these regions due to production of insufficient food grains and at times food supplies are not easily available. Foods obtained from WEFs serve as 'buffer food' rescuing lives during food shortages and famines (Chakravarty et al. 2016).

Adansonia digitata L. (Malvaceae) is commonly known as baobab tree native to Africa. Baobab is a multi-purpose tree which offers protection and provides food, clothing and medicine as well as raw material for many useful items. The fruit pulp, seeds, leaves, flowers, roots, and bark of baobab are edible and they have been studied by scientists for their useful properties. The fruit pulp has very high vitamin C, calcium, phosphorus, carbohydrates, fibers, potassium, proteins and lipid content, which can be used in seasoning as an appetizer and also make juices. Seeds contain appreciable quantities of phosphorus, magnesium, zinc, sodium, iron, manganese, whereas they have high levels of lysine, thiamine, calcium and iron. Baobab has numerous biological properties including antimicrobial, anti-malarial, diarrhoea, anaemia, asthma, antiviral, anti-oxidant and anti-inflammatory activities amongst others. Phytochemical investigation revealed the presence of flavonoids, phytosterols, amino acids, fatty acids, vitamins and minerals (Rahul et al. 2015). Leaves are digitate, normally having 5 leaflets when mature. The leaflets have entire margins and are elliptic to obovate-elliptic, with acuminate apex and a decurrent base. Mature leaf size may reach a diameter of 20 cm. The flowers bloom during the wet season and the dry season as well. They are very large and suspended on long peduncles. The fruit is bottle or cucumber shaped and develops 5–6 months after florescence. The internal fruit pulp is split into mealy agglomerates that enclose several reniform seeds (Sidibe and Williams 2002). *A. digitata* fruits contain 50% more calcium than spinach, is high in anti-oxidants, and has three times the vitamin C of an orange. It is sometimes called a superfruit. In Kordofan and Darfour states the leaves mixed with onion slides and peanut paste and eaten as a salad, in other African countries the fruit dissolved in milk or water and used as a drink. The seeds also produce edible oil. In 2008, the European Union approved the use and consumption of baobab fruit as an ingredient in smoothies and cereal bars (http://en.wikipedia.org/wiki/Advisory_Committee_on_Novel_Foods_and_Processes). The baobab fruit pulp is probably the most important foodstuff. It can be dissolved in water or milk as in some Sudanese areas where young children shepherds used to mixed baobab fresh fruit with goat milk to have thick delicious porridge known as *ummartabo*. In other African countries the produced liquid is then used as a drink, a sauce for food, a fermenting agent in local brewing, or as a substitute for cream of tartar in baking (Sidibe and Williams 2002).

4.5 Conclusion

The wild fruit species thought of sacred, and are measure necessary within the diet. They are significantly utilized throughout the amount of seasonal food shortages; they're usually the sole fruit supply and that they are usually high in nutrients. Some households are concerned within the promoting of wild fruits and receive a little financial gain from sales. Wild fruit trees might play a very important role in conservation as a result of they assist maintain contour ridges. They are also necessary for shade. It's shocking, that wild fruit trees are given therefore very little attention, each in research and extension. More studies and extension on wild fruit trees ought to be a priority; with time this valuable resource might disappear.

References

Asfaw Z, Tadesse M (2001) Prospects for sustainable use and development of wild food plants in Ethiopia. Econ Bot 55(1):47–62

Badimo D, Lepetu J, Teketay D (2015) Utilization of edible wild plants and their contribution to household income in Gweta Village, central Botswana. Afr J Food Sci Technol 6(7):220–228

Chakona G, Shackleton C (2017) Minimum dietary diversity scores for women indicate micronutrient adequacy and food insecurity status in South African Towns. Nutrients 9(8):87–94

Chakravarty S, Bhutia KD, Suresh CP, Shukla G, Pala NA (2016) A review on diversity, conservation and nutrition of wild edible fruits. J Appl Nat Sci 8(4):2346–2353

Deb D, Sarkar A, Barma BD, Datta BK, Majumdar K (2013) Wild edible plants and their utilization in traditional recipes of Tripura, Northeast India. Advan Biol Res 7(5):203–211

Eltahir BA, Gebauer J (2004) Non-timber forest products: Opportunities and constraints for poverty reduction in Nuba Mountains, South Kordofan, Sudan. A paper presented at the conference on International Agricultural Research for Development, Deutscher Tropentag 2004, Berlin, October 5–7, 2004

FAO, WFP, IFAD (2012) The state of food insecurity in the world 2012. Economic growth is necessary but not sufficient to accelerate reduction of hunger and malnutrition. Rome, Italy.

http://en.wikipedia.org/wiki/Advisory_Committee_on_Novel_Foods_and_Processes

Rahangdale DR, Rahangdale SS (2014) Potential wild edible plant resources from Maharashtra: future prospects for their conservation and improvement. Life Sci Leaflets 57:73–85

Rahul J, Jain MK, Singh SP, Kamal RK, Naz AA, Gupta AK, Mrityunjay SK (2015) *Adansonia digitata* L. (baobab): a review of traditional information and taxonomic description. Asian Pac J Trop Biomed 5(1):79–84

Sawian JT, Jeeva S, Lyndem FG, Laloo RC (2007) Wild edible plants of Meghalaya, North-East India. Indian J Nat Prod Resour 6(5):410–426

Seyoum Y, Teketay D, Shumi G, Wodafiras M (2015) Edible wild fruit trees and shrubs and their socioeconomic significance in Central Ethiopia. Ethnobot Res Appl 14:183–197

Sidibe M, Williams JT (2002) Baobab. Adansonia digitata. International Centre for Underutilised Crops, UK

Chapter 5
Economic of Wild Fruits and It Is Contribution to Rural People

Elsedig Elbadawi Ahmed Awad and Abdalbasit Adam Mariod

Abbreviation

FAO	Food and Agriculture Organization
GDP	Gross domestic product
KFA	Kenya's Forests Act
TSS	Total Soluble Solids
TTA	Titratable Acidity
WEFs	Wild edible fruits

5.1 Introduction

Around 75, 000 types of plants worldwide are accepted to be edible. These species are broadly accessible however are essential just in situ. The national per capita supplies are essentially contributed by just 103 types of food plants (90%) and of these, staples are just 20 to 30 species (Chakravarty et al. 2016). More than 7000 species of Wild edible plants (WEPs) are registered worldwide (Grivetti and Ogle 2000). About 1000 species were well known in the Americas, 1200 species in Africa and 800 species in Asia. Around 3000 tropical fruits worldwide were accounted for as still unexploited. Around 400 wild consumable fruit species were accounted for from Kenya (Hassan et al. 2005), 700 from Tanzania, 150 from Eastern Madagascar, 110 from Swaziland, 300 from Cameroon and 105 from Uganda, 123 from southern Yunnan in China, 33 from Nepal and 11 from Pakistan. Very rich diversity of wild fruit species was reported from Nigerian lowland rainforest where in species richness of a few hectares is higher than the entire vegetation of Europe (Adekunle and Oyerinde 2004). In Ethiopia, the

E. E. A. Awad (✉)
Department of Management, Economic and Social Research Bureau, Ministry of Higher Education and scientific research, Khartoum, Sudan

A. A. Mariod
Indigenous Knowledge and Heritage Center, Ghibaish College of Science & Technology, Ghibaish, Sudan

A. A. Mariod (ed.), *Wild Fruits: Composition, Nutritional Value and Products*,
https://doi.org/10.1007/978-3-030-31885-7_5

wealth of WEP species is huge and 38–182 eatable fruit species accounted for to be expended in Ethiopia (Addis et al. 2013). The wild palatable fruit species reported in India from Himalayas are in excess of 675 species (Chettri et al. 2005).

Wild fruits assume a fundamental job in the family pay and food security of numerous networks in the rustic zone through the world. In addition, wild fruits contain imperative supplements (Carbohydrates, protein, minerals and so on) and basic nutrients, which are vital, particularly for the developing children who are inclined to hunger and related illnesses. A portion of the fruits has moved toward becoming articles of trade in the nearby, national and worldwide markets, in this way, adding to family incomes and food security (FAO 2012). In this way, wild fruits collecting empowers provincial networks to broaden their income sources. In addition, fruit gathering is most imperative cash income, create in dry season and starvation time (Gebauer et al. 2002).

There is a lot of work have been done all over the world in wild fruits and record of several hundred wild edible plants, which not only satisfy hunger of the people but have been proved nutritious too. So wild fruits can be utilized as substitutes of cereals to diminish hunger and reduce destitution. The region exists in savannah and semi-arid locales, past investigations uncovered that these areas were a vital source for the wild fruits accumulation and promoting (Eltahir, and Gebauer 2004).

Wild palatable plants are characterized as plant species gathered in the wild to be devoured as food or drink. Wild eatable plants have been an essential piece of human diet regimen since ancient time and around the globe. Forest trees make critical commitments to the sustenance, security of rural societies, supply fundamental supplement, particularly when food sources inaccessible. Nevertheless, sharing of wild fruit business in local societies livelihoods and consumption isn't yet very much reported (Olawoye 1996).

The present review has portrayed the wild edible plants in the setting of their assorted variety, conventional information, preservation rehearses and dietary composition from the accessible secondary literature. Writers feel there is still degree to join increasingly relevant factors for clarifying more varieties implanted with local individuals' impression of qualities and use of these wild edible fruit.

5.2 Who Use Wild Edible Plants

The United Nations embraced the Millennium Declaration of September 2009 to improve the worldwide living conditions through diminishing destitution and hunger. However, extensive quantities of individuals are as yet living in absolute penury and are denied of an honorable living. In such rough conditions nature's free present as wild eatable foods are profiting the subordinate communities. Wild edible plants (WEPs) are the species those are neither cultivated nor domesticated, anyway are accessible in their wild regular living space and utilized as sources of these plants have assumed a huge role in the improvement and progress of mankind's history all through the ages and globe. These wild eatable plants have assumed a critical role in providing food and nourishing necessities of poor people in numerous local groups of the world. These wild edibles can be promoted just when they are analyzed for their nutritional and medical advantages with major or broadly utilized cultivated plants. The social, religious, and belief of the provincial societies is deficient without these wild edible plants. Domestication of these wild edible plants can increase their utilization and their conservation as well.

5.3 Income Generation from Wild Fruits

Unavoidable food uncertainty and neediness of a significant part of the world drives poor societies to collect normal assets as methods for producing more money and facing family unit needs. Wild eatable plants are an especially normal and compelling adapting technique utilized to increment socio-ecological resilience in Sub-Saharan Africa, where horticultural frameworks are frequently touchy to natural annoyances and precariousness. Shumsky et al. (2014) gathered wild fruits from Eastern Province, Kenya farming regions to govern-oversaw peaks with changing degrees of progress and lawfulness. Their numerous contextual analysis examines the formal woods controls and land residency rights, just as local enforcement and comprehension of those rules, so as to get it their effect on the capacity of poor people to utilize wild edible fruits as an adapting system. These authors recommended that broad confusion, trust issues and a solid spotlight on the trade of wild fruits are constraining the possible contribution of wild fruits to food security and expanded socio-ecological versatility. They distinguished various approach changes and expansion programs that could better help local communities depending on wild fruits for subsistence purposes to enhance their versatile capacity.

The cropping of wild species, especially of fish and timber, represents an imperative commitment to the worldwide economy (FAO 2010). At a local dimension, the poor population in developing nations depend vigorously on woods assets for their income (Angelsen et al. 2014). Wild cropping, if controlled reasonably, can be a chance to scale up conservation efforts. This is especially vital for developing nations where arrangements coordinating improvement and preservation are basic (Hutton and Leader-Williams 2003).

Leão et al. (2017) studied 87 instances of wild collecting of vertebrates and plants in developing nations to comprehend the conditions impacting its sustainability. They utilized irregular timberland and calculated relapse to locate the most essential indicators among a scope of organic, demographic and financial markers, and to detail the anticipated impacts for the maintainability of harvest. Species flexibility, Gross domestic product (GDP) per capita and destitution head check ratio were the most strongest indicators of demographic. Species strength was decidedly identified with maintainability of harvest, though GDP per capita and poverty proportion were negatively related. Harvesting a species with low strength is bound to be unsustainable when harvested in a middle-income, high poverty proportion nation than in a low-income, low poverty proportion nation. Their examination featured the risks of wild populace harvest under the difficult economic conditions intrinsic in developing nations, and visualize where endeavors are most expected to accomplish maintainable harvest.

The outcomes propose that there are serious correspondence and understanding breakdowns with respect to the use of Kenya's Forests Act (KFA), where subsistence gatherers are hindered because of strict requirement that targets small-scale harvesters and considerable boundaries to KFA creation for the local society. The emphasis on commercialization of wild products by foresters, extension workers and private landowners additionally add to negative effects on more poor family, for example, loss of access to wild fruits assets on public and private lands and decreased advantages when neighborhood elites are better arranged to exploit esteem option exercises and markets. Moreover, focusing on wild fruits for money collection limits the education and outreach programs, that may some way or another advantage local clients and adds to a conviction that wild fruit harvesters have ulterior thought processes in entering collection zones. The outcomes got through this research add to a more noteworthy comprehension of the resource access conditions that exist in rural, semi-arid Kenya and advise feasible food security arrangement as the conventional land residency frameworks change in private possession, State-oversaw forests and community claimed assets. Future strategy investigations ought to think about how KFA could be corrected to more readily support national and universal food security objectives. Specifically, strategy and research endeavors to more readily support the practical utilization of wild fruits for subsistence purposes will probably result in enhanced family food security and expanded socio- ecological resilience in the rural communities of arid and semi-arid Kenya (Shumsky et al. 2014).

5.4 Wild Fruits Supported Rural People Income

Rural people group that need flexibility and are profoundly sensitive to environmental tend to used some strategic to increased their income. Wild edible fruits (WEFs), defined by the Food and Agriculture Organization (FAO) as "plants that grow spontaneously in self-maintaining populations in natural or semi-natural ecosystems and can exist independently of direct human action", have been recognized as an especially imperative way that households in rustic can diminish their sensitivity to environmental change while additionally adjusting to less good conditions. The key characteristics of WEFs are as follows: (Shumsky et al. 2014).

1. They are locally accessible and their utilization depends on customary ecological learning
2. They are a low-input, minimal cost choice for expanding nutrition and diminishing the need to spend restricted cash
3. They give more noteworthy advantages to defenseless population (poorer family units, ladies, and kids, who are regularly influenced by climate occasions

4. They add to livelihoods and are accessible during times of the dry season or conflict-driven.
5. They endure water stress superior to anything their domesticated relatives, having a "natural strength to quick environmental change, which is frequently lacking in exotic species.

Precise examinations of human reliance on forests and natural assets have been challenging, because of heterogeneous approach. Particular Forestry Modules have been created, with the objective of filling current data gaps concerning the economic significance forests and wild items in family welfare and rural livelihoods. Bakkegaard et al. (2017) displayed their outcomes from a pilot assessment of the forestry modules in West Kalimantan, Indonesia, they demonstrated that the Forestry Modules perform well in extricating the normal data: mean per capita forests and wild item income shifts as per the geological "forest gradient". Fundamentally, in the timberland rich upstream village, mean wood and wild item income and mean backwoods related wage and business incomes surpasses current mean agrarian pay measurements for West Kalimantan and mean non-farming rural family unit earnings in the most minimal and lowest bracket. Utilization of forest items and significance as an adapting methodology was higher in the most upstream village, where sale of forest products in the midst of shock was progressively more marked in the most downstream village.

Bakkegaard et al. (2017) investigation has demonstrated that the Forestry Modules do perform well in separating the expected data as indicated by the "forest gradient". What's more, the pilot testing underlines the significance of separating information collected at various spatial and time scales, for example community vs household and money versus subsistence, catch of occasional items versus precision of a year review - contemplations that may go plain in different sectors. These authors affirmed that, the Forestry Modules expected to think about the harmony between study execution at the national scale, and the significance of catching subtlety at different scales. In this manner, the Forestry Modules mean to guarantee that the specificities of forest related exercises and commitments can be caught while ensuring their relevance crosswise over many forest settings. Their deliberate usage will augment the viability and portrayal of forest and wild item conditions in national financial information and add to molding proper national strategy that mirrors the circumstances of family units in forest regions (Bakkegaard et al. 2017). Chakravarty et al. (2016) reviewed the accessible data with respect to the assorted variety, conservation, indigenous knowledge and physico-chemical attributes of wild consumable fruits. These authors confined their research work to regions with severe poverty. They revealed that, wild edible fruits additionally have the potential through specific conservation and domestication, which can add to the upkeep of plant biodiversity. These authors detailed a need to recognize/perceive these fruits and their value added products included items in the nearby or national or global market. It is additionally vital to examine the market condition for these fruits compared with elective potential outcomes, for example, intriguing fruits or agricultural crops.

5.5 The Role Wild Edible Fruits in Food Production

Wild edible plants (WEPs) assume an imperative role in food processing and keeping up biological community administrations, particularly in Sub-Saharan Africa (Grivetti and Ogle 2000). These wild food assets supplement energy and micronutrients, improve the flavor of staple foods and broaden food sources (Arnold and Perez 2001). This is particularly valid in poor families and in rural territories where they can decrease spending of constrained cash assets on energy, shelter, food and medicinal needs (Shackleton and Shackleton 2004). Numerous WEPs can be assembled without financial expense and don't require costly information sources, machinery or preparing, which means beginning investment in production is certainly not a boundary to effective results (Jama et al. 2008).

Ndabikunze et al. (2011), assessed baobab powder as a potential option for pectin in jam making. They looked at its gelling capacity and potential use to that of commercial pectin in the processing of jam from various indigenous fruits accessible in Tanzania. Jams were processed from indigenous fruits got from different zones of Tanzania including, Smelly-berry (*Vitex mombassae*), Wild loquat (*Uapaca kirkiana*) and Marula plum (*Sclerocarya birrea*) utilizing lemon extract, commercial pectin and baobab (*Adansonia digitata* L.) powder as sources of pectin. Their discoveries of this examination explain that, total soluble solids (TSS) extended from 11.6% in Adansonia digitata L. to 16.9% in Uapaca kirkiana. Titratable acidity (TTA) was most astounding in Adansonia digitata L. (2.27%) and most minimal in Uapaca kirkiana (0.05%). Pectin content in fruits was most astounding in Adansonia digitata L. (2.56%) and most reduced in Vitex mombassae (0.12%). Jams processed without utilization of pectin were inferior in quality. Conventional pectin and baobab powder gave jams which did not differ essentially in moisture content, TSS and TTA. Ndabikunze et al. (2011) revealed that, the investigated wild fruits do fluctuate in useful attributes required for jam processing. The considered fruits contain low amount of pectin aside from A. digitata L., which had high pectin and acidity. The pectin found in baobab (*A. digitata* L.) powder was effectively used in the processing of other fruit jams and ended up being a potential substitute for costly conventional pectin. Commercial pectin produced from being expensive isn't effectively accessible in rural regions, where these fruits are found in substantial amounts. For instance in Tanzania, 1 kg of commercial pectin was accessible at US$ 50.00 (May, 2009). The indigenous natural fruits are less expensive, when compared with grown fruits. If the farmers, in locations where these fruits are found, are prepared to have adequate training, they can undoubtedly use them to process fruit items for commercial reason to obtain income and for their own utilization. Along these lines, indigenous fruit can add to the improvement of food and food security. It is in this manner suggested that, the utilization of baobab powder as a substitute for conventional pectin ought to be advanced. In any case, an inside and out examination of chemical changes of the fruits from development to ripening should be seriously considered so as to enhance the preparing method and limit losses experienced amid ripening and storage. In addition the standard commitment of WEPs to family food security, they are likewise an imperative method for dealing with mechanism amid times of food insecurity directed by dry season, political instability and precarious goods markets, because of the accessibility of various species over the datebook year (Fentahun and Hager 2009) also, moderately high tolerance to water pressure (Addis et al. 2005). In spite of the fact that WEPs are a vital dietary source in provincial territories, their commitment to food security is regularly overlooked by policy-makers, prompting formal strategies on access, extraction and sale that can need comprehension of local conditions (Shackleton and Shackleton 2004).

5.6 Other Uses of Wild Edible Fruits of Economic Importance

Antioxidants are materials found in few amounts that restricts oxidation in a chain reaction, ascorbic acid is an antioxidant that reduces free radical attack by scavenging oxyradicals. Numerous wild consumable fruits like berries are accounted for as rich sources of cancer prevention agents like vitamins B, E, and C (Abu Bakar et al. 2009; Ikram et al. 2009). Wild eatable fruits like mulberry, Ficus roxburghii and Capparis zeylanica are rich in vitamin C (Deshmukh and Waghmode 2011). Carotenoids announced from numerous wild palatable fruits like ocean buckthorn and mulberry are the chemo-defensive agents or are cancer prevention agents and are in charge of appearance and attractiveness quality of natural fruits (Venkatesh and Chauhan 2011). Anthocyanin is a polyphenolic dye of flavonoid group that grants orange, red, blue and violet color to numerous wild eatable natural fruits like chokeberry and bilberry (Chandra et al. 1992). Cyanidin 3-glucoside is the most well-known anthocyanin found in these fruits that potential defensive impacts against illness (Fukumoto and Mazza 2000).

5.7 Conclusion

Regardless of the difficulties, there are incredible open doors for Wild Fruits on the planet. Numerous universal tourists are keen on attempting local foods and supporting local organizations, and Wild Fruits are a normal fit for this specialty. Wild fruits are at present selling to Air line organizations (Air Botswana), which is their biggest client, so travelers are acquainted with their items on their flights. Wild fruits are investigating new markets in the EU, the US and Japan, from where there have gotten item enquiries.

References

Abu Bakar MF, Mohamed M, Rahmat A, Fry J (2009) Phytochemicals and antioxidant activity of different parts of bambangan (Mangifera pajang) and tarap (Artocarpusodoratissimus). Food Chem 113:479483

Addis G, Urga K, Dikasso D (2005) Ethnobotanical study of edible wild plants in some selected districts of Ethiopia. Hum Ecol 33(1):83–118

Addis, G., Asfaw, Z., Woldu, Z. (2013). Ethnobotany of wild and semi-wild edible plants of Konso ethnic community, South Ethiopia. Ethnobot Res Appl 11:121–142

Adekunle VAJ, Oyerinde OV (2004) Food potential of some indigenous wild fruits in lowland rainforest ecosystem of South West Nigeria. J Food Technol 2:125–130

Angelsen A, Jagger P, Babigumira R, Belcher B, Hogarth NJ, Bauch S, Börner J, Smith-Hall C, Wunder S (2014) Environmental income and rural livelihoods: a global-comparative analysis. World Dev 64(Supplement 1):S12–S28

Arnold JEM, Perez MR (2001) Can non-timber forest products match tropical forest conservation and development objectives? Ecol Econ 39(3):437–447

Bakkegaard RK, Hogarth NJ, Bong IW, Bosselmann AS, Wunder S (2017) Measuring forest and wild product contributions to household welfare: testing a scalable household survey instrument in Indonesia. Forest Policy Econ 84:20–28

Chakravarty S, Bhutia KD, Suresh CP, Shukla G, Pala NA (2016) A review on diversity, conservation and nutrition of wild edible fruits. J Appl Nat Sci 8(4):2346–2353

Chandra A, Nair MG, Lezzoni A (1992) Evaluation and characterization of the anthocyanin pigments in tart cherries (Prunus cerasus L.). J Agric Food Chem 40:967–969

Chettri N, Sharma E, Lama SD (2005) Non timber forest produces utilization, distribution and status in a trekking corridor of Sikkim, India. Lyonia 8:89–101

Deshmukh BS, Waghmode A (2011) Role of wild edible fruits as a food resource: traditional knowledge. Int J Pharm Sci Res 2:919–924

Eltahir BA, Gebauer J (2004) Non-timber forest products: opportunities and constrains for poverty reduction in Nuba Mountains, South Kordofan, Sudan. A paper presented in the conference on international Agricultural Research for Development, Deutscher Tropentag 2004, Berlin, October 5–7, 2004

FAO (2010) Global forest resources assessment 2010. In: FAO forestry paper No. 163. Organization of the United Nations, Rome

FAO, WFP and IFAD (2012) The state of food insecurity in the world 2012. Economic growth is necessary but not sufficient to accelerate reduction of hunger and malnutrition. FAO, Rome

Fentahun M, Hager H (2009) Exploiting locally available resources for food and nutritional security enhancement: wild fruits diversity, potential and state of exploitation in the Amhara region of Ethiopia. Food Sec 1(2):207–219

Fukumoto LR, Mazza G (2000) Assessing antioxidant and pro-oxidant activity of phenolic compounds. J Agric Food Chem 48:35973604

Gebauer J, El-siddig K, Elbert G (2002) Baobab (Adansonia digitata L.): a review on a multipurpose tree with promising future in the Sudan. Gatenbauwissenschaft 67(4):S. 155–S. 160. ISSN0016-478X

Grivetti LE, Ogle BM (2000) Value of traditional foods in meeting macro and micro nutrient needs: the wild plant connection. Nutr Res Rev 13:31–46

Hassan R, Scholes R, Ash N (2005) Current state and trends. Millennium ecosystem assessment. Island press, Washington DC

Hutton JM, Leader-Williams N (2003) Sustainable use and incentive-driven conservation: realigning human and conservation interests. Oryx 37:215–226

Ikram EHK, Eng KH, Jalil AMM, Ismail A, Idris S, Azlan A, Nazri HSM, Diton NAM, Mokhtar RAM (2009) Antioxidant capacity and total phenolic content of Malaysian underutilized fruits. J Food Compos Anal 22:388–393

Jama B, Mohamed A, Mulatya J, Njui A (2008) Comparing the Big Five: a framework for the sustainable management of indigenous fruit trees in the dry lands of East and Central Africa. Ecol Indic 8(2):170–179

Leão TCC, Lobo D, Scotson L (2017) Economic and biological conditions influence the sustainability of harvest of wild animals and plants in developing countries. Ecol Econ 140:14–21

Ndabikunze BK, Masambu BN, Tiisekwa BPM, Issa-Zacharia A (2011) The production of jam from indigenous fruits using baobab (*Adansonia digitata* L.) powder as a substitute for commercial pectin. Afr J Food Sci 5(3):168–175

Olawoye JE (1996) Sociological issues in sustainable forest management. Ghana J Forest 3. pp 7,

Shackleton C, Shackleton S (2004) The importance of non-timber forest productsin rural livelihood security and as safety nets:a review of evidence from South Africa. S Afr J Sci 100:658–664

Shumsky SA, Hickey GM, Pelletier B, Johns T (2014) Understanding the contribution of wild edible plants to rural social ecological resilience in semi-arid Kenya. Ecol Soc 19(4):34. https://doi.org/10.5751/ES-06924-190434

Venkatesh KR, Chauhan S (2011) Biochemical constituents of different parts of mulberry genotypes. Int J Agric Sci 3:90–96

Chapter 6
Domestication of Indigenous Fruit Trees

Abdel Moneim Elhadi Sulieman and Abdalbasit Adam Mariod

Abbreviations

IFT Indigenous fruit tree
WAC World Agroforestry Center

6.1 Introduction

Plants are principal to all life on earth, giving, nourishment and insurance from organisms ranging from bacteria to large mammals. Multipurpose plants have different economical, ecological, natural and social perspectives (Pye-Smith 2010). Plants that are fundamentally utilized as food are accepted to have extra advantages to health, or that likewise have therapeutic properties (Cunninghum 2001). Among the indispensable jobs of plants in numerous parts of common life more remote than prescription and sustenance, they are pivotal for family apparatuses, fills, scavenge, material for development and cordage. In rural villages, home garden fences quite often contain live plants which give extra advantages to families (Asfaw and Tadesse 2001). Such multipurpose plants which give significant services are presently getting rare in zones where natural and human effects are exceptionally articulated except if protection systems are inspected (FAO 2008).

Wild food plants are normally considered as all plant assets that are neither developed nor domesticated, yet utilized as a wholesome enhancement to the nearby individuals. They are accessible in their common natural habitats, for example, for-

A. M. E. Sulieman (✉)
Department of Biology, Faculty of Science, University of Hail,
Ha'il, Kingdom of Saudi Arabia
e-mail: am.sulieman@uoh.edu.sa

A. A. Mariod
Indigenous Knowledge and Heritage Center, Ghibaish College of Science & Technology,
Ghibaish, Sudan

A. A. Mariod (ed.), *Wild Fruits: Composition, Nutritional Value and Products*,
https://doi.org/10.1007/978-3-030-31885-7_6

ests savannah and other shrubbery arrive regions (Asfaw and Tadesse 2001; Lulekal et al. 2011). Teketay et al. (2010) likewise portrayed that wild food plants will be planted with consumable parts that additionally develop normally on farmlands. However, wild species vary from domesticated ones, since the breeding system of the latter has been changed through hereditary or phenotypic selection that they end upward upon the continued human help for survival (Cunningham 2001). Generally, utilization of wild edibles depends on their harvest yield, on nearby markets where food gets and different incomes. In addition, this is progressively normal in food uncertain territories of the country and it likewise differs among poor, medium, and moderately well off family units (FAO 2005).

Domestication of wild fruit trees contains knowing and characterizing of species diversification and control of the hereditary; growing of good breed plants and helpful growing of the species in controling agroecosystems (Simons 2003). Vegetative propagation strategies such as establishing of stem cuttings, uniting and marcotting warrant early fruiting and certification that the perfect characteristics of better mother trees are passed to the posterity. Effective undertakings on domestication of IFTs, for instance, in Cameroon, show that fruit cultivation and processing have significant effects on rural advancement and changing individuals' lives (Fig. 6.1).

Fig. 6.1 *Crataegus submollis* wild fruit tree. (Source: https://commons.wikimedia.org/wiki)

6.2 Potentials of Fruit Tree Domestication

Domestication is made by man, including determination forms showing the objective species progressively helpful to individuals by expanding the production or potentially the nature of whatever product they obtain. The WAC characterizes domestication as "the socio-economic and biophysical processes engaged with the distinguishing proof, characterization, choice, augmentation, and development of high-esteem tree species in controling systems" (Garrity 2008). Generally, local fruit tree species have as it were experienced deliberate determination to a certain point, and a large portion of them are described as being undomesticated or semi-domesticated in the genetic from worldwide.

Genetic enhancement possibly are characterized as a procedure that changed attributes over the ages by modification of the targeted genes (Namkoong et al. 1988). The process include recognizable proof and the choice of the prevalent population and their appealing phenotypes, followed by investigating and choice (Eriksson et al. 2006). Essential stages include:

1. The screening and activation stage, where populations and individual with wanting phenotypic qualities are recognized and chosen
2. The testing stage where the people are assessed and chosen;
3. The rearing stage where unrivaled genotypes are crossed.
4. The repetitive stage where descendants crossed and assessed over progressive ages.

On a fundamental level, such repetitive breeding exercises can proceed over numerous ages, managing and improving breeding population (Namkoong et al. 1988).

Indeed, even solid determination force in a populace seeming phenotypic changeability will conceivably prompt extensive increments if the heritability is of satisfactory size, in light of the way that the gain depends upon the item between choice power, heritability and inside populace variety (Eriksson et al. 2006). Learning of the heritability and the phenotypic fluctuation of different fruit characters is in this way basic to make sense of which qualities justify deciding for.

6.3 Importance of Plant Domestication

Tree domestication offers a bunch of alternatives for a revived reforestation and agroforestry programs. It could likewise influence biodiversity and hereditary asset preservation, and community forestry. The Land Grant encounter attested that indigenous learning framework is a rich vault of data on characteristic assets the executives. The absence of formal logical writing need not be a constraining component for the utilization of indigenous tree species. It has been proven that nearby individuals are "researchers uncommonly". In any case, good judgment and essential ranger service science standards should be utilized to connect data holes. Tree domestication projects will likewise be fruitful if participatory methodology is embraced. This isn't just in leading on-cultivate preliminaries yet more essentially in giving secure land residency to the upland ranchers. Tree domestication will surely be basic intending to the tree species needs of agriculturists for enhanced upland development and better personal satisfaction. Small preparing groups, particularly of ladies, advantage from enhanced fruit product development and help to decrease post-harvest losses. In any case, there is a high-unexploited potential for redesigning, business improvement and salary age through preparing of both intriguing and wild or tamed indigenous fruits. For example, a feasibility examination of little scale concentrated juice preparing undertakings determined a potential net benefit of around 28% of the gross production value in Malawi (Jordaan et al. 2008).

Fruit tree growing gives extraordinary increase for farmers income if they are:-

(i) connected to business sectors to diminish input costs and enhance prices for their products,
(ii) trained in best on-farm control of fruit trees; and
(iii) in developing an enhanced, high esteem varieties and species, which suit present and future market requests. When farmers approach made strides united planting material, they can expect a generally speedy come back from their new trees as joined trees will begin fruiting a few years in the wake of planting.

6.4 Nutritional Value of Wild Fruit Tree

Nutrient composition varied among selected wild fruit species. Some of the wild fruit species are richer in vitamin C and sugar. High sugar content in most wild fruit species indicated that they could be utilized as an energy source for humans, especially during periods where conventional energy foodstuffs such as maize or sorghum are scarce. Many studies have demonstrated that the wild food plant species have high healthful potential, and their dietary benefits are more noteworthy than that of some green developed crops. As for their mineral substance, numerous wild plants may offer a superior nourishing potential. For instance *Tamarindus indica* and *Passiflora edulis* contained the highest concentration of ascorbic acid, which is only comparable to oranges (270 g/100 g). A higher abundance of Fe and vitamin C are found in some of the wild fruits. Splittstoessr (1990) found that 60 mg vitamin C, 750 mg P and 10 mg Fe are required daily by humans. Therefore, 254 mg *Tamarindus indica* and 250 g *Passiflora edulis* could meet the daily recommended vitamin C, Fe, and P requirements. *Passiflora edulis* with dietary fiber of

38.27 mg/100 g could assume an essential job in diminishing the dangers of numerous disarranges, for example, blockage, diabetes, cardiovascular infections, and over weight (Spiller 2001). *Vitellaria paradoxa* had the highest fat content. Fats and oils provide more calories/gram than any other nutrient and allow for the absorption of vitamins A, D and E (Gullick 1997). Numerous wild fruits contain kernels, which are rich in lipids and from where considerable amounts of oil can be separated for cooking purposes. High fat foods are particularly especially imperative for kids who require energy-dense foods for growth. Wild grains, seeds and kernels provide significant amounts of calories, proteins and oils. The calorific values of wild grains are frequently more noteworthy than those of the cultivated varieties and they tend to be more balanced than cereals when the overall nutritional value is taken into consideration.

6.5 Domestication of Indigenous Fruit Tree (IFT) in Tropical Africa

In the 1990s, researchers from the WAC led reviews in West Africa, Southern Africa and the Sahel to build up which indigenous trees were most esteemed by neighborhood individuals. Exactly 6000 agriculturists reacted to the overview in Cameroon, Gabon, Ghana and Nigeria. They were anticipating that individuals should point to monetarily imperative timber species, however, what they esteemed most were indigenous fruit trees.“ Although there were a few varieties in inclinations, both inside and between nations, a moderately modest number of trees – including shrub mango (*Irvingia gabonensis*), African plum (*Dacryodes edulis*), African nut (*Ricinodendron heudelotii*) and severe kola (*Garcinia kola*) – were especially prominent. In spite of the fact that they were generally normal in woodlands, and as wild trees on homesteads, they were relatively obscure to science. “We knew their natural names, yet we knew their organic names, yet that was pretty much all,” says Ebenezar Asaah, a tree researcher at the WAC. “We had no clue to what extent it took them to achieve development and create organic product, and we didn't know anything about their regenerative conduct. Top picks for the future local individuals, interestingly, knew a decent arrangement about these trees, as they and their precursors had been collecting and preparing their foods grown from the ground for ages. They had likewise drilled an extremely fundamental type of domestication by choosing and eating the most delicious assortments of African plum and a couple of different animal varieties, planting their seeds and exchanging the seedlings. In any case, this was an erratic and informal approach to domesticate wild plants. As local individuals discovered significantly more about these species than the researchers did, the WAC chose to work with them to create predominant assortments. The participatory tree domestication program was propelled in 1998. This has included rural communities selecting, spreading and overseeing trees as per their very own necessities, in association with researchers, metro specialists and business organizations. Projects, for example, these are normally arranged towards nearby markets and incorporate the utilization of both indigenous learning and hereditary determination dependent on logical standards. In West and Central Africa, the program's first assignment was to build up which qualities were most refreshing in the towns. Most ranchers said a similar thing: they needed trees that create expansive, sweet natural product, develop at an early age, and are generally less in stature. One of the ordinary methods for growing new harvest assortments includes sexual generation. For instance, researchers may choose they need to join the sweetness of the organic product delivered by variety A with the extensive size of natural product created by Variety B. This is accomplished through cross-fertilization, and the subsequent mixture may then be back-crossed with one of the parent assortments to deliver offspring with an extensive variety of changeability got from the two guardians. Among this changeability there might be people with exactly the size and taste of natural product wanted. This type of plant reproducing works especially well for yearly harvests, yet numerous assortments of apple, pear and orange have likewise been produced along these lines. Nevertheless, it is an exceptionally tedious method for creating new tree assortments. Unrivaled assortments of indigenous organic product trees were required now if they somehow happened to acquire advantages to small holders the quick future.

6.6 Old Procedures, New Methodology for Domestication

Some trees produce sour fruit, which are left to spoil; others deliver sweet fruit, which are accumulated by villagers before daybreak. The span of the fruit may change as well, with the biggest African plums being multiple times the mass of the littlest. Training tries to catch and duplicate the trees with attractive attributes, in this manner exploiting the varieties found in nature. With the assistance of local farmers and gatherers, the researchers could distinguish the individual trees, which had the ideal attributes. They gathered germ plasm, as vegetative material, and this was utilized to set up predominant 'increases' at research locales and nurseries. A basic piece of the procedure included a chronicle accurately where each example originated from, utilizing a GPS, and marking the parent tree. The researchers at that point continued to inquire about the ideal methods for engendering unrivaled trees with the goal that expansive quantities of indistinguishable duplicates would be accessible in a moderately brief time frame. In the meantime, they started to prepare agriculturists in methods, for example, establishing, joining and marketing, which would empower them to create their own predominant assortments. At first, numerous agriculturists saw the methods with doubt. Farmers also received training in how to clone predominant trees by taking cuttings – extraordinary compared to other methods for delivering huge quantities of hereditary indistinguishable plants – and how to make marcots, a training that includes stripping bark far from a branch and deceiving it into creating roots while it is as yet joined to the parent tree. When the roots show up, the branch can be chopped down and planted. For a few animal varieties, marcotting turned out to be a phenomenal method for building up clones from explicitly develop wild trees. These could then be utilized as a source of cuttings or scions for establishing and joining. Marcotting likewise lessens the time it takes a tree to achieve development and prove to be fruitful. "There is a colloquialism round here that on the off chance that you plant the nut of the cola tree (Cola species), you will pass on before the primary reap. It is as yet not realized to what extent it takes a wild cola tree to achieve development – presumably 20 years or more. In any case, by utilizing marcots, agriculturists can raise cola trees that organic product after only 4 years. Additionally, these are predominate assortments, an imperative thought in a nation where numerous individuals tumble to their demise when reaping the natural product from tall trees. Certain species, for example, shrub mango and severe kola have demonstrated hard to marcot, however are most effectively duplicated by joining. It is moderately simple, then again, to take marcots from the African plum, however, it has just experienced an indefatigable procedure of research that the researchers have decided the strategies that outcome in elevated amounts of fruitful establishing and nursery foundation. Information such as this has assumed an essential job in helping agriculturists to spread wild species in their own nurseries and on their homesteads.

6.7 Domesticated Fruits for Health and Food Security

Inadequacy of iron and nutrient is predominant in many parts of the world. Consumption of nutrient A – around 50 million African kids are in danger of insufficiency – is viewed as Africa's third most noteworthy general medical issue after HIV/AIDS and jungle fever (Agudo 2004). Vitamin C from fruit, then again, is fundamental for engrossing iron, an imperative mineral that is available in noteworthy amounts in green leafy vegetables. Indigenous fruits add to the nutrient and mineral supply of neighborhood communities, e.g. baobab (*Adansonia digitata*) for vitamin C, marula (*Sclerocarya birrea*) for vitamin A and white cross berries (*Grewia tenax*) for iron. A child could cover 100 per cent of its vitamin C requirement by eating only about 10 g of baobab pulp a day (http://www.hort.purdue.edu; Lukmanji et al. 2008).

Concerning iron, utilization of 40–100 g white cross berries coats very nearly 100% of the everyday iron necessity of a kid under 8 years of age. Notwithstanding micronutrients, fruits for example, tamarind (*Tamarindus indica*) and baobab contribute a lot of energy supply because of their sugar content. However, information on the nutrient contents of numerous indigenous fruits are either inaccessible or questionable. The high inconstancy of supplement substance given in the literature (http://www.hort.purdue.edu; Lukmanji et al. 2008) might be caused by utilizing diverse techniques for analysis, but also by the fact that a very high various naturally occurs among different populations of the same species as long as the species is undomesticated.

Tree crops such as fruit trees are contributing not exclusively to nutrition security, yet moreover to food security. Because of their wide and significant establishing rooting systems, fruit trees are less delicate to droughts as appeared differently in relation to yearly stable crops and give a gather notwithstanding when the staple harvests fail. Not just amid dry seasons, and in addition especially amid the pre-reap times of yearly staples portrayed by food deficiencies, the fruits of some IFT species might be prepared for collect to fill in as crisis food or to be sold, subsequently adding to food and nourishment security.

6.8 Wild Domesticated Fruits for the Future

The practice over the previous time demonstrates that the sharing way to deal with tree domestication has a lot to give. The domestication programs depend on basic, low cost agricultural procedures and they expand on conventional employments of tree items, for instance as foods catalysts and drugs. They have a relatively quick effect by diminishing destitution and enhancing human welfare. Participatory domestication urges ranchers to receive different cultivating frameworks and diminish their reliance on products, for example, cocoa and coffee. It additionally recognizes local ownership of germplasm, in spite of the fact that the global property rights system, in its present shape, still can't seem to give lawful assurance for locally developed varieties of superior fruit tree. The tree domestication program has demonstrated that it is so critical to include local communities. Participatory tree domestication is a self improvement way to deal with research that completes things rapidly, and maintains a strategic distance from a considerable lot of the issues which happen when examine spending plans are controlled by public sector." for the time being, the WAC will keep on advancing participatory tree domestication in new territories inside Cameroon, Ghana, Nigeria and DRC, and it would like to stretch out its exercises to different nations in West and Central Africa, including Liberia, Sierra Leone, Guinea and Gabon. Its scientists will likewise proceed to produce and test new varieties of indigenous fruit tree, including those that deliver fruits in the off-season, similar to the Noel cultivar of the African plum. Ranchers who move the products of this cultivar in December can get multiple times more than those moving their plums in August, when the majority of trees are fruiting. During the previous decade, the WAC has gathered more than 600 accessions of African plum, and many increases of a few different indigenous fruit trees. Utilizing sub-atomic markers and quantitative genetics, researchers will set up the level of genetic variation between the diverse increases. In the future, farmers will be urged to plant a genetically diverse scope of trees. In-reproducing may diminish efficiency and the absence of genetic variation departure from homesteads may make trees more disease prone." So far, explore in West and Central Africa has concentrated on fewer trees, yet reviews demonstrate that there are somewhere around 30 different types of indigenous fruit tree which are normally utilized by local individuals. During the coming years, a significant number of these will be liable to the participatory domestication approach developed over the previous decade. An extraordinary assorted variety of tree crops and a significantly more intricate, increasingly practiced (Pye-Smith 2010).

6.9 Selected Common African Wild Fruits

6.9.1 *Oil Palm (*Elaeis guineensis*)*

Oil palm (*Elaeis guineensis*) is widely distributed in tropical Africa, cannot tolerate full shade, but prefers disturbed habitats. West African origins, but has spread throughout tropical Africa. It is a tree. The oil palm is presently a standout amongst the most financially imperative palms in Africa. It has a walnut-size fruit bunched in pods, with a stringy mash, rich in oil (which is energy, fatty acids, and a great source of Vitamin A). Inside the husk is a hard-shelled seed containing a palatable portion (eaten by chimps and individuals) (Fig. 6.2).

The species still develops wild, and in addition being cultivated and planted by individuals. The wild form developing in Congo, gives 9% of the aggregate caloric intake, for instance (Ngalle et al. 2014).

Fig. 6.2 Oil Palm (*Elaeis guineensis*). (Source: https://commons.wikimedia.org)

6.9.2 Ziziphus mauritiana

Well known *Ziziphus*species are found in southwestern Asia, the Mediterranean region, and western Africa, India, Afghanistan, China, Malaysia, Australia, Pacific regions, Brazil and the United States. It can shape thick stands and wind up intrusive in a few zones, including Fiji and Australia and has turned into a genuine natural weed in Northern Australia. It is a quickly developing tree with a medium life expectancy that can rapidly reach up to 10–40 ft tall (Fig. 6.3).

Ziziphus mauritiana is a sharp, evergreen bush or little tree up to 15 m high, with a trunk of 40 cm or more in width; spreading crown; stipular spines and many hanging branches. The fruit is of variable shape and size. It is oval, obovate, elliptical or round, and that can be 1–2.5 in (2.5–6.25 cm) long, contingent upon the assortment. The tissue is white and fresh. At the point when marginally under ready, this organic product is a to some degree succulent and has a beguiling smell. The organic product's skin is reflexive, thin, however tight. It is the most usually found in the tropical and sub-tropical districts (Azam-Ali et al. 1968).

The plant contains crude protein, fat, fiber, ash, calcium, phosphorus, magnesium, potassium, sodium, chlorine, sulphur, they also contain ceryl alcohol and the alkaloids, protopine and berberine, quercetin, kaempferol, sitosterol, stigmasterol, lanosterol, diosgenin. The leaves contain flavonoids, tannins and holosides, mucilages, sterol, triterpenoids, cardiotonic glucosides, and leucoanthocyanes. Fresh fruits contain protein, fat, fiber, carbohydrates, reducing sugars, non-reducing sugars, ash, calcium, phosphorus, iron, carotene, thiamine, riboflavin, niacin, citric acid, ascorbic acid, fluoride, pectin. The fresh fruits also contain some malic and oxalic acid and quercetin (Sulieman et al. 2012; Kundu et al. 1989). The fruits are an imperative hotspot forbirds, which eat the entire fruit, expanding the seeds in the best conditions for germination (or nithochory). Secondly, seed dispersalis done by mammals or fishes. The fruit is vitality rich because of the far reaching measure of sugar it contains. It is produced and eaten fresh, dry, and in jam. It is additionally included as a base in suppers and in the make of sweets. The leaves can be either

Fig. 6.3 *Ziziphus mauritiana.* (Source: https://commons.wikimedia.org)

deciduous or evergreen relying upon species, and are sweet-smelling. The fruit works as a remedy, diuretic, and emollient. The leaves are astringent and febrifuge. Likewise, said to advance hair development. The dried fruits are used in many medicinal uses e.g. anticancer, sedative, stomachache, styptic and tonic. They are utilized to clear the blood and help digestion. They are utilized directly to treat chronic fatigue, loss of appetite, diarrhea, and anemia. The root is utilized in the treatment of dyspepsia (Devi et al. 1989; Azam-Ali et al. 2012) (Fig. 6.4).

Ziziphus mauritiana contains many important compounds such as triterpenes, cyclopeptide alkaloids and flavonoids that have been have anti- inhibitory effects on some important enzymes (Lee et al. 2004).

6.9.3 Balanites aegyptiaca *(Laloab)*

Locally known in Sudan as Lalob or Hegleeg, it has other names, such as: Indian almonds or lupus. It is a large old tree, and the parts used are the fruits and leaves, and usually fruits are harvested after maturity, and is similar to dry dates in its form. The native habitat of a plant that passes slaves: India, Pakistan, Iran and Sudan. Its contents include chloric acid, ephesian materials, resinous materials, fixed oils, and achene ether materials.

The fruit has many therapeutic uses: in a clinical study at the University of Khartoum Hospital proved the effect of the fruits of this plant on the hepatitis B

Fig. 6.4 Ripped *Ziziphus mauritiana* - Indian Jujube. (Source: https://commons.wikimedia.org/w/)

Fig. 6.5 *Balanites aegyptiaca* tree and fruit. (Source: https://commons.wikimedia.org/w/)

virus. Another study in India on the leaves of this plant found that the leaf extract reduces liver damage caused by chemicals inhaled from factories such as carbon tetrachloride and others. It helps in cure of colon cancer since it increases the movement of the colon and intestine, and is used as a natural laxative for the stomach that prevent constipation (Fig. 6.5).

6.9.4 Hyphaene thebaica *(Doum)*

The palm of the doum is an African origin, cultivated in Egypt since ancient times and abounds in Upper Egypt and the oases of the sea and the country of Nuba and Sudan. The Doum is classified as an ornamental tree, which was planted in the gardens, is slow growing and the nucleus is wooden. The old doum trees were used as supports for the roofs of houses, because the "corona" insect cannot kill it. It also

Fig. 6.6 *Hyphaene thebaica* **(Doum)** tree and fruits. (Source: https://commons.wikimedia.org/w/)

makes palm fronds, baskets, mats, mats and dishes, while the Pharaohs' lineage was made of palm tree fibers, and the length of the rope was about 300 cubits (Fig. 6.6).

The tree is similar to the camel. It bears the thirst and the extreme temperature rise, and it is very widespread in the area west of the Nile in Aswan Governorate, especially in the villages of Kubaniya and Fars in Egypt and thought Sudan.

The fruits of dom are rich in vitamins and minerals, the most important is vitamin A, which works on the strength of sight, and contains vitamin B1, B2, which work to strengthen the nerves, and contains antioxidants that delay the symptoms of aging, and is also very useful for patients with blood pressure, Because Doum regulates blood pressure and makes it at the normal level. Doum also reduces cholesterol in the blood, and can be taken as a drink cold or hot, which is fond of children who eat it moist, and works to whitening tooth strengthening.

Doum contains many vitamins such as vitamin A, B1, C, B2, B3, E and some minerals such as phosphorus, potassium, calcium, potassium, and therefore regulates blood pressure. For man, Doum helps to treat erectile dysfunction and increases the number of sperm. Doum contains a large percentage of fiber that gives the feeling of fullness and fullness, and helps digestion and get rid of digestive problems such as diarrhea and it works to lose weight and helps to burn excess fat when eating a healthy diet for the loss of excess weight. Beside these, Doum has many other uses such as:

1. Increases body immunity.
2. Helps treat sunburn.
3. Treatment of oral ulcers.
4. Enter into the treatment of hemorrhoids.
5. Helps in the treatment of asthma.
6. Works on teeth whitening.
7. Strengthens memory.
8. Protects against some types of cancer.
9. Reduces exposure to premature aging.
10. Helps to treat some skin diseases.
11. It treats infertility and helps with pregnancy.
12. Nourishes the scalp and helps germinate hair in the case of baldness.

*6.9.5 African Fan Palm (*Borassus aethiopum*)*

African fan palm (*Borassus aethiopum*) which give individuals a wide convenience on natural, nutritional, little scale, creating and producing and culturally plans (Bassir 1968; Sokpon et al. 2004; Sakandé et al. 2004). The tree roots are utilized as hostile to asthmatic. Leaves are utilized in numerous experts items. The petioles are utilized as fencing, kindling and the hypocotyls are consumable (Sokpon et al. 2004). The male blossoms are utilized as to manure soils and as feed (Ouinsavi et al. 2011). They are considered as a fantastic grain and their supplements content are like those of groundnut and cowpea (Atta 1997). The bole of the palm is utilized in carpentry for house and scaffold building. Its wood is exceptionally impervious to termites and organisms assaults and to atmosphere varieties (Ouinsavi et al. 2011).

African fan palm fruits are very rich in dietary fiber (about 500 grams per fruit) and have a strong odor to contain turpentine oil. They are consumed raw or cooked, preferably used with rice. The raw fruit nucleus contains albumin, with a sweet and refreshing taste. The immature seeds can be eaten and contain sweet jelly with a tasty taste. The tree buds are also used as delicious food. Fresh sap is used as a yeast or vinegar (http://hacen.net/%) (Fig. 6.7).

The fruit is a hard shell after ripening, and the hard shell can be removed. It contains a distinctly fibrous pulp, forming a layer on the nut. Each palm bear 8–15 clusters of fruit with a total of about 80 fruits per year. Fruits and young leaves are used as feed for livestock as well. The parts of the Boras are used in traditional

Fig. 6.7 African fan palm (*Borassus aethiopum*) tree and seed. (Source: https://commons.wikimedia.org/w/)

medicine, from which a libido is extracted. The flowers help heal the hoarseness of the sound and the leaves are used to stop the bleeding.

6.9.6 Tinas grewia

Known locally in Sudan as Gudem, which is a small fruit, has the size of a small chickpeas, orange in color, mixed with bright red color, grown in western Sudan, and in some areas of Yemen. The tree leaves are utilized as a powder for the treatment of bacteria and fungus. For their use, it is utilized as an effective insecticide, and the tree as a whole uses good and nutritious feed for the animal, the veins and roots treat jaundice. It is utilized in several forms, including juice, often in Ramadan and warm growth. The price of Gedim until recently was cheap and within reach, but after the fame gained by the latter of the great benefits for patients with anemia, it has increased the turnout, so that doctors themselves recommend it as the best option in nutrition. There is no doubt that Sudan is a country of wonders that is full of contradictions. At a time when it finds vegetarian properties in the rich, its people find a peak in poverty. Gudem plant grows in all environments of Sudan to the urban ones. And do not exaggerate if we say that a cup of cold Gudem juice in the early morning helps greatly in pushing the production wheel and a huge reduction in the bill of treatment.

6.9.7 *Tamarind (***Tamarindus indica***)*

Tamarind belongs to the legume family. It is the core of the horny fruits of a tree growing evergreen tree growing at a height of up to three meters and leaves clustered flower cluster, yellow and solid wood color reddish-brown fruits are horns and uses brown meat acid taste that encapsulates the seeds and when the fruit is still collecting straws. The tamarind is a large, evergreen trees with a length of 10 meters and up to 30 meters. They are perennial trees. They are about 200 years old. It is most likely that the origin of the tamarind tree in the Eastern African Equatoria region has been introduced to India for a long time and its cultivation flourished there. It was known by ancient Egyptians and Greece since the fourth century BC. The tamarind tree gives about 200 kg of fruits every year. The fruits of tamarind are irregular horns, about 17 centimeters long and 2.5 centimeters wide, with a number of seeds (1–10 seeds) of smooth solid structure surrounded by pulp. If you leave the fruit after maturity on the tree, it may remain a full year without spoiling.

The tamarindis fruits has many benefits, starting from the tree itself, it is a tropical ornamental plants often planted to beautify the gardens and to give shade. All its parts have been utilized: wood, bark, fiber, leaves and fruit. Tamarind has been used as a medical herb since the time of the silk and the pulp of tamarind fruit is prescribed and licensed in the British and American Pharmacopoeia and most other food constitutions in the world. Although there have been several cases of medical benefits for medicinal preparations containing pulp, flowers and tamarind leaves, what has been proven effective is the anti-scurvy agent, the tamarind-laxative agent and the diuretic agent in the fluid in the fiber. Their importance to the human being lies in the nutritional aspect (Sulieman et al. 2015; Saha et al. 2010) (Fig. 6.8).

Small plants, leaves and soft flowers are consumed as vegetables in the salads, soup, and cherries. The soft, unfinished centuries are used as spices for cooking rice, fish and meat. The fruit of the finished fruit is an important agricultural product with great economic value in many regions of the world. In the countries of Asia, the pulp is sold in the form of briquettes as a delicious food known for dates and compressed figs. During storage, there is a change in the strength and color of the pulp. After it was dry, the color of the dark becomes soft and black and color. This change is due to the work of bactinic enzymes and the nature of the gyroscopic pulp (absorption of moisture from the atmosphere) and contain a high proportion of sugar and acid. Tamarind pulp is an important ingredient in popular products such as Worcestershire

Fig. 6.8 Tamarind tree and fruit. (Source: https://wikimedia.org)

sauce, barbeque sauce, fruit source in the preparation of jams and jelly, and a number of drinks, including soda drinks. In tropical America, the pulp is used in refreshing beverages, including soft drinks. The seeds of Tamarind are used to a limited extent as food, but their industrial uses are many and varied, most notably the manufacture of glue seed.

*6.9.8 African Ebony (***Diospyros mespiliformis***)*

Diospyros mespiliformis belongs to Ebenaceae family. The plant is one of the largest and most majestic berry-bearing trees found in Africa, sub-tropical regions, and generally common to river and water-rich areas. It is height is 25 meters. It is found

Fig. 6.9 African Ebony tree+ fruits. (Source: https://commons.wikimedia.org)

in swamp semi-deciduous and evergreen forest, up to about 1000 m in elevation, will in general keep away from the wettest backwoods. It's anything but a typical tree in natural forests, grown alone or in a small group. The most well known part is the heart wood used as timber, which is dark brown with black strips. It is utilized for substantial ground surface, ship building, vehicle bodies, knife handles, and music instruments. The tree bark and sap utilized as antibacterial and antifungal medicine (Fig. 6.9).

The roots, leaves and fruits contain tannin and act as an antibiotic, and are utilized for conventional therapeutic purposes, treating various disorders such as skin eruptions, acne and sores. People, animals and birds eat the ready yellow-green berries. The berries are about 2 cm in distance across, green at first, and end up yellow when ready. Boiled bark yields is utilized as a dark color.

*6.9.9 Carob Tree (*Ceratonia siliqua*)*

The Carob name is of Arabic origin. It is an evergreen tree belonging to the family (Caesalpinioideae) of the legume (Latin: Fabaceae), live naturally in the Mediterranean region, including the Arab countries. Carob seeds are significantly equal in size and weight. Carob tree extended tree evergreen shadows reach a

Fig. 6.10 *Ceratonia siliqua* carob tree, unripe and dried pods. (Source: https://commons.wikimedia.org/w)

height of between 15 and 17 meters. At the age of 18, the leg thickness is 85 cm (Craig and Nguyen 1984). Leaves are composed of a long neck with 6–10 opposite leaves, a leaf shape. The edge is rounded at the top, the leaves are dark green leather with a length (2.5–6.25 cm) (Fig. 6.10).

The carob tree is an economically promising tree in the long run. The agricultural production of carob seeds is about 330,000 tons per year. Spain (especially the regions of Valencia, Catalonia and the Balearic Islands) is one of the most important regions in the cultivation of this tree. Production is estimated at 45% of the world production, followed by Italy with 16%, Portugal 9%, Morocco 7.5%, Cyprus 6%, then Greece 5%, then Turkey 4%. Smaller quantities are produced in Algeria, Tunisia, Israel and the rest of the Mediterranean.

It contains 60% protein and plenty of cholesterol-free oils. The carob fruits contain vitamins A, B, B, B, B, B, and D) and important metal elements such as potassium, calcium, iron, phosphorus, manganese, barium, copper, nickel, magnesium, etc., and oxalic acid, which inhibits the absorption of calcium and other mineral elements, Facilitate the process of intestinal absorption of these minerals and benefit greatly, and is characterized Bactin fruits not caused by the emergence of symptoms of allergy (Burg 2007).

*6.9.10 Moringa (*Moringa oleifera*)*

Moringa oleifera Lam. (Moringa) is a most well known cultivated tree belonging to Moringaceae family (Fuglie 1999). The tree is cultivated in all sub-tropical zones. This tropical quickly developing tree is impervious to dryness. Moringa is distributed in Tanzania, Nicaragua, Malawi, Brazil, Niger, Indonesia and Senegal (Optima of Africa, Ltd.) The ground seed powder in Sudan is used for the purification drinking water in rural areas (Jahn and Dirar 1979). The Moringa tree can adapt to any environment characterized by its high drought tolerance. It does not need much water and is sufficient for rainwater, so it grows in the mountains and deserts; it grows in arid, warm, semi dry, dry, temperate and warm areas. It is also an evergreen tree used to improve soil properties and is utilized in several other areas such as nematode control, animal feed and beekeeping. It is successfully cultivated on canal and waterway bridges, in household gardens and on farms, and does not suffer from pests or diseases unless planted in areas with bad drainage. Moringa grow speedy, with a height of more than 2 meters in less than 2 months and more than three meters in less than 10 months.

Moringa leaves are rich in beta-carotene, vitamin A and C, iron, protein, potassium and phosphorus. They form an integral food in some regions of Africa. They are also used as a dietary supplement for people with HIV in some African countries. Vitamins. The powdered leaves are used as a seasoning for meals. Experiments have shown that the addition of Moringa leaves in the diet of lactating women has led to increased milk yield, and leaf juice lowers high blood pressure, and is effective in the production of urine (https://www.ruoaa.com). The leaves contain seven times the vitamin C found in oranges, three times the banana content of potassium, four times the milk content of the calcium, four times the carrot content of vitamin A and the weaker the milk content of the protein. Leaf leaves can be eaten either fresh or cooked like spinach, as can be dried and ground in a powder image that can be added to sauces or soups. Moringa stalks are used as firewood in rural communities. The bark produces a substance used in some pharmaceutical industries and is also used in the treatment of diarrhea. Moringa roots are a treatment for rheumatism in some areas (Fig. 6.11).

The Moringa plant is used to purify the water of the remaining seeds after extracting the oil by adding it to the water reservoirs. This is because it has the properties of collecting and depositing the water impurities in a trap-like manner. It works to purify the water from impurities and bacteria at the same time. Which has contaminated water and no special purification plants in Africa to reduce epidemics and diseases. *M. oleifera* offers an elective answer for the utilization of costly compound coagulants. The tests with these natural coagulants gave separated water turbidity not exactly or relatively equivalent to 1NTU (Nephelometric Turbidity Unit) and in this manner met the turbidity criteria for drinking water according to WHO guides (Yousif et al. 2013).

Fig. 6.11 The Moringa plant, fruits and flowers. (Source: https://commons.wikimedia.org/w/)

References

Agudo A (2004) Measuring intake of fruit and vegetables, Background paper for the Joint FAO/WHO workshop on fruit and vegetables for health, 1–3 September 2004, Kobe, Japan. Available at: http://www.who.int/dietphysicalactivity/publications/f&v_intake_measurement.pdf. Accessed July 2012

Asfaw Z, Tadesse M (2001) Prospects for the sustainable use and development of wild food plants in Ethiopia. Econ Bot 55:47–62

Atta S (1997) The Palmyra (*Borassus aethiopum*). Flowering, biomass and contribution of nutrients, Report, UICN/ PAIGLR/MH/E, Niger, p 24

Azam-Ali S, Bonkoungou E, Bowe C, deKock C, Godara A, Williams JT, Bassir O (1968) Some Nigerian wines. West Africa J Bio App Chem 10:42–45

Azam-Ali S, Bonkoungou E, Bowe C, deKock C, Godara A, Williams JT (2012) Ber and other jujubes, Ziziphus species. Fruits for Gupta et al., IJPSR. 3(3):818–821. ISSN: 0975-8232 Available online on www.ijpsr.com

Bassir O (1968) Some Nigerian wines. W Afr J Biol Chem 10:42–45

Burg B (2007) Good treats for dogs cookbook for dogs: 50 home-cooked treats for special occasions. Quarry books. CABI, Oxford shire, p 28

Craig WJ, Nguyen TT (1984) Caffeine and theobromine levels in cocoa and carob products. J Food Sci 49(1):302–303, 305. https://doi.org/10.1111/j.1365-2621.1984.tb13737.x

Cunningham AB (2001) Applied ethnobotany: people, wild plant use and conservation. Earthscan Publications Ltd, London/Sterling

Devi KR, Atluri JB, Reddi CS (1989) Pollination ecology of *Zizyphus mauritiana* (Rhamnaceae). Proceedings of the Indian Academy of Sciences. Plant Sci 99(3):223–239

Eriksson G, Ekberg I, Clapham D (2006) Introduction to forest genetics, 2nd edn. SLU, Uppsala

FAO (2005) Building on gender, agrobiodiversity and LK: a training manual. Publishing

FAO (2008) http://www.fao.org/3/a-i0100e.pdf

Fuglie LJ (1999) The miracle tree: *Moringa oleifera*: natural nutrition for the tropics. Church World Service, Dakar. 68 pp; revised in 2001 and published as The Miracle Tree: The Multiple Attributes of Moringa, 172 pp.

Garrity D (2008) Foreword. In: Akinnifesi FK, Leakey RRB, Ajayi OC, Sileshi G, Tchoundjeu Z, Matakala P, Kwesiga F (eds) Indigenous fruit trees in the tropics: domestication, utilization and commercialization. Cambridge, MA, Wallingford

Gullick (1997) Characterization of chimeric proteins constructed from human epidermal growth factor (EGF) and the Drosophila EGF-receptor antagonist argos. Int J Oncol 10(4):677–682. https://doi.org/10.3892/ijo.10.4

http://hacen.net/

http://www.fineli.fi/, Accessed 20 July 2012

http://www.hort.purdue.edu/newcrop/

https://commons.wikimedia.org

https://www.ruoaa.com/2013/09/moringa-tree-seeds-leaves-oil-benefits.html

Jahn SA, Dirar H (1979) Studies on natural water coagulants in the Sudan, with special reference to *Moringa Oleifera* seeds. S Afr Water Res commission 5(2):90–95

Jordaan DPS, Akinnifesi FK, Ham C, Ajayi OC (2008) The feasibility of small scale indigenous fruit processing enterprises in Southern Africa. In: Akinnifesi FK, Leakey RRB, Ajayi OC, Sileshi G, Tchoundjeu Z, Matakala P, Kwesiga FR (eds) Indigenous fruit trees in the tropics: domestication, utilization and commercialization. CAB International, Wallingford, UK, in association with the World Agroforestry Centre, Nairobi, pp 273–287

Kundu AD, Barik BR, Mandal DN, Dey AK, Banerji A (1989) Zizybernalic acid, a penta cyclic triterpenoid of *Ziziphus jujuba*. Phytochemistry 28(11):3155–3158

Lee JT, Connor-Appleton S, Haq AU, Bailey CA, Cartwright AL (2004) Quantitative measurement of negligible trypsin inhibitor activity and nutrient analysis of guar meal fractions. J Agric Food Chem 52:6492–6495

Lukmanji Z, Hertzmark E, Mlingi N, Assey V, Ndossi G, Fawzi W (2008) Management service, Rome

Lulekal E, Asfaw Z, Kelbessa E, Van Damme P (2011) Wild edible plants in Ethiopia: a review on their potential to combat food insecurity. Africa Focus 24(2):71–121

Namkoong G, Kang HC, Brouard JS (1988) Tree breeding: principles and strategies. Monographs on theoretical and applied genetics, vol 11. Springer, New York

Ngalle HB, Bell JM, Ngando-Ebongue GF, Eman-Evina H, Ntsomboh GN, Nsimi-Mva A (2014) Morphogenesis of oil palm (*Elaeis guineensis* Jacq.) fruit in seed development. J Life Sci 8:946–954

Ouinsavi C, Gbémavo C, Sokpon N (2011) Ecological structure and fruit production of African fan palm (*Borassus aethiopum*) populations. Am J Plant Sci 2:733–743. https://doi.org/10.4236/ajps

Pye-Smith C (2010) The fruits of success: a programme to domesticate West and Central Africa's wild fruit trees is raising incomes, improving health and stimulating the rural economy. World Agroforestry Centre, United Nations Avenue, Gigiri, Nairobi

Saha P, Chowdhury S, Gupta S, Kumar I, Kumar R (2010) Assessment on the removal of malachite green using tamarind fruit shell as biosorbent. Clean (Weinh) 38:437–445

Sakandé J, Nacoulma OG, Nikiema JB, Compo M, Bassene E, Guissou IP (2004) Effect of male inflorescence of *Borassus aethiopum*. Médécine d'Afrique Noire 51(5):280–282

Simons AJ (2003) Concepts and principles of tree domestication. In: Simons AJ, Beniest J (eds) Tree domestication in agroforestry. World Agroforestry Centre, Nairobi

Sokpon N, Houankoun E, Ouinsavi C, Yabi J (2004) Importance Socio-Economique du Rônier (*Borassus aethiopum* Mart): Différent Usages et Commercialisation de Quelques Sous Produit au Bénin," Communication PrésenTée à L'atelier Scientifique Sud-Centre, Abomey-Calavi, 2004. Source: https://commons.wikimedia.org/w/

Spiller A (2001) Preispolitik für ökologische Lebensmittel: Eine neo institutionalistische Analyse. Agrarwirtschaft 50:451–461

Splittstoesser WE (1990) Vegetable growing handbook. Avi Publishing Company Inc Tanzania food composition Tables. MUHAS-TFNC, HSPH, Dar es Salaam Tanzania, Westport

Sulieman AE, Yousif HB, Elkhalifa EA (2012) Production and quality assessment of vinegar prepared from Nabag (*Zizyphus spina christi*). Sudan J Agric Res 20:49–60

Sulieman AE, Alawad SM, Osman MA, Abdelmageed EA (2015) Physicochemical characteristics of local varieties of tamarind (*Tamarindus indica* L), Sudan. Int J Plant Res 5(1):13–18

Teketay D, Senbeta F, Maclachlan M, Bekele M, Barklund P (2010) Development of wild food plants in Ethiopia. Econ Bot 55:47–62

Yousif NN, Elamin HB, Ali HA, Sulieman AME (2013) Utilization of *Moringa oleifera* seeds in purification of tap water in Khartoum State, Sudan. J Microbiol Res 3(5):171–175

Chapter 7
Contribution of Wild Edible Fruits to Rural Peoples Income

Rania Mustafa Abdalrhman Ahmed

7.1 Introduction

The wild edible plants (WEPs) can be define as the plant species can be utilized as food sources and are never cultivated and never domesticated. The essential source of food for the many early hunter gathering communities they were woodland, forests, wild plants and animals (Alemayehu et al. 2015) and are still present for food dependent communities. Edible wild plants play an important role in making a living ensuring food security and livelihoods for countless families and communities around the world (Bell 1995; Guijt et al. 1995; Molla et al. 2011). They are considered a part of the diet of farm households on a daily basis and during the periods of food shortage. Edible wild plants have been showed to have a number of important dietary elements. Like, vitamins, minerals, proteins, and higher fat contents than cultivated species. They contain niacin, iron, zinc, iodine, calcium, thiamine, riboflavin, and vitamins A and C (Ohiokpehai 2003). Hence, edible wild plants are providing nutritional balance in the diet (Ohiokpehai 2003). Furthermore the exhaustion of edible wild plants is an important part of the strategy adopted by people in order to survive and serve as a major source of food during harsh environments and periods moreover, during times of stress such as failure, pest attack, and drought (Guinand and Lemessa 2001; Neudeck et al. 2012). Moreover the demand for the marketing of edible wild plants is increasing due to increased demand for fruit in urban centers and the lack of alternative economic options for rural populations (Mithöfer and Waibel 2003). Economically about 15 million people in sub-Saharan Africa has been Large access to cash income from forest-related activities (Oksanen and Mersmann 2003).

R. M. A. Ahmed (✉)
Industrial Research & Consultancy Centre, Pharmaceutical Industries Section, Khartoum, Sudan

A. A. Mariod (ed.), *Wild Fruits: Composition, Nutritional Value and Products*,
https://doi.org/10.1007/978-3-030-31885-7_7

7.2 Nutritive Value of Edible Wild Fruits

A few studies in nutrient constituent of edible wild bulbs *Fritillaria camschatcensis* report to be a good source of carbohydrates. Wild *Parkia biglobosa* is wild flowers used as food which show considerable quantities of retinol equivalent, Bambuseae are rich in protein, phosphorus and calcium, but they have small amounts of vitamin C, on the other hand, tubers contained high amounts of available carbohydrates and phosphorus. While *Hibiscus sabdariffa* contain amounts of magnesium, calcium, zinc and iron. The roots and rhizomes is similar in nutrient composition, but ash content is higher in rhizomes than in roots (Guil et al. 1996).

7.3 Wild Edible Plants Habitat

The climate plays important role of type edible plant. If the place humidity, the wild edible plant places are sunnier. Wild edible plants grow close to water sources if living region is more dry.

7.3.1 Berries and Cherries Fruit Edible Plant

The berries and cherries taste good to eat. Especially silver berry is very tasty food at eating, its leaves show silvery appearance and very delicious to eat. Wild cherries can be found on trees during winter season (Fig. 7.1).

7.3.2 Nuts Underneath the Trees

Fresh nuts are generally wet and easy to digest. There are so many acorns under the oak trees. The fruit trees usually found at places which are sunnier, such as the roadsides, forest edges, and canal banks rich with fruiting.

7.4 Conservation and Diversity

75,000 species of plant worlds – wide are believed to be edible. Till now 2,50,000–3,00,000 higher plant species are documented, although a few hundred species are documented, just only hundred species cultivated and thousand are gathered from the wild (Sekeroglu et al. 2006). The national population contributes significantly to only 103 species of food plants by 90%. Of these species, only 20% of

Fig. 7.1 Wild edible plant fruit, (**a**) Berries, (**b**) nuts, (**c**) wild apple, and (**d**) wild mulberries

species are stable (Heywood 1999). Current studies have proven, About 1200 species were identified in Africa, 1000 species were recognize in America, and 800 species were found in Asia (Verheij and Coronel 1991; Heywood 1999). There are about 3000 tropical fruits all over the world Remain untapped (Di Castri and Younes 1996). The promotion and localization of these wild fruit species improve nutritional value In protecting them from the loss of the wild environment (Brownrigg 1985).

7.5 Physiochemical Properties

Physico-chemical characterization of wild edible fruits very important will help in the picking of convenient wild species which can then be developed through hybridization for improvement of a cultivated variety. Several analytical reports concerning to these wild fruits from different parts of the world which are necessary for commercial and nutritional values (Muok et al. 2000; Zatylny et al. 2005; Paull and Duarte 2011).

Acidity and Total soluble solid (TSS) are an important characteristic, giving information about potential acidity and sugar content of fruits. Sugar provides the main source of energy that is found to be high in many edible wild fruits. Acidity indicated the total number of acid molecules. TSS, sugar and acidity of many wild edible fruits have been studied from a different country from the world (Sundriyal and Sundriyal 2001; Ercisli and Orhan 2007; Iqbal et al. 2010; Deshmukh and Waghmode 2011; Kumari and Kumari 2011). Fat and protin studied varied from 2.0–5.0% and 0.4–18% respectively (Sundriyal and Sundriyal 2001; Deshmukh and Waghmode 2011). Essential amino acid are also present in the plant (Cameron et al. 1976). other physico-chemical properties such as antioxidants substance reduce damage chain (Fischer-Nielsen et al. 1992). Some studied wild edible fruits contain antioxidant e.g. vitamin, pro vitamin A. β-carotene (Beekwilder et al. 2005; Netzel et al. 2007; Ikram et al. 2009).

7.6 Utilization of Wild Edible Fruits

WEPs have been a source of edible food which had provided the community with food and income (Heywood 1999; Grivetti and Ogle 2000). The economic demand and traditional knowledge system of a community affects the utilization of WEPs in a specific area (Edeoga et al. 2003; Tabuti et al. 2004; Sawian et al. 2007; Jama et al. 2008). Most of the fruit purchased by the community is located near the forest (Styger et al. 1999; Pender and Hazell 2000; Harris and Mohammed 2003; Afolayan and Jimoh 2009). Some reports strongly show that the WEPs can significantly substantiate the global food basket in today's era of climate change (Nazarudeen 2010).

7.7 Case Study in Economic Value of Wild Fruit

The wild plants are one of the most important means used for food deficit in the dry season of the rural families in Sudan. For example, the wild fruit in Kordofan contributed to increase the family income and marketing and house hold food security, like *adansonia digitata, Banalities aegyptiaca, Grewia tenax.*

The wild plant should be considered for both economic and nutritional value, And then they can be exchanged or sold in the markets in addition to collecting and selling firewood, Women also collect and sell firewood and use it as charcoal to cook food, the sale of wild plants Such as *Moringa oleifera*, lalop, ardeeb (*Tamarindus indica*), lulu or shea Nuts (*Vitellaria paradoxa*) in the market. They can be kept longer compare with vegetables. in the markets of Akobo and Kapoeta The oil tin of Lalop 20 SSP As in North Bahr El Ghazal trade is more like where Some shops were established to sell lalop seeds by a 50 kg sack for 300 SSP each. (Dragicevic 2017). Kudera has been sold to1 or 2 SSP per handful. There are some plants sold in the villages for their nutritional value and distinctive smells for example, in Dengjok, Akobo key Plant (Nue) (Elmola and Ibrahim).

In another study The baobab plant was found to be has numerous nutritional and medicinal uses in Africa (von Maydell 1986; Van Wyk and Gericke 2000). the baobab tree result shows its useful (Owen 1970). The fruit pulp is used in preparing hot and cool drinks in rural areas further more become a popular ingredient in ice products in urban areas (Gebauer et al. 2002). Botswana is one of the semi-arid states located in two thirds of Kalahari Sandfield. Rain falls clearly in November to March (Govedich et al. 2010). Although the area is barren, the country is rich in about 150 wild and non-edible wild plants (Flyman and Afolayan 2006). Studies indicate that there is 1 year of every 3 years in which agriculture is failing in Botswana due to lack of rainfall and general lack of surface water (Fend and Brinkhurst 2010; Govedich et al. 2010) During these drought years, rural communities in Botswana are using local food plants as a strategy to cope with hunger to maintain their livelihoods (Twyman 2001).

Future economic potential of cultivated crops in compare to Edible wild plants study showed the highest in minerals concentrations and vitamins (Smith et al. 1996). The vitamin C content of more than 2500 mg kg–1 is considered to be high compare to other fruits. The use of baobab fruit could raise nutritional standards and also endues the market availability and trade of baobab powder. Currently baobab fruits are already an important source of both income and food for certain tribes living in Central and South Sudan (Sidibé et al. 1996). In the Economic valuation of non-marketed wild edible plants in Thailand in family of Karen held 190 trips in the forest, spending 14.59 days, costing US $ 30.79, and local wages averaging US $ 2.11 per daod crops from the market was found to cost 531 kg of wild plants. The Karen family costs $ 302.40 per family per year. For this study, they reached several conclusions. For Karen, eating edible wild plants appear to be more effective way. To get $ 302.40 a year, each household unit must work for 143.32 days of work, 14.59 days, with an increased opportunity cost., Economically, gathering of edible wild plants leads to decrease the value of the purchase of commercial crops. Thus, edible wild plants replace commercial crop (Delang 2005). Gweta Village is an a good example for economic utilization of EWPs. This rural area rich with a variety of ranging from leafy vegetables and fruits utilized by the population. The studies in gwrea revealed on documenting and synthesizing indigenous information explain how to utilized the EWPs and evaluate their contribution to household food supply and income in Gweta Village. Central Botswana. Data were gathered through direct household survey of forty five households and five key informants. Twenty four species were present that belong to thirteen families, utilized as source of beverage and food. Most food supply for EWPs include *Scloreocarya birrea, Grewia flava, Corchorus olitorius, Grewia bicolor, Corchorus olitorius, Amaranthus thunbergii, Cleome gynandra*. They are the common source of food and income, during the harvested season everyone distributed in the Village. About 52% of the respondents were engaged in the sale of nine EWPs, and the sale contributed between BWP 50.00 to more than BWP 400.00 per week. The research showed that women are main users and collectors of EWPs and children are participated in regular harvesters in small quantities for consumption as a snack. Furthermore, the study showed that elephants are the major threat to diminishing of EWPs in the Village (Badimo et al. 2015).

7.8 Conclusions

This article aims to review available information on diversity, conversation, local knowledge and the physico-chemical properties of edible fruit. The research was confined to poverty-stricken areas. So as to achieve rural development in order to improve living conditions. More areas of research should be absorbed and more scientific research should be conducted to determine the nutritional composition of wild edible fruits for their many benefits compared to crops and fruits., Further reflection on edible fruits around the world should be studied in the forefront by involving indigenous communities. It is concluded that edible wild plants can be important nutrient contributors in the diet, during the period of famine and healthy foods. To know the contribution of edible wild food resources more information is required on nutrient constituents in raw foods as well as on the bioavailability of specific nutrients. Much effort should be made in the area in the coming years. This will be necessary, as a proportion of the population increase in recent addition to the fact that agriculture is very expensive. There is still an enormous work and effort to be done to improve, induce and market EWPs. Awareness and give full information to the communities about the value and benefits of EWPs should be a priority in the rural economy directly conduct to the constancy with the cost-benefit and analyses before allocation of land either for arable agriculture or residential plots. This is important in the protection and conservation of valuable EWPs in various localities.

References

Afolayan A, Jimoh F (2009) Nutritional quality of some wild leafy vegetables in South Africa. Int J Food Sci Nutr 60(5):424–431

Alemayehu G, Asfaw Z, Kelbessa E (2015) Plant diversity and ethnobotany in Berehet District, North Shewa Zone of Amhara region (Ethiopia) with emphasis on wild edible plants. J Med Plants Stud 3(6):93–105

Badimo D, Lepetu J, Teketay D (2015) Utilization of edible wild plants and their contribution to household income in Gweta Village, Central Botswana. Afr J Food Sci Technol 6:220–228

Beekwilder J, Jonker H, Meesters P, Hall RD, van der Meer IM, Ric de Vos C (2005) Antioxidants in raspberry: on-line analysis links antioxidant activity to a diversity of individual metabolites. J Agric Food Chem 53(9):3313–3320

Bell J (1995) The hidden harvest. In seedling, the quarterlynewsletter of Genetic Resources Action International (GRAIN)

Brownrigg LA (1985) Home gardening in international development: what the literature shows, including an annotated bibliography, and inventories of international organizations involved in home gardening and their projects. League for International Food Education, Washington, D.C

Cameron B, Chasseaud L, Conway B, Fox N, Taylor T (1976) Absorption and disposition of econazole nitrate after application to the skins and vaginas of rabbits. Arzneimittelforschung 26(11):2054–2059

Delang CO (2005) Economic valuation of non-marketed wild edible plants in Thailand. Environ Conserv 32(4):285–287

Deshmukh B, Waghmode A (2011) Role of wild edible fruits as a food resource: traditional knowledge. Int J Pharm Life Sci 2(7):919

Di Castri F, Younes T (1996) Biodiversity, science and development: towards a new partnership. CAB International, Oxford

Dragicevic H (2017) Everything except the soil: understanding wild food consumption during the lean season in South Sudan

Edeoga H, Okwu D, Mbaebie B (2003) Minerals and nutritive value of some Nigerian medicinal plants. J Med Aromat Plant Sci 25:1010–1015

Elmola SAF, Ibrahim AH, Tropentag (2015) Contribution of wild fruits to household income and food security among small scale farmers in West Kordofan State-Central-west of Sudan. International Research on Food Security, Natural Resource Management and Rural Development, Berlin, Germany.

Ercisli S, Orhan E (2007) Chemical composition of white (Morus alba), red (Morus rubra) and black (Morus nigra) mulberry fruits. Food Chem 103(4):1380–1384

Fend SV, Brinkhurst RO (2010) Contributions towards a review of the genus Rhynchelmis Hoffmeister (Clitellata: Lumbriculidae). Zootaxa 2407:1–27

Fischer-Nielsen A, Poulsen HE, Loft S (1992) 8-Hydroxydeoxyguanosine in vitro: effects of glutathione, ascorbate, and 5-aminosalicylic acid. Free Radic Biol Med 13(2):121–126

Flyman M, Afolayan A (2006) The suitability of wild vegetables for alleviating human dietary deficiencies. S Afr J Bot 72(4):492–497

Gebauer J, El-Siddig K, Ebert G (2002) Baobab (Adansonia digitata L.): a review on a multipurpose tree with promising future in the Sudan. Gartenbauwissenschaft 67(4):155–160

Govedich FR, Bain BA, Moser WE, Gelder SR, Davies RW, Brinkhurst RO (2010) Annelida (Clitellata): Oligochaeta, Branchiobdellida, Hirudinida, and Acanthobdellida. In: Ecology and classification of North American freshwater invertebrates. Academic Press: San Diego, pp 385–436, Pretoria, 352

Grivetti LE, Ogle BM (2000) Value of traditional foods in meeting macro-and micronutrient needs: the wild plant connection. Nutr Res Rev 13(1):31–46

Guijt I, Hinchcliffe F, Melnyk M (1995) The hidden harvest: the value of wild resources in agricultural systems: a summary. IIED, London, pp 28

Guil JL, Torija ME, Giménez JJ, Rodríguez-García I, Giménez A (1996) Oxalic acid and calcium determination in wild edible plants. J Agric Food Chem 44(7):1821–1823

Guinand Y, Lemessa D (2001) Wild-food plants in Ethiopia: reflections on the role of wild foods and famine foods at a time of drought. Potential Indig Wild Foods 22:31

Harris FM, Mohammed S (2003) Relying on nature: wild foods in northern Nigeria. AMBIO J Hum Environ 32(1):24–30

Heywood VH (1999) Use and potential of wild plants in farm households. Food and Agriculture Organization of the United Nations. Rome, Italy

Ikram MA, Seshadri S, Bis JC, Fornage M, DeStefano AL, Aulchenko YS, Debette S, Lumley T, Folsom AR, Van Den Herik EG (2009) Genomewide association studies of stroke. Food & Agriculture Org, pp 113

Iqbal M, Mir K, Munir M (2010) Physico-chemical characteristics of different mulberry cultivars grown under agro-climatic conditions of Miran Shah, North Waziristan (Khyber Pakhtunkhwa), Pakistan. J Agric Res (03681157) 48(2):209

Jama B, Mohamed A, Mulatya J, Njui A (2008) Comparing the "Big Five": a framework for the sustainable management of indigenous fruit trees in the drylands of East and Central Africa. Ecol Indic 8(2):170–179

Kumari S, Kumari S (2011) Documentation and nutritional analysis of underutilized fruits and vegetables of Himachal Pradesh. CSKHPKV, Palampur

Mithöfer D, Waibel H (2003) Income and labour productivity of collection and use of indigenous fruit tree products in Zimbabwe. Agrofor Syst 59(3):295–305

Molla EL, Asfaw Z, Kelbessa E, Van Damme P (2011) Wild edible plants in Ethiopia: a review on their potential to combat food insecurity. Afr Focus 24(2):71–121

Muok B, Owuor B, Dawson I, Were J (2000) The potential of indigenous fruit trees: results of a survey in Kitui District, Kenya. Agrofor Today 12(1):13–16

Nazarudeen A (2010) Nutritional composition of some lesser-known fruits used by the ethnic communities and local folks of Kerala. Indian J Tradit Knowl 9:398
Netzel M, Netzel G, Tian Q, Schwartz S, Konczak I (2007) Native Australian fruits—a novel source of antioxidants for food. Innov Food Sci Emerg Technol 8(3):339–346
Neudeck L, Avelino L, Bareetseng P, Ngwenya BN, Motsholapheko M (2012) The contribution of edible wild plants to food security, dietary diversity and income of households in Shorobe Village, northern Botswana. Ethnobot Res Appl 10:449–462
Ohiokpehai O (2003) Promoting the nutritional goodness of traditional food products. Pak J Nutr 2(4):267–270
Oksanen T, Mersmann C (2003) Forests in poverty reduction strategies-An assessment of PRSP processes in Sub-Saharan Africa. Forests in poverty reduction strategies: capturing the potential. EFI Proceedings
Owen J (1970) The medico-social and cultural significance of Adansonia digitata (baobab) in African communities. Afr Notes 6(1):26–36
Paull RE, Duarte O (2011) Tropical Fruits (Agriculture). CABI, pp 436
Pender J, Hazell P (2000) Promoting sustainable development in less-favored areas. International Food Policy Research Institute (IFPRI), Washington, D.C
Sawian JT, Jeeva S, Lyndem FG, Mishra BP, Laloo RC (2007) Wild edible plants of Meghalaya, North-east India. Natural Product Radiance 6(5):410–426
Sekeroglu N, Ozkutlu F, Deveci M, Dede O, Yilmaz N (2006) Evaluation of some wild plants aspect of their nutritional values used as vegetable in eastern Black Sea region of Turkey. Asian J Plant Sci 5(2):185–189
Sidibé M, Scheuring J, Tembely D, Sidibé M, Hofman P, Frigg M (1996) Baobab-homegrown vitamin C for Africa. Agrofor Today 8(2):13–15
Smith GC, Dueker SR, Clifford AJ, Grivetti LE (1996) Carotenoid values of selected plant foods common to southern Burkina Faso, West Africa. Ecol Food Nutr 35(1):43–58
Styger E, Rakotoarimanana J, Rabevohitra R, Fernandes E (1999) Indigenous fruit trees of Madagascar: potential components of agroforestry systems to improve human nutrition and restore biological diversity. Agrofor Syst 46(3):289–310
Sundriyal M, Sundriyal d C (2001) Wild edible plants of the Sikkim Himalaya: nutritive values of selected species. Econ Bot 55(3):377
Tabuti J, Dhillion S, Lye K (2004) The status of wild food plants in Bulamogi County, Uganda. Int J Food Sci Nutr 55(6):485–498
Twyman C (2001) Natural resource use and livelihoods in Botswana's wildlife management areas. Appl Geogr 21(1):45–68
Van Wyk B-E, Gericke N (2000) People's plants: a guide to useful plants of Southern Africa. Briza Publications, Pretoria, pp 352
Verheij E, Coronel R (1991) Plant resources of South-East Asia. No. 2: edible fruits and nuts
von Maydell, H.-J. (1986). Trees and shrubs of the Sahel, their characteristics and uses
Zatylny A, Ziehl W, St-Pierre R (2005) Physicochemical properties of fruit of 16 saskatoon (Amelanchier alnifolia Nutt.) cultivars. Can J Plant Sci 85(4):933–938

Chapter 8
The Impact of Wild Fruits to a Better Life Worldwide

Suzy Munir Salama

Abbreviations

Ca	Calcium
Cu	Copper
Fe	Iron
K	Potassium
Mg	Magnesium
Mn	Manganese
P	Phosphorus
WHO	World Health Organization
Zn	Zinc

8.1 Introduction

Plant foods especially wild edible fruits and vegetables contain plenty of health-promoting nutrients such as vitamins and minerals, and biologically active compounds that are sufficient enough to reduce the incidence of chronic diseases (Phillips et al. 2014). Based on United States Dietary Guidelines Advisory Committee 2015, one of the dietary patterns recommended for reducing cardiovascular disease is the intake of meals containing less saturated trans-fats, sodium and red meat, and high vegetables and fruits (Jenkins et al. 2018; Millen et al. 2016). In rural regions, wild fruits play crucial role in preventing diseases and maintain the health of indigenous tribes (Borek 2017). Healthy diet should contain the right proportions of nutrients to lower the incidence of health disorders and mortality related to malnutrition (Bvenura and Sivakumar 2017). The regular intake of fruits, vegetables and fibre-rich food stuffs significantly lower the risk of many health ailments

S. M. Salama (✉)
Department of Molecular Medicine and Biomedical Science, Faculty of Medicine, University of Malaya, Kuala Lumpur, Malaysia
e-mail: suzymunir@um.edu.my

A. A. Mariod (ed.), *Wild Fruits: Composition, Nutritional Value and Products*,
https://doi.org/10.1007/978-3-030-31885-7_8

(Wilson et al. 2017). Based on recent report from World Health Organization (WHO), many people changed their lifestyle shifting to processed fatty food that is rich in energy and sugars/salts (Fernández-Ruiz et al. 2017). Allen et al. (2006) published that about 80% of people from both sexes and many developing countries utilize less than ½ kg of the daily suggested amount of fruits and vegetables (Allen et al. 2006).

Medical Scientists defined dietary fibre as indigestible oligosaccharides and polysaccharides that are found plenty in fruits and vegetables. Recent studies published that dietary fibre plays important role in maintaining health of the human body particularly gut and heart (Slavin 2013). Due to the role of dietary fibre in preventing the risk of many diseases such as colon cancer and cardiovascular diseases, Nutritionalists suggest adults to have daily consumption of 25 g dietary fibers *via* the intake of whole grain meals, wild edible fruits and vegetables (Bvenura and Sivakumar 2017).

Many studies showed that oxidative stress and the imbalance between oxidants and antioxidant machinery of the body play key role in initiating many health problems such as autoimmune, neurodegenerative and heart diseases, liver cirrhosis and cancer (Kumagai et al. 2003; Mohanty and Cock 2012; Na et al. 2014). Borek, 2017 reported that vitamins E and C as natural dietary antioxidants are considered as immunity-enhancing factors in cancer therapy via relieving the side effects caused by chemotherapy and radiotherapy in some patients (Borek 2017). The longevity of Mediterranean people compared to Europeans and Americans may be referred to their high intake of dietary antioxidant-rich vegetables and fruits (Wilson et al. 2017). Dietary antioxidants from fruits and vegetables e.g. Phenolic compounds, flavonoids, anthocyanins and carotenoids have attracted the attention of many researches for their pivotal role in preventing and treating many diseases as indicated by the outcomes of their pre-clinical and clinical studies (Bvenura and Sivakumar 2017).

In this chapter, we present some of the nutritional and phytochemical profiling of selected wild fruits together with their impact on the health of human worldwide. Figure 8.1 illustrates schematic diagram for the wild fruits selected from each continent.

8.2 Wild Fruits from Africa

Rural Africans depend mainly on wild fruits as their main food source. In dry African regions where cultivation is difficult and poverty is problematic, people consider wild fruits very important for maintaining their health (Fernández-Ruiz et al. 2017). Poor nutrition is one of the biggest problems in Africa where people records the highest incidence of protein, iron, vitamin A and micronutrients deficiencies in the world manifesting many health disorders such as anemia and heart diseases (Allen et al. 2006). Recently, researchers started to focus on the nutritional and pharmacological activities of wild fruits in Africa analyzing their chemical

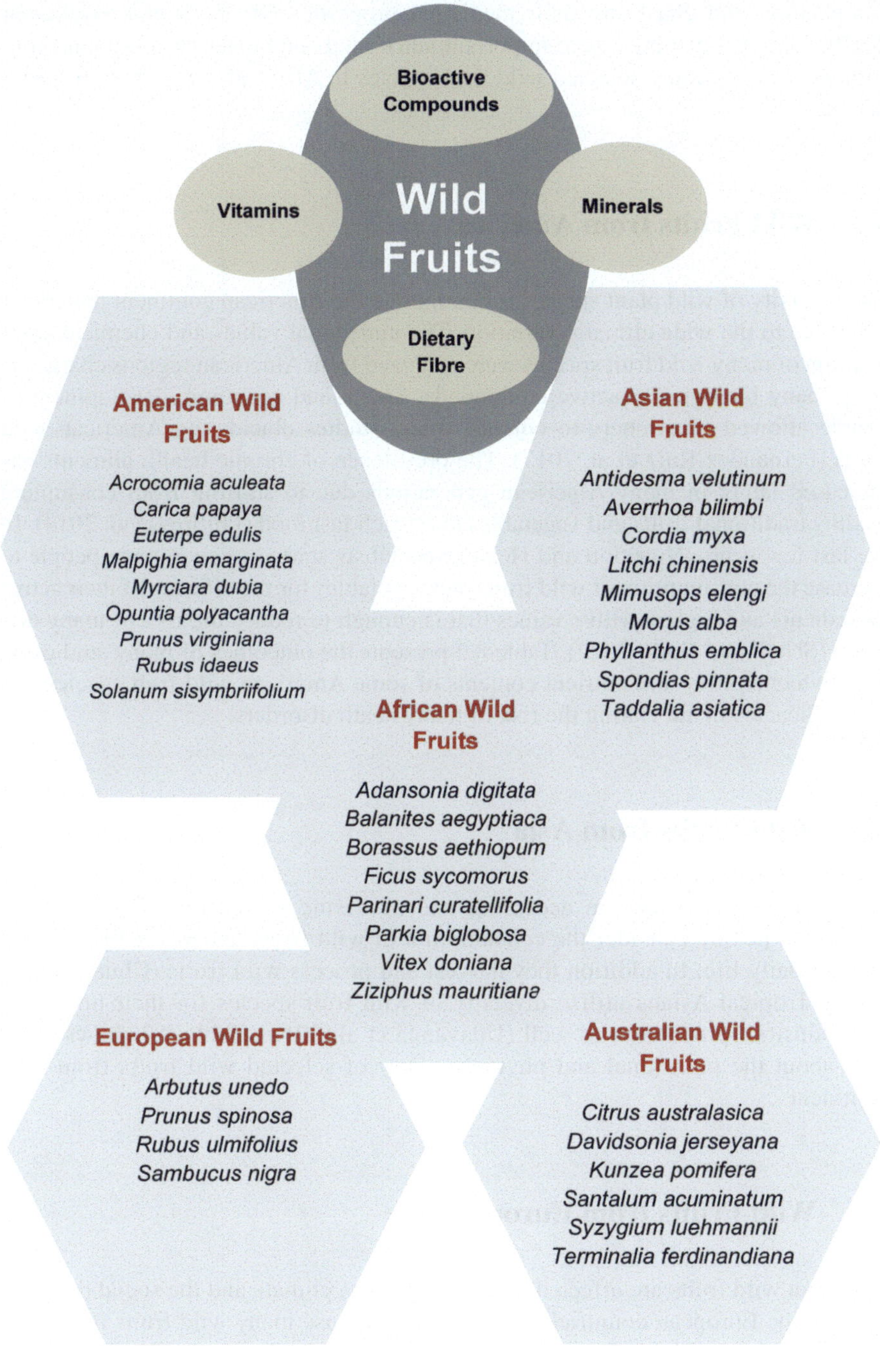

Fig. 8.1 Selected wild fruits from different continents

composition and clarifying their medicinal properties (Bvenura and Sivakumar 2017). Table 8.1 displays some important nutritional and bioactive compound constituents of commonly selected wild fruit species in Africa along with their health benefits.

8.3 Wild Fruits from America

The diversity of wild plant species is common in the American continent than other areas due to the wide climatic variation. The nutritional values and chemical composition of many wild fruit species were analyzed from American regions characterizing many biologically active compounds. Social and economic development of people allowed researchers to conduct many studies elucidating American wild fruits (Fernández-Ruiz et al. 2017). The prevalence of chronic health ailments has increased lately in many American generations due to shifting from consuming healthy traditional fruits and vegetables to fat-rich fast food (Phillips et al. 2014). In the last few years, Nutrition and Health Specialists started to encourage people to increase the consumption of wild fruits and vegetables for the efficacy of their active ingredients and high nutritive values that is enough to reduce the risk of many diseases (Schell and Gallo 2012). Table 8.2 presents the outcomes of many studies on the phytochemistry and nutrient contents of some American wild fruit species and their efficacies in alleviating the risk of many heath disorders.

8.4 Wild Fruits from Asia

Asian population varies in accepting and accessing wild fruits. For example, Philippian people consider the consumption of wild fruits and vegetables as part of their daily life. In addition they harvest and process wild fruits (Chua-Barcelo 2014). Tropical Asians utilize diversity of wild fruit species for their home use and industrial production as well (Udayanga et al. 2013). Table 8.3 shows some data about the nutritional and phytochemistry of selected wild fruits from Asia continent.

8.5 Wild Fruits from Europe

European wild fruits are affected by the geography, climate and the social development of the European countries. People of Europe use many wild fruits in making desserts and jams either for selling their products or family use such as *Sambucus nigara* and *Prunus spinosa* (Łuczaj et al. 2013) while other fruits are used in making liqueurs such as *Rubus ulmifolius* (Pardo-de-Santayana et al. 2007). Table 8.4 records some wild fruit species from Europe and the benefits of their phytochemistry in health.

Table 8.1 Nutritional and bioactive contents of some selected African wild fruits, and their medicinal properties

Botanical name	Nutritional importance	Bioactive compounds	Medicinal property	Reference
Adansonia digitata L.	Rich in vitamin C, iron, calcium and dietary fibre	Flavonoids and polyphenols	Antioxidant antimicrobial, anti-inflammatory and anti-anaemic	De Caluwé et al. (2010), Rahul et al. (2015)
Balanites aegyptiaca L.	Rich in vitamin C and iron	Flavonoids, saponins, glycosides and phenolics	Cardioprotective, antioxidant, anticancer, antibacterial, antiviral, anti-inflammatory, hepatoprotective and antidiabetic	Chothani and Vaghasiya (2011)
Borassus aethiopum Mart.	Rich in vitamin C	Flavonoids, saponins, glycosides and phenolics	Anti-atherogenic, antibacterial, anti-inflammatory, antioxidant, cardioprotective and antidiabetic	Peprah et al. (2018)
Ficus sycomorus L.	Rich in dietary fibre	Triterpenoids, phenolic acids, flavonoids and chalcones	Antimicrobial, cardioprotective, antioxidant, ant-inflammatory and anticancer	Hossain (2018)
Parinari curatellifolia Planch. ex Benth.	Rich in dietary fibre	Polyphenols, alkaloids, flavonoids, anthraquinones and glycosides	Hypolipidemic, hepatoprotective, antiatherogenic and antioxidant	Manuwa et al. (2017)
Parkia biglobosa (Jacq.) R.Br. ex G.Don	Rich in vegetable protein	Polyphenols, glycosides and triterpenoids	Antibacterial, antidiabetic, anti-hypertension, anti-snake venom, antioxidant, hepatoprotective and anti-inflammatory	Gnansounou et al. (2018), Nkafamiya et al. (2007)
Vitex doniana sweet	Rich in vitamins, minerals and fibre	Flavonoids, phenolic compounds and essential oils	Anti-inflammatory, antitumor, antioxidant antidysentry, hypoglycemic, antimicrobial and antiplasmodial	Moke et al. (2018)
Ziziphus mauritiana Lam.	Rich in vitamins, A, B and C, calcium and iron	Tannins, flavonoids and polyphenols	Hypolipidemic, antidiabetic, antioxidant, anti-inflammatory and accelerate wound healing	Abubakar et al. (2018), Verma et al. (2018)

Table 8.2 Nutritional and bioactive contents of some selected American wild fruits, and their medicinal properties

Botanical name	Nutritional importance	Bioactive compounds	Medicinal property	Reference
Acrocomia aculeata Lodd. ex Mart.	Rich in dietary fibre	Carotenoids, tocopherols, β-carotene and phenolic compounds	Hypoglycemic, antioxidant, anti-inflammatory	Arena et al. (2018), Canavaciolo et al. (2015), Coimbra and Jorge (2011), Nunes et al. (2018)
Carica papaya L.	Rich in vitamins A, C, E and minerals Mg, K, Fe	Carotenoids, flavonoids, alkaloids, enzymes, and lycopene	Antioxidant, anticancer, anti-inflammatory, hypoglycemic, hypolipidemic, antifungal, antibacterial, antifertility, anti-sickling, anti-Helmintic, antihypertensive, immunomodulatory and accelerates wound healing	Shahid and Fatima (2018)
Euterpe edulis Mart.	Rich in fibre and vitamin C	Anthocyanins, flavonoids, polyphenols and unsaturated fatty acids	Anti-inflammatory, antioxidant, antipoliferative and hepatoprotective	Almeida Morais et al. (2014), Marques Cardoso et al. (2015)
Malpighia emarginata DC.	Rich in vitamin C	Flavonoids, anthocyanins, carotenoids	Antioxidant, antihyperglycemic, antihyperlipidemic, anti-photoaging, antimicrobial, antitumor, anti-inflammatory, antigenotoxic, hepatoprotective	Belwal et al. (2018)
Myrciara dubia (Kunth)	Rich in vitamin C	Polyphenols, pro-anthocyanins, tanins	Antioxidant, antihypertensive, anti-inflammatory, antigenotoxic and antiobesity	Anhê et al. (2019), Aride et al. (2018), dos Santos Garcia et al. (2018), Serrano et al. (2018)
Opuntia polyacantha Haw.	Rich in calcium	Pro-anthocyanins, carotenoids, betanins, polysaccharides	Neuroprotective, antiulcer, hypoglycemic, anti-inflammatory, hepatoprotective, antioxidant and immunomodulatory	Schepetkin et al. (2008)
Prunus virginiana L.	Rich in dietary fibre and minerals, Ca, K, Mn and Zn	Anthocyanins and phenolic compounds	Antioxidant and protects against eye diseases	Aladedunye et al. (2014), Fang (2015), Grover and Samson (2014)

(continued)

Table 8.2 (continued)

Botanical name	Nutritional importance	Bioactive compounds	Medicinal property	Reference
Rubus idaeus L.	Rich in dietary fibre and vitamin C	Anthocyanins, phenolic, compounds	Anti-inflammatory, anti-obesity, antidiabetic and antioxidative	Noratto et al. (2017), Surya et al. (2018), Szymanowska et al. (2018), Zou et al. (2018)
Solanum sisymbriifolium Lam.	Rich in minerals, Ca, K, Mn and Zn	Triterpenoids, steroidal glycosides, flavonoids and alkaloids	Antitumor, antineoplastic, anticancer, antioxidant, antiviral, hypotensive and nematicidal	Fernández-Ruiz et al. (2017), Pestana et al. (2014)

Table 8.3 Nutritional and bioactive contents of some selected American wild fruits, and their medicinal properties

Botanical name	Nutritional importance	Bioactive compounds	Medicinal property	Reference
Antidesma velutinum L.	Rich in dietary fibre, vitamin E and minerals Ca, Mg, P	Phenolic compounds, quercetin and caffeic acid	Antioxidant and antibacterial	Fernández-Ruiz et al. (2017), D'Annibale et al. (2011), Panda et al. (2017), Shajib et al. (2013)
Averrhoa bilimbi L.	Rich in dietary fibre	Flavonoids, phenolic compounds, quercetin	Hypoglycemic, hypolipidemic, antidiabetic, hypotensive, antioxidant, antiulcer and anti-inflammatory	Fernández-Ruiz et al. (2017), Jagessar et al. (2018), Kurup and Mini (2017), Suluvoy et al. (2017)
Cordia myxa L.	Rich in minerals Mg, Cu, Fe, Zn and selenium	Phenolic compounds and flavonoids	Antipyretic, antiobesity, antioxidant and antidiabetic and antimicrobial	Nasab et al. (2017), Padhi and Singh (2017)
Litchi chinensis Sonn.	Rich in fibre, vitamins B, C, E, K and minerals Mg, Ca, Zn and selenium	Anthocyanins, polyphenols and carotenoids	Antiobesity, antidiabetic, antioxidant, hepatoprotective, anti-inflammatory and antitumor	Alexander-Aguilera et al. (2019), Emanuele et al. (2017), Hu et al. (2018)
Mimusops elengi L.	Rich in Ca	Flavonoids, phenolic compounds, alkaloids and tannins	Antimicrobial, antihelmintic, antibacterial, gastroprotective, cardiotonic, hypotensive and protects gum from bleeding	Fernández-Ruiz et al. (2017), Mathur and Vijayvergia (2017)

(continued)

Table 8.3 (continued)

Botanical name	Nutritional importance	Bioactive compounds	Medicinal property	Reference
Morus alba L.	Rich in minerals K, Ca, Mg, Fe Cu and selenium	Anthocyanins, polyphenols, flavonoids, alkaloids, glycoside derivatives	Anticholesterol, antidiabetic, antioxidant, antiobesity. Hypolipidemic, hepatoprotective, neuroprotective and cardioprotective	Zhang et al. (2018)
Phyllanthus emblica L.	Rich in minerals, and vitamin C	Tannins, phenolics, flavonoids, gallic acid and emblicol	Antioxidant, antibacterial, anti-inflammatory, cardioprotective, anti-Alzheimer, hypocholesterolemic, and anticancer	Gao et al. (2018), Rahmatullah et al. (2009)
Spondias pinnata Kurz.	Rich in vitamin E	Flavonoids, triterpenoids, phenolic compounds, syringic acid, essential amino acids and tannins	Antioxidant, antiulcer, anti-inflammatory, hepatoprotective, photoprotective, anti-arthritite, thrombolytic, analgesic, antipyretic, antimicrobial, antihypertensive, hypoglycemic, laxative, anti-helmintic and antipsychotic	Fernández-Ruiz et al. (2017), Sameh et al. (2018)
Taddalia asiatica Baill.	Rich in minerals, Fe, Cu, Mn	Essential oils, cumarins, triterpenoids, alkaloids	Antibacterial, antiviral, anti-diarrheal, anti-malarial, antidiabetic, cadioprotective, antitumor, anti-inflammatory and antioxidant	Fernández-Ruiz et al. (2017), Kariuki et al. (2013)

8.6 Wild Fruits from Australia

Indigenous Australians are depending on wild fruits as important source of their food since thousands of years ago. For example, *Kunzea pomifera* fruits were used by native Australians as a source of flour making home cakes and also for trading the surplus. Native people of Australia used to gather seasonally ripened fruits and vegetables, and eat healthy diets free from sugars and preservatives. Additionally they store those seasonal fruits for their consumption in the seasons when those fruits are not available (Clarke and Jones 2018). At the pharmacology level, researchers have raised their reports that Australian wild plants used by native people contain many medicinal compounds (Clarke 2008). Table 8.5 views some of the health-promoting ingredients of selected wild fruit species from Australia.

Table 8.4 Nutritional and bioactive contents of some selected European wild fruits, and their medicinal properties

Botanical name	Nutritional importance	Bioactive compounds	Medicinal property	Reference
Arbutus unedo L.	Rich in vitamin C, energy and fibre	Phenolic compounds, anthocyanins, tocopherols and carotenoids flavonoids	Neuroprotective, cardioprotective, gastroprotective, anticancer	Fonseca et al. (2015), Ruiz-Rodríguez et al. (2011)
Prunus spinosa L.	Rich in minerals and fibre	Flavonoids, phenolic acids and anthocyanins	Cytotoxic and apoptotic to cancer cells, antioxidant and treating gastrointestinal disorders	Meschini et al. (2017), Rop et al. (2009)
Rubus ulmifolius Scott.	Rich in vitamin C, iron and dietary fibre	Flavonoids, polyphenols, carotenoids and anthocyanins	Antilipidemic, antiglycemic, antioxidant, anti-inflammatory and antimicrobial	Surya et al. (2018), de Souza et al. (2018)
Sambucus nigra L.	Rich in calcium, iron and other minerals, and fibre	Anthocyanins, phenolic compounds, and triterpenoids	Antiproliferative, antidiabetic, and antioxidant	Divis et al. (2015), Gleńsk et al. (2017), Ho et al. (2017)

Table 8.5 Nutritional and bioactive contents of some selected American wild fruits, and their medicinal properties

Botanical name	Nutritional importance	Bioactive compounds	Medicinal property	Reference
Citrus australasica F.Muell.	Rich in dietary fibre, vitamin C and minerals K, Ca, Mg and Cu	Polyphenols, terpenoids, sesquiterpenes and volatile metabolites	Hypotensive, anti-inflammatory, antidiabetic, antioxidant, anticancer, hepatoprotective and antimicrobial	Delort et al. (2015), Donkersley et al. (2018), Mohib et al. (2018)
Davidsonia jerseyana F.Muell. ex F.M.Bailey	Rich in dietary fibre, minerals Ca, Zn, Cu, Fe and vitamin C	Polyphenols, pro-anthocyanins and flavonoids	Antibacterial, antiproliferative, antioxidant, anticolon cancer and antihepatocarcinoma	Sakulnarmrat et al. (2014), Sirdaarta (2016), Williams et al. (2016)
Kunzea pomifera F.Muell.	Rich in dietary fibre, minerals Ca, Zn, Cu, Fe and vitamin C	Triterpenoids, tannins, flavonoids, saponins, anthocyanins and phenolic compounds	Antioxidant, anticancer, antiproliferative, anti-inflammatory and immunomodulatory	Sirdaarta (2016)

(continued)

Table 8.5 (continued)

Botanical name	Nutritional importance	Bioactive compounds	Medicinal property	Reference
Santalum acuminatum A.DC.	Rich in dietary fibre, minerals Ca, Zn, Cu, Fe and vitamin C	Chlogenic acid-rich polyphenols, phenolic compounds pro-anthocyanins and flavonoids	Cardioprotective, antiobesity, antioxidantant and pancreatic lipase inhibitor	Sakulnarmrat et al. (2014)
Syzygium luehmannii (F.Muell.)	Rich in dietary fibre, minerals Ca, Zn, Cu, Fe and vitamin C	Phenolic compounds, flavonoids, sesquiterpenes, tannins and terpenoids	Antimicrobial, antioxidant, antibacterial, antiseptic, antiproliferative, antiobesity and anti-inflammatory	Williams et al. (2016), Murhekar et al. (2017), Wright et al. (2016)
Terminalia ferdinandiana Exell	Rich in dietary fibre, minerals Ca, Zn, Cu and Fe in particular, and very rich in vitamin C	Polyphenols, flavonoids, tannins, cardiac glycosides and carotenoids	Anticancer, anti-inflammatory, antiproliferative, antiobesity and antioxidant	Mohanty and Cock (2012), Sirdaarta (2016)

8.7 Conclusion

Although many wild fruits are endangered in many regions around the world due to changes in the climatic conditions, pressure from over-population and many human activities, but many wild fruits are still under regular consumption from indigenous people particularly children and pregnant women for their considerable health benefits. In addition, many wild fruits are under research to elucidate more beneficial compounds for better life worldwide through treating micro and macronutrient deficiencies and health problems such as heart, autoimmune and Alzheimer's disease. For that purpose, financial and technological facilities are required from the concerned governments and organizations to support researchers in elucidating more bioactive compounds from wild fruits for better and healthy life.

References

Abubakar SM, Umar SA, Alexander I, Abubakar N, Abdulazeez MA, Sule MS (2018) Evaluation of hypoglycaemic, hypolipidemic, and non-toxic effect of hydro-methanolic extracts of *Ziziphus mauritiana*, *Ziziphus spina* christi fruit and glibenclamide on allatoxin induced diabetic rats. J Drug Deliv Ther 8(3):82–92

Aladedunye F, Przybylski R, Niehaus K, Bednarz H, Matthäus B (2014) Phenolic extracts from *Crataegus* × *mordenensis* and *Prunus virginiana*: composition, antioxidant activity and performance in sunflower oil. LWT-Food Sci Technol 59(1):308–319

Alexander-Aguilera A, Aguirre-Maldonado I, Antolín JR, Toledo LN, Rodríguez IS, Otero MGS (2019) Effect of *Litchi chinensis* on adipose and hepatic tissues in rats with obesity and non-alcoholic fatty liver disease (NAFLD). J Saudi Soc Agric Sci 18(3):235–240. https://doi.org/10.1016/j.jssas.2017.06.002

Allen LH, De Benoist B, Dary O, Hurrell R, World Health Organization (2006) Guidelines on food fortification with micronutrients. World Health Organization, Geneva

Almeida Morais C, Oyama LM, de Oliveira JL, Carvalho Garcia M, de Rosso VV, Sousa Mendes Amigo L et al (2014) Jussara (*Euterpe edulis* Mart.) supplementation during pregnancy and lactation modulates the gene and protein expression of inflammation biomarkers induced by trans-fatty acids in the colon of offspring. Mediat Inflamm 2014:11

Anhê FF, Nachbar RT, Varin TV, Trottier J, Dudonné S, Le Barz M et al (2019) Treatment with camu camu (*Myrciaria dubia*) prevents obesity by altering the gut microbiota and increasing energy expenditure in diet-induced obese mice. Gut 68(3):453–464. https://doi.org/10.1136/gutjnl-2017-315565

Arena A, Jorge B, Silva M, de Barros A, Fernandes A, Nóbrega R et al (2018) *Acrocomia aculeata* oil: beneficial effects on cyclophosphamide-induced reproductive toxicity in male rats. Andrologia 50:e13028

Aride P, Oliveira A, Batista R, Ferreira M, Pantoja-Lima J, Ladislau D et al (2018) Changes on physiological parameters of tambaqui (*Colossoma macropomum*) fed with diets supplemented with Amazonian fruit camu camu (Myrciaria dubia). Braz J Biol 78(2):360–367

Belwal T, Devkota HP, Hassan HA, Ahluwalia S, Ramadan MF, Mocan A et al (2018) Phytopharmacology of Acerola (Malpighia spp.) and its potential as functional food. Trends Food Sci Technol 74:99–106

Borek C (2017) Dietary antioxidants and human cancer. J Restor Med 6(1):53–61

Bvenura C, Sivakumar D (2017) The role of wild fruits and vegetables in delivering a balanced and healthy diet. Food Res Int 99:15–30

Canavaciolo VLG, Perez RCS, Ferreiro RMM, Guerra YP, Yera AO, Leyes EAR, et al (2015) Compounds from the fruits of acrocomia crispa and acrocomia aculeata for use against oxidative stress and inflammation. Google Patents, US 2015/0190358 A1

Chothani DL, Vaghasiya H (2011) A review on *Balanites aegyptiaca* Del (desert date): phytochemical constituents, traditional uses, and pharmacological activity. Pharmacogn Rev 5(9):55–62

Chua-Barcelo RT (2014) Ethno–botanical survey of edible wild fruits in Benguet, Cordillera administrative region, the Philippines. Asian Pac J Trop Biomed 4:S525–SS38

Clarke P (2008) Aboriginal healing practices and Australian bush medicine. J Anthropol Soc S Aust 33:3–38

Clarke PA, Jones DS (2018) Aboriginal culture and food-landscape relationships in Australia: indigenous knowledge for country and landscape. In: Zeunert J, Waterman T (eds) Routledge handbook of landscape and food. Routledge, London/New York, pp 41–60

Coimbra MC, Jorge N (2011) Proximate composition of guariroba (*Syagrus oleracea*), jerivá (*Syagrus romanzoffiana*) and macaúba (*Acrocomia aculeata*) palm fruits. Food Res Int 44(7):2139–2142

D'Annibale A, Maraldo K, Sechi V, Larsen T, Strandberg B, Cortet J, et al (eds) (2011) Response of the soil mesofauna trophic structure to elevated CO_2 and GM barley. In: 13th Nordic soil zoology symposium and PhD course, Lammi Biological Station, Finland, 25–29 Aug 2011

De Caluwé E, Halamová K, Van Damme P (2010) *Adansonia digitata* L.: a review of traditional uses, phytochemistry and pharmacology. Afrika Focus 23(1):11–51

de Souza AV, Vieites RL, Gomes EP, da Silva Vieira MR (2018) Biochemical characterization of blackberry fruit (Rubus sp) and jellies. Aust J Crop Sci 12(4):624–630

Delort E, Jaquier A, Decorzant E, Chapuis C, Casilli A, Frérot E (2015) Comparative analysis of three Australian finger lime (*Citrus australasica*) cultivars: identification of unique citrus chemotypes and new volatile molecules. Phytochemistry 109:111–124

Divis P, Porizka J, Vespalcová M, Matejicek A, Kaplan J (2015) Elemental composition of fruits from different black elder (*Sambucus nigra* L.) cultivars grown in the Czech Republic. J Elem 20(3):549–557

Donkersley P, Silva FW, Carvalho CM, Al-Sadi AM, Elliot SL (2018) Biological, environmental and socioeconomic threats to citrus lime production. J Plant Dis Prot 125(4):339–356

dos Santos Garcia VA, Borges JG, Maciel VBV, Mazalli MR, das Graças Lapa-Guimaraes J, Vanin FM et al (2018) Gelatin/starch orally disintegrating films as a promising system for vitamin C delivery. Food Hydrocoll 79:127–135

Emanuele S, Lauricella M, Calvaruso G, D'Anneo A, Giuliano M (2017) *Litchi chinensis* as a functional food and a source of antitumor compounds: an overview and a description of biochemical pathways. Nutrients 9(9):pii: E992. https://doi.org/10.3390/nu9090992

Fang J (2015) Classification of fruits based on anthocyanin types and relevance to their health effects. Nutrition 31(11–12):1301–1306

Fernández-Ruiz V, Morales P, Ruiz-Rodríguez BM, TorijaIsasa E (2017) Nutrients and bioactive compounds in wild fruits through different continents. In: Ferreira ICFR, Morales P, Barros L (eds) Wild plants, mushrooms and nuts: functional food properties and applications. Wiley, Chichester, pp 263–314

Fonseca DF, Salvador ÂC, Santos SA, Vilela C, Freire CS, Silvestre AJ et al (2015) Bioactive phytochemicals from wild *Arbutus unedo* L. berries from different locations in Portugal: quantification of lipophilic components. Int J Mol Sci 16(6):14194–14209

Gao Q, Li X, Huang H, Guan Y, Mi Q, Yao J (2018) The efficacy of a chewing gum containing *Phyllanthus emblica* fruit extract in improving oral health. Curr Microbiol 75(5):604–610

Gleńsk M, Czapińska E, Woźniak M, Ceremuga I, Włodarczyk M, Terlecki G et al (2017) Triterpenoid acids as important antiproliferative constituents of European elderberry fruits. Nutr Cancer 69(4):643–651

Gnansounou SM, Iskandar S, Robin M, Brunel J-M, Dahouenon E, Piccerelle P (2018) Dialium guineense Willd. Parkia biglobosa (Jacq.) R. Br. Ex Benth. and Tamarindus indica L.: review of known and synergetic bioactive compounds. J Med Plants 6(3):103–111

Grover AK, Samson SE (2014) Antioxidants and vision health: facts and fiction. Mol Cell Biochem 388(1–2):173–183

Ho GTT, Kase ET, Wangensteen H, Barsett H (2017) Phenolic elderberry extracts, anthocyanins, procyanidins, and metabolites influence glucose and fatty acid uptake in human skeletal muscle cells. J Agric Food Chem 65(13):2677–2685

Hossain MA (2018) A review on *Ficus sycomorus*: a potential indigenous medicinal plant in Oman. J King Saud Univ Sci. https://doi.org/10.1016/j.jksus.2018.07.002

Hu B, Li J, Wang D, Wang H, Qin Y, Hu G et al (2018) Transcriptome profiling of *Litchi chinensis* pericarp in response to exogenous cytokinins and abscisic acid. Plant Growth Regul 84(3):437–450

Jagessar R, Rodrigues A, Prasad K, Husain A, Kanhai V, Bernarai B (2018) An investigation of the hypoglycemic effect of the aqueous extract of the fruits of *Psidium guajava*, *Averrhoa bilimbi* and the peel of *Tamarindus indica* in normoglycemic guinea pigs. World J Pharm Pharm Sci 7(4):77–101

Jenkins DJ, Spence JD, Giovannucci EL, Kim Y-I, Josse R, Vieth R et al (2018) Supplemental vitamins and minerals for CVD prevention and treatment. J Am Coll Cardiol 71(22):2570–2584

Kariuki HN, Kariuki J, Mwonjoria K (2013) Ethnomedicinal, phytochemical and pharmacological profile of genus *Toddalia*. Phytopharmacology 4(2):259–268

Kumagai S, Nobuhara Y, Saegusa J (2003) Oxidative stress and autoimmune diseases. J Jpn Soc Intern Med 92(6):1096–1103

Kurup SB, Mini S (2017) *Averrhoa bilimbi* fruits attenuate hyperglycemia-mediated oxidative stress in streptozotocin-induced diabetic rats. J Food Drug Anal 25(2):360–368

Łuczaj Ł, Köhler P, Pirożnikow E, Graniszewska M, Pieroni A, Gervasi T (2013) Wild edible plants of Belarus: from Rostafiński's questionnaire of 1883 to the present. J Ethnobiol Ethnomed 9(1):21

Manuwa TR, Akinmoladun AC, Crown OO, Komolafe K, Olaleye MT (2017) Toxicological assessment and ameliorative effects of *Parinari curatellifolia* alkaloids on triton-induced hyperlipidemia and atherogenicity in rats. Proc Natl Acad Sci India Sect B Biol Sci 87(2):611–623

Marques Cardoso L, Dias Novaes R, De Castro CA, Azevedo Novello A, Gonçalves RV, Ricci-Silva ME et al (2015) Chemical composition, characterization of anthocyanins and antioxidant potential of *Euterpe edulis* fruits: applicability on genetic dyslipidemia and hepatic steatosis in mice. Nutr Hosp 32(2):702–709

Mathur R, Vijayvergia R (2017) Determination of total flavonoid and phenol content in *Mimusops elengi* Linn. Int J Pharm Sci Res 8(12):5282–5285

Meschini S, Pellegrini E, Condello M, Occhionero G, Delfine S, Condello G et al (2017) Cytotoxic and apoptotic activities of *Prunus spinosa* Trigno ecotype extract on human cancer cells. Molecules 22(9):pii: E1578. https://doi.org/10.3390/molecules22091578

Millen BE, Abrams S, Adams-Campbell L, Anderson CA, Brenna JT, Campbell WW et al (2016) The 2015 dietary guidelines advisory committee scientific report: development and major conclusions. Adv Nutr 7(3):438–444

Mohanty S, Cock IE (2012) The chemotherapeutic potential of *Terminalia ferdinandiana*: phytochemistry and bioactivity. Pharmacogn Rev 6(11):29–36

Mohib M, Afnan K, Paran TZ, Khan S, Sarker J, Hasan N et al (2018) Beneficial role of citrus fruit polyphenols against hepatic dysfunctions: a review. J Diet Suppl 15(2):223–250

Moke LE, Koto-te-Nyiwa Ngbolua GN, Bongo L, Messi M, Noté OP, Mbing JN et al (2018) *Vitex madiensis* Oliv.(Lamiaceae): phytochemistry, pharmacology and future directions, a mini-review. J Pharmacogn Phytochem 7(2):244–251

Murhekar S, Wright MH, Greene AC, Brownlie JC, Cock IE (2017) Inhibition of Shewanella spp. growth by *Syzygium australe* and *Syzygium luehmannii* extracts: natural methods for the prevention of fish spoilage. J Food Sci Technol 54(10):3314–3326

Na J, Sirdaartaa J, Cocka I (2014) The anti-proliferative properties of Australian plants with high antioxidant capacities against cancer cell lines. Pharmacogn Commun 4(4):71–82

Nasab FY, Hajinezhad MR, Hashemi H, Miri A (2017) Comparison the effects of *Trigonella foenum*-graecum L. seed and *Cordia myxa* fruit extracts on diabetes-induced memory impairment in rats. Int J Basic Sci Med 2(4):184–188

Nkafamiya I, Modibbo U, Manji A, Haggai D (2007) Nutrient content of seeds of some wild plants. Afr J Biotechnol 6(14):1665–1669

Noratto GD, Chew BP, Atienza LM (2017) Red raspberry (*Rubus idaeus* L.) intake decreases oxidative stress in obese diabetic (db/db) mice. Food Chem 227:305–314

Nunes ÂA, Buccini DF, Jaques JA, Portugal LC, Guimarães RC, Favaro SP et al (2018) Effect of *Acrocomia aculeata* kernel oil on adiposity in type 2 diabetic rats. Plant Foods Hum Nutr 73(1):61–67

Padhi M, Singh S (2017) Surface sterilization for reducing microbial contamination in *In Vitro* propagation of lasora (*Cordia myxa* Roxb.) using nodal segments. Int J Curr Microbiol App Sci 6(8):836–842

Panda SK, Padhi L, Leyssen P, Liu M, Neyts J, Luyten W (2017) Antimicrobial, anthelmintic, and antiviral activity of plants traditionally used for treating infectious disease in the Similipal Biosphere Reserve, Odisha, India. Front Pharmacol 8:658. https://doi.org/10.3389/fphar.2017.00658

Pardo-de-Santayana M, Tardío J, Blanco E, Carvalho AM, Lastra JJ, San Miguel E et al (2007) Traditional knowledge of wild edible plants used in the northwest of the Iberian Peninsula (Spain and Portugal): a comparative study. J Ethnobiol Ethnomed 3(1):27

Peprah M, Apprey C, Larbie C, Asamoah-Boakye O (2018) Phytochemical constituents of flour and composite bread from African palmyra (*Borassus aethiopum*) fruit from Ghana. Eur J Med Plants 23(1):1–7

Pestana M, Rodrigues M, Teixeira L, Abrantes IDO, Gouveia M, Cordeiro N (2014) *In vitro* evaluation of nematicidal properties of *Solanum sisymbriifolium* and *S. nigrum* extracts on *Pratylenchus goodeyi*. Nematology 16(1):41–51

Phillips KM, Pehrsson PR, Agnew WW, Scheett AJ, Follett JR, Lukaski HC et al (2014) Nutrient composition of selected traditional United States northern plains native American plant foods. J Food Compos Anal 34(2):136–152

Rahmatullah M, Noman A, Hossan MS, Rashid M, Rahman T, Chowdhury MH et al (2009) A survey of medicinal plants in two areas of Dinajpur district, Bangladesh including plants which can be used as functional foods. Am Eurasian J Sustain Agric 3(4):862–876

Rahul J, Jain MK, Singh SP, Kamal RK, Naz A, Gupta AK et al (2015) *Adansonia digitata* L. (baobab): a review of traditional information and taxonomic description. Asian Pac J Trop Biomed 5(1):79–84

Rop O, Jurikova T, Mlcek J, Kramarova D, Sengee Z (2009) Antioxidant activity and selected nutritional values of plums (*Prunus domestica* L.) typical of the White Carpathian mountains. Sci Hortic 122(4):545–549

Ruiz-Rodríguez B-M, Morales P, Fernández-Ruiz V, Sánchez-Mata M-C, Camara M, Díez-Marqués C et al (2011) Valorization of wild strawberry-tree fruits (*Arbutus unedo* L.) through nutritional assessment and natural production data. Food Res Int 44(5):1244–1253

Sakulnarmrat K, Srzednicki G, Konczak I (2014) Composition and inhibitory activities towards digestive enzymes of polyphenolic-rich fractions of Davidson's plum and quandong. LWT-Food Sci Technol 57(1):366–375

Sameh S, Al-Sayed E, Labib RM, Singab AN (2018) Genus *Spondias*: a phytochemical and pharmacological review. Evid Based Complement Alternat Med 2018:13

Schell LM, Gallo MV (2012) Overweight and obesity among North American Indian infants, children, and youth. Am J Hum Biol 24(3):302–313

Schepetkin IA, Xie G, Kirpotina LN, Klein RA, Jutila MA, Quinn MT (2008) Macrophage immunomodulatory activity of polysaccharides isolated from *Opuntia polyacantha*. Int Immunopharmacol 8(10):1455–1466

Serrano A, Ros G, Nieto G (2018) Bioactive compounds and extracts from traditional herbs and their potential anti-inflammatory health effects. Medicines 5(3):76. https://doi.org/10.3390/medicines5030076

Shahid S, Fatima U (2018) Pharmacological activities of *Carica papaya* Linn. J Basic Appl Sci 14:210–216

Shajib MTI, Kawser M, Miah MN, Begum P, Bhattacharjee L, Hossain A et al (2013) Nutritional composition of minor indigenous fruits: cheapest nutritional source for the rural people of Bangladesh. Food Chem 140(3):466–470

Sirdaarta JP (2016) Phytochemical study and anticancer potential of high antioxidant Australian native plants. Downloaded from https://research-repository.griffith.edu.au/handle/10072/365567

Slavin J (2013) Fiber and prebiotics: mechanisms and health benefits. Nutrients 5(4):1417–1435

Suluvoy JK, Sakthivel K, Guruvayoorappan C, Grace VB (2017) Protective effect of *Averrhoa bilimbi* L. fruit extract on ulcerative colitis in wistar rats via regulation of inflammatory mediators and cytokines. Biomed Pharmacother 91:1113–1121

Surya MI, Suhartati S, Ismaini L, Lusini Y, Destri D, Anggraeni D et al (2018) Fruit nutrients of five species of wild raspberries (Rubus spp.) from Indonesian mountain's forests. J Trop Life Sci 8(1):75–80

Szymanowska U, Baraniak B, Bogucka-Kocka A (2018) Antioxidant, anti-inflammatory, and postulated cytotoxic activity of phenolic and anthocyanin-rich fractions from polana raspberry (*Rubus idaeus* L.) fruit and juice—in vitro study. Molecules 23(7):1

Udayanga D, Manamgoda DS, Liu X, Chukeatirote E, Hyde KD (2013) What are the common anthracnose pathogens of tropical fruits? Fungal Divers 61(1):165–179

Verma RK, Pandey M, Indoria MD, Singh R, Suthar S (2018) Phytochemical investigation and pharmacological evaluation of leaves of *Ziziphus mauritiana* for wound healing activity in albino rats tropical. J Pharm Life Sci 5(1):08–18

Williams D, Chaliha M, Sultanbawa Y (2016) Australian native plants: anti-obesity and anti-inflammatory properties. Google Books. CRC Press, Boca Raton, pp 237–294

Wilson DW, Nash P, Buttar HS, Griffiths K, Singh R, De Meester F et al (2017) The role of food antioxidants, benefits of functional foods, and influence of feeding habits on the health of the older person: an overview. Antioxidants 6(4):pii: E81. https://doi.org/10.3390/antiox6040081

Wright MH, Lee CJ, Pollock CE, Greene AC, Cock IE (2016) Growth inhibitory activity of selected high antioxidant Australian *Syzygium* species against the food poisoning and tissue necrotic pathogen Clostridium perfringens. Pharmacogn Commun 6(2):93–99

Zhang H, Ma ZF, Luo X, Li X (2018) Effects of mulberry fruit (*Morus alba* L.) consumption on health outcomes: a mini-review. Antioxidants 7(5):pii: E69. https://doi.org/10.3390/antiox7050069

Zou T, Wang B, Yang Q, de Avila JM, Zhu M-J, You J et al (2018) Raspberry promotes brown and beige adipocyte development in mice fed high-fat diet through activation of AMP-activated protein kinase (AMPK) α1. J Nutr Biochem 55:157–164

Chapter 9
The Utilization of Multi-Purpose Wild fruits in Central Sudan, because gum belt is Part of Central Sudan

Hagir Mahagoub Elnasri

Abbreviations

FAO	Food and Agriculture Organization
NTFPs	Non timber forest products
NWFPs	Non-wood forest products
WFs	Wild fruits
WHO	World Health Organization
IUCN	International Union for Conservation of Nature

9.1 Introduction

Multi-purpose wild plants are essential and as the largest flora on the earth, they store carbon and release oxygen, stabilize the soil and provide life to the world's wild animals. Also they supplies us with the resources for shelter and tools. Wild plants products including Non-wood forest products (NWFPs) producing from multi purposes trees had received a very high consideration because they have an important usages all over the world during the last years, and the most important NWFPs is the Wild Fruits (WFs). Forestry Commission at the United Kingdom's defines NTFPs as "all biological resources found in woodlands not including timber", and in Forest Harvest, part of the Reforesting Scotland project, defined as "resources supplied by woodlands – with the exception of conventional harvested timber". These definitions include manage and wild game, insects and fishes,. Commonly NTFPs are grouped into many categories such as floral greens, foods, flavors, decorative, saps and resins, medicinal plants, and fragrances and fibers. The term NTFP comprises literally all tangible harvest originated from forests, excluding

H. M. Elnasri (✉)
Department of Forest Management, University of Khartoum, Khartoum, Sudan

A. A. Mariod (ed.), *Wild Fruits: Composition, Nutritional Value and Products*,
https://doi.org/10.1007/978-3-030-31885-7_9

timber, and including a broad varieties of products, such as rattan, nuts, essential oils, resins, latex, spices, seeds, fruits, game, leaves, fish, eggs, birds and honey (Ros-Tonen and Wiersum 2005). In this chapter, the expression NTFP is used to incorporate trees and shrubs fruits from wild source.

Based on these special and diverse meanings of wild natural forest harvest may be conceived of as products that are derived from traditional nature-analogue production systems, or wilderness areas, or from species that are not acculturated in the sense of being subject to institutionalized practices. In a number of policy documents the concept of wild harvest is interpreted as referring to products that are derived from wild species that take place in self-maintaining populations in natural or semi-natural ecosystems and can survive independently and separately from human action (Heywood 1999). This reflects in EU Regulation 2092/91 on organic production and labeling that defines the gathering of wild harvest as concerning the collection of edible parts or plants which are growing naturally in natural areas, forests, and farming areas. The World Health Organization (WHO), the World Wildlife Fund (WWF) and the International Union for the Conservation of Nature (IUCN),adhere to worldwide standard for sustainable collection of aromatic plants and wild medicinal that refers to the practice as the collection of non-cultivated native, inhabitant or naturalized resource reserve from its natural environment habitat (Censkowsky et al. 2007). Thus, wild harvests are generally consider as derive from autonomously growing plants that may occur not only in untouched wilderness areas, but moreover in human being modified and adapted environments such as managed forests and even agricultural land.

The objectives of the chapter were directed towards highlight the importance of wild fruits their distribution and local uses and benefits the production, marketing channels and utilization of wild fruits and attempted to explore the promotion of their potential and participation of the neighborhood communities in sustainable management approaches in addition to the future perspectives analyzes the constraints and challenges confronting their sustainability, and concluded by some recommendations.

9.2 Distribution

Sudan is considered one of the largest countries in Africa with an area which include diverse ecological zones from the desert in the Northern parts to the tropical humid rainforest in the Southern part (Ngambi 2013). The vast area of occupied at least 37% of the total land of the in Sudan, and which are habitats for many various plant types (El-Amin 1990), there are diverse ecotypes which bear a variety of fruit shapes and qualities in different areas of the Sudan. Wild fruits are extremely significant in multiple crop system where multi-use shrubs and trees are linked and coupled with seasonal and perennial foodstuff crops and domestic animals. The successful agroforestry systems are rely on overall high yield of biomass, good quality nutritional value of the edible wild crops and a wide variety of harvest. Livelihood system within the central Sudan including Gum belt from the East to the West is varies. Primarily, they depend on both traditional mode of agriculture (west) and mechanized farming (east and central) and followed by the animal rearing, daily labour and wild fruits collection.

9.3 Production of Wild Fruits

NWFPs including wild fruits are obtaining from about 3000 types in the country and form a significant resource of foodstuff and income for local community mainly to those who living closest to the forests (Mitra 2000). Generally there are two important systems for the collection and production of wild fruits the first system is depend on family labours (Nafeer) in both sand and vertisol (mud) soils, and the collection is done through participation and hired seasonal labour, in this system the productive areas are varies according to the land tenure, family size and customary use right and responsibility (smallholder producers), the second system is clearly found and practiced by the local communities living in or around the reserved forests as a privileges by the laws and legislation (large areas)

9.4 Women and Wild Fruits

In several communities women are responsible for household activities that involve forest-based foods (wild fruits) and fuel wood.

Forests in addition to trees out-side forests contribute to the livelihoods of more than 1.6 billion citizens (FAO 2010).Poor households particularly depend on multi-use trees (WFs) which provide essential nutrition and food fodder, medicine, construction materials, fuel and, nonfarm income. WFs are mainly important in relieving "famine period" in the agricultural cycle; they offer seasonal employment and barrier against risk and households emergency. Moreover the poor, tend to have additional access to the natural woodland (forest) than other natural capital and few land right elsewhere. Within deprived household, gender asymmetry in tenure position and access to fruitful and productive resources such as cause women to rely for income and nutrition on NWFPs disproportionately (FAO 1995), women specifically those who are poor, heavily depends on ordinary common resources for meeting their survival requirements (Kumar et al. 2000).

9.5 Characteristic of Wild Fruits

In General non-wood forest products (NWFPs) have received a very high awareness and attention because they have important usages all over the world during the previous few years, and the forests wild fruit are one of which. The information on the different wild fruits physical properties and chemical composition is still limited.

Different studies were conducted to study the characterization of some of indigenous wild fruits. Many studies were focus on both physical properties such as colour, dimensions including width thickness and length, weight, seed, peels and edible part, and physico-chemical analysis for the components percentages of the edible portions including moisture, crude fibre, ash, fats, crude protein, ascorbic acid, carbohydrates, in addition to energy value, sugars, ß-carotene and pectin, unfortunately the study covered only four wild fruits (*Hyphaene thebaica* L, *Grewia tenax and Randia geipaeflora, Naucleae latifolia,*). The results approved that the wild fruits constituents of a good quantities of crude protein, ash and crude fiber. In addition to that they contain excellent quantities of ascorbic acid, high percentages of carbohydrates, also they contain high values of energy which ranged from 310 to 372 Kcal/g and adequate amounts of ß-carotene. Different studies was carried out for a numerous wild fruits, and concluded that wild fruits have a high vitamins and minerals e.g., *Grewa tenax* is a good source of iron for children under 8 years old (Kehlenbeck et al. 2013). *Tamarindus indica* and *Adansonia digitata* are important sources of energy because they contain high sugar percent (FAO 2013).

Three different studies carried by and El-Amin (1990), Vogt (1995), and Abdel Rahman (2014); Confirmed that the weight of godeim fruit which was obtained from 100 seeds weighed 14.0 g. 52.20% of this weight is edible part and peels together, consequently and 47.80% is seeds, moisture content for the forest fruits was ranging from 5.47% (doum) to 60.52% (karmadoda), the (doum) value was higher than the findings of FAO (2006) for African doum which was 4.00%. While, the karmadoda value was not coping with the result (70.1%) reported by Kuria (2005). The differences in moisture content are influenced by cultivation and post-harvest conditions (Bates et al. 2001). The protein content is ranging from 3.80% to 7.71% for doum and kirkir, respectively. The doum value was similar and equal to the African doum value (3.80%), while the minimum recommended target of dietary protein for adults is approximately 70 g/day (Layman 2003). Moreover, kirkir could be exploited for protein supplementation since the plants are inexpensive source of protein. The changes in protein contents however, indicate variations in metabolic activity during the different development stages (Wills et al. 1981). These results are clear evidence that the researches are focused on specific wild fruits. Diverse and difference regions of the Sudan proofs that there is considerable variation within species. The clear huge and enormous variation in the fruits and seeds includes shape, color and size. One of studies which concerned about the variation in physiochemical characteristics of fruits was done for wild *Ziziphus spina–christi* "Nabag" from three different sites and the result has proved that there are significant differences between the three sites and among trees in the same site regarding fruit diameter (Fig. 9.1).

In regard to the mineral content, the results indicated that the important constituents of fruit pulp are calcium, phosphorus, manganese potassium magnesium, sodium, and iron with an average which was significantly different between sites, while ascorbic content ranged between 29 and 44 mg/100 g. Some nutritional constituents of "Nabag" fruit meet and sometimes exceed body requirements (BR) of the human, determined by FAO and WHO (1974). For example, 100 gram of "Nabag" fruit pulp contains 264 mg of vitamin C (Ascorbic acid), while the daily

Fig. 9.1 *Ziziphus spina–christi* "Nabag" *Capture by the Author's Camera*

recommended dose is 64 mg, and iron content accounted for 13.7 mg, while BR was 14 mg (Abdalla and Elzebeir 2014). *Balanites aegyptiaca* which is 'Lalob' is another example for the clear differences in fruits; the fruits of the *Balanites aegyptiaca* in Sinnar area are yellow and longish in compare to the red round fruits from Kordofan state (Fig. 9.2). Also there are taste differences; Sinnar fruits are sweeter.

Forest trees producing wild fruits require more investigation in the area of nutritional value, as the determination of abundant elements and amino acids profile It is high time to conduct researches for the proper assessment and maximum utilization of edible forest fruits and possible use of fresh products as table and processed items; emphasizing the populating, nutritional and economic value (Abdel Rahman 2011).

9.6 Marketing Channels

9.6.1 Current Situation

Forests products can be very important commercially at a local or region level, where they are trade in villages shop and marketor they are sent to the bigger cities and towns. In the Sudan the marketing of wild fruits like other NTFPs naturally grown in remote areas and their marketing process is starting the collection and

Fig. 9.2 *Balanites aegyptiaca (Laloub) Captured by the Author*

gathering operations of the product at the households level and at a village level (Primary Producers) as a whole. The collection is not organized and the products were sold at very low rate and valued, village traders buy the wild fruits collected from the primary producers (collectors or gatherers), and then sell it to the whole traders in their village (daily and weekly market) or to the village's consumer. Again whole village traders sells it to the city traders sells it to the industry sectors, whole sellers and city consumers depending upon the marketability whole sellers has three main channels first is city consumers and the second is exported it and the third was the industry sectors Typical marketing channels Fig. 9.3 (Elnasri 2010). Subsequently, wild fruits are passed through long channels (middlemen or traders) to the consumer and end users (Dash 2016). Internationally some of the forests wild products

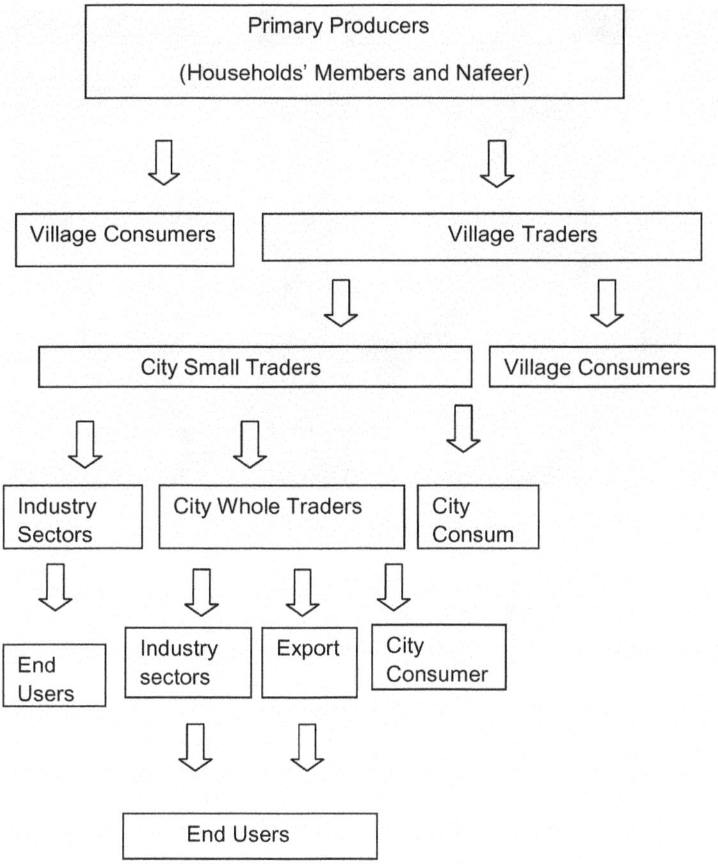

Fig. 9.3 Typical market channel for wild fruits modified from Elnasri (2010)

are valuable and precious export and sell as overseas commodities, they include, turpentine bamboos, resin and gums, honey, various oils, tanning materials, medicinal plants, species, leaves and bark, rattan the long thin stem of a climbing palm (mainly calamu sp), has become an important export for Indonesia, Malayisa, Portugal, Morocco, the Philippines, and other Mediterranean countries which export great quantity of cork derivative from the cork oak, Quercussuber, the Democracy of Korea has build up trade in edible forest fungi an exported it, while the gum Arabic from the gift of the God *Acacia senegal* trees is an ancient and the most important Sudan's exported tangible products (Elnasri 2010), recently and for the coming future, the most promising wild fruit in the world is Tabaldi (Baobab) (Figs. 9.4, and 9.5).

Fig. 9.4 *Adansonia digitata* (Tabaldi). *Captured by the Author*

Fig. 9.5 *Adansonia digitata* (Tabaldi). *Captured by the Author*

9.6.2 Benefits, Uses and Utilization

Over thousands of years millions all over the world relied on forests as a resource for their livelihood, and remained food forests (Powell 2011; Vinceti 2008), insects and wildlife as important components in the diet of rural areas (FAO 2013), As well as, it is good sources of nutrients, protein and fibre than common fruits and vegetables for rural population (Johns and Sthapit 2004; Leakey 1999). Kumar and Mitra 2000 stated that non-timber forest products contribute a high profitable return than timber. But unfortunately about sixty percent of the collection are unrecorded while they are consumed or battered by 15 million of people who are living in and around, the best example is the tendu leaf (*diospyrosme lanoxylon*) which is used locally for manufacturing bidi cigarette, this practice of leaves collection and processing is estimated provides employment as part-timer for up to 0.5 million of women.

Trees and shrubs are providing a multitude of useful products especially in semi-arid areas (Gebauera et al. 2002). With a new innovative understanding of the indigenous wild trees values in the provision and meetings of the nutritional needs and food security, these wild plant trees will be given high attention, because they are found naturally and traditionally managed in Africa forests and wooded lands (Ofori 2014).

FAO 2001 stated that "Africa has faced with the severe crisis and problem of not being capable to nourish and feed its inhabitants people or supplies them with fuel wood". As a result of frequent crop failure in arid and semi-arid areas, the local people suffering from the poor nutrition of local people (Waterlow et al. 1998; Maxwell 1991). That is why it is very important to discover and locate other resources for receiving sufficient and adequate food for the increasing population (Gebauera et al. 2002). The uses of fruits from forest plants species played a significantly important role in the diet of people in the savanna belt of Sudan particularly throughout food shortages and famines in Kordufan and Darfur states during the years 1983–1984. During this time, the forest wild fruits were the main ingredients given as a gesture of hospitality and generosity (Abdel Muti 2002). Sudan is one of the poor developing countries if not the poorest. As the case with all developing countries in the world, the rural communities depend on forest products. Forests provide food (wild fruits, leaves, roots…etc) in addition to other essential goods to meet fundamental requirements at the community level. Moreover, wild forests trees and shrubs lands provide ecological stability necessary for continuous supply for food production and (Ikhlas 2018).

Forest products for rural communities can generate income and employment, also the urban poor in Sudan dominantly uses wood for fuel the principal structural material for construction of shelter and housing. Wild fruits used intensively because it could be acquired at little cost, often no more than the cost and effort of its collections. There are tremendous benefits that can be gained from the forests, when the environment is not endangered (Adger et al. 2004).

9.6.3 Utilization of Wild Fruits

Wild Fruits which comprise of plants for medicinal purposes, animal fodder, food, fibers, dyes, and other necessities, for example people of Gorotire village in southern para state (the kayapo)'in Brazil, they utilize over 98% of the 120 species occurring inside the savannah local scrub (campo cerrado). the kayapo' even prepare from litter, termite and ant nests as a planting medium and they take it to the woodland where they plant useful wild species (Anderson and Posey 1989).

Fruits which are widely and extensively eaten as new and fresh include 'Medaika' which is a very mouth-watering and delicious fruit. The succulent juicy, yellow plum- similar to fruit is eaten fresh and it can be store for short time only. Several wild fruits are very accepted and popular amongst kids; they harvested them and eat them as 'snacks'. The children pick and choose them from the tree and directly eaten them. On the other hand, the majorities of the wild fruits are dried up and have very good storage ability. Regularly they are often store for some months during the dry period. From the pulp of the fruit local beverages are prepared in a diverse different ways. Gebauera et al. (2002) stated that a amount number of under-utilized wild species fruits which produce edible fruits and have different purposes for protecting. Many fruits of these indigenous species have been identified with in the Central Sudan. These species have a significant wide range of benefits and uses (Table 9.1). Beside the edible fruits, multipurpose trees provide a different and important other forest products. The species have immense great potential for future advanced development. on the other hand, comprehensive scientific researches, studies and trials are needed. (Figs. 9.4–9.6)

Dirar (1993), stated that seed of wild fruit are grounded and are very important ingredient elements for the thickening soup or even to make bread, and oil is extract from the dehydrated and crushed kernel of 'Lalob' (*Balanites aegyptiaca*) and is used in various ways for preparing food. Table 9.2 shows the wild fruit trees and their valuable uses of Seeds.

Table 9.1 An important wild fruits in Sudan, local name, uses and observations

Scientific (Latin) name	Local name	Use	Observations
Cordia africana	Gambil	Fresh fruits	–
Sclerocarya birrea	Homeid	Fresh fruits	–
Ximenia americana	Medaika	Fresh fruits	Very delicious fruit
Diospyros mespiliformis	Joghan	Fresh fruits	Popular among children
Pithecellobium dulce	Tamarhindi	Fresh fruits	Popular among children
Tamarindus indica	Aradaib	Local beverages	Good storage capacity
Grewia tenax	Gudeim	Local beverages	Good storage capacity
Adansonia digitata	Gunguleiz	Local beverages	Good storage capacity
Balanites aegyptiaca	Lalob	Local beverages	Good storage capacity
Ziziphusspina-christi	Nabag	Local beverages	Good storage capacity
Borassusa ethiopum	Dalaib	Fruits, nice smell	Palm, better air condition
Hyphaene thebaica	Doum	Sweet fruits	Palm, good storage capacity
Naucleae latifolid	Karmadoda	Fruit, medicine	Good storage after treatment
Randia geipaeflora	Kirkir	Fruit	Food
Naucleae latifolid	Gaggaq	Sweet fruit	Storage capacity, food
Acaci nailotica	Garad	Medicine, tannin	High storage capacity

Fig. 9.6 Sweets from *Tamarindus indica* (Aradaib) and *Adansonia digitata* (Tabaldi). Auther's Camera

Table 9.2 Valuable uses of some wild fruits seeds

No	Latin name	Local name	Uses of seeds
1	*Adansonia digitata*	Gungulaise	Seeds for thickening soups and breads,
2	*Tamarindus indica*	Aradaieb	Thickening soups and breads
3	*Sclerocarya birrea*	Humaid	Thickening soups and breads
4	*Cordia africana*	Gembail	Thickening soups and breads
5	*Balanites aegyptiaca*	Lalob	Oil from dried kernels used in food
6	*Hyphaene thebaica*	Doum	Buttons, pearls and carving

9.7 Resources Improvement and Management

According to the FAO definition: "Forest management deals with the overall economic administrative, technical, social, legal, and scientific aspects related to natural and established and planted forests". It implies various different degree of deliberate human being involvement, ranging from action that aimed at safe guard and maintain the forest (woodland) ecosystem and the environment functions, to favouring given economically valuable species or socially groups of species to improve of the production of environmental services and goods. Management of forest sustainably will guarantee and ensure that values derivative from the forest meet the present—day current needs and at the same time they ensure their continuous accessibility and contribution to long-term development advance t needs (FAO 1993a, b).

In customary traditional management of forest, it mainly and primarly been managed as a wood produc unit (major products) and the other yields and products have traditionally been referred to as NWFPs (minor forest products) and the supply of the minor products was not given high concern or priority. Only in latest recent time that appropriate interest and suitable attention has been given paid to weight the significance of these minor products, particularly in their local limited context where they considerably have received more precieous important valuable than the major wood obtained from forests. Now it is becoming accepted in many cases that recognition of social importance and the economical values of these other minor forest products may be the key in the active participation and involvement of indigiounous people in forests management. One of the most successful and effective ways of sustainable management of trees and forest resources is Self-help management by rural people. Yet, great large of forests formations are under the threat of overuses by inhabitant groups, including local communities. On the one hand, local rural people value forest tree products and functions considerably; on the other hand, the local people are often unwilling and reluctant to invest in forest tree management or may still jeopardize forest management. Consciousness-raising element of projects on the importance of forest trees is often ineffective. The actual real reason for the lack of local people participation is those outsiders misperceive local people constraints to and attitudes towards involvement and participation. Donors and planners need to have more additional confidence in local people's ability to contribute and participate in all stages of activities, planning, Implementation and monitoring and evaluations (Elagab 2016). The feeling of ownership rights and the guarantee of benefits and profits only will encourage local community to take on long-term management forests. Local communities' collaboration is considered necessary and needed if both kinds of forest as well as trees are to be more effectively managed on national, communal and private lands. Participation is local women and men (insiders) implementing wise trees and forest management practices with adequate sufficient support, or in partnership with outsiders (e.g. foresters).

Partnership or joint management is particularly significant model concept regarding and concerning forest resources, since most forests are legally controlled by forest service's governments. Outsiders can support local management by providing or granting access to forest resources, removing impediments to receiving benefits from their management inputs, strengthening the local organizations' situation or that long-term benefits go to those whom an age wisely, providing outside technical formation, etc. Significantly, this definition implies a change of roles for both foresters and farmers in that participation means insiders are doing and responsible from some or all of the administration and managing, and services from the forest are really acting in a service role. For this to happen, both farmers and foresters must see rewards from participatory management. In many places, farmers are not participating in this way. They are deforesting unsuitable land for farming or over using the forest resource base in a way that undermines future production of either trees or other plant or animal crops. Furthermore, rural people do not all want the same outputs. Community level resource conflicts can undermine equity and sustainability. Foresters, too, frequently take on roles other than service roles. However, the authors cited above maintain that lack of participation is frequently an institutional problem and does not stem from a lack of interest or understanding. It is, therefore, important to clarify the term "institution" and identify priority institutional opportunities to increase local participation. Political scientists have a useful definition of institutions for this context: institutions are groups or organizations which have its own set of rules and laws that cover predictable and expected behavior, sanction for breaking the rules and rewards for behaving in the agreed prescribed manner (Thomson 1992). Organizations may codify these rules, however, in local settings, although not necessarily codified, they are understood by groups or group members In community forestry activities, there are three sets of institutions: at national level, including the national policy-makers who decide land tenure and other regulations and donors who often establish sanctions and rewards and exert pressure; forestry and other field-level organizations (e.g. extension services) that administer policies and have codified as well as non-codified rules of behavior; communities or community based groups whose members expect a certain type of behavior towards, among other things, land, trees and management and use of forest.

Natural resources including land, water, animals, soil, wood and plants are of vital paramount significance value to the livelihood of millions of people however in many cases, the indigiounous people did not participated and involved in the management plan of resources. The participatory approach management (the engagement of the community in the decision making to guarantee and ensure fair equitable and sustainability of using the environment) is the best system to improve the status of resources is not adopted although the local people have their own local valuable and rich indigenous knowledge (Messerschmidt and Hammett 1997).

It is recognizing e that local people livelihoods rely on the health (physical condition) and productivity of their own landscapes, and the actions of them as stewards of the land playing a critical significant role in maintenance of this health and productivity, so conservation and proper management requires active involvement of local community (Dash 2016). In Sudan forest policy 1986 and Forest Act 1989, the rights and privileges over wild fruits have been given to the local communities living near and round reserved forests the privilege of collecting the wild fruits and dead trees and fallen branches and since they benefit from the reserved forest they shoulders the role of conservation and protection. An important study conducted by Kumar et al. (2000) confirmed that more than 50% of the respondent preferred formation of association in the villages and camps and about 40% agreed to the formation of villages and camps committees improvement and management can be by native administration was considered by 5.4% of the respondents.

Management approach for NWFPS is base on the application of suitable and appropriate logical scientific methodologies which must focused on the levels of the organization, which encompass the essential processes, structure, functions and the interactions amongst organisms and the environment.

9.7.1 Suggestion for Improvements

All systems of wild fruits production depend on and affected by the type of land use systems (traditional agricultural system, mechanized agriculture, reserved forests, natural range) to improve the current situation some suggestion are below:

1. **Production**
   - Increase the rate of production for the stabilized and sustainable products.
   - Increase and maximize the production horizontally by
   - Mapping of the productive area and expanding it by new plantations according to management plan (afforestation program) and rehabilitate the existing areas (reforestation program) productive wild fruit trees. and planted area in the production
   - Adoption of new technologies such as water harvesting techniques for establishment of provision of water.
   - construct and establishment of camps for production operations (collection, cleaning, packages, guarding) for marketing transporting
2. **Policy Statement**
   Regarding the role of the government and it's institutions toward the natural resources is on a high concern but in this chapter the priority will be specifically given to the latest national forest policy issued one in 2006 and revised 2015. The vision of forestry resources will be utilized and used in efficient, rational, and sustainable manner according to values and in response to the requirements and needs of the communities of the Sudan, thus creates job and trade opportunities will help in poverty alleviation, achievement of food security and bring about improvements of the physical environment of country. The Forest Policy National Goals include, Population welfare, Governance Create a greener Sudan, Peoples' participation Maintaining competitiveness, Land utilization and tenure conflict resolution, Conservation of Biodiversity and Development of jobs and income generation program. The Specific objectives are focused on governance, on maintaining forestry competitive advantage through; Forests and Wood Industries, Energy, oil and forestry. Non-wood forest products and Greening Sudan through addressing issues of Desertification & Deforestation and on Land Tenure and Conservation of Biodiversity (Ministry of Agriculture and Forest, 2015).
3. **Act, legislation and human resources**
   This is what is formulated by the Forest National Corporation (FNC) experts through participatory approach for all who are at stake and it is promising, but if we want to develop and promote this sector to achieve its fruitful objectives it must be given a priority, adopted and implemented by the government and backed by financial, act, legislation, and trained human resource supports.

9.8 Wild Fruits Future Perspectives

To increasing the rate of production of wild fruit there are some important considerations must be achieved:

- Adoption of scientific technologies regarding the time and the methods of collection and harvesting
- Empower-producers through micro-finance and solving the problems of low price.
- Training the producers for the adoption and applying new technologies and on the best practices regarding the production chain of wild fruits.
- Development and improvement of infrastructures in the production area.
- Revision, formulation and implementation of policies to protect the wild fruit trees and their owners from the illicit forests violation the strategic objective is to provide sustainable production through rational and sustainable management
- Training producers in good practices regarding wild fruits production chain.
- Development and improving the infrastructures in the production area
- Adoption and implementation of policies and legislation to protect trees.
- Expansion of productive tree covered areas.

9.9 Challenges Confronting and Facing Wild Fruits Production

- **Institutional constraints**:

  Policies, price, finance, and Inefficient marketing links (small quantities, remote and isolated home and production areas).
  Insufficient credits (limited and mistime)

- Socioeconomic aspect:

  Households area, productive area, infrastructure (carbon market) Insufficient practices in production and marketing (poor post harvest handling, cleaning, and packing) and the wild fruit reputation is poor (low value, low quality).

- **Environmental**: Drought, desertification, climate change.
- **Opportunities**:
  - Under utilization of the resource potential large area from the dry forest in central Sudan is underutilized only few fruit trees species are utilized commercially in country.
  - Loss of substantial value added due to marketing and exporting of unprocessed and semi processed form of wild products.

- Contribution of wild fruits to household income versus different sources.
- Accumulation of indegiounous knowledge and experience (long period of collection of wild fruits) producing age.

9.10 Recommendations

- Value added increment through the processing and semi processing industries (increase marginal benefits)
- Transparency, improvement of marketing and establishment of free zones
- Wild fruits advertising for local uses and as organic produce
- Benefiting from the environmental value of the wild fruit trees for carbon market.
- Comply with the international market requirement such as certification of products
- The wild fruits production could be increased to satisfy the demand through looking deeply into the constraints that face the sector and find out solutions related to wild fruits supply

Suggested important priority theme for further investigation:

1. In depth survey and mapping of the local native wild fruit trees reservoir in Sudan using remote sensing and geographical information system.
2. Establishment of nurseries for seedlings production.
3. Establishment plots and pilot plantations for monitoring and evaluation of under-utilized fruit species
4. Development and improvement of processing and marketing channels for indigenous wild fruits.
5. Identify the reasons that led producers to ignore the production of wild fruits, and give the producers priorities to enhance programs to raise the productivity due to their experience

References

Abdalla AMA, Elzebeir HBG (2014) Variation of physicochemical characteristics of *Ziziphus spina–christi* fruits from three geographical sites in Sudan. M.Sc thesis. Faculty of Forestry. University of Khartoum, Khartoum, Sudan

Abdel Muti OMS (2002) Nutritive value of wild plants of the Sudan. Arab J Food Nutrition 3(3):6–67

Abdel Rahman NA (2011) A study of some Sudanese edible forest fruits and their nectars. Ph.D thesis. Sudan Academy of Sciences, Khartoum, Sudan

Abdel-Rahman NA, Awad II, Babiker ELB (2014) Characterization of some Sudanese Edible Forest Fruits. J Agri-Food Appl Sci 2(2):39–44, 28 February, 2014 E-ISSN: 2311–6730. Available online at http://jaas.blue-ap.org/

Adger WN, Brooks N, Bentham G, Agnew M, Eriksen S (2004) New indicators of vulnerability and adaptive capacity. Technical Report Number 7. Tyndall Centre for Climate Change Research, University of East Anglia, Norwich

Anderson AB, Posey DA (1989) Management of a tropical scrub savanna by the korotire kayapo of Brazil. Adv Econ Bot 7:159–173. Quoted in FAO, 1993. Sustainable Management of Forest in the Tropic and Subtropics for Non-Wood Forest Products. FAO Forestry Paper, Rome

Bates RB, Morris JR, Gandall PG (2001) Chapter III. Priciples and practices of small and medium-scale fruit juice processing: fruit morphology and composition. In: Agriculture and consumer protection. FAO, Rome

Censkowsky U, Helberg U, Nowack A, Steidle M (2007) Overvieew of world production and marketing of organic wild collected products. International Trade Centre UNCTAD/WTO and International Federation of Organic Agriculture Movements (IFOAM), Geneva

Dash BB (2016) NTFPs collection and marketing pattern in tribal doninated villages of Keonjhar District in Odisha. Intercont J Mark Res Rev. ISSN:2321-0346 - ONLINE ISSN:2347-1670 - PRINT -IMPACT FACTOR :1.590

Dirar HA (1993) The indigenous fermented foods of the Sudan. CAB International, Wallingford

Elagab K (2016) Prospect of community forestry in ELsuki locality SinnarState, Sudan. M.Sc thesis. Faculty of Forestry. University of Khartoum

El-Amin MH (1990) Trees and shrubs of the Sudan. 1st published, pp: 106, 400, 451. Ithaca Press, England

Elnasri HM (2010) Certification of gum Arabic in the Sudan: an approach to sustainable management. Ph.D thesis. Faculty of Forestry. University of Khartoum

FAO and WHO Handbook on Human Nutritional Requirements (1974) Nutrition Reviews. 33(5):147–157. https://doi.org/10.1111/j.1753-4887.1975.tb07110.x Published: 01 May 1975

FAO (1993a) Cosecha de hongos en la VII region de Chile. Estudio Monografico de Explotacion Forestal, 2. FAO. Rome. Ensuring sustainable management of wildlife resources. The Case of Africa. Technical Papers. FAO. Rome

FAO (1993b) Conservation of genetic resources in tropical forest management:principles and concepts FAO Forestry paper 107. Rome. Publication Division, FAO. Vialedelle Terme di Caracalla. 00100 Rome Italy

FAO (1995) Report of the international Expert Consultation on Non Wood Forest Products. Yogyakarta, 17–27. Appendix 4.1.2 Non Wood Forest Products and Nutrition. Food and Nutrition Division, FAO, Rome

FAO (2001) Die Tragödie des Hungers hält an: Rund 815 Millionen Menschen unterernährt. Weltberichtzu Hunger und Unterernährung der VereintenNationen (in English)

FAO (2006) Composition and characteristics of selected palm products. FAO Corporate Document Repository. Forestry Department. Tropical Palm. FAO. Rome, Italy

FAO (2010) Global forest resources assessment 2010: progress towards sustainable forest management. FAO, Rome. Available at: www.fao.org/docrep/013/i1757e/i1757e.pdf

FAO (2013) International Conference on the role of forests in the Food Security and Nutrition. Forestry Department. FAO. May 13–15. Rome, Italy

Gebauera J, El-Siddigb K, Eberta G (2002) The potential of under-utilized fruit trees in Central Sudan. Conference on International Agricultural Research for Development. Deutscher Tropentag. Witzenhausen, October 9–11. 2002

Heywood V (1999) Use and potential of wild plants in farm households. FAO Farm Systems Management Series No. 15. FAO. Rome

Ikhlas OSE (2018) A participatory monitoring and evaluation of plantation in Setait locality, Sudan. M.Sc thesis. Faculty of Forestry. University of Khartoum

Johns T, Sthapit BR (2004) Biocultural diversity in the sustainability of developing country food system. Food Nutr Bull 25(2):143–155

Kehlenbeck K, Asaah E, Jammadass R (2013) Diversity of indigenous fruit trees and their contribution to nutrition and livelihoods in sub Saharan Africa: examples from Kenya and Cameroon. In: Fanzo J, Hunter D et al (eds) Diversifying food and diets: using agricultural biodiversity to improve nutrition and health issues in agricultural biodiversity. Earthscan, London, pp 257–269

Kumar N, Saxena NC, Alagh Y, Mitra K (2000) India: alleviation poverty through forest development. Evaluation country case study. The World Bank, Washington, DC

Kuria SG, Wanyoike MM, Gachuiri CK, Wahome RG (2005) Nutritive value of important range forage species for camels in Marsabit District, Kenya. Trop Subtrop Agroecosyst 5:15–24

Layman DK (2003) The role of leucine in weight loss diets and glucose homeostasis. J Nutr 133(1):261–267
Leakey RRB (1999) Potential for novel food products from agrofoeretry trees. Food Chem 64:1–14
Maxwell S (1991) To cure all hunger: food policy and food security in Sudan. Short Run Press, London
Messerschmidt DA, Hammett AL (1997) Local knowledge of alternative forest resources. J Sustain For 7:1-2, 21-55. https://doi.org/10.1300/J091v07n01_02
Ministry of Agriculture and Forests (2015) Forests National Corporation Forest Policy, Legal and Institutional Framework Sudan's Forests' Policy
Powell B, Hall J and Johns T. (2011) Forest Cover, Use and Dieatry Intake in East Usambara Mountains, Tanzania. International Forestry Review. 13(3):305–317
Ros-Tonen MAF, Wiersum KF (2005) The scope for improving rural livelihoods through non-timber forest products: an evolving research agenda. For Trees Livelihoods 15:129–148
Vogt K (1995) A field worker's guide to the identification. Propagation and Uses of: Common Trees and Shrubs of Dry land Sudan
Vinceti B, Eyzaguirre P andohns T. (2008) The Nutritional Role of Forest Plant Food for Rural Cmmunities. In C. J. P. Colfer, ed. Human Health and Forests: A global Overview of Issue, Practice and Policy. London: Earthscan (Chapter 4).
Waterlow JC, Armstrong DG, Fowden L (1998) Feeding a word population of more than eight billion people. Oxford University Press, New York
Wills RHH, Lee TH, Graham D, McGlasson WB, Hall EG (1981) Post-harvest: an introduction to the physiology and handling of fruits and vegetables. AVI Publishing, Co., Westport, CT

Chapter 10
Wild Fruits: Medicinal Use and Health Benefits

Abdalbasit Adam Mariod

10.1 Introduction

Wild fruits assume a crucial role in the household income and food security of numerous communities in the rural societies through the world. In addition, wild fruits contain fundamental supplements (Carbohydrates, protein, minerals) and basic nutrients, which are critical, particularly for the developing youngsters who are inclined to lack of healthy food and related infections. A portion of the fruits has progressed toward becoming articles of business in the local, national and universal markets, thus adding to household incomes and food security. Wild fruits are the capacity supply of human fitness due its active compounds this is answerable for its diverse pharmacological activities (FAO 2012).

As the lack of traditional claims, various elements of *Muntingia calabura* have been utilized to cure one of a kind sorts of illnesses. In Peruvian folklore medication, the flowers and bark are utilized as an antiseptic and to lessen swelling in the lower extremities, even as the leaves, both boiled or steeped in water, are utilized to lessen gastric ulcer and swelling of the prostate gland, and to relieve headache and cold (Mahmood et al. 2014).

The plant *Annona squamosa* Linn traditionally known as Custard apple possesses potent bioactive principals in all its parts. In Mexico the juice is utilized for chills and fever. The mash was found to have mutagenic property. Natural product and natural product juice are taken for worms and parasites, to cool fevers, and as an astringent for loose bowels and loose bowels. The pulverized seeds are utilized against inner and outside parasites, head lice, and worms. The bark takes off, and roots are considered narcotic, ulcer treatment and a nerving tonic and a tea is made for different clutters towards those impacts. The roots contain acetogenins demonstrated to have anti-carcinogenic impact by repressing DNA union. Takes off are

A. A. Mariod (✉)
Indigenous Knowledge and Heritage Center, Ghibaish College of Science & Technology, Ghibaish, Sudan

A. A. Mariod (ed.), *Wild Fruits: Composition, Nutritional Value and Products*,
https://doi.org/10.1007/978-3-030-31885-7_10

utilized to treat insanity, blacking out spells. The leaf decoction is utilized within the treatment of cold, hack, intestinal diseases and causticity condition. Bark decoction is utilized in the runs. Roots are utilized in loose bowels. Smashed takes off is utilized in inner and outside wounds and bubbles, the leaf decoction is utilized in gastritis. The seed extricate have appeared, noteworthy anti-tumor exercises against human hepatoma cells in vitro and in vivo, illustrating a potential for creating the extricate as a novel anti-liver cancer sedate. Extricates of Annona squamosa seeds have noteworthy antitumour action in vivo against AD-5 tumor (Bhattacharya and Chakraverty 2016). In conventional medication, Opuntia ficus indica has been utilized for the treatment of burns, wounds, edema, hyperlipidemia, weight and catarrhal gastritis. Alcoholic extricates are demonstrated for anti-inflammatory, hypoglycemic, and antiviral purposes (Kaur et al. 2012).

10.2 Medicinal Uses of Wild Fruits

Specific parts of one-of-a-kind species of Grewia are applied as human beings medicinal drug in the distinctive portion of globes. Its root bark is applied as a cure for ailment. 50% Ethanolic extricate of airy components of *G. Asiatica* appeared hypotensive motion while the fluid extricate of the stem bark is special to be antidiabetic. Its seed extricate and seed oil shown antifertility. The herbal product is astringent and stomachic. The unripe natural fruit reduces inflammation, respiration, cardiac and blood disarranges, in addition to in fever lessening. MA mixture of the bark is given as a remedy for diarrhea. Ether extricate of *G. bicolor* is utilized for treating postulant pores and skin lesions. *Grewia bicolor* is part of Sudanese traditional medicinal drug, and is utilized in the treatment of the pores and skin accidents. Unique parts of *G. hirsuta* are utilized in cerebral pain, eye lawsuits, bruises and cholera while ethanolic extricate of the stem bark exhibited antiviral. The leaves are beneficial in the nostril and eye sicknesses, treating spleen enlargement, piles, rheumatism and relieving joint pain at the same time as the roots are utilized in diarrhea, dysentery and as a dressing for wounds. The plant *G. microcos* is utilized for treating indigestion, eczema and itch, small pox, typhoid fever, dysentery and syphilitic (Goyal 2012).

10.2.1 Antibacterial Activity

Fangling et al. (2019) analyzed the polyphenols profiles of the methanol extricates of *Armeniaca Sibirica* L. kernel skins (AKS) utilizing fluid chromatography-electrospray ionization mass spectrometry. The polyphenols emphatically diminished Fe3 + and shown great scavenging action towards 2,2′-azino-bis(3-ethylbenzothiazoline-6-sulphonic acid free radicals, 1,1-diphenyl-1-picrylhydrazyl free radicals, hydroxyl radicals, superoxide anions and hydrogen peroxide.

The AKS polyphenols controlled oxidant stretch in HepG2 cells by down controlling diminished glutathione, up directing oxidative glutathione, malondialdehyde and progressed oxidation protein items, and decreased cell practicality to actuate apoptosis of HepG2 cells in vitro. The AKS polyphenols appeared solid antibacterial action against *E. coli*, *Staphylococcus aureus*, *Bacillus subtilis*, *Acetobacter aceti* and *Bacillus cereus*. Subsequently, the antioxidant, inhibitory impact on HepG2 cells and the antimicrobial movement of the AKS polyphenols were unmistakable an and worthy of further consideration for medical industry applications (Fangling et al. 2019).

Anthocyanins of wild blueberries repressed the development of *Listeria monocytogenes*, *Staphylococcus aureus*, *Salmonella enteritidis* and *Vibrio parahaemolyticus*. After a treatment of anthocyanins, the layer penetrability and digestion system of four pathogens were evaluated. When food borne pathogens were treated with anthocyanins for 2 h, the nucleic acid spillage and protein discharge expanded, which illustrated that anthocyanins might devastate the cell film. Besides, the entire protein substance and chemical movement diminished after treatment. The solid antimicrobial impact of Chinese wild blueberry anthocyanins gives application potential within the field of food security (Sun et al. 2018).

Four wild berries were tried for antibacterial action against seven strains of Salmonella typhimurium separated from poultry chain. Candidates were basically screened utilizing the disk-agar strategy for antibacterial activity against E.coli ATCC 25922 and appeared a tall antibacterial action of the restorative herbs tried. The extricates of the wild berries displayed a direct action against E. coli (ATCC 25922) with (12, 8 and 15 mm of diameter) (Ayachi et al. 2009).

10.2.2 Antidiabetic Activity

Crowberry fruit products are advantageous in lessening the hazard of type-2 diabetes. Rough ethyl extricate of crowberry fruit products and its fractions were tried for α-glucosidase and α-amylase inhibitory exercises. To recognize the potential of crowberry extract as an antidiabetic supplement, the ethyl acetic acid derivation fraction shown a better inhibitory impact (IC50 0.5 ± 0 μg/ml) than did the other fractions, and the hexane fraction appeared the least inhibitory impact (IC50 > 100 μg/ml). Moreover, starch corruption by pancreatic α-amylase was hindered by 200 μg of the ethyl acetic acid derivation and n-butanol fraction, showing that the critical restraint of α-glucosidase and α-amylase exercises by the ethyl acetic acid derivation fraction (versus the other fractions) is due to the nearness of polyphenolic compounds, which are known to be dissolvable in polar solvents (Hyun et al. 2018).

Alamin et al. (2015) illustrated that watery extricates of *Tinospora bakis*, *Nauclea latifolia*, *Randia nilotica* and *Mitragyna inremis* at 400 mg/kg essentially brought down ($P < 0.05$) blood glucose levels in diabetic rats. These extricates demonstrated to have critical ($P < 0.05$) antihyperglycemic impact and have the capacity to redress the metabolic unsettling influences related with diabetes. The fluid extricates of these four plants have restorative esteem in diabetes and related complications and hence supporting the conventional employments of these plants in Sudanese conventional medication.

Furthermore Mishra and Garg (2011) investigated in vivo hypoglycemic Also antidiabetic achievable for methanolic extricate about tree grown foods mash for *Feronia elephantum* in glucose stacked animals and alloxan precipitated diabetic animals. Previously, each those styles *Feronia elephantum* diminished the blood glucose stage when contrasted with diabetic deal with gathering What's more push an enormous hypoglycemic Furthermore antidiabetic action. Then again the potency of the herb might have been When considerably short of what that about standard medication regardless metformin. *Feronia elephantum* Corr. Methanolic extricate Moreover decreased the value for body weight passing clinched alongside general What's more alloxan provoked diabetic animals. The results from claiming this look into printed the vicinity of a incredible antidiabetic feasible for methanolic extricate from claiming *Feronia elephantum* Corr. Clinched alongside alloxan achieved diabetic rats.

10.2.3 Antimalarial Activity

Annona squamosa compounds showed moderate activity against a chloroquine-sensitive strain and a chloroquine-resistant strain of *Plasmodium falcifarum* (Bhattacharya and Chakraverty 2016). Mzena and Swai (2018) showed that the ethyl acetate, methanolic and chloroform extracts of *C. metuliferus* and *L. kituiensis* significantly ($p < 0.05$) inhibited parasitemia in a dose-dependent manner and prevented loss of body weight at the dose levels of 600 mg/kg and 1500 mg/kg, respectively. In addition, the extracts prolonged the mean survival time of *P. berghei*-infected mice compared to the non-treated control. The plant extracts did not show reduction of PCV except at the low dose of 300 mg/kg. The highest suppression was recorded at the dose level of 1500 mg/kg. At this dose, *C. Metuliferus* in chloroform, methanol and ethyl acetate extracts had percentage suppression of 98.55%, 88.89% and 84.39%, respectively, whereas *L. kituiensis* in ethyl acetate, chloroform and methanolic extracts exhibited suppression of the pathogens of 95.19%, 93.88% and 74.83%, respectively (Mzena and Swai 2018).

10.2.4 Anti-tumor Activity

Antitumor bioassay was performed to show the biological activities of eight different wild fruits (*Viburnum opulus* L. (guelder rose), *Viburnum lantana* L. (wayfaring tree), *Cornus mas* L. (cornelian cherry), *Pyracantha coccinea* Roemer (firethorn), *Rubus caesius* L. (dewberry), *Crataegus tanacetifolia* (Lam.) Pers (tansy-leaved thorn), *Crataegus monogyna* Jacq. (hawthorn) and *Rosa canina* L. (dog rose)] grown in Turkey. Antitumor activity was evaluated with potato disc tumor induction assay. Best antitumor activity was obtained with cold water extract of fresh fruits of *R. caesius* (100% inhibition). Cold or hot ethanol extracts of fresh *V. lantana* fruits (90.5% and 95.2%, respectively), cold water extract of fresh *C. monogyna* fruits (85.7%) and hot ethanol extracts of fresh *C. tanacetifolia* fruits (71.4%) also exhibited strong tumor inhibition (Turker et al. 2012).

10.3 Antioxidant Activity

Chalise et al. (2010) studied fifteen fruits many times used via the ethnic population in Nepal for their antioxidant recreation and whole polyphenol content (TPC). Among them, *Terminalia bellirica, Terminalia chebula, Phyllanthus emblica* and *Spondias pinnata* had been the most robust antioxidants as compared with nutrition C based totally on the 1,1-diphenyl-2-picryl hydrazyl radical assay. These fruits additionally contained high TPCs. *Spondias pinnata* used to be determined to be extra effective (16% radical scavenging recreation at 5 microg/ml) than nutrition C (5% radical scavenging activity at 5 microg/ml). Shan et al. (2019) assessed polyphenolics content and antioxidant attainable in edible wild fruits used as food and to treat a variety of ailments by the inhabitants of Himalayan area of Pakistan. The fruits of 20 plant species were evaluated using well known protocols, whereas records on medicinal makes use of was gathered via semi-structured interviews. Comparatively, *Prunus domestica* and *Rubus ellipticus* fruits exhibited absolute best ranges of phenolics and flavonols contents in acetone extract. Nevertheless, flavonoids had been most in the water extract of *Rosa moschata*. Contrary, *Duchesnea indica* fruit depicted considerable conceivable to scavenge DPPH and H2O2 radicals at 94.66 ± 8.89% in acetone extract and 83.54 ± 9.37% in water extract, while acetone extract of *Rubus ellipticus* had maximum attainable to minimize ferric ions (133.66 ± 15.00 μM GAE/100 g FW). Additionally, total antioxidant potential used to be easiest in the acetone extract of *Berberis lycium* fruit (332.08 ± 21.90 μM AAE/100 g FW). The relationships between polyphenolics and antioxidant pastime printed synergistic role of secondary metabolites in the prevention of diseases. These authors two revealed that wild fruits ate up by using the nearby communities of Himalayas are rich in fitness recommended phytochemicals and hold significant viable to treat continual diseases, specifically related with free radicals. Their findings revealed that the studied samples have good sized conceivable to hunt free radicals and are rich in herbal antioxidants, particularly phenolics compounds. These fruits are not only used to deal with various fitness disorders, however ought to also make a contribution extensively to the prevention of degenerative illnesses (Shan et al. 2019).

The polyphenols in Fuji' apples and crabapple extracts have extra plentiful phenols and greater whole polyphenols. Crabapple extracts possessed greater antioxidant activity than apple by way of DPPH and ABTS analysis. All fruit extracts exhibited inhibitory consequences on proliferation in unique cancer cells. These outcomes confirmed that crabapples, especially red crabapples, have high-quality possible as a wholesome food, as they are rich in phenolic compounds with excessive antioxidant and anti-proliferative activities to most cancers cells (Han et al. 2019). Cosmulescu et al. (2017) investigated the total phenolic, complete flavonoids, phenolic compounds, the mineral content, and antioxidant activity of fruit extracts of seven wild species (*Crataegus monogyna* Jacq., *Prunus spinosa* L., *Rosa canina* L., *Hippophaë rhamnoides* L., *Rubus fruticosus* L., *Prunus padus*, Cornus mas L.). Their results indicated extensive variations ($p < 0.05$) in the total phenolics and whole flavonoids content, between the seven analyzed species. This recreation used to be linked to a large extent to the type of person phenolic compounds and to a lesser extent to the total phenolic compounds.

10.4 Conclusion

Different types of wild fruits have proven several biological activities, such as antioxidant, antimicrobial, anti-inflammatory, anticancer, and anti-acetylcholinesterase activities. Some wild fruits have extra than one bioactivity. The consumption and utilization of some wild fruits have been increasing, and some wild fruits have been developed into purposeful foods. In addition, the toxicological assessment of some wild fruits is also imperative for protected human consumption.

References

Alamin MA, Yagi AI, Yagi SM (2015) Evaluation of antidiabetic activity of plants used in Western Sudan. Asian Pac J Trop Biomed 5(5):395–402

Ayachi A, Alloui N, Bennoune O, Yakhlef G, Daas Amiour S, Bouzid W, Zoughlache D, Boudjellal K, Abdessemed H (2009) Antibacterial activity of some fruits; berries and medicinal herb extracts against poultry strains of Salmonella. Am Eurasian J Agric Environ Sci 6(1):12–15

Bhattacharya A, Chakraverty R (2016) The pharmacological properties of Annona squamosa Linn: a review. Int J Pharm Eng 4(2):692–699

Chalise JP, Acharya K, Gurung N, Bhusal RP, Gurung R, Skalko-Basnet N, Basnet P (2010) Antioxidant activity and polyphenol content in edible wild fruits from Nepal. Int J Food Sci Nutr 61(4):425–432. https://doi.org/10.3109/09637481003591590

Cosmulescu S, Trandafir I, Nour V (2017) Phenolic acids and flavonoids profiles of extracts from edible wild fruits and their antioxidant properties. Int J Food Prop 20(12):3124–3134

Fangling Q, Lu Y, Cairui L, Cong L, Yehua S (2019) Phenolic composition, antioxidant and antibacterial properties, and in vitro anti-HepG2 cell activities of wild apricot (*Armeniaca Sibirica* L. Lam) kernel skins. Food Chem Toxicol 129:354–364

FAO, WFP and IFAD (2012) The State of Food Insecurity in the World 2012. Economic growth is necessary but not sufficient to accelerate reduction of hunger and malnutrition. FAO, Rome

Goyal PK (2012) Phytochemical and pharmacological properties of the genus Grewia: a review. Int J Pharm Pharm Sci 4(4):72–78

Han M, Li G, Liu X, Li A, Li H (2019) Phenolic profile, antioxidant activity and anti-proliferative activity of crabapple fruits. Hortic Plant J 5:155–163

Hyun TK, Ra JH, Han SH, Kim JS (2018) Antioxidant, antimicrobial, and antidiabetic activities of crowberry fruits. Indian J Pharm Sci 80(3):489–495

Kaur M, Kaur A, Sharma R (2012) Pharmacological actions of *Opuntia ficus indica*: a review. J Appl Pharm Sci 2:1

Mahmood ND, Nasir NLM, Rofiee MS, Tohid SFM, Ching SM, Teh LK, Salleh MZ, Zakaria ZA (2014) *Muntingia calabura*: a review of its traditional uses, chemical properties, and pharmacological observations. Pharm Biol 52(12):1598–1623

Mishra A, Garg GP (2011) Antidiabetic activity of fruit pulp of *Feronia elephantum* Corr. Pharm J 3(20):27–32

Mzena, T., Swai, H., Chacha, M. (2018). Antimalarial activity of Cucumis *metuliferus* and *Lippia kituiensis* against *Plasmodium berghei* infection in mice Res Rep Trop Med; 9: 81–88

Shan S, Huang X, Shah MH, Abbasi AM (2019) Evaluation of polyphenolics content and antioxidant activity in edible wild fruits. Bio Med Res Int 2019, Article ID 1381989, 11 pages. https://doi.org/10.1155/2019/1381989

Sun X-h, Zhou T, Wei C-h, Lan Wq, Wu VCH (2018) Antibacterial effect and mechanism of anthocyanin rich Chinese wild blueberry extract on various food borne pathogens. Food Control 94:155–161

Turker AU, Yildirim AB, Karakas FP (2012) Antibacterial and antitumor activities of some wild fruits grown in Turkey. Biotechnol Biotechnol Equip 26(1):2765–2772

Chapter 11
Adansonia digitata: Phytochemical Constituents, Bioactive Compounds, Traditional and Medicinal Uses

Muneer E. S. Eltahir and Mohamed E. O. Elsayed

11.1 The Plant Distribution and Botanical Description

Baobab (*Adansonia digitata* L.) distributes in sub-Saharan Africa (Wickens and Lowe 2008), and specifically is found in west, east and southern Africa in the drier plant communities of the Sudano-Zambesian lowlands where annual rainfall is 200–800 mm annually (Venter 2012). It is one of eight species of baobabs in the genus Adansonia (Malvaceae subfamily Bombacoideae). Six species occur in Madagascar, one in Australia and one in mainland Africa (Baum 1995). In natural landscapes, baobab densities and not size-class distributions differed significantly between vegetation types and soil types (Venter and Witkowski 2011). Baobabs in Sudan are most frequently found in the southern part of the country (Gebauer et al. 2016), and in the western part as well (Eltahir 2017). They thrive on sandy and rocky soils, from the short-grass savannah (El Amin 1990) to the deciduous savannah woodlands (Wickens and Lowe 2008). They often occur as widely spaced individuals or small groups of individuals scattered over large areas. Baobabs are also common on mountain slopes such as the Jebel edDair in central Sudan and also in the Nuba Mountains (Wiehle et al. 2014). They are also reported in the eastern foothills of the Jebel Marra massif (Wickens 1982). Along wadis and in depressions, where water is being collected during the rainy season, baobabs are found even in the very dry northern parts of Darfur and Kordofan, on heavy soils such as the flat clay plains around Habila in the Nuba Mountains (Gebauer et al. 2016). According to ElAmin (1990), baobabs form belts in Kordofan, Darfur, Blue Nile in Sudan and Upper Nile and Bahr El Ghazal in South Sudan.

M. E. S. Eltahir (✉)
Department of Gum Arabic Researches, Institute of Gum Arabic Research and Desertification Studies, University of Kordofan, Elobeid, Sudan

M. E. O. Elsayed
Department of Biochemistry and Food Science, University of Kordofan, Elobeid, Sudan

A. A. Mariod (ed.), *Wild Fruits: Composition, Nutritional Value and Products*,
https://doi.org/10.1007/978-3-030-31885-7_11

Baobab tree is a massive deciduous tree, up to 20–30 m tall with a diameter up to 2–10 m at adult age (Rahul, J. et al. Asian Pac J Trop Biomed 2015; 5(1): 79–8480). Figure 11.1 tree (https://commons.wikimedia.org) Baobabs is characterized by swollen trunks and palmately compound leaves. The trunks consist of soft, fibrous wood that can store water. Adult trees begin each season by producing simple leaves followed by 2–3-leaflets leaves; mature leaves (20 cm diameter with about 5–9 leaflets) appear later (Rahul, J. et al. Asian Pac J Trop Biomed 2015; 5(1): 79–8480).The leaves of juvenile trees are simple and gradually change to 5–7 foliate compound leaves as the tree gets older. Flowers are borne in the axils of the leaves and comprise a single, large, odoriferous white flower made up of both male and female reproductive parts. The fruits are large, ovoid, and covered in a yellow/green velvety indumentum. The pericarp is woody and indehiscent. Seeds are renal form, embedded in a soft dry matrix (Baum 1995). The native African populations commonly use the baobab fruit as famine food to prepare decoctions, sauces and natural refreshing drink, due to its possession of nutritional properties. The pulp is therapeutically employed as febrifuge, analgesic, anti-diarrhea/ anti-dysentery and for treatment of smallpox and measles (Silvia et al. 2002), Fig.11.1 (fruit, pulp, leaves and seeds of baobab tree).

11.2 Phytochemical Constituents and Bioactive Compounds

Several ethno botany studies have been conducted to characterize the bioactive constituents and biological, pharmacological properties. However, the major interest on baobab product relies in its richness in amino acids (AA) and dietary fibres content. In particular baobab fruit pulp represents the most important natural sources of AA, while the leaves are characterized by the presence of pro vitamin A. In this regards, the baobab fruit pulp can be considered a highly valuable source containing vitamin C ranging from 2.8 to 3.0 g kg^{-1}, in comparison to the fruits that are generally considered as the best source of AA reaches approximately six times of its content in orange. Regarding other parts of the plant, the highest level of pro vitamin A was detected in the young leaves, especially when they are used as dried material (Silvia et al. 2002). The content of pro vitamin A, expressed as retinol equivalents, is between 9 and 27 mg kg^{-1} depending on the method of leaf drying. The combination of small leaves and shade drying pushes the pro vitamin A content up to 27 mg retinol Equivalents per gram of dried leaf powder. To the best of our knowledge in the regards of antioxidant activity, previous investigations (Silvia et al. 2002) were conducted only on fresh leaves, not considering any other parts of the plant.

The baobab fruit pulp is contained in a very resistant external capsule named epicarp whereas the internal ripe fruit which termed as endocarp, is split in small floury, dehydrated and powdery slices that enclose multiple seeds and filaments, the red fibers that subdivide the pulp in segments.

Fig. 11.1 Baobab products: flowers (**a**), fruit (**b**) fibre inside pulp (**c**), pulp (**d**), seeds (**e**), leaves (**f**) leaf powder (**g**), fibres (**h**). (Source: field visits 2017, 2018 by the authors)

The ripe fruit pulp appears as naturally dehydrated, powdery, whitish colored and with a slightly acidulous taste, and its separation from the shell only needs of a single mechanical process without any extraction, concentration or chemical treatment (Sokeng et al. 2019; Nour et al. 1980). This ensures to the pulp the characteristic of a slightly processed food.

Baobab pulp is found to be rich in ascorbic acid (AA) or vitamin C, the leaves are rich in good quality proteins and minerals, and the kernels are rich in fat. Most essential amino acids are present in the leaves. The pulp and the leaves also exhibit antioxidant activity (Chadare et al. 2008).

In virtue of high antioxidant capacity bounded to the characteristic of a slightly processed food, it seems reasonable to consider the baobab fruit pulp as new valuable ingredient for food and/or nutritional application in the promotion of health (Silvia et al. 2002).

It was noted that baobab fruit pulp has very high vitamin C content (280–300 mg 100 g^{-1}) repeated in the previous line, which is seven to ten times more than oranges (51 mg 100 g^{-1}) Such study demonstrated that the consumption of 40 g of baobab pulp provided 100% of the recommended daily intake of vitamin C in pregnant women (19–30 years). The ascorbic acid content was evaluated in the fruit of A. digitate and it was found to contain 337 mg 100^{-1} of ascorbic acid (Rahul et al. 2015). Sidibé and Williams (2002) recommended that baobab leaves should be stored as whole leaves rather than ground leaf powder in order to preserve the high vitamin content (USDA 1998; Sidibé and Williams 2002). Although baobab is mostly regarded as a fruit-bearing forest tree, it is a multipurpose and widely-used species with medicinal properties, numerous food uses of various plant parts, and bark fibers that are used for a variety of purposes. Centuries ago the products were traded: it was well known in Cairo markets in the sixteenth century (Rahul et al. 2015). The leaves are particularly rich in calcium (307–2640 mg 100 g^{-1} dw), and they are known to contain good quality proteins with a chemical score of 0.81. The whole seeds and the kernels have a relatively high lipid content which cotain 11.6–33.3 mg 100 g^{-1} dw and 18.9–34.7 mg 100 g^{-1} dw, respectively. The pulp and leaves exhibit antioxidant properties with a higher activity in the pulp than in the leaves. Reported nutrient contents of different baobab parts showed a large variation, which may have arisen from various factors (Chadare et al. 2008). Previously published biochemical analyses revealed that baobab's edible parts (pulp, leaves and seeds) are rich in micronutrients including vitamin C, provitamin A, minerals (Yazzie et al. 1994; Nordeide et al. 1996; Sidibé et al. 1996; Barminas et al. 1998; Sena et al. 1998; Codjia et al. 2001; Sidibé and Williams 2002; Tembo 2008), procyanidin B_2, epicatechin and organic acids (Tembo et al. 2019). According to Ismail et al. (2019) the application of LC-MS-QTOF-MS enabled identification of 46 phytochemicals in baobab fruit pulp. Hydroxycinnamic acids derivatives, hydroxybenzoic acid derivatives, flavonols, and flavan-3-ols (monomeric and polymeric proanthocyanidins), triterpenoid and steroid glycosides saponins were identified in baobab fruit extract. Proanthocyanidins, phenolic acids, flavonols and saponins were found to be the most common phytochemicals in baobab fruit pulp. Therefore, by considering all the results, 80% acetone was determined as the best solvent for the extraction of phytochemicals in baobab fruit pulp and its usage.

Baobab seed oil served as one ingredient which rapidly gained popularity on global markets and is highly regarded by researchers in the field of ethnobotany and ethnophar-macology particularly for cosmetic benefit (Komane et al. 2017). Regarding the effects of a powder of the fruit of baobab on serum lipids There was a notable reduction in the levels of total cholesterol (120.06% vs. 49.06%, $P < 0.001$) and triglycerides (78.13% vs. 57.44%, $P < 0.001$) in the intervention group compared with the control group (Gadoura et al 2017).

The extracted oil from baobab seeds kernel contains active anti-oxidants i.e. vitamin A and E which makes it a useful active ingredients (moisturizer) in cosmetic products which helps to fight the skin from ageing Modiba et al. (2014). Modiba et al. (2014) reported an optimum biodiesel yield (96 wt.%) obtained from baobab seed kernel oil. Baobab seed oil contains 75–81% of β-sitosterol followed by campesterol (6–6.3%) (Chadare et al. 2008). β-sitosterol, campesterol, and stigmasterol are the main sterols in plants and constitute bioactive compounds that can decrease the plasma/serum levels of lipids and lipoprotein lipids (Li and Sinclair 2002).

The yields of the hydro-distilled the essential oil (EO) from the leaf and stem-bark of *Adansonia digitata* were: 0.302% and 0.403%; while the principal chemical constituents of EO were: hydrocarbons, alkene alcohol, cyclic ketonic ether, terpenoids, amides, esters. Tetramethyl-2-hexadecen-1-ol (26.31%), 8-dimethyl-2-(1-methylethenyl) (8.20%), tetracosan (6.54%), heptacosane (5.81%) and tetratetracontane (5.59%) were dominant compounds in the leaves EO. Even though, the major compounds of the stem-bark EO were octadecane (9.30%), cyclopentane (8.81%), 1-octadecanesulphonyl chloride (8.73%), heptadecane (8.50%), ecosane (8.34%), and tetracosan (7.12%) (Kayode et al. 2018). They concluded that the essential oil of baobab leaves and stem-bark have antimicrobial and cytotoxic properties, as a result of the presence of various chemical constituents. Their findings supported the use of essential oils as natural fungicide for the treatment of fruits and applications as a potential source of sustainable eco-friendly botanical fungicides.

Based on Komane et al. (2017), occlusivity wipe-off test indicated an increased moisture hydration ($p < 0.001$) and decreased transepidermal water loss (TEWL) when baobab oil was applied. Baobab possesses hydrating, moisturising and occlusive properties when topically applied to the skin. They concluded that baobab seed oil could be a valuable functional ingredient for cosmeceutical applications.

11.2.1 The Proximate Analysis of the Leaves

Abiona et al. (2015) investigated the proximate analysis and some phytochemicals of baobab leaves, they reported that it contained protein (13.6%), fat (2.71%), sugar which (0.01%), ash (4.08%), crude fibre (2.45%), moisture content (78.2%), and vitamin C (14.98 mg 100 g^{-1}). Sokeng et al. (2019) reported from the quantitative NMR spectroscopic analysis indicate that sucrose accounts for about 18% of the total amount of detected sugars in leaves, whereas glucose represents about 35% in both α and β isomers, and β-fructopyranose accounts for 34.5%. They also found that amount of myo-inositol was found to be higher in leaves than in pulp, at about 10%, with both α and β galactose isomers accounting for 1.5% of the total amount of sugars detected in leaves. All phytochemicals screened were found to be present.

The seed was found to be a good source of energy, protein, and fat. Both the kernel and the pulp contain substantial quantities of calcium, potassium, and magnesium. Amino acid analyses revealed high glutamic and aspartic acid contents and the sulfur-containing amino acids was the limited amino acids. The fatty acid profile showed that oleic and linoleic acids were the major unsaturated fatty acids, whereas palmitic was the major saturated fatty acid. 0.1 M NaOH was found to be the most effective solvents to solubilize the seed protein. The protein was more soluble at alkaline than acidic pH, with the lowest solubility at pH 4.0 (Osman 2004).

11.2.2 Chemical Composition of Sudanese Baobab Fruit Pulp

Concerning the results of the Sudanese baobab pulp (Table 11.1), it was found that calcium (Ca) content was higher (558.5 mg 100 g^{-1}) than those reported in other country not seen in Table 11.1. In this regard, many authors from Africa such as Saka and Msonthi (1994), Glew et al. (1997), Lockett et al. (2000) and Osman (2004) who reported that Ca contents was 115.6 mg 100 g^{-1}, 341 mg 100 g^{-1}, 211 mg 100 g^{-1} and 295 mg 100 g^{-1} in the same order. The variations in nutrient and mineral contents of baobab pulp based on fruit shapes, may suggest the fact that there are ecotypic and phenotypic variations between trees which ultimately affect the biochemical properties of the baobab pulp. The fruit pulp has a very high vitamin C content as well as carbohydrate with exception of protein and fat which were extremely low. This rich vitamin C in baobab fruits may fill important gaps in food deficiency in the rural areas where more vitamins producing trees are lacking (Gurashi 2015). According to Abdalla et al. (2010) the Sudanese baobab fruit pulp from different regions of the country proved to be rich in ascorbic acid (vitamin C), Ca and P with average values of 358.44, 393.55 and 91 mg 100 g^{-1}, respectively, it also showed high protein content (5.2%). Extraction, filtration and spray drying methods adopted reduced the above mentioned nutrients, except Phosphorus. Solubility of the sprayed powder

Table 11.1 Chemical composition of Sudanese baobab fruit pulp (mg 100 g^{-1})

Reference	Carbu-hydrate	Fat	Ash	Crude fiber	Crude protein	Moisture
Nour et al. (1980)	86.2	0.2	5.3	5.7	2.6	6.7
Obizoba and Amaechi (1993)	73.4	5.1	2.4	–	19.1	19.9
Saka and Msonthi (1994)	79.5	4.3	5	8.3	3.1	13.2
Odetokun (1996)	45.2	4.3	2	6.2	10.9	6.2
Lockett et al. (2000)	70.0	0.4	5.7	11.2	2.2	10.55
Murray et al. (2001)	46.6	0.7	5.1	4.51	4.7	4.7
Manfredini et al. (2002)	30	0.2	–	44	5.3	–
Osman (2004)	76.2	0.3	4.5	5.4	3.2	10.4
Gurashi (2015)	69.55	2.02	5.09	4.71	11.46	7.21

was significantly ($p \leq 0.05$) reduced as a result of treatment with gum Arabic, sugar and Carboxy Methyl Cellulose (CMC). Dispersibility was significantly ($p \leq 0.05$) improved by concentration with sugar, significantly ($p \leq 0.05$) deteriorated as a result of CMC treatment and not affected by addition of gum Arabic. Furthermore, wettability was significantly ($p \leq 0.05$) reduced due to addition of both CMC and gum Arabic (Abdalla et al. 2010).

11.3 Traditional and Medicinal Uses

The baobab is a multi-purpose tree with products having numerous food uses and medicinal properties, and a fibrous bark that is used for various applications (Sidibé and Williams 2002; Codjia et al. 2001; Wickens 1982). The pulp of the fruit, the seeds and the leaves are all utilized and are essentially wild gathered foods. They are consumed daily by rural populations in Africa and are also commercialized. The tuberous tap root of seedlings and young saplings are also eaten, especially during times of famine (Chadare et al. 2009).

Baobab leaves, bark, roots, pulp and seeds are used for multiple medicinal purposes in many parts of Africa and were found to show interesting pharmacological properties including antioxidant, prebiotic-like activity, anti-inflammatory, analgesic, antipyretic activity, anti-diarrhoea, anti-dysentery activity and excipient. Now, bioavailability and pharmacokinetics study of the plant need to be carried out for the assurance of safety reasons (Kaboré et al. 2011).

Abiona et al. (2015) studied the antimicrobial activity of the aqueous extract of the baobab leaves, they found it active against the tested organisms except klebsiellaepneumonae, and *Staphylococcus aureus* was more sensitive to the antimicrobial activity of the extract.

The various parts of the plant (leaves, bark and seeds) are used as a panacea, that is, to treat almost any disease and specific documented uses include the treatment of malaria, tuberculosis, fever, microbial infections, diarrhoea, anaemia, dysentery, toothache, etc. also the leaves and fruit pulp are used as febrifuge as well as an immune stimulant (Rahul et al. 2015). Traditionally, the whole seeds are soaked and boiled for 2–3 hours. Afterwards, the seeds are individually and manually dehulled. Next, the kernels are dried (Chadare et al. 2009). In India, it is reported that baobab pulp is used externally with buttermilk for the relief of diarrhoea and dysentery, while the young leaves are crushed and used to treat painful swellings (USDA 1998). In some countries in West Africa, the leaves, fruit pulp and seeds are the main ingredients in sauces, porridges and beverage. Recently, baobab has been referred to as a "super fruit" based on its nutritional profile (e.g. vitamin, fatty acid, mineral) (Rahul et al. 2015).

The antimicrobial activity of the essential oil of the leaf and stem-bark of *Adansonia digitata* against post-harvest decay of tomato fruits stored at room temperature (27 ± 2 °C) indicated significant ($P < 0.05$) reduction of bacterial and fungal loads at 2000 ppm when compared with control. The lethality assay of leaf and stem-bark EOs showed significant toxicity against brine shrimps, with LC50 values of 1750 μg/ml and 1932 μg/ml, respectively (Kayode et al. 2018).

References

Abdalla AA, Mohammed MA, Mudawi HA (2010) Production and quality assessment of instant baobab (*Adansonia digitata* L.). Adv J Food Sci Technol 2(2):125–133

Abiona DL, Adedapo Z, Suleiman MK (2015) Proximate analysis, phytochemical screening and antimicrobial activity of baobab (*Adansonia digitata*) leaves. IOSR J Appl Chem 8(5):60–65

Barminas JT, Charles M, Emmanuel D (1998) Mineral composition of nonconventional leafy vegetables. Plant Foods Hum Nutr 53:29–36

Baum DA (1995) A systematic revision of *Adansonia* (Bombacaceae). Ann Mo Bot Gard 82:440–470

Chadare FJ, Hounhouigan JD, Linnemann AR, Nout MJR, van Boekel MAJS (2008) Indigenous knowledge and processing of *Adansonia digitata* L. food products in Benin. Ecol Food Nutr 47(4):338–362

Codjia JTC, Fonton Kiki B, Assogbadjo AE, Ekue MRM (2001) Le baobab (*Adansonia digitata*), une espèce à usage multiple au Bénin. Cecodi/Cbdd/Veco/Snv/Fsa, Cotonou, Bénin

El Amin HM (1990) Trees and Shrubs of the Sudan. Ithaca Press Exeter, UK

Eltahir MES (2017) Demographic structure and factors affecting regeneration of baobab (*Adansonia digitata L.*) Populations in some Parts of Kordofan, Sudan. Ph.D. Dissertation, University of Kordofan, Sudan

Gadoura MO, Khidira HB, Adama I, Gasimb GI (2017) Effects of a Powder of the fruit of *Adansonia digitata* (Tabaldia, Gongolase, or baobab tree) on serum lipids, Journal of Herbal Medicine 8:14–16

Gebauer J, Adam YO, Cuní Sanchez A, Darr D, Eltahir MES, Fadl KEM, Fernsebner G, Frei M, Habte T-Y, Hammer K, Hunsche M, Johnson H, Kordofani M, Krawinkel M, Kugler F, Luedeling E, Mahmoud TE, Maina A, Mithoefer D, Munthali CRY, Noga F, North R, Owino WO, Prinz K, Rimberia FK, Saied A, Schüring M, Sennhenn A, Spath MA, Taha MEN, Triebel A, Wichern F, Wiehle M, Wrage Mönnig N, Kehlenbeck K (2016) Africa's wooden elephant: the baobab tree (*Adansonia digitata* L.) in Sudanand Kenya: a review. Genet Resour Crop Evol. https://doi.org/10.1007/s10722-015-0360-1

Glew HR, Van der Jaget JD, Lockette TC, Grivetti EL, Smith CG, Pastuszyn A, Milsom M (1997) Amino acid, Fatty Acid and Mineral Composition of 24 indigenous plants of Burkino Faso. J Food Comp Anal 10:205–217

Gurashi NA (2015) Ecotypic and Phenotypic Variations in *Adansonia digitata L.* (Baobab) From Blue Nile and North Kordofan States, Sudan, Ph.D. Dissertation, University of Khartoum, Sudan

Ismail BB, Pu Y, Guo M, Ma X, Liu D (2019) LC-MS/QTOF Identification of Phytochemicals and the Effects of Solvents on Phenolic Constituents and Antioxidant Activity of Baobab (*Adansonia digitata*) fruit pulp. foodchem

Kaboré D, Sawadogo-Lingani H, Diawara B, Compaoré CS, Dicko MH, Jakobsen M (2011) A review of Baobab (*Adansonia digitata*) products: effect of processing techniques, medicinal properties and uses. Afr. J Food Sci. 5:833–844

Kayode RM, Azubuike CU, Laba SA, Dauda AO, Balogun MA, Ajala SA (2018) Chemical composition and anti-microbial activities of the essential oil of Adansonia digitata stem-bark and leaf on post-harvest control of tomato spoilage. LWT 93:58–63

Komane BM, Vermaak I, Kamatou GPP, Summers B, Viljoen AM (2017) Beauty in Baobab: a pilot study of the safety and efficacy of Adansonia digitata seed oil. Revista Brasileira de Farmacognosia 27:1–8

Li D, Sinclair AJ (2002) Macronutrient innovations: the role of fats and sterols in human health. Asia Pac J Clin Nut 11:155–162

Lockett CT, Calvert CC, Grivetti LE (2000) Energy and micronutrient composition of dietary and medicinal wild plants consumed during drought. Study of rural Fulani, Northeastern Nigeria. Int J Food Sci Nutr 51:195–208

Manfredini S, Vertuani S, Buzzoni V (2002) *Adansonia digitata*.Il baobab farmacista. L'integratore nutrizionale 5:25–29

Modiba E, Osifo P, Rutto H (2014) Biodiesel Production from Baobab (*Adansonia digitata* L.) Seed Kernel Oil and its Fuel Properties. Ind Crops Prod 59:50–54

Murray SS, Schoeninger MJ, Bunn HT, Pickering TR, Marlett JA (2001) Nutritional compositionof some wild food plant foods and honey used by Hadza foragers of Tanzania. J Food Comp Anal 14:3–13

Nordeide MB, Hatloy A, Folling M, Lied E, Oshaug A (1996) Nutrient Composition and Nutritional Importance of green leaves and wild food resources in an agricultural district, Koutiala, in Southern Mali. Int J Food Sci Nutr 47(6):455–468

Nour AA, Magboul BI, Kheiri NH (1980) Chemical composition of baobab fruit (*Adansonia digitata* L.). Trop Sci 12:383–388

Obizoba IC, Amaechi NA (1993) The effect of processing methods on the chemical composition of baobab (*Adansonia digitata* L.) pulp and seed. *Ecol Food Nutr* 29:199–205

Odetokun SM (1996) The nutritive value of baobab fruit (*Adansonia digitata*). La Rivista Italiana delle Sostanze Grasse 73:371–373

Osman MA (2004) Chemical and nutrient analysis of baobab (*Adansonia digitata*) fruit and seed protein solubility. Plant Foods Hum Nutr 59:29–33

Rahul J, Jain MK, Singh SP, Kamal RK, Anuradha Naz A, Gupta AK, Mrityunjay SK (2015) *Adansonia digitate* L. (baobab) a review of traditional information and taxonomic description. Asian Pac J Trop Biomed 5(1):79–84

Saka JDK, Msonthi JD (1994) Nutritional-value of edible fruits of indigenous wild trees in Malawi. Forest Ecol Manag 64:245–248

Sena LP, Vanderjagt DJ, Rivera C, Tsin ATC, Muhamadu I, Mahamadou O, Millson M, Pastuszyn A, Glew RH (1998) Analysis of Nutritional Components of Eight Famine Foods of the Republic of Niger. Plant Foods Hum Nutr 52(1):17–30

Sidibé M, Scheuring JF, Tembely D, Sidibé MM, Hofman P, Frigg M (1996) Baobab – Homegrown vitamin C for Africa. Agrofor Today 8:13–15

Sidibé M, Williams JT (2002) Baobab (*Adansonia digitata*). International Centre for Underutilized Crops, UK p. 100

Silvia V, Elena B, Valentina B and Stefano M (2002) Antioxidant capacity of *Adansonia digitate* fruit pulp and leaves, ACTA Phytotherapeutica 2

Sokeng AJT, Anatoly PS, Arianna DL, Jianbo X, Luisa M, Donatella C, Maria D (2019) Metabolite characterization of powdered fruits and leaves from *Adansonia digitata* L. (baobab): a multimethodological approach. Food Chem 272:93–108

Tembo DTW, Holmes MJ, Marshall LJ (2019) Effect of thermal treatment and storage on bioactive compounds, organic acids and antioxidant activity of baobab fruit (*Adansonia digitata*) pulp from Malawi. J Food Comp Anal 58:40–51

Tembo DTW (2008) Optimisation of vitamin C extraction and physicochemical properties of *Adansonia digitata*, *Parinari Curatellifolia*, *Strychnos Cocculoides* and *Ziziphus Mauritiana* fruits of Malawi. MSc thesis. University of Malawi, Chancellor College, Zomba, Malawi

United States Department of Agriculture (USDA) (1998) Food industry red book. Nutrient tables. US Government Printing Office, Washington, DC

Venter SM, Witkowski ETF (2011) Baobab (*Adansonia digitate* L.) fruit production in communal and conservation landuse types in Southern Africa. Forest Ecol Manag 261: 630–639

Venter SM (2012) The ecology of baobabs (*Adansonia digitate* L.) in relation to sustainable utilization in northern Venda, South Africa. Ph.D. Dissertation, University of the Witwatersrand, Johannesburg, South Africa

Wickens GE (1982) The baobab—Africa's upside-down tree. Kew Bull 37:173–209

Wickens GE, Lowe P (2008) The Baobab trees: Pachycauls of Africa, Madagascar and Australia. Springer, Berlin

Wiehle M, Prinz K, Kehlenbeck K, Goenster S, Mohamed SA, Finkeldey R, Buerkert A, Gebauer J (2014) The African baobab (*Adansonia digitate* L.)—genetic resources in neglected populations of the Nuba Mountains, Sudan. Am J Bot 101:1489–1507

Yazzie D, Van der Jaget DJ, Pastuszen A, Okolo A, Glwu RH (1994) The Amino Acid and Mineral Content of Baobab (*Adansonia digitate* L.) Leaves. J Food Comp Anal 7:189–193

Chapter 12
Annona squamosa: Phytochemical Constituents, Bioactive Compounds, Traditional and Medicinal Uses

Rasheeda Hamid Abdalla Ahmed and Abdalbasit Adam Mariod

12.1 Introduction

From early time plants have been one of the basic sources of medications. The interest of using medicinal plant in treatment of many aliments is increased, this because of the emergence of multidrug resistant of many microbes and the harmful side effects of the synthetics drugs this beside its high cost. Numerous lifesaving drugs utilized in the armamentarium of current prescription are given by herbs (Goyal et al. 2007) *Annona squamosa* Linn which belong to the family Annonaceae is one of the fundamental therapeutic plants, ordinarily called "custard apple". It has been accounted for to have numerous pharmacological exercises and is utilized in customary applications (Raj et al. 2009). Therefore a lot of research was done on the different parts of the plant because of the occurrence of precious annonaceous acetogenins, which are generally utilized for the treatment of numerous illnesses (Kaleem et al. 2008). It is known with different names in the different language like Custard apple in English and Sharifa in Hindi, it is distributed in all over India and other tropical countries (Morton 1987).

R. H. A. Ahmed (✉)
National Public Health Laboratory, Ministry of Health, Khartoum, Sudan

A. A. Mariod
Indigenous Knowledge and Heritage Center, Ghibaish College of Science & Technology, Ghibaish, Sudan

A. A. Mariod (ed.), *Wild Fruits: Composition, Nutritional Value and Products*,
https://doi.org/10.1007/978-3-030-31885-7_12

12.2 Botanical Description

Annona squamosa Linn is also known with other various names such as common-seureuba, sugar apple, gishta, sweet apple, sweetsop (Wunderlin and Hansen 2008). *Annona squamosa* is a little tree that grows up to 3–8 m, with wide, sporadically branches of light dark colored bark having flimsy leaves that happen independently, estimating 5–17 cm long and 2–6 cm in width (Pandey et al. 2014). Flowers are greenish-yellow, on slim bristly stalks, sepals pointed, bushy, green, around 16 mm long; 3 external petals elliptical, thick and adjusted at the tips, yellow-green, somewhat hairy, inside light yellow and keeled with a purplish or ruddy spot at the thin, stamens various, white, under 16 mm long; ovary light green, styles white on the raised axis (Orwa et al. 2009). The round or heart-molded greenish yellow, matured fruit is pendulous on a thickened stalk. The pulp is white-tinged yellow, palatable, and sweetly fragrant. Every carpel contains an elliptical, gleaming and smooth, dull dark colored, 1.3–1.6-cm seed (Pandey et al. 2014). Stem is cylindrical with characteristic odour and bitter taste. Outer side thick cork cells are found upon maturation (Anshuman and Raja 2016; Bhattacharya, and Chakraverty 2016) (Fig. 12.1).

Fig. 12.1 Mature *Annona squamosa* tree with fruits. (https://commons.wikimedia.org)

12.3 Plant Distribution

Annona squamosa (sugar apple) is constantly developed in tropical South America yet not regularly in Central America, in all respects as often as possible in southern Mexico, West Indies, Bahamas, Bermuda and once in a while southern Florida just as in Jamaica, Puerto Rico, Barbados and in dry locales of Australia. It was developing in Indonesia from the get-go in the seventeenth century and has been generally received in southern china, tropical Africa, Egypt and marshes of Palestine. Development is most broad in India where the tree is exceedingly prominent. The sugar apple is a standout amongst the most essential organic products in the inside of Brazil (Morton 1987). It is traditionally cultivated in north eastern parts of Thailand (Intaranongpai et al. 2006).

12.4 Biochemical Analysis

Anuragi et al. (2016) studied the proximate analysis of Annona, they reported 73.9% moisture content, 1.4% ash content, 23.2% total carbohydrates, 16.6% total soluble sugars, 7.8% reducing sugars, 3.3% fiber content, 1.9% protein content, 0.3% phenol content, 0.22% titratable acidity, 31.5% ascorbic acid content, and 25.5% seed oil content. Andrade et al. (2001) investigated the Brazilian *Annona squamosa* fruit pulp chemical composition they reported the amounts of sugars as 58% of dry mass. They reported very low triglyceride and a considerable amount of 0.25% for diterpenoid compound kaur-16-en-18-oic acid detected in the lipid fraction. These authors investigated the essential oil of the fruit pulp and they confirmed the major compounds as α-pinene (25.3%), sabinene (22.7%) and limonene (10.1%).

12.5 Medicinal Uses

12.5.1 Antibacterial Activity

The emergence of multidrug resistance by many organisms and the side effects of the synthetic antibiotics triggered the search for antimicrobial agents from the plant source to control microbial infections. All the distinctive pieces of the plant demonstrated affectability to numerous life forms. The Leaves of *Annona squamosa* are utilized in local prescription for disease treatment and aversion and they have been appeared to have antibacterial activity in ongoing investigations (Dholvitayakhun et al. 2016). Vohra et al. (1975) reported that seed extracts of *A. squamosa* show antibacterial action. Oil of *A. squamosa* also showed a clear sensitivity against *Bacillus subtilis* and *Staphylococcus aureus* (Chavan et al. 2006). Many studies disclosed that the solvents which is used in the extractions of phytochemicals from *Annona squamosa*, affect its efficiency as antimicrobial. A study by Padhi et al. (2011) investigated three extracts (methanol, petroleum, and aqueous) of the leaf of *A. squamosa* and *A. reticulate* against three Gram-positive bacteria and five Gram-negative bacteria, by using agar cup and broth dilution methods. It was shown that *B. subtilis, S. epidermidis, S. aureus* and *V. alginolyticus* were most susceptible to the extracts among the tested bacteria. But, *S. typhi* didn't depict any zone of inhibition. Therefore, the methanol extract revealed the highest inhibition zone to bacterial growth then came petroleum ether and aqueous extracts for both *A. squamosa* and *A. reticulata* leaf solvent ether, chloroform water, petroleum ether, chloroform and alcohol leaves extracts were subjected to antibacterial screening, the three late extracts reveal the greatest zone of inhibition. Within these extracts the petroleum ether extracts showed significant results. Therefore comparing to the control group all the outcomes was critical for various parameters in wound healing activity (Shenoy et al. 2009). Another study assessed the antibacterial activity of different solvent extracts of *A. squamosa* against different Bactria spp. four of the species showed weak activity (El-Chaghaby et al. 2014). *Annona squamosa* bark is also active against bacteria *Bacillus coagulans* and *Escherichia coli* are more sensitive to methanol extract of stem bark than other bacteria (Kachhawa et al. 2012). It was found that *Neisseria gonorrhoeae, Campylobacter jejuni, Staphylococcus aureus, Streptococcus pneumoniae, Bacillus cereus*, and *Listeria monocyto-genes* were susceptible to the *A. squamosa* leaf extract in vitro. This has been confirmed by Shokeen et al. (2005), and Yusha'u et al. (2011) using *A. squamosa* leaf extract. Salman and Senthilkumar (2015) reported that, *S. sobrinus* is more susceptible to the *A. squamosa* bark extract than *S. mutans,* but there is no much differences within the species. They also found that *S. mutans* and *S. sobrinus* are susceptible to 5–50 mg/ml concentration of the bark extract of *A. squamosa.* The susceptibility of *S. mutans* and *S. sobrinus* to A. squamosa bark extract made its useful as antibacterial agent and it can be used to prevent dental caries, this lead to suggestion of addition of the extract to the tooth paste or mouth wash so as to prevent the tooth decay. Number of *A. squamosa* phytochemical compounds showed antibacterial activity, such as linalool, borneol, eugenol, farnesol, geraniol, tannins, phenolic compounds, polyphenols, annotemoyin-1, annotemoyin-2, squamocin and cholesteryl glucopyranoside, and flavonoids, which showed strong antibacterial activity. In addition to silver nanoparticles (AgNPs) that biologically synthesized from *A. squamosa* (Bulut and Rapid 2009), has antibacterial action (Patel and Kumar 2008). The antibacterial activity of *A. squamosa* leaf extracts suggested by Dholvitayakhun et al. (2016) to be due to disrupting cell wall formation. The change in shape of *B. cereus* after treated with inhibitory concentrations of *A. squamosa* extract, strongly supported the suggestion of the cell wall disruption by the plant (Dholvitayakhun et al. 2016).

12.5.2 Antidiabetic

Demand of diabetic patients to utilize natural products has increased as the result of the adverse impacts of insulin and oral hypoglycaemic drugs (Kameswara and Appa 2001). The extracts of *Annona squamosa* leaves, seeds and roots were found to have antidiabetic and hypoglycaemic effects (Tomar and Sisodia 2012; Gupta et al. 2005). A few clans in India take a blend of 4–5 recently risen youngleaves alongside five grains of dark pepper (*Piper nigrum*) promptly in the first part of the day for the treatment of diabetes. Proceeded with treatment with this mixture ensured up to 80% positive results (Topno 1997). Supplementation with the aqueous extract of *A. squamosa* is helpful in controlling the blood glucose level, improves the plasma insulin, lipid digestion and is valuable in keeping diabetic complexities from lipid peroxidation and cancer prevention agent frameworks in exploratory diabetic rats. Panda and Kar (2007) investigated the effect of ethanolic extract of *A. squamosa* L. in the Alloxan induced diabetic control rats. They found a considerable decrease in blood glucose level when compared to diabetic control group. However, the addition of *A. squamosa* (300 mg/kg) water extract orally to diabetic rats for 30 days reduced blood glucose, urea, uric acid and creatinine significantly; nevertheless, they elevated the actions of insulin, C -peptide, albumin, albumin/globulin ratio beside restoring all marker enzymes nearly to the control levels (Kaleem et al. 2008). *A. squamosa* leaf extract was also found to decrease blood triacylglycerol and total cholesterol levels in diabetic animals (Gupta et al. 2008). Kalidindi et al. (2015) reported that the leaves extract of *A. squamosa* possesses anti-diabetic action because it induces the insulin release from the pancreatic islets, raises the consumption of glucose in muscle, as well as inhibits the glucose productivity from the liver. This effect comes from the fact that this plant contains quercetin which has antidiabetic action. Shirwaikar et al. (2004) suggested that the critical antidiabetic action of A. squamosa water concentrate is because of the presence of flavonoids and adhesive in the plant. Additionally, quercetin-3-O-glucoside segregated from *A. squamosa* leaf was found to restrain glucose 6 phosphatase action in the liver and to bring down blood glucose level (Panda and Kar 2007). The mechanism by which *A. squamosa* brings about its hypoglycemic action in diabetic rat was suggested by (Shirwaikar et al. 2004) to be by potentiating the insulin impact of plasma by expanding either the pancreatic secretion of insulin from the current beta cells or by its discharge from the bound structure.

12.5.3 Antimalarial Activity

The different parts of *Annona squamosa* exhibits antimalarial activities. In vitro *Annona squamosa* leaf methanol extract delineated the strong hindrance against the chloroquine-sensitive strain 3D7 and chloroquine-resistant strain Dd2 of threatening plasmodium, additionally its stem skin methanol concentrate had moderate inhibitory impact against Dd2 (Johns et al. 2011). Kamaraj et al. (2012) also reported that the bark extract displayed IC50 of 30 μg/ml against blood stage of Plasmodium falciparum. The N-Nitrosoxylopine, Roemerolidine and Duguevalline separated from *Annona squamosa* leaf extricate are known alkaloids in charge of antimalarial properties (Johns et al. 2011). The noteworthy activity appeared by the concentrates of *Annona squamosa* suggested that the plant may had emphatically kill insects especially mosquitoes, in this manner giving a promising source of larvicidal agents. Brazilian plant demonstrated larvicidal impact against *Aedes adopictus* and *C. quinquefascinits* and against *Anopheles stephensi*. Present larvicidal action result underpins the reports and exhibited that concentrate of Annona species are potential anti-mosquito agents (Magadula et al. 2009.

12.5.4 Therapeutic Potentials of Annona squamosa Linn

The plants are essential wellsprings of drugs since the start of human progress. *Annona squamosa* is widely used in traditional medicine. It possesses effective bioactive principals in all its parts (Ranjan and Sahai 2009). The distinctive pieces of Annona squamosa, for example, fruits, seeds, leaves and barks have been utilized to treat diverse aliments. The pharmacological potential of Custard apple has many uses in medicine. Furthermore custard apple has number of uses in food and it is a source of industrial products. Recently, extensive research were done on the pharmacological properties of various pieces of *Annona squamosa* and has effectively disengaged and distinguished dynamic constituents in charge of therapeutic potential. Uses of various pieces of *Annona squamosa* plant are given in Table 12.1.

12.6 Food Uses

Custard apple is a multipurpose tree with edible fruits. Its chemical constituents have shown the potentially useful source of nutraceutical and flavoring agents. This fruit is soft and juicy and has a sweet taste. Furthermore, Custard Apple is widely used at homes, hotels and restaurants to make salads. Spray dried apple powder is extensively used to make juice, mango flavored, yoghurt and in many other applications (Zahid et al. 2018). It's also used in making of ice creams and milk beverages (Pandey and Barve 2011), or eaten fresh (Vanitha et al. 2010). It is a mixture of cooked rice and sitaphal in specific proportions with added flavor with cardamom (Kaur et al. 2015). It is also used to make wine (Leal 1990). The higher quality oil can be extracted from the seeds (Mariod et al. 2012). Delicious products such as jam and squash can also be made from the fruit pulp (Haq and Hughes 2002).

Table 12.1 Uses of different parts of *Annona squamosa* plant

Pharmacologyical action	Plant part	Active ingredients	Reference
Anti-head lice effect	Seed	Oleic acid and triglyceride	Kumar et al. (2010)
	Leaf	Linalool, Borneol, Eugenol, Farnesol, Geraniol and Flavonoids	Padhi et al. (2011)
	Leaf	Flavonoids	(Patel and Kumar (2008)
Antimicrobial activity	Seeds	Ent-kauranes, Acetogenins, essential oils and Benzylisoquinolines alkaloids and diterpene	Kumar et al. (2010)
		Annotemoyin-1, Annotemoyin-2, squamocin and cholesteryl glucopyranoside	Rahman et al. (2005)
		16-hentriacontanone (palmitone) and 10- hydroxy-16-henriacontanone while squamocin A and G, and squamostatin A	Dang et al. (2011)
Antimicrobial activity	Seeds	2,2-azinobis- (3-ethylbenzothiazoline-6-sulphonate) (ABTS) 1,1-diphenyl, 2-picryl hydrazyl (DPPH)	Shirwaikar et al. (2004); Baskar et al. (2007)
	Seeds	1, 1- diphenyl-2-picrylhydrazyl, nitric oxide, and hydrogen peroxide	Kalidindi et al. (2015) and El-Chaghaby et al. (2014); Chandrashekar and Kulkarni (2011)
	Leaves	Quercetin-3-O glycoside	Panda et al. (2007)
	Leaves	5,7,4′ trihydroxy-6,3′ dimethoxyflavone 5-O-α-l-rhamnopyranoside (THDMF-Rha)	Panda et al. (2015)
Antimicrobial activity	Leaves	Ascorbic acid	Kothari and Seshadri (2010)
Antimicrobial activity	Leaves	Tocopherol	Luzia and Jorge (2012)
Antimicrobial activity	Seeds	O-methylarmepavine and C37 trihydroxy adjacent bistetrahydrofuran acetogenins	Vila-Nova et al. (2011)
Anti-fungal activity	Seeds	Annonaceous acetogenins and alkaloids asimicin and bullatacin	Nandagaon and Kulkarni (2012)
Antioxidant – activity	Leaf	Squadiolins A and B and squafosacin B	Liaw et al. (2008)
Antioxidant – activity	Fruit	Annosquamosin A, B and C	Sun et al. (2012)
Antioxidant – activity	Seed	Acetogenin squamotacin	Hopp et al. (1996)

(continued)

Table 12.1 (continued)

Pharmacologyical action	Plant part	Active ingredients	Reference
Antioxidant – activity	Bark	12,15-cissquamostatin-A and bullatacin	Chen et al. (2012)
	Bark	Squamoxinone-D	Miao et al. (2016)
	Seed	Annosquacin A, B and C, annosquatin A and B, and squamostanin A, B and D, squamostolide, bullatacin and uvarigrandin A	Chen et al. (2011, 2012); Yang et al. (2009)
	Pericarp oil	Two entkaurane diterpenoids, ent-kauran-16-en-19-oic acid and ent-kauran-15-en-10-oic acid	Chen et al. (2017)
	Bark	Mosin A, B and C, annoreticuin-9-one, squamotacin, bullacin B, tetrahydrosquamone and bullatacinone	Hopp et al. (1998)
Antioxidant – activity	Leaf seeds	Lanuginosine	Nakano et al. (2013)
Antitumour activity	Leaf, pericarp	*Squamocin P* and *annosquatin III*	Ma et al. (2017)
		asimicin and *bullatacin*	Nandagaon and Kulkarni (2012)
Renoprotective activity	Leaf, pericarp	5,7,4′ trihydroxy-6,3′ dimethoxy-flavone 5-O-α- Irhamnopyranoside (THDMF-Rha), acetogenins	Panda and Kar (2015); Singh and Singh (2001)
Mosquitocidal activity	Seeds stem	The ent-kaurane diterpenoids 'ent-Kaur-16-en-19-oic acid' and '16alpha-hydro- 19-al-ent-kauran-17-oic acid' 18-acetoxy-ent-kaur-16-ene	Chavan et al. (2011)
Molluscicidal activity	Bark	18-acetoxy-ent-kaur-16-ene	Chavan et al. (2011)
	Leaf	A kaurane diterpenoid 16β, 17-dihydroxy-entkauran- 19-oic acid	Wu et al. (1996)
	Seed	(+)-O-methylarmepavine, Nmethylcorydaldine and isocorydine	Yadav et al. (2012)
	Leaves, bark	Glycosaminoglycan	Ponrasu and Suguna (2014)
	Fruit	Glycosaminoglycan	Ponrasu and Suguna (2014)
	Fruit	A cyclic octapeptide, cyclosquamosin B	Iizuka et al. (2006); Morita et al. (2006)
	Seeds	The ent-kaurane diterpenoids 'ent-Kaur-16-en-19-oic acid' and '16alpha-hydro- 19-al-ent-kauran-17-oic acid'	Yang et al. (2002)
Vasorelaxant activity	Stem	Carophyllene oxide	Chavan et al. (2010)
Anti-platelet activity	Bark	18-acetoxy-ent-kaur-16-ene	Chavan et al. (2011)
Analgesic & Anti-inflammatory	Bark	A kaurane diterpenoid 16β, 17-dihydroxy-entkauran-19-oic acid	Wu et al. (1996)

(continued)

Table 12.1 (continued)

Pharmacologyical action	Plant part	Active ingredients	Reference
Analgesic & anti-inflammatory	Fruit	Acidic heteropolysaccharide known as GASP3–3-I isolated	Ren et al. (2017)
Antiviral activity	Fruit	Quercetin-3-O-glucoside	Panda et al. (2007)
Antidiabetic	Leaf	L-lysine and L-proline,	Singh et al. (2010)
		1-(4-β-D glucopyranosyloxyphenyl) – 2 – (β – D -glucopyranosyloxy)-ethane	Yadav et al. (2011)
	Twigs	(+)-O-methylarmepavine, Nmethylcorydaldine and isocorydine	Yadav et al. (2012)
Antiulcer	Twigs	Glycosaminoglycan	Ponrasu and Suguna (2012, 2014)
	Leaves	EtOAc fractions	Rahman et al. (2005)
Wound healing activity	Leaf	N-Nitrosoxylopine, roemerolidine and Duguevalline	Ponrasu and Suguna (2012)

12.7 Other Uses

Annona squamosa leaves yield an excellent oil rich in terpenes and sesquiterpenes, mainly B-caryophyllene, which finds limited use in perfumes (Bhattacharya and Chakraverty 2016). The leaves also provide ingredients used to make dyes, stains, inks, tattoos and mordants (Singh 2011). The strongest bark of the plant is used for carrying burdens in the Amazon Rainforest (Raj et al. 2009), and for wooden implements, such as tool handles and pegs. The wood is valued as firewood. Furthermore it is used as ornamental plant and it is cultivated along with banana plantation (Oliveira et al. 2010). A. squamosa seed oil was reported to be use in the soap and plasticizer industry as well as in alkyd manufacturing, the seeds are acrid and poisonous (Morton 1987).

References

Andrade E, Zoghbi MGB, Maia JGS, Fabricius H, Marx F (2001) Chemical characterization of the fruit of *Annona squamosa* L. occurring in the Amazon. J Food Comp Anal 4(2):227–232. https://doi.org/10.1006/jfca.2000.0968

Anuragi H, Dhaduk HL, Kumar S, Dhruve JJ, Parekh MJ, Sakure AA (2016) Molecular diversity of *Annona* species and proximate fruit composition of selected genotypes. 3 Biotech 6(2):204. https://doi.org/10.1007/s13205-016-0520-9

Baskar R, Rajeswari V, Kumar TS (2007) In vitro antioxidant studies in leaves of Annona species. Indian J Exp Biol 45(5):480–485

Bhattacharya A, Chakraverty R (2016) The pharmacological properties of *Annona squamosa Linn:* a review. Inter J Pharm Eng 4(2):692–699

Bulut E, Rapid OM (2009) Facile synthesis of silver nanostructure using hydrolysable tannin. Ind Eng Chem Res 48:5686–5690

Chandrashekar C, Kulkarni VR (2011) Isolation, characterizations and free radical radical scavenging activity of *Annona squamosa* leaf. J Pharm Res 4:610–611

Chavan MJ, Shinde DB, Nirmal SA (2006) Major volatile constitute of *Annona squamosal* L. bark. Nat Prod Res 20:1–2

Chavan MJ, Wakte PS, shinde DB (2010) Analgesic and anti-inflammatory activity of Carophyllene oxide from *Annona squamosa* L bark. Phytomedicine 17(2):149–151

Chavan MJ, Wakte PS, Shinde DB (2011) Analgesic and anti-inflammatory activities of 18-acetoxy-ent-kaur-16-ene from *Annona squamosa L.* bark. Inflammopharm 19:111–115

Chen Y, Chen JW, Li X (2011) Cytotoxic bistetrahydrofuran annonaceous acetogenins from the seeds of *Annona squamosa*. J Nat Prod 74:2477–2481

Chen Y, Xu S, Chen JW, Wang Y, Xu X, Fan N, Li X (2012) Antitumor activity of *Annona squamosa* seeds extract containing annonaceous acetogenin compounds. J Ethnopharmacol 142:462–466

Chen YY, Peng CX, Hu Y, Bu C, Guo SC, Li X, Chen Y, Chen JW (2017) Studies on chemical constituents and anti-hepatoma effects of essential oil from *Annona squamosa L.* pericarp. Nat Prod Res 31:1308–1308

Dang QL, Kim WK, Nguyen CM, Choi YH, Choi GJ, Jang KS, Park MS, Lim CH, Luu NH, Kim J-C (2011) Nematicidal and antifungal activities of Annonaceous acetogenins from *Annona squamosa* against various plant pathogens. J Agric Food Chem 59(20):11160–11167. https://doi.org/10.1021/jf203017f

Dholvitayakhun A, Trachoo N, Narkkong N, Cushnie TPT (2016) Using scanning and transmission electron microscopy to investigate the antibacterial mechanism of action of the medicinal plant Annona squamosa Linn. J Herb Med 7:31–36

El-Chaghaby GA, Ahmad AF, Ramis ES (2014) Evaluation of the antioxidant and antibacterial properties of various solvents extracts of *Annona squamosa* L. leaves. Arab J Chem 7:227–233

Goyal BR, Goyal RK, Mehta AA (2007) Phyto-pharmacology of *Achyranthes aspera*: a review. Pharmacogn Rev 1(1):143–150

Gupta RK, Kesari AN, Watal G, Murthy PS, Chandra R, Maithal K, Tandon V (2005) Hypoglycaemic and antidiabetic effect of aqueous extract of leaves of *Annona squamosa* (L.) in experimental animal. Curr Sci 88(8):1244–1254

Haq N, Hughes A (2002) Fruits for the future in Asia. International Centre for Underutilized Crops, University of Southampton. Available from http://www.cropsforthefuture.org/publication/proceeding/Fruits_for_the_Future_in_Asia.pdf

Hopp DC, Zeng L, Gu ZM, McLaughlin JL (1996) Squamotacin: an Annonaceous acetogenin with cytotoxic selectivity for the human prostate tumor cell line (PC-3). J Nat Prod 59(2):97–99

Hopp DC, Alali FQ, Gu ZM, McLaughlin JL (1998) Three new bioactive bis-adjacent THF-ring acetogenins from the bark of *Annona squamosa*. Bioorg Med Chem 6:569–575

https://commons.wikimedia.org

Iizuka SSH, Choo T, Chan CY, Takeya KL, Kobayashi K (2006) Vasorelaxant activity of cyclic peptide, cyclosquamosin B, from Annona squamosa. Bioorg Med Chem Lett 16(17):4609–4611

Intaranongpai J, Chavasiri W, Gritsanapan W (2006) Anti-head lice effect of Annona squamosa seeds. Southeast Asian J Trop Med Public Health 37(3):532–535

Johns T, Windust A, Jurgens T, Mansor SM (2011) Antimalarial alkaloids isolated from *Annona squamosa*. Phytopharm 1:49–53

Kachhawa JBS, Sharma N, Tyagi S, Sharma KK (2012) Screening of stem bark methanol extract of *Annona squamosa* for antibacterial activity. Inter J Cur Pharm Res 4(1):48–50

Kaleem M, Medha P, Ahmed QU, Asif M, Bano B (2008) Beneficial effects of *Annona squamosa* extract in streptozotocin-induced diabetic rats. Singapore Med 49(10):800

Kalidindi N, Thimmaiah NV, Jagadeesh NV, Nandeep R, Swetha S, Kalidindi B (2015) Antifungal and antioxidant activities of organic and aqueous extracts of *Annona squamosa* Linn. leaves. J Food Drug Anal 4:795–802

Kamaraj C, Kaushik NK, Mohanakrishnan D, Elango G, Bagavan A, Zahir AA, Rahuman AA, Sahal D (2012) Antiplasmodial potential of medicinal plant extracts from Malaiyur and Javadhu hills of South India. Parasitol Res 111:703–715

Kameswara RB, Appa RCH (2001) Hypoglycemic and antihyperglycemic activity of alternifolium Walp. Seed extracts in normal and diabetic rats. Phytomedicine 8:88–93

Kaur R, Kaur K, Kaur P, Singh I (2015) Sitapha L: unexplored therapeutic potential. Asian J Res Chem Pharm Sci 3(4):129–141

Kothari V, Seshadri S (2010) Antioxidant activity of seed extracts of *Annona squamosa* and *Carica papaya*. Nut Food Sci 40:403–408

Kumar AJ, Rekha T, Devi SS, Khannan M, Jaswanth A, Gopal V (2010) Insecticidal activity of ethanolic extract of *Annona squamosa*. J Chem Pharm Res 2(5):177–180

Liaw CC, Yang YL, Chen M, Chang FR, Chen SL, Wu SH, Wu YC (2008) Mono-tetrahydrofuran annonaceous acetogenins from *Annona squamosa* as cytotoxic agent and calcium ion chelators. J Nat Prod 71(5):764–771

Luzia DMM, Jorge N (2012) Soursop (*Annona muricata L.)* and sugar apple (*Annona squamosa L.):* antioxidant activity, fatty acids profile and determination of tocopherols. Nut Food Sci 42:434–441

Magadula JJ, Innocent E, Otieno JN (2009) Mosquito larvicidal and cytotoxic activities of 3 Annona species and isolation of active principles. J Med Plant Res 3(9):674–680

Mariod AA, Elkheir S, Ahmed YM, Matthaus B (2012) *Annona squamosa* and *Catunaregam nilotica* seeds, the effect of the extraction method on the oil composition. J Amer Oil Chem Soc 87(7):763–769

Miao Y, Xu X, Yuan F, Shi Y, Chen Y, Chen J, Li X (2016) Four cytotoxic annonaceous acetogenins from the seeds of *Annona squamosa*. Nat Prod Res 30:1273–1279

Morita H, Iizuka T, Choo CY, Chan KL, Takeya K, Kobayashi J (2006) Vasorelaxant activity of cyclic peptide, cyclosquamosin B, from *Annona squamosa*. Bioorg Med Chem Lett 16(1):4609–4611

Morton JF (1987) Fruits of warm climates (Sugar Apple), Published by Julia F. Morton 20534 SW 92 Ct. Miami, FL. 33189, USA. Miami, FL, pp 69–72

Nakano D, Ishitsuka K, Kamikawa M, Matsuda M, Tsuchihashi R, Okawa M, Okabe H, Tamura K, Kinjo J (2013) Screening of promising chemotherapeutic candidates from plants against human adult T-cell leukemia/lymphoma. J Nat Med 67:894–903

Nandagaon VS, Kulkarni AR (2012) *In vitro* antioxidant and cytotoxicity activity of aqueous and alcoholic extracts of *Annona Squamosa*. Res J Pharm Biol Chem 3(4):479

Oliveira LC, Hankerson SJ, Dietz JM, Raboy BE (2010) Key tree species for the golden-headed lion tamarin and implications for shade-cocoa management in southern Bahia, Brazil. Anim Conserv 13:60–70

Orwa C, Mutua A, Kind R, Jamnadass R, Anthony S (2009) Agroforest tree Database: a tree reference and selection guide version 4.0. http://www.worldagroforestry.org/sites/treedbs/treedatabases.asp

Padhi LP, Panda SK, Satapathy SN, Dutta SK (2011) In vitro evaluation of antibacterial potential of *Annona squamosa* Linn. and *Annona reticulata* L. from Similipal Biosphere Reserve, Orissa, India. J Agri Technol 7(1):133–142

Panda S, Kar A (2007) Anti-diabetic and antioxidative effects of *Annona squamosa* leaves are possibly mediated through quercetin-3-O-glucoside. Biofactors 31(3–4):201–210. https://doi.org/10.1142/S0192415X17500501.

Panda S, Kar A (2015) Protective effects of 5,7,4′-trihydroxy-6,3′dimethoxy-flavone 5-O-I-rhamnopyranoside, isolated from *Annona squamosa* leaves in thyrothoxicosis and in hepatic lipid peroxidation in rats. Bioorg Med Chem Lett 25:5726–5728

Pandey N, Barve D (2011) Pharmacological and phytochemical review on *Annona squamosa*. Int J Res Pharm Bio Med Sci 2(4):1404–1414

Pandey VK, Giri IC, Prakashdeep SS, Srivastava A (2014) Pharmacognostical and physiochemical study on the leaves of *Annona squamosa* Linn. Int J Res Pharm Sci 4:8–12

Patel JD, Kumar V (2008) *Annona squamosa* L. phytochemical analysis and antimicrobial screening. J Pharm Res 1(1):34–38

Ponrasu T, Suguna L (2012) Efficacy of *Annona squamosa* on wound healing in streptozotocin-induced diabetic rats. Int Wound J 9:613–623

Ponrasu T, Suguna L (2014) Efficacy of *Annona squamosa L* in the synthesis of glycosaminoglycans and collagen during wound repari in Streptozotoxin induced diabetic rats. Bio Med Res Inter, 10 pages, Article ID 124352

Rahman MM, Parvin S, Ekramul Haque M, Islam ME, Mosaddik MA (2005) Antimicrobial and cytotoxic constituents from the seeds of *Annona squamosa*. Fitoterapia 76:484–489

Raj DS, Vennila JJ, Aiyavu C, Selvam KP (2009) Hepatoprotective effect of alcoholic extract of *Anona squamosa* leaves on experimentally induced liver injury in swiss albino mice. Inter J Integr Biol 5(3):182–186

Ranjan R, Sahai M (2009) Coumarinolignans from the seeds of *Annona squamosa* Linn. E-J Chem 6(2):518–522

Ren YY, Zhu ZY, Sun HQ, Chen LJ (2017) Structural characterization and inhibition on α-glucosidase activity of acidic polysaccharide from *Annona squamosa*. Carbohydr Polym 174:1–12

Salman HA, Senthilkumar R (2015) Antibacterial activity of *Annona squamosa*, and *Annona reticulata* Against clinical isolates of *Mutans streptococci* the causative agents of dental caries. Asian J Pharm Clin Res 8(4):152–155

Shenoy C, Patil PMB, Kumar R (2009) Antibacterial and wound healing activity of the leaves of *Annona squamosa*. Res J Pharmacog Phytochem 1(1):44–50

Shirwaikar A, Rajendran K, Kumar CD (2004) In vitro antioxidant studies of *Annona squamosa* leaves. Indian J Exp Biol 42(8):803–807

Shokeen P, Ray K, Bala M, Tandon V (2005) Preliminary studies on activity of *Ocimum sanctum*, *Drynaria quercifolia*, and *Annona squamosa* against *Neisseria gonorrhoeae*. Sex Transmit Dis 32:106–111

Singh, S., Manvi, F.V., Nanjwade, B., Nema, K.R. (2010). Antihyperlipidemic Screening of polyherbal formulation of Annona squamosa and Nigella sativa. Inter. J. Toxicol. Pharmacol. Res. 2(1):1–5.

Singh S (2011) Pharmacological screening of combined extract of *Annona squamosa* and Nigella sativa. Inter J Pharma Biol Sci 2(2):1183–1189

Singh A, Singh DK (2001) Molluscicidal activity of the custard apple (*Annona squamosa* L.) alone and in combination with other plant derived molluscicides. J Herbs Spi Med Plants 8(1):23–29. https://doi.org/10.1300/J044v08n01_04

Sun L, Zhu H, Gan L, Mo J, Feng F, Zhou C (2012) Constituents from the bark of *Annona squamosa*and their anti-tumor activity. Zhongguo Zhong Yao Z Zhi 37:2100–2104

Tomar RS, Sisodia SS (2012) Antidiabetic activity of *Annona squamosa* L. in experimental induced diabetic rats. Inter J Pharm Biol Arch 3:1492–1495

Topno KK (1997) Plants used by tribal of Chotanagpur against diabetes. Botanica 47:99–101

Vanitha V, Umadevi KJ, Vijayalakshmi K (2010) *In vitro* assessment of alcoholic leaf extracts of *Annona squamosa* and *Aegle marmelos*. Bioscan 5(2):225–229

Vila-Nova NS, Morais SM, Facao MJ, Machado LM, Becilaqua CM, Costa IR, Brasil NV, Andrade Junior HF (2011) Leishmanicidal activity and cytotoxicity of compounds from two *Annonacea* species cultivated in Northeastern Brazil. Rev Soc Bras Med Trop 44:567–571

Vohra SB, Ishwar K, Naqvi SAH (1975) Phytochemical, pharmacological, antibacterial and antiovulatory studies on *Annona squamosa*. Planta Med 28:97–100

Wu YC, Hung YC, Chang FR, Cosentino M, Wang HK, Lee KH (1996) Identification of *ent*-16β,17-dihydroxykauran-19-oic acid as an anti-HIV principle and isolation of the new diterpenoids annosquamosins A and B from *Annona squamosal*. J Nat Prod 59:635–637

Wunderlin R, Hansen B (2008) Synonyms of *Annona squamosa*. Atlas Fla Vasc Plants 4:24–27

Yadav DK, Singh N, Dev K, Sharma R, Sahai M, Palit G, Maurya R (2011) Anti-ulcer constituents of *Annona squamosa* twigs. Fitoterapia 82(4):666–675. https://doi.org/10.1016/j.fitote

Yadav DK, Singh N, Dev K, sharma R, Sahai M, Palit G, Maurya R (2012) Anti-ulcer constituents of *Annona squamosa* twigs. Fitoterapia 82:666–675

Yang YL, Chang FR, Wu CC, Wang WY, Wu YC (2002) New ent-kaurane diterpenoids with antiplatelet aggregation activity from Annona squamosa. J Nat Prod 65(10):1462–1467

Yang HJZN, Chen JW, Wang MY (2009) Two new cytotoxic acetogenins from *Annona squamosa*. J Asian Nat Prod Res 11:250–256

Yusha'u M, Taura DW, Bello B, Abdullahi N (2011) Screening of Annona squamosa extracts for antibacterial activity against some respiratory tract isolates. Int Res Pharm Pharmacol 1:237–241

Zahid M, Mujahid M, Singh PK, Farooqui S, Singh K, Parveen S, Arif M (2018) *Annona squamosa* Linn. (custard apple): an aromatic medicinal plant fruit with immense nutraceutical and therapeutic potentials. IJPSR 9(5):1745–1759

Chapter 13
Vangueria madagascariensis: Phytochemical Constituents, Bioactive Compounds, Traditional and Medicinal Uses

Sara Elkheir Mustafa, Rafiah Awad Mohamed, and Abdalbasit Adam Mariod

13.1 Plant Description

Vangueria madagascariensis (VM) J.F. Gmelin, family Rubiaceae is African small tree (Mahomoodally and Dilmohamed 2016). It contains over 50 species, are mainly tropical forested plants and comprise mostly of trees and shrubs (Mongrand et al. 2005; Ramalingum and Mahomoodally 2014). Ramalingum and Mahomoodally (2014) stated that VM is a profusely branched tree of 1.5–15 m tall (Fig. 13.1), with plane grey bark. Leaves, with stems 5–10 mm long, are opposed and dark green. The plant bears small green to yellow flowers in groups. Fruits are pretty much round, smooth, and glossy with measurement 2.5–5 cm, differing from green to caramel red (Fig. 13.2). They seem greenish when unripe, changing to tanish-red when ready, and contain 4–5 wooded seeds up to 1.6 cm long (Fig. 13.3) (El Ghazali et al. 2003; Elamin 1990).

Vangueria madagascariensis played an essential role in the diet of Sudanese people, especially during food scarcities and famines. During this time, the forestry fruits were the main ingredients given as an indication of hospitality (Abdel Muti 2002). Abdel-Rahman et al. (2014) indicated that VM fruits are palatable, and con-

S. E. Mustafa (✉)
College of Agriculture and Veterinary Science, Department of Food Science and Human Nutrition, Qassim University, Buraydah, Saudi Arabia

R. A. Mohamed
Old Age Home, Ministry of Social Development, Khartoum, Sudan

A. A. Mariod
Indigenous Knowledge and Heritage Center, Ghibaish College of Science & Technology, Ghibaish, Sudan
e-mail: aalnadif@uj.edu.sa

A. A. Mariod (ed.), *Wild Fruits: Composition, Nutritional Value and Products*,
https://doi.org/10.1007/978-3-030-31885-7_13

Fig. 13.1 *Vangueria madagascariensis* tree. (Source: https://commons.wikimedia.org)

Fig. 13.2 *Vangueria madagascariensis* fruits. (Source: https://commons.wikimedia.org)

Fig. 13.3 *Vangueria madagascariensis* seeds

sist of of many nutrients such as protein, carbohydrate, oil and minerals and (Mustafa 2011) stated that V.M seed contained high protein and oil. Recently researches started to be much concern about assessment of the nutritional and medical values of desert fruits in Sudan.

This multibranched bush or tree is mainstream as a wellspring of both wood and charcoal additionally the tree wood is reasonable for building structure, apparatus handles and cutting (Pino et al. 2004). Fentahun and Hager (2009) announced that, Vangueria madagascariensis wild fruits were wealthy in significant supplements and was accessible all year in Amhara locale of Ethiopia with vital cover on occasion of intense sustenance and supplement lack. *Vangueria madagascariensis* is a native medicinal food plant, is used as aantibacterial and ant diabetics in Africa and has many folk medicine use in various countries (Ramalingum and Mahomoodally 2014).

13.2 *Vangueria madagascariensis* Distribution

Vangueria madagascariensis found in Africa south of the Sahara with one species scattering in Madagascar (*Vangueria madagascariensis*). It is widely distributed in Sudan, Darfur and Southern Kordofan state. It develops normally in swamp timberlands and volcanic fiery debris soils all through Africa and Asia. The plant can likewise be found in the Republic of Mauritius (including Rodrigues Island), Seychelles, India, Northern Australia, Singapore and Trinidad (Ramalingum and

Mahomoodally 2014) VM is a multipurpose tree in Sudan locally known as Kirkir. El Amin (1990) reported that, *Vangueria madagascariensis* reported the appearance of the tree in Kassala, Blue Nile, Kordofan, Bahr El Ghazal and Equatoria (Abdelmuti 1991).

13.3 *Vangueria madagascariensis* Chemical Composition

The proximate wholesome content of the VM fruit per 100 g was recognized depend on the pulp to fruit ratio of 47.5% (Ramalingum and Mahomoodally 2014). Mustafa et al. (2017) reported chemical composition of VM fruits showed that the *Vangueria madagascariensis* fruit contains 6.7% protein. Kirkir fruit contains 0.66% fat content, and 11.9% as crude fiber content and 4.9% ash content.

13.3.1 Mineral Content

The mineral content of *V. madagascariensis* fruit. Potasium content was found to be the highest one 310.9 mg/l which was lower than that reported by Ramalingum and Mahomoodally (2014) (521 mg) followed by magnesium 66.1 which was higher than that reported by Ramalingum and Mahomoodally (2014) (39 mg), sodium 8.5 which was higher than that reported by Ramalingum and Mahomoodally (2014) (28 mg), iron 0.253 which was lower than that studied by Ramalingum and Mahomoodally (2014) (1.1 mg) and the lowest was zinc 0.025 mg\l and cupper 0.028 .

13.3.2 Oil Content of the V. madagascariensis *Seed*

The characteristics of oil recovered from seeds of *V.madagascariensis* have been investigated. The seed oil (Fig. 13.4) content was found very high (43.6%). The oil contained plant sterols and tocopherols (Mustafa 2011; Mustafa et al. 2017).

13.3.3 Physiochemical Properties of V. madagascariensis *Seed Oil*

The refractive index of *Vangueria madagascariensis* seed oils was 1.475, the acid value was 0.46, while the peroxide value was 1.0, with saponification value of 181.4 KOH/g (Mustafa 2011). Mustafa (2011) reported linoleic (63.4%), as the most prevailing fatty acid in *Vangueria madagascariensis* oil, followed by oleic acid (10.5%), then palmitic (9.8%) and stearic (5.4%) acids.

Fig. 13.4 *Vangueriamadagascariensis* seed oil

13.4 *Vangueria madagascariensis* Main Bioactive Components

The search for bioactive compounds with antioxidant activity hasgreater than beforein recent times as they play an essential therapeutic role in human disease. Introductory phytochemical study of the leaves and stems show the presence of alkaloids, terpenes, cyonogenetic heterosides as well as phenols, tannins, and saponosides which may probably be in charge of its reported antimicrobial impacts (Mustafa et al. 2017).

The alcoholic extract of leaves and stem of VM have been found to contained β-sitosterol, stigmasterol, palmitic acid, scopoletin, *p*-coumaric acid, protocatechuic acid, esculetin, ethyl-1-O-glucosyl-4-O-(E) caffeoylquinate, kaempferol-3-O-rhamnoside, 7-O-rutinoside, and vanillic acid (Ramalingum and Mahomoodally 2014).

Different *V. madagascariensis* parts are used as both food and herbal medicines as these parts have several phytochemical compounds and micronutrientsrequired for human nutrition and health. Earlier research indicated that the medicinal and nutritional properties of edible fruits collected from the wild enable local communities to use such plant resources as traditional medicines, at the same time increase their nutritional options, micronutrients, diet, and vitamins .

The value of medicinal drugs derived from plants, other natural health products, nutraceuticals, and functional foods are being promoted throughout the world as an another strategy for disease risk decrease and reduction in health care costs (Maroyi 2018).

Mustafa et al. (2017) analyzed methanolic leaf, seed and bark extracts of *Vangueria madagascariensies* for their phenolic content, flavonoids and antioxidant activity. Their results revealed higher total phenolic and flavonoid levels. Methanolic leaf extract showed high potential free radical scavenging activity with IC50 value of 7.8 μg/ml. Compare to methanolic seed extract that showed IC50 value of 62.5 μg/ml. The highest phenolic and flavonoid content were observed from bark and leaf extract which were 170.4 mg/g plant extract as GAE and 298.8 RE/g of extract respectively. Methanolic leaf extract of VM also examined for their antioxidant activity by using the oxygen radical absorbance capacity, their data revealed that 72.71 ± 0.89 (μM of Trolox) compared with the positive control; Quercetin (58.97 ± 0.02).

13.5 Food Uses

The fruit pulp is a good source of both macrominerals and trace elements such as potassium, zinc, calcium, magnesium, chromium, phosphorus, copper, manganese, and iron. Mustafa (2011) stated that VM seed contained high protein and oiland several amino acids and fatty acids have been identified from the seed of *V. madagascariensis*, and these include the essential amino acids such as lysine, threonine, histidine, leucine, phenylalanine, valine, isoleucine, and methionine. Research done by Mariod et al. (2017) revealed that the contribution of conditionally essential amino acids such as tyrosine, arginine, glycine, and cysteine and nonessential amino acids such as glutamic acid, aspartic acid, and serine was close to 50% (5.9 g out of 14.2 g/100 g) of the total amino acids identified from the species. The aggregate sum of the fundamental amino acids were analyzed in Kirkir seeds by Mustafa (2017), their results revealed 8.319 g/100 proteins. Mustafa (2011) reported an oil content of *Vangueria madagascariensis* seeds as 40.0% and protein content as 22.2%, with palmitic, oleic, linoleic as main fatty acids. The fruit of V M is edible when ripe, it is yellowish brown when mature with 4–5 seeds (El Ghazali et al. 2003; El Amin 1990).

Abdelmuti (1991) reported that, fruits were very sweet when ripe with 8 mm in diameter and weighing 5.5–6.5 g. The reddish fruits were eaten fresh or dried, the fruit contained 5.9% protein, 1.2% fat, 12.30% crude fiber, 4.9% ash and 75.7% carbohydrates content, while Abdel-Rahman (2011) reported that, kirkir fruit contains 7.71% protein, 2.35% fat, 18.9% crude fiber, 3.6% ash content, 67.4% carbohydrate. Abdel-Rahman et al. (2014) indicated that the VM fruits were edible and comprised of many nutrient such as protein, carbohydrate, oil and minerals. The fruit pulp is suitable for concentrated juice and Abdel-Rahman (2011) studied VM fruits nectars and their data showed that kirkir contain 2.2% polyphenols (tannins), 169.7 mg/100 g ascorbic acid (vitamin C), and 115.6 1IU β-carotene.

The dried fruits are effectively rehydrated by putting in water. Ready fruits are hand communicated and the juice given to lacating moms to discharge stomach collics and made into porridge in Ramadan in Sudan, likewise it is eaten as starvation foods (Abdelmuti 1991). The delicate tissue of the ready fruit has a satisfying somewhat acid flavor. It is normally eaten crisp, in jelly and refreshments and it has a wonderful chocolate-like flavor. The fruit is likewise used to add flavor to beer

(Orwa et al. 2009). Pino et al. (2004) studied the aroma of VM from Cuba using GC and GC/MS. They reported 74 compounds, of which 2-furfural and hexadecanoic acid were found to be the major constituents.

Abdel-Rahman et al. (2014) analyzed the edible portion of *V. madagascariensis* for it is proximate analysis sugars, ascorbic acid, ß-carotene pectin and energy value. Their results reported that the fruits were rich of vitamin C (156. mg/100 g), protein, fibre and ash. In the same time these authors stated that VM fruit contained high amounts of carbohydrates and an adequate amounts of ß-carotene (115.61 IU vitamin A/100 g). With energy value ranged from 310 to 372 Kcal/100 g .

Vangueria fruits in Malawi have been found to contain at 26.5% dry matter; 78.1% total carbohydrate, 10.2%, 2.6% fat, 5.7% crude protein. Ascorbic acid content is 16.8 mg per 100 g fresh weight (Saka 1995). In Botswana, an ascorbic acid content of 4.7 mg per 100 g has been reported for fruits with 64.4% moisture (Leakey 1999).

13.6 Medicinal Uses

Vangueria madagascariensis is considered as a major African medicinal plant. Roots and bark of *Vangueria madagascariensis* are utilized in folk therapy; e.g., in Tanzania the root extract is utilized to treat worm infections (Orwa et al. 2009).The seeds, bark, leaves, fruits, roots, and stem bark of *V.* madagascariensis are exploited in monotherapeutic or multitherapeutic applications in Eritrea, Kenya, Madagascar, Mauritius, Sudan, and Tanzania. Bark, fruit, leaf, and root maceration of *V.* madagascariensis is taken by mouth for diabetes in Madagascar, Mauritius, and Sudan. Bark, leaf, root bark, and stem bark infusion of *V. madagascariensis* is taken by mouth for bloody diarrhoea in Tanzania, dysentery in Mauritius, and stomach problems in Kenya. Root bark and root infusion of *V.* madagascariensis is taken by mouth for intestinal worms in Eritrea and Tanzania (Maroyi 2018).

Jain and Srivastava (2005) reported that, *Vangueria madagascariensis* leaf bath is given for skin diseases and abscesses. Kotowaroo et al. (2006) investigated the possible impacts of *Vangueria madagascariensis* on starch decomposition using alpha-amylase in vitro.

Vangueria has gotten logical consideration for its broad ethnomedicinal applications around the world. Generally developed for its sweet fruits, this plant has likewise acquired noteworthy commitment the African Materia Medica for its antimicrobial properties since long time. In vitro investigation demonstrated that VM has antimicrobial potential against *Staphylococcus aureus* (ATCC 25923) and *Escherichia coli* (ATCC 25922). In vitro information uncovered the nearness of various bio-constituents with pluri potential mechanism of activity which may be in charge of its restorative temperance (Ramalingum and Mahomoodally 2014). VM leaf and fruit, removes displayed incredible antibacterial and anti-infection potentiating activity against tried antibiotics in vitro (Mahomoodally and Dilmohamed 2016). Maroyi, (2018) revealed that *Vangueria madagascariensis* was utilized as herbal drug against anti diabetes, antigastro intestinal and for skin illnesses.

References

Abdel Muti OMS (1991) Biochemical and Nutritional Evaluation of Famine Foods of the Sudan. Ph. D. Thesis. University of Khartoum, Sudan

Abdel Muti OMS (2002) Nutritive value of wild plants of the Sudan. Arab J Food Nutr 3(3):6–67

Abdel-Rahman NA (2011) A study of some Sudanese edible forest fruits and their nectars. Ph.D. Thesis, Sudan Academy of Sciences. Khartoum, Sudan

Abdel-Rahman NA, Awad I, Elshafe'a ELBB (2014) Characterization of some Sudanese edible forest fruits. J Agri-Food Appl Sci 2(2):39–44

El Amin HM (1990) Trees and shrubs of the Sudan, Biddles LTd Guidforde & kings' Lynn, I edn. Ithaca Press, Exeter

El Ghazali GEB, Abdalla WE, Khalid HE, Khalafalla MM, Ahamad AA (2003) Medicinal plants of the Sudan part V medicinal plants of the Ingassana Area. Ministry of Science &Technology Sudan Currency printing Press, Khartoum

Fentahun MT, Hager H (2009) Exploiting locally available resources for food and nutritional security enhancement: wild fruits diversity, potential and state of exploitation in the Amhara region of Ethiopia. Food Sci 1:207–219

Jain SK, Srivastava S (2005) Traditional uses of some Indian plants among islanders of the Indian Ocean. Indian J Trad Knowl 4:345–357

Kotowaroo MI, Mahmoodally MF, Gurib-Fakim A, Subratty AH (2006) Screening of traditional antidiabetic medicinal plants of Mauritius for possible alpha-amylase inhibitory effects invitro. Phytother Res 20(3):228–231

Leakey RRB (1999) Potential for novel food products from agroforestry trees: a review. Food Chem 66:1–14

Mahomoodally MF, Dilmohamed S (2016) Antibacterial and antibiotic potentiating activity of *Vangueria madagascariensis* leaves and ripe fruit pericarp against human pathogenic clinical bacterial isolates. J Trad Compl Med 6:399–403

Mariod AA, Mirghani MES, Hussein I (2017) *Vangueria madagascariensis* (Rubiaceae) as new oil source. In: Mariod AA, Mirghani MES, Hussein I (eds) Unconventional oilseeds and oil sources. AcademicPress, London, pp 167–178

Maroyi A (2018) *Vangueria madagascariensis* fruit tree: nutritional, phytochemical, pharmacological, and primary health care applications as herbal medicine. Scientifica 2018:Article ID 4596450, 17 pages,. Hindawi. https://doi.org/10.1155/2018/4596450

Mongrand S, Badoc A, Patouille B, Lacomblez C, Chavent M, Bessoule J (2005) Chemotaxonomy of the Rubiaceae family based on leaf fatty acid composition. Phytochemistry 66:549–559

Mustafa SE (2011) Evaluation of oil components of *Annona squamosa* (Gishta) and *Vangueria madagascariensis* (Kirkir) seeds and antioxidant activity of their methanolic extracts , Ph.D Thesis, Sudan University of Science and Technology, Khartoum, Sudan

Mustafa SE (2017) The potential of *Vangueria madagascariensis* as a source for un-conventional oil in Sudan. Agri Biolo Sci J 3(4):28–32

Mustafa SE, Mariod AA, Ahmed YM, Abdelwahab SI (2017) Antioxidant activity and cytotoxicity study of *Vangueria madagascariensis* leaf, bark and seed cake methanolic extracts. Aust Basic Appl Sci 11(3):64–70

Orwa C, Mutua A, Kindt R, Jamnadass R, Simons A (2009) Agro forest tree database: a tree reference and selection guide version 4.0 (http://www.worldagroforestry.org/af/treedb/)

Pino J, Marbot R, Vazquez C (2004) Volatile components of the fruits of *Vangueria madagascariensis* J. F. Gmel. F grown in Cuba. J Ess Oil Res 16(4):318–320

Ramalingum N, Mahomoodally MF (2014) *Vangueria madagascariensis* (Rubiaceae)-an under-utilised African traditional medicinal food plant with potential application. J Intercult Ethnopharmacol 3(1):45–48

Saka JDK (1995) The nutritional value of edible indigenous fruits:present research status and future directions. In: Maghembe JA, Ntupanyama Y, Chirwa PW (eds) Improvement of indigenous fruit trees of the Miombo Woodlands of Southern Africa, Nairobi, pp 50–57

Chapter 14
Grewia tenax (Guddaim): Phytochemical Constituents, Bioactive Compounds, Traditional and Medicinal Uses

Abdel Moneim Elhadi Sulieman and Abdalbasit Adam Mariod

14.1 Introduction

Underused wild fruits species have exceptional money related, dietary benefit just as potential wellspring of appealing quality properties. They give food for human, grub for creature and drug and other natural gets by to human and their animals. These fruits have particularly incredible adjustment abilities to various biotic and a biotic anxieties like warmth resilience, saltiness/alkalinity resilience, dry season resistance and protection from different infections (Grivetti and Ogle 2000; Dev et al. 2017).

Grewia tenax is a deciduous tropical bush or tree, distributed in different climates. The wild bush is the principle source of the developing, growing commercial interest for the fruit. It is viewed as a prime contender for taming and commercialization as new yield for the semi – arid areas of numerous African and African and Asiatic countries (El-Siddig et al. 2003). It happens in a substantial zone, recovers well and is generally secured during clearing and supported by farmers.

The genus *Grewia* contains roughly 150 species of little trees, and bushes, distributed in the tropics and sub-tropics worldwide and is the only genus in the family that yields eatable fruits. *Grewia* species, are well-known for their nutritional and medicinal uses. Despite their diverse use, they have suffered notable disregard, as is evident from the lack of literature on this plant (Bayer et al. 2003; Ullah et al. 2012). Naturally, it can withstand ecological pressure all the more effectively contrasted

A. M. E. Sulieman (✉)
Department of Biology, Faculty of Science, University of Hail, KSA, Ha'il, Saudi Arabia
e-mail: am.sulieman@uoh.edu.sa

A. A. Mariod
Indigenous Knowledge and Heritage Center, Ghibaish College of Science & Technology, Ghibaish, Sudan
e-mail: aalnadif@uj.edu.sa

A. A. Mariod (ed.), *Wild Fruits: Composition, Nutritional Value and Products*,
https://doi.org/10.1007/978-3-030-31885-7_14

and yearly harvests and in this manner make a vital commitment to supportable generation without requiring costly contributions of water or compost. The bush can produce fruits three times each year, if there is adequate rain. If the main crop gather is postponed its fruits help to fill the food gap. Fruits are palatable. The ripen fruits are gathered and eaten crude. The fruits are sweet and might be eaten either as a whole or bit and just the sweet juice is swallowed (El-Siddig et al. 2003).

Fruits might be beat, dried and stored. Generally, children gather and expend the fruit in ordinary occasions, however, adults may consume too in food deficiency periods. The fruit is an essential monetary product, where locally, it has food and medicinal utilizations while globally utilized and exported as food and pharmaceutical enterprises. It has been the subject of much worldwide enthusiasm in research and development as it may be the solution of worldwide standing issues such as iron deficiency anemia (El-Siddig et al. 2003).

14.2 Description and Geographical Distribution

Grewia tenax is a three meters tall, erect tree, with grey stem bark. 3–5 shaggy on the two sides leaves. Blooms are white, infrequently cream to yellow in color, 5.0–35 mm long seeming like a jointed pedicel. Sepals 8–16 mm long, greenish outwardly, Petals white elongated to barely lanceolate, 5–12 mm long. Fruit is drupe generally 1–4 lobed, 8–12 mm wide, the lobes 6–8 mm long, 5–6 mm wide, orange, yellow or reddish tinge, shimmering, glabrous (Fig.14.1) (Gebauer et al. 2007a; Whitehouse et al. 2001).

Fig. 14.1 *Grewia tenax* plant. (Source: https://commons.wikimedia.org)

G. tenax commonly found in parched and semiarid plains, piedmont plain, marshes and mountains up to 1250 m over sea level furthermore, in districts with mean annual rainfall of 200–1000 mm. Plant inclines toward disintegrated rocky, sandy, gravelly, stony and lateritic soils. The normal territory of the plant is tropical forest in association with other wild species (Alrikain 2004; Sharma and Patni 2012).

G. tenax is astoundingly dry spell and occurs in the driest savannas at desert edges and zones of higher rainfall, where it creates in shrubberies on termite hills in generally regularly flooded areas. In the Sahel, it creates in rough places on slopes and inclines, in areas with 100–600 mm of rain annually (Von Maydell 1986). *G. tenax* is widespread in Africa from the Transvaal and South West Africa to Ethiopia and Arabia in the North – East and through West Africa to Senegal. It is just found in the driest kinds of wood land or semi-desert scrub. Geographically distributed in most African countries, India, and Pakistan.

14.3 Taxonomy of the Plant

Grewia tenax is scientifically classified as follows

Kingdom: Plantae, Division: Angiospermae, Sub-division: Dicotyledons, Class: Polypetalae

Series: Thalamiflorae, Order: Malvales, Family: Malvaceae (formerly: Tiliaceae), Subfamily: Grewioideae, Genus: Grewia.

14.4 Chemical Composition

Many researchers have investigated the chemical composition of *Grewia* spp. *Grewia tenax* is wealthy in supplements and they could be imperative supporters of improving the nourishing substance of provincial and urban individuals in many African and Asiatic countries (Elhassan and Yagi 2010). Data in Table 14.1 indicate the proximate chemical composition, while the data in Table 14.2 indicate tannins, phytic acid and ascorbic acid contents of fruits of four *Grewia* species (Sati and Ahmed 2018).

Grewia spp have varying amounts of amino acids which fulfill the requirements of the FAO/WHO standard protein. the fruit contains 817 mg/100 g potassium, 20.8–22.3 mg/100 g iron, so the fruits are utilized as an iron supplement for anemic children. *G. tenax* has higher value of reducing sugar (13.8%) and starchcontent (44.4%) when compared with other *Grewia* spp. (Elhassan and Yagi 2010).

G.tenax fruit is arich source of carbohydrates, protein,vitamins and minerals and the species is nutritionally balanced. Fruits are a good source of nutritional compo-

Table 14.1 Chemical composition of four *Grewia* spp.

Parameter Species	Ash (%)	Moisture (%)	Protein (%)	Fat (%)	Fiber (%)	Carbohydrate (%)
G.tenax	3.555B	12.92A	6.700C	0.40393B	25.50B	51.287
G.villosa	3.577B	7.805C	8.400B	0.4037B	13.79C	66.040
G.flavescens	3.400B	9.505B	10.54A	0.4041B	31.87A	44.330
G.mollis	4.303A	8.000C	10.50A	1.631A	33.33A	42.300
Mean	3.709	9.559	9.035	0.621	26.123	
sig	**	***	***	***	***	
SE+	0.13	0.23	0.37	0.02	0.54	
C.V%	6.13	4.10	7.06	5.03	3.60	
LSR(p = 0.05)	0.4294	0.7389	1.201	0.05854	1.771	

Source: Sati and Ahmed (2018)
, * *Significant at 0.01 and 0.001 probability levels, respectively*

Table 14.2 Tannins, phytic and ascorbic acids contents of fruit of four *Grewia* species

Parameter Species	Tannin (%)	Phytic acid (%)	Ascorbic acid (%)
G.tenax	0.3333A	0.08903B	0.03573C
G.villosa	0.2777C	0.08940B	0.04880B
G.flavescens	0.2055D	0.1777A	0.01950D
G.mollis	0.3277B	0.09007B	0.05870A
Mean	0.286	0.112	0.041
sig	***	***	***
SE+	0.00	0.00	0.00
C.V%	5.83	1.93	6.96
LSR(p = 0.05)	0.001883	0.001883	0.001883

Source: Sati and Ahmed (2018)
, * *Significant at 0.001 probability levels*

nents and essential nutrients, including minerals and amino acids, and have functional properties (Abdelmuti 1991; Hamed 1995; Rahamtalla 1999; Aboagarib et al. 2014) reported the following chemical compositions for *Grewia tenax*, the data are shown in Table 14.3.

14.5 Phytochemical Properties of *Grewia tenax*

Various parts of different species exhibit different medicinal importance and yet to be phytochemically investigated. Phytochemical study of the ether extract of the leaves showed that it contains diterpenes, glycosides and fats; chloroform extract contains alkaloids and glycosides, while ethanolic extract contains triterpenoids, sterols, flavonoids, saponins and tannins (Patil et al. 2011) (Fig. 14.2).

Table 14.3 Chemical composition of *G. tenax* as reported by some authors (1986–1999)

Parameter (%)	Boutros (1986)	Abdelmuti (1991)	Hamed (1995)	Aboagarib et al. (2014)
Moisture	5.3	4.5	5.3	11.72 ± 0.08
Ash	–	–	–	4.12 ± 0.03
Protein	8.0	6.3	7.50	7.68 ± 0.05
Fiber	14.3	8.1	9.5	9.41 ± 0.15
Carbohydrates%	85	80.7	71.5	66.59 ± 0.04
Sucrose	–	–	16	
Fe	0.0031	0.0074	0.0077	8.222 ± 1.05
K	–	–	1.45	856.25 ± 1.76
P	–	–	0.08	
Na				
Mg				22.135 ± 0.16
Cu				0.708 ± 0.02
Cr				0.063 ± 0.01
Zn				2.107 ± 0.05
Mn				1.034 ± 0.17

Source: Boutros (1986); Abdelmuti (1991); Hamed (1995); Aboagarib et al. (2014)

Fig. 14.2 *Grewia tenax* plant flower and ripening fruit. (Source: https://en.wikipedia.org/wiki/Grewia)

The prevailing phytochemical of *Grewia tenax* fruits are sugar, starch, tannins, phenolic mixes, flavonoids, steroids, ascorbic acid, phenolic mixes and proteins (Gupta et al. 2006; Sharma and Patni 2012).

Total phenol (mg GAE/g) and antioxidant activity (DPPH, ABTS and FRAP) of *Grewia tenax* were the highest in seeds which were 3.25, 4.439,4.108 and

4.595; respectively. Plant-derived compounds have been identified to prevent and treat of cancer, such as resveratrol, lycopene and astaxanthin, and phenolic acids (Li et al. 2012; Deng et al. 2012). Studies have proposed that antioxidants from fruits and their residues can diminish the danger of cancer and related mortality; and expending foods rich in polyphenols may prompt a lower occurrence of cancers. In addition, the antioxidants' potential and functional properties of nutrients from various natural sources have been investigated, essentially to replace the utilization of synthetic antioxidants in food products that can be a health hazard (Suhail et al. 2012).

G. tenax possesses the highest reducing potential as well as antioxidant capacity. *G. tenax* seed also possesses the maximum amount of phenolic content in aqueous extract, the highest antioxidant activity is due to the presence of high phenolic compounds in the species. The total phenolic content was 10.67 μg/ml, and and total flavonoids were 32.7 μg/ml in hydroalcoholic extract of *G. tenax*. The presence of antioxidants, phenolic compounds and flavonoids indicated that this species can be used for therapeutic usage at a very lower cost, and with minimum side effects in comparison to other commercial drugs available in the market (Basri et al. 2014).

14.6 Traditional Uses of *Grewia tenax*

Distinctive parts of *Grewia tenax* plant are customarily utilized for treating diverse human and livestock illnesses in some parts of India. fruit pulp is used to treat swelling in the body. For this reason seeds are separated from whole fruits and remaining pulp is connected remotely on the influenced parts by delicate application. Root bark and root powder are utilized for looseness of the bowels and diarrhea. Root and root bark are dried in the shade and squashed into fine powder (El Ghazali et al. 1997). This 5–10 gram powder boiled in water and taken orally for 3–4 days for dysentery remedy. Roots are likewise utilized in treating the issues associated with female regenerative systems. The green leaves and delicate twings are perused by the animals' given subsequent to slashing amid lean period when other fodder source are rare. Seeds and green leaves used at the season of the animal delivery (Aboagarib et al. 2014; Elhassan and Yagi 2010).

The fruits of *G. tenax* are amazing for making juice and squash, which are viewed as exceptionally nutritious drinks by indigenous people. A restoring summer drink arranged from fruits, called *phalsay ka sharb* at is accessible in sustenance stores and is accepted as a heart tonic. The fruit juice might be sustained with different supplements to additionally upgrade its dietary commitment to the diet. The juice has a low glycemic index and may be taken to oversee diabetes, since carbohydrates in low glycemic index foods digested gradually.

14.7 Medicinal Uses

Leaves and twigs of G. tenax are crucial constituents of rural drug for the treatment of numerous disarranges, for example, trachoma, tonsillitis contaminations and are utilized as a poultice to treat swelling (EL Ghazali et al. 1994; El Ghazali et al. 1997). Because of its high iron substance, fruits of *G. tenax* are regularly used in exceptional foods for pregnant ladies and kids experiencing sickness. *G. tenax* plant is utilized for the treatment and avoidance of iron lack frailty. Porridge, called Nesha, is set up by cooking fruit pulp of *G. tenax* and millet flour given to lactating moms (Gebauer et al. 2007b). Ointment of whole plant extract applied locally for hard tissue repair and bark glue of *G. tenax* can be connected as plaster (El Ghazali 1997). *G. tenax* fruit powder blended with milk is given for the treatment of bone break and swelling (Shekhawat and Batra 2006). All through screening work with various plant extracts for their carcinolytic exercises, *G. tenax* concentrate was found to have antitumour exercises when mixed in rodents bearing Rhabdomyosarcoma. The effect of fluid concentrate of *G. tenax* fruit was tried on the assortment of in vitro iron absorption. The incubation of freshly prepared rat everted gut sac in Ringer medium containing FeSO4 within the sight of concentrate at various concentrations supports essentially the iron transfer from the mucous side toward the serous one. Limit of iron absorption was reported within the sight of watery extract at 10 mg/ml and 5 min of incubation time in stomach, duodenum and jejunum.

14.8 Miscellaneous Uses

Food The fruits utilized by individuals and animals contain a considerable measure of iron and can be made into an a refreshing beverage. Fruit storage can be extended by drying. The dead leaves are eaten, however just while they stay on the plant. Its fruits are as refreshment in summer season. A beverage is set up by extracting the fruit 8 hours, hand-squeezing, sieving, and adding sugar.

Fodder Young leaves are utilized by animals, they are marginally agreeable toward the finish of dry seasons, and have genuinely great feed esteem.

Fuel The branches are utilized as firewood, and can be used in charcoal making.

Fiber Ligno-Cellulosic Fibre with good tensile strength is made by the bark, which is used to make ropes and for binding purposes in house construction.

Timber *G. tenax* wood is utilized in making weapons, for example, clubs, bows, bolts and for other general purposes. Toxic substance: An adhesive bark planning is utilized by ladies against hair vermin (Sharma and Patni 2012).

References

Abdelmuti OM (1991) Biochemical and nutritional status of famine foods of the Sudan. Ph.D Thesis in Fac. Agric. University of Khartoum, Khartoum, Sudan

Aboagarib EAA, RuijinYang XH, Siddeeg A (2014) Chemical compositions, nutritional properties and volatile compounds of Guddaim (*Grewia Tenax*. Forssk) Fiori fruits. J Food Nutr Res 2(4):187–192

Alrikain AEMS (2004) The yield and characteristics of gudeim (*Grewia tenax*) juice. Dissertation, University of Khartoum, Khartoum, Sudan

Basri DF, Heng KY, Meng CK, Ghazali AR (2014) Screening of antioxidant phytoextracts of Canarium odontophyllum (Miq.) leaves in vitro. IOSR Journal Of Pharmacy. 4(12), 01–06

Bayer C, Kubitzki K (2003) Malvaceae, the families and genera of vascular plants. Flowering plants, Dicoyledons. Springer, Berlin

Boutros JZ (1986) Sudan food composition tables, 2nd edn. National Chemical Laboratories, Ministry of health, Khartoum

Deng GF, Xu XR, Li S, Li F, Xia EQ, Li HB (2012) Natural sources and bioactivities of resveratrol. Int J Modern Biol Med 1:1–20

Dev R, Sharma GK, Singh T, Dayal D, Sureshkumar M (2017) Distribution of Grewia species in Kachchh Gujarat, India: taxonomy, traditional knowledge and economic potentialities. Int J Pure App Biosci 5(3):567–574

El Ghazali GEB, El Tohami MS, El Egami AAB. Medicinal plants of the Sudan (1994) In: Medicinal plants of the White Nile provinces. Khartoum University Press, Khartoum

El Ghazali GEB, El Tohami MS, El Egami AAB, Abdalla WS, Mohammed MG (1997) Medicinal plants of Sudan, part III Medicinal plants of the eastern Nuba Mountains. Khartoum University Press, Khartoum

El-Siddig K, Ebert G, Inanaga S (2003) Drought tolerance of *Grewia tenax* - a potential new small fruit for the Sudan. Deutscher Tropentag, October 8–10, Gottingen, Germany

Elhassan GM, Yagi SM (2010) Nutritional composition of Grewia species (Grewia tenax (Forsk.) Fiori, G. flavescens Juss and G. villosa Willd) fruits. Adv J Food Sci Technol 2(3):159–162

Gebauer J, Patzelt EA, Hammer EK, Buerkert EA (2007a) First record of *Grewia tenax* (Forssk.) Fiori in northern Oman, a valuable fruit producing shrub. J Genet Resour Crop Evol 54:1153–1158

Gebauer J, El-Siddig K, El Tahir BA, Salih AA, Ebert G, Hammer K (2007b) Exploiting the potential of indigenous fruit trees: *Grewia tenax* (Forssk.) Fiori in Sudan. J Genet Resour Crop Evol 54:1701

Grivetti LE, Ogle BM (2000) Value of traditional foods in meeting macro-and micronutrients needs: the wild plant connection. Nutr Res Rev 13:31–46

Gupta MK, Sharma PK, Ansari SH, Lagarkha R (2006) Pharmacognostical evaluation of *Grewia asiatica* fruits. Int J Plant Sci 1:249

Hamed KA (1995) The industrial utilization of Guddiem (*Grewia tenax*). M.Sc. Thesis, University of Gezira, Gezira: Sudan

https://commons.wikimedia.org/wiki/Grewia_tenax

https://en.wikipedia.org/wiki/Grewia

Li F, Xu XR, Li S, Deng GF, Wu S, Li HB (2012) Resources and bioactivities of lycopene. Int J Food Nutr Safety 1:15–31

Patil P, Patel MM, Bhavsar CJ (2011) Preliminary phytochemical and hypoglycemic activity of leaves of *Grewia asiatica* L. Res J Pharm Biol Chem Sci 2:516–520

Rahamtalla (1999) Consumption patterns and Anti -anemic properties of Guddiem (Grewia tenax) M. Sc. Thesis University of Khartoum, Khartoum: Sudan.

Sati NME, Ahmed FAM (2018) Botanical over view and chemical composition of someGrewia spp."Gudeim plant" in Sudan. Open Sci J 3:1–12

Sharma N, Patni V (2012) *Grewia tenax* (frosk.) Fiori.- a traditional medicinal plant with enormous economic prospective. Asian J Pharm Clin Res 5(3):28–32

Shekhawat D, Batra A (2006) House hold remedies of Keshavraipatan tehsil in Bundi district, Rajasthan. Indian J Tradit Knowl 5(Suppl 3):362–367

Suhail N, Bilal N, Khan HY, Hasan S, Sharma S, Khan F (2012) Effect of vitamins C and E anti-oxidant status of breast-cancer patients undergoing chemotherapy. J Clin Pharm Ther 37:22–26

Ullah W, Uddin G, Siddiqui BS (2012) Ethnic uses, pharmacological and phytochemical profile of genus *Grewia*. J Asian Nat Prod Res 14:186–195

Von Maydell HJ (1986) Trees and shrubs of Sahel – their characteristics and uses. GTZ 6MBH, Eschborn

Whitehouse S, Varni JW, Seid M, Cooper-Marcus C, Ensberg MJ, Jacobs JJ Mehlenbeck RS (2001) Evaluating a children's hospital garden environment: Utilization and consumer satisfaction. J Environ Psychol, 21:301–314

Chapter 15
Ziziphus spina-christi: Analysis of Bioactivities and Chemical Composition

Ahmed S. Hussein

Abbreviations

IZ	Inhibition zone
MBC	Minimum Bactericidal Concentration
MIC	Minimum Inhibitory Concentration
ZSC	*Ziziphus spina-christi*

15.1 Introduction

15.1.1 Plant Description and Distribution

The genus *Zizyphus* belongs to family (Rhamnaceae) and comprises nearly 100 species of deciduous or evergreen trees and shrubs distributed worldwide. Is native and well known to large area of Africa stretching from Mauritania to the Red Sea (Motamedi et al. 2014; Izuagie et al. 2012). The plant is known to be very heat resistant as well as drought tolerant (Paroda and Mal 1989). *Zizyphus spina-christi* (L.) Willd is an armed shrub or evergreen tree called Christ's thorn (Al-Wakeel 2008; Anthony and Dweck 2005) with edible sweet fruits (Fig. 15.1). It has common Arabic names such as Sidr and Nabeq or Nabaj (fruit) and many local names such as Nibs, Jabat, Zejzaj, Zefzoof, Ardeg, Ghosl, and Kanar. *Ziziphus spina-christi* sometimes can reach a height of 20 m and a 60 cm diameter; bark light-grey, branches whitish brown, often zig-zagging, thinly hairy when young. It is a most important cultivated native tree species of Arabia with historical, medicinal and religious benefits (Sameera and Mandakini 2015; Yossef et al. 2011; Adzu et al. 2002).

Recently medicinal plants have generally gained more interest as potential sources of pharmaceutical agents and/or as sources of precursors in drug

A. S. Hussein (✉)
Department of Chemistry, Faculty of Sciences and Arts-Alkamil, University of Jeddah, Jeddah, Saudi Arabia

A. A. Mariod (ed.), *Wild Fruits: Composition, Nutritional Value and Products*,
https://doi.org/10.1007/978-3-030-31885-7_15

Fig. 15.1 *Ziziphus spina-christi* (**a**) tree (**b**) fruits (**c**) seeds separated from the fruits (**d**) kernel separated from their seeds

development. Based on the World Health Organization (WHO) reports, greater than 80% of the populations within developing countries relay on herbal and other traditional medicines for their primary healthcare (Alhakmani et al. 2014). *Ziziphus* plants have been widely used in the folklore medicine for curing many diseases including malaria, digestive disorders, skin infections, liver complaints, diarrhea, urinary problems, diabetes, fever, pneumonia, dysentery, scorpion stings, cough, constipation, weakness, hepatic diseases, obesity, skin infections, and insomnia intestinal worms bronchitis, heal wounds, sores, against ringworm, sex diseases and ulcers, tuberculosis, anaemia, heal swellings, antispasmodic, purgative and against gonorrhea (Ads et al. 2017; Karar et al. 2016; Motamedi et al. 2014; Alhakmani et al. 2014; Mishra and Bhatia 2014; Goyal et al. 2012; Youssif et al. 2011). In traditional Chinese medicine, *Ziziphus* is believed to nourish the heart, supplement to the liver blood, and for treatment of irritability, insomnia and heart palpitations (Ads et al. 2017).

15.2 Chemical Composition

Generally, separation, identification and quantification of the major and minor chemical constituents of the plant are of great interest. Extensive studies have been carried out on *Ziziphus spina-christi* to determine its major and minor chemical constituents (Ikram and Tomlinson 1976; Aynehchi and Mahoodian 1973; Younes et al. 1996; Mahran et al. 1996). The main phytoconstituents of the plant were alkaloids, flavonoids, saponins, tannins, terpenoids, peptide, cyclopeptide, sterols, betulinic acid, triterpenoidal, glycosides, triterpenic acids, quercetin, kaempferol, and phloretin derivatives (Suleiman 2015; Asgarpanah and Haghighat 2012; Abu-Taleb et al. 2011; Guo et al. 2011; Pawlowska et al. 2008; Shahat et al. 2001; Tripathi et al. 2001; Mahran et al. 1996) and proanthocyanidins, and chlorogenic acids (Karar et al. 2016)). In addition to zizyphine-F, jubanine-A, amphibine-H, spinanine-A (a 14-membered cyclopeptide alkaloid of the amphibine-B type) have also been previously isolated from the SZC stem bark (Dweck 2005; Abdel-Galil and El-Jissry 1991; Ismail 1998). About 57 different phenolic compounds were also detected in ZSC. Among these phenolic compounds 52 of them are found in high (regioisomeric) levels in the methanol/water extract of *Z. spina-christi* leaves (Karar et al. 2016). The leaves also contain some sugars such as lactose, glucose, galactose, arabinose, xylose and rhamnose. The main flavonoids include quercetin 3-Orhamnoglucoside 7-O-rhamnoside and rutin and the highest amount reported (0.66%) was in the leaves (Brantner and Males 1999). The main chemical constituents found in the oil include alpha-terpineol (16.4%), linalool (11.5%) and neutral hydrocarbons including n-pentacosane forms (81%). Generally, fatty acids and their esters are the main components of ZSC fruits, and among all, dodecanoic acid is the main component (Said et al. 2010) Some methyl esters including methyl palmitate, methyl stearate, methyl myristate. Beta-sitosterol, oleanolic acid and maslinic acid were also isolated from the plant leaves. Many aromatic hydrocarbons (2-phenyl undecane, 3-phenyl undecane, 4-phenyl undecane, 5-phenyl undecane, 6-phenyl tridecane, 2-phenyl dodecane, 3-phenyl dodecane, 4-phenyl dodecane, and 6-phenyl dodecane) and volatile compounds (m-cymene, crypton, α-pinene, farnesan, (+)-sabinene and α-bergamotene) have also been exploited in ZSC fruit extract (EL-Hefny et al. 2018)

The proximate chemical composition (%) of *Ziziphus spina-christi* fruits, seeds stem, leaves and oil is shown in Tables 15.1, 15.2, 15.3, and 15.4. The values of moisture content in fruits, seeds, and leaves were (5.4–16.8%), (4–14.89%), and (7.72%), respectively (Ahamed 2016; Duke 1985; Adekunle and Adenike 2012; Ibrahim et al. 2015; Ahmed et al. 2015). It is obvious that major constituents of ZSC are polyphenols, carbohydrates, minerals, oil, fatty acid, and vitamins. The great number of these constituents were detected in the fruits, followed by seed, stem and finally kernel oil. The greatest number of the different constituents found in ZSC

Table 15.1 The proximate chemical composition of *Ziziphus spina-christi* fruits, seeds, stem and leaves

Component	Fruits	Seeds	Stem	Leaves	Reference
Moisture (%)	–	4 ± 0.76	–	–	Ahamed (2016)
	–	4.20 ± 0.15	–	–	Ahmed et al. (2015)
	5.4	–	–	–	Ahmed and Sati (2018)
	7.4 ± 0.2	–	–	–	Yossef et al. (2011)
	9.3	–	–	–	Duke (1985)
	13.30 ± 0.16	–	–	–	Singh et al. (2012)
	16.83 ± 0.03	14.89 ± 0.13	–	–	Adekunle and Adenike (2012)
	–	–	–	7.72 ± 0.01	Ibrahim et al. (2015)
Protein (%)	–	4.8	–	–	Duke (1985)
	5.40 ± 0.13	–	–	–	Singh et al. (2012)
	–	18.6	–	–	Saied et al. (2008)
	–	31.9 ± 0.61	–	–	Ahamed (2016)
	–	38.79 ± 0.33	–	–	Ahmed et al. (2015)
Protein (crude) (%)	4.1	–	–	–	Ahmed and Sati (2018)
	4.7 ± 0.8	–	–	–	Yossef et al. (2011)
	4.8	–	–	–	Duke (1985); Berry-Koch et al. (1990)
	4.8–5.6 (dry)	–	–	–	Dweck (2005)
	8.23 ± 0.03	24.07 ± 0.04	–	–	Adekunle and Adenike (2012)
Lipid (%)	–	28.5%	–	–	Saied et al. (2008)
Fat (%)	0.66 ± 0.05		–		Singh et al. (2012)
	0.9	–	–	–	Ahmed and Sati (2018); Berry-Koch et al. (1990); Duke (1985)
	0.9 (2.1) (dry)	–	12.11 ± 1.4	–	Dweck (2005)
	0.94 ± 0.3	–	–	–	Yossef et al. (2011)
Crude fat (%)	1.94 ± 0.05	1.24 ± 0.02	–	–	Adekunle and Adenike (2012)
Fiber (%)	2.5	–	–	–	Ahmed and Sati (2018)
	–	21.3 ± 1.8	–	–	Ahamed (2016)
Fibre (crude) (%)	–	3.84 ± 0.42	–	–	Ahmed et al. (2015)
	4.1 (dry)	–	–	–	Dweck (2005)
	6.09 ± 0.02	18.73 ± 0.02	–	–	Adekunle and Adenike (2012)
Ash (%)	–	2.88 ± 0.83	–	–	Ahmed et al. (2015)
	–	3.5 ± 0.01	–	–	Ahamed (2016)
	3.79 ± 0.4	–	–	–	Yossef et al. (2011)
	4.11 ± 0.11	–	–	–	Singh et al. (2012)
	4.3	–	–	–	Ahmed and Sati (2018)
	4.4	–	–	–	Duke (1985)
	7.92 ± 0.01	5.82 ± 0.03	–	–	Adekunle and Adenike (2012)
	–	–	–	5.78 ± 0.05	Ibrahim et al. (2015)

(continued)

Table 15.1 (continued)

Component	Fruits	Seeds	Stem	Leaves	Reference
Ash insoluble (%)	4.4 (dry)	–	–	–	Dweck (2005)
Acid insoluble ash (%)	–	–	–	5.05 ± 0.03	Ibrahim et al. (2015)
Water Soluble ash (%)	–	–	–	1.99 ± 0.08	Ibrahim et al. (2015)
Water extractive (%)	–	–	–	4.51 ± 0.02	Ibrahim et al. (2015)
Alcohol extractive	–	–	–	3.43 ± 0.00	Ibrahim et al. (2015)

were reported by Adekunle and Adenike (2012). Table 15.1 illustrates that the high percentages were exhibited with carbohydrates (up to 82.7% in fruits), protein (up to 38.8% in seeds), oil content (27.8–28.89% in seeds). Both fruits and seeds are also rich in calories, minerals, fatty acids, and vitamins. The protein is more concentrated in seed (4.8–38.79 ± 0.33%) followed by fruits (4.8–8.23%) (Ahamed 2016; Ahmed et al. 2015; Duke 1985; Adekunle and Adenike 2012; Ikram and Tomlinson 1976). Fat was found to be mainly concentrated in stem (12.11%) and minor quantities in seed (0.9–1.9%) (Ahamed 2016; Duke 1985; Adekunle and Adenike 2012; Ikram and Tomlinson 1976). The fibre quantity represents about 18.7–21.3% of ZSC seed while ash (4.4–8%), (3.5–8.5%), and (2–5.8%) in fruit, seed, and leaves, respectively (Ahamed 2016; Ikram and Tomlinson 1976; Adekunle and Adenike 2012; Ahmed et al. 2015).

From the data depicted in Table 15.2, it is clear that ZSC contains many polyphenols, rich in calories (314 calories/100 g fruits) (Duke 1985), and carbohydrates (58.02 ± 1.18% in fruits, and 35.25 ± 0.06% in seeds) (Adekunle and Adenike 2012), besides minor quantities of many minerals (Ca, Mg, Na, K, Mn, S, Zn, Fe, Cu, Mn, Pb, and P). Both minerals and carbohydrates are concentrated in fruit and seeds. However, the concentration of the minerals are low but represent relatively valuable nutrients. The presence of these calories, minerals and carbohydrates highly indicate the potential of SZC fruits and seeds for nutritional uses and applications. It can also be clearly seen that ZSC seeds have a relative quantity of oil ranging from 27.8 ± 1.69% (Ahamed 2016) to 28.89 ± 0.58% (Ahmed et al. 2015).

Most interestingly, SZC fruits represent a valuable natural source for many amino acids and vitamins as illustrated in Table 15.3. It is obvious that amino acids of high percentages in fruits are glutamic acid (17.6%) and aspartic acid (15.1%) (Ikram and Tomlinson 1976), while in seeds are glycine (17.1%), histidine (12.3%), proline (11.3%) and serine (10.8%). The percentages of the vitamins are generally

Table 15.2 Proximate analysis of polyphenols, carbohydrates and minerals in *Ziziphus spina-christi* fruits, seeds and stem

Component	Fruits	Seeds	Stem	Reference
Rutin (%)	0.0159	–	0.325	Ghafoor et al. (2012)
Apigenin (%)	0.0095	–	0.123	Ghafoor et al. (2012)
Quercetin (%)	0.00	–	0.0181	Ghafoor et al. (2012)
p-Coumaric acid (%)	0.00	–	0.253	Ghafoor et al. (2012)
Chlorogenic acid (%)	0.0124	–	0.0217	Ghafoor et al. (2012)
Syringic acid (%)	0.0061	–	0.0276	Ghafoor et al. (2012)
T. phenolic (mg/g gallic)	7.55 ± 0.33	–	–	Yossef et al. (2011)
Calories	314	–	–	Duke (1985); Berry-Koch et al. (1990)
Calcium, Ca (%)	0.14	–	–	Duke (1985); Berry-Koch et al. (1990)
	0.452	0.347	–	Adekunle and Adenike (2012)
	0.61 (dry)	–	–	Dweck (2005)
	0.85	–	–	Ahmed and Sati (2018)
Magnesium, Mg (%)	0.12 (dry	–	–	Dweck (2005)
	0.036	0.192	–	Adekunle and Adenike (2012)
	0.50	–	–	Ahmed and Sati (2018)
Sodium, Na (%)	0.01 (dry)	–	–	Dweck (2005)
	0.055	0.051		Adekunle and Adenike (2012)
Potassium, K (%)	0.13 (dry)	–	–	Dweck (2005)
	0.0679	0.0447	–	Adekunle and Adenike (2012)
Sulphur, S (%)	0.04 (dry)	–	–	Dweck (2005)
Zinc, Zn (%)	0.0009 (dry)	–	–	Dweck (2005)
	0.0569	0.0223	–	Adekunle and Adenike (2012)
Iron, Fe (%)	0.002 (dry)	–	–	Dweck (2005)
	0.003	–	–	Duke (1985); Dweck (2005); Abdelmuti (1991)
	0.0377	–	–	Ahmed and Sati (2018)
	0.040	0.0585	–	Adekunle and Adenike (2012)
Copper, Cu (%)	0.00123	0.00257	–	Adekunle and Adenike (2012)
	0.005 (dry)	–	–	Dweck (2005)
Manganese, Mn (%)	0.00019	0.00014	–	Adekunle and Adenike (2012)
	0.0013 (dry)	–	–	Dweck (2005)
Lead, Pb (%)	0.00067	0.00065	–	Adekunle and Adenike (2012)
Phosphorus, P (%)	0.0347	0.0427	–	Adekunle and Adenike (2012)
	0.28	–	–	Ahmed and Sati (2018)
Sulphur, S (%)	0.00004	–	–	Dweck (2005)

(continued)

Table 15.2 (continued)

Component	Fruits	Seeds	Stem	Reference
Carbohydrates (%)	–	11.5 ± 2.6	–	Ahamed (2016)
	–	16.12 ± 0.35	–	Ahmed et al. (2015)
	58.02 ± 1.18	35.25 ± 0.06	–	Adekunle and Adenike (2012)
	76.50 ± 0.27	–	–	Singh et al. (2012)
	80.6	–	–	Saied et al. (2008); Abdelmuti (1991)
	82.7	–	–	Ahmed and Sati (2018)
	83.18 ± 0.29	–	–	Yossef et al. (2011)
Starch (%)	21.8 (dry)	–	–	Saied et al. (2008); Dweck (2005); Abdelmuti (1991)
Sugar (%)	14.16	–	–	Orwa et al. (2009)
D-fructose (%)	16.0 (dry)	–	–	Saied et al. (2008); Dweck (2005); Abdelmuti (1991)
Fructose	0.16 ± 0.02	0.15 ± 0.01	–	Adekunle and Adenike (2012)
D-glucose (%)	9.6 (dry)	–	–	Saied et al. (2008); Dweck (2005); Abdelmuti (1991)
Sucrose (%)	21.7	–	–	Ahmed and Sati (2018)
	21.8 (dry)	–	–	Dweck (2005); Abdelmuti (1991)
Dextrose (%)	0.15 ± 0.01	0.15 ± 0.01	–	Adekunle and Adenike (2012)
Hydrated maltose	0.26 ± 0.02	0.25 ± 0.01	–	Adekunle and Adenike (2012)
Anhydrous lactose	0.19 ± 0.01	0.20 ± 0.01	–	Adekunle and Adenike (2012)
Hydrated lactose	0.23 ± 0.01	0.20 ± 0.01	–	Adekunle and Adenike (2012)
Total reducing sugar	0.15 ± 0.01	0.16 ± 0.01	–	Adekunle and Adenike (2012)
	22.6	–	–	Ahmed and Sati (2018)
Oil content (%)	–	27.8 ± 1.69	–	Ahamed (2016)
	–	28.89 ± 0.58	–	Ahmed et al. (2015)

low (<0.03%) except that reported by Orwa et al. (2009), who reported a value of 1.6% for Vitamin C.

Table 15.4 displays the physicochemical characteristics of ZSC kernel oil as well as the fatty acids in both seeds and kernel oil. These physicochemical characteristics include refractive index, density, viscosity, saponification value, unsaponifiable, acid value, proxide value, iodine index, total caroteniods, total chlorophyll, tocopherols, total phenolic compound, and some fatty acids contents. From the table it is noticeable that kernel oil has relatively low density (0.89–0.9 g/cm^3) and high viscosity (37.3%). The investigated fatty acids include hexanoic acid, capric acid, caprylic acid, palmitoleic acid, palmitic acid, oleic acid, linoleic acid, linolenic acid, arachidic acid, gadoleic acid, behenic acid, dodecanoic acid, lauric acid and myristic acid (Ahamed 2016; IUPAC 1992; Azangueu Mafo et al. 2013; Nazif 2002; Said et al. 2010). Among all the high percentages were reported with gadoleic acid (12.91%), palmitic acid (18.03%), and linoleic acid (35.44%) (Azangueu Mafo et al. 2013). Ahamed (2016) has also detected many fatty acids

Table 15.3 Amino acids and vitamins in *Ziziphus spina-christi* fruits and seeds

Component	Fruits	Seeds	Reference
Ascorbic acid (%)	0.00864		Ahmed and Sati (2018)
	0.014		Singh et al. (2012)
	0.03		Duke (1985); Berry-Koch et al. (1990)
Glutamic acid (%)	17.6		Dweck (2005)
Aspartic acid (%)	15.1		Dweck (2005)
	–	0.122	Nazif (2002)
Alanine (%)	3.4		Dweck (2005)
		7.45	Nazif (2002)
Arginine (%)	3.4		Dweck (2005)
		4.85	Nazif (2002)
Glycine (%)	3.1		Dweck (2005)
		17.1	Nazif (2002)
Histidine (%)	0.9		Dweck (2005)
	–	12.3	Nazif (2002)
Cysteine (%)	0.5		Dweck (2005)
Isoleucine (%)	2.3		Dweck (2005)
	–	1.15	Nazif (2002)
Leucine (%)	3.9		Dweck (2005)
	–	4.75	Nazif (2002)
Lysine (%)	2.3		Dweck (2005)
	–	1.92	Nazif (2002)
Niacin (%)	0.0037		Duke (1985); Berry-Koch et al. (1990)
Methionine (%)	0.4		Dweck (2005)
		0.747	Nazif (2002)
Phenylalaine (%)	2.2		Dweck (2005)
	–	1.75	Nazif (2002)
Proline (%)	5.3		Dweck (2005)
		11.3	Nazif (2002)
Riboflavin (%)	0.00013		Duke (1985); Berry-Koch et al. (1990)
Serine (%)	3.9		Dweck (2005)
	–	10.8	Nazif (2002)
Threonine (%)	2.2		Dweck (2005)
Tyrosine (%)	1.8		Dweck (2005)
	–	1.15	Nazif (2002)
Valine (%)	3.1		Dweck (2005)
	–	4.42	Nazif (2002)
Thiamin (%)	0.00004		Duke (1985); Berry-Koch et al. (1990)
Vitamin B1 (%)	0.00004		Dweck (2005)
Vitamin B2 (%)	0.00013		Dweck (2005)
Vitamin C (%)	0.03		Dweck (2005)
	1.6		Orwa et al. (2009)

Table 15.4 The physicochemical characteristics of *Ziziphus spina-christi* kernel oil and fatty acids of kernel oil and fruit

Component	Kernel oil	Reference
Physicochemical characteristics		
Refractive index	1.467 ± 0.00	Ahamed (2016)
Density (g/cm^3)	0.891 ± 0.002	IUPAC (1992)
	0.928 ± 0.00	Ahamed (2016)
Viscosity (poise)	37.31 ± 0.02	Ahamed (2016)
Saponification value (mg KOH/g)	126.55 ± 4.58	IUPAC (1992)
	159.6 ± 0.99	Ahamed (2016)
Unsaponifiable (%)	11.96 ± 1.76	IUPAC (1992)
Acid value (mg KOH/g)	4.7 ± 1.15	Ahamed (2016)
Acid value (mg/g oil)	6.50 ± 0.41	
Peroxide value (mg/kg)	0.8 ± 0.28	Ahamed (2016)
Iodine index (gI2/100 g)	7.93 ± 0.43	IUPAC (1992)

Fatty acids profile	Fruit	Kernel oil	Reference
Hexanoic acid (C6:0) (%)	–	nd	Azangueu Mafo et al. (2013)
Capric acid (C10:0) (%)	–	3.75	Azangueu Mafo et al. (2013)
Caprylic acid (%)	0.106	–	Nazif (2002)
Palmitoleic acid (C16:1) (%)	–	6.07	Azangueu Mafo et al. (2013)
Palmitic acid (C16:0) (%)	–	18.05	Azangueu Mafo et al. (2013)
	10.0	–	Nazif (2002)
Oleic acid (C18:1) (%)	–	8.87	Azangueu Mafo et al. (2013)
Linoleic acid (C18:2) (%)	–	35.44	Azangueu Mafo et al. (2013)
	45.5	–	Nazif (2002)
Linolenic acid (%)	20.0	–	Nazif (2002)
Arachidic acid (C20:0) (%)	–	nd	Azangueu Mafo et al. (2013)
	4.59	–	Nazif (2002)
Gadoleic acid (C20:1) (%)	–	12.91	Azangueu Mafo et al. (2013)
Behenic acid (C22:0) (%)	–	nd	Azangueu Mafo et al. (2013)
Dodecanoic acid (%)	–	22.4	Said et al. (2010)
Lauric acid (%)	0.129	–	Nazif (2002)
Myristic acid (%)	0.689	–	Nazif (2002)

in ZSC seeds including heptanoic acid, octanoic acid, 5-octadecenic acid, 5-octadecenic-6-octadecenoic acid, (Z)-, dodecannoic acid, Cis-5-dodecenoic acid, entadecanoic acid, 9-hexadecenoic acid, (Z)-, hexadecanoic acid, Cis-10-heptadecenoic acid (Z,Z), heptadecanoic acid, 9,12-octadecadienoic acid (Z,Z)-, 9-octadecenoic acid, (Z)-, 11,14-eicosadienoic acid, oxiraneoctanoic acid, 3-octyl-,11-eicosenioc acid, heneicosanoic acid, 3-docosenoic acid, tricosanoic acid, etracosanoic acid (all as methyl esters). Moreover, the author has also detected butlated hydroxytluene, methyl tetradecanoate, methyl hexadec-9-enoate, methyl stearate, methyl 10-trans,12-cis-Octadecadienaoate, methyl 5,13-docosadienoate, ethyl 18-methylnonadecanoate, methyl 20-methyl-heneicosanoate, and squalene (Ahamed 2016).

15.3 Bioactivities of *Ziziphus spina-christi*

For many years plants have been considered as valuable sources of food and plenty of medicinally active agents and many modern drugs have been successfully derived from natural plants (Ahmad et al. 2016; Shahidi 2009; Tukan et al. 1998). The genus *Zizyphus* common medicinal properties include hypoglycemic, hypotensive hepatoprotective, anti-inflammatory, antibacterial, antimicrobial, antifungal, antioxidant, antitumour, antidiabetic, liver protective agent, antihyperglycaemic agent and an immune system stimulant and (Panduraju et al. 2009; Abalaka et al. 2010; Mohammed et al. 2013; Singh et al. 2012; Asgarpanah and Haghighat 2012; Said et al. 2006; Hafiz and Mubaraki 2016; Abdel-Zaher et al. 2005; Ali et al. 2001; Shahat et al. 2001).

The extensive investigation carried out on phytochemical, pharmacological and microbiological activities of *Ziziphus spina-christi* led to a fact that the plant possess multi-biologically active compounds. These active compounds include alkaloids (spinanine A, tanines), sterols like β sitosterol, flavonoids (rutin and quercetin derivatives), triterpenoids, sapogenins and saponins (betulinic acid) (Farmani et al. 2016; Brantner and Males 1999; Godini et al. 2009; Abalaka et al. 2010). It has been reported that flavonoids can serve as antioxidant (Bors and Saran 1987), antimicrobial (Cowan 1999), anti-inflammatory agents (Moroney et al. 1988), can retard platelet aggregation (Van Wauwe and Goossens 1983), and mast cell histamine release (Amellal et al. 1985). Phenolic compounds are powerful chain breaking agents, and therefore, they serve as good antioxidants (Liu et al. 2011). Likely, the antioxidant phenolics have been suggested to prevent the progression of the diseases and cancers such as prostate, breast, lung, colon and rectal cancers (Kris-Etherton et al. 2002). The most investigated bioactivities using *Ziziphus spina Christi* fruits, leaves, bark and roots extracts are summarized in Table 15.5.

15.4 Antibacterial Activity

Most likely, the plant fruits contain polyphenolics which can serve as potential source of physiologically active natural antioxidant and other therapeutic agents (Ahmad et al. 2016; Kaur and Kapoor 2001; Vinson et al. 2001). A traditional practice has been carried out by Muslims using the boiled water extracts of ZSC leaves for cleaning of dead bodies before burial suggesting antibacterial properties of ZSC (Karar et al. 2016). It is reported that aromatic hydrocarbons and their substitutions have a good antibacterial activity (Harami et al. 2010; Yanping et al. 2010; Jahangirian et al. 2013; Salem et al. 2014; Islam et al. 2008). Decane has presented good antifungal and antibacterial activity (Gholamreza et al. 2012).

Table 15.5 Summary of some bioactivities of *Ziziphus spina-cristi*

Part of plant used	Type of extract	Antibacterial activity	Anti-diabetic activity	Antifungal activity	Anti-proliferative activity	Antioxidant activity	Reference
Fruits	Butanol extract, saponins glycoside, christinin-A	–	A significant increase in serum insulin levels in diabetic rats	–	–	–	Glombitza et al. (1994); Adzu et al. (2002)
	Ethanol extract	High activity against *Pseudomonas aeruginosa*	–	–	–	–	Temerk et al. (2017)
	n-hexane extracts	–	–	–	–	Antioxidant activity with IC_{50} values of 5.5 ± 0.1 and 4.1 ± 0.1 μg/mL	EL-Hefny et al. (2018)
	n-hexane extracts	Highest activity against B. pumilus, with an IZ value of 13.3 mm	–	–	–	–	EL-Hefny et al. (2018)
	Fruit oil	Good activity against P. carotovorum	–	–	–	–	EL-Hefny et al. (2018)
	Ethanol extract	–	–	–	–	Antioxidant activity with IC_{50} value of 120 μg/mL against FL-cells	Ali et al. (2001)
	Water and ethyl acetate extracts	–	–	–	–	Have no antioxidant activity	Ali et al. (2001)

(continued)

Table 15.5 (continued)

Part of plant used	Type of extract	Antibacterial activity	Anti-diabetic activity	Antifungal activity	Anti-proliferative activity	Antioxidant activity	Reference
	Methanol extract	No antibacterial activity against the Gram-negative bacteria Moderate antibacterial activities against the Gram-positive bacteria	–	–	–	–	Abdallah (2017); Ali et al. (2015)
	Methanolic extract	–				A strong antioxidant activity with 51% DPPH radical scavenging	Singh et al. (2012)
	Hydro alcoholic extract (500 mg/kg)	–	A significant decrease in the level of blood glucose A concomitant significant increase in the serum insulin level, in diabetic dogs ($p < 0.05$)	–	–	–	Avizeh et al. (2010)
	Aqueous extract of the pulp fruit	Antibacterial activity against Gram-negatives (Escherichia coli)	–	Antifungal potency against Candida albicans	–	–	Tom et al. (2009)

Leaves	Ethanol 70% extract	–	Highest ant hyperglycemic activity ($P < 0.01$)	–	–	–	Nesseem et al. (2008)
	Ethanol and methanol extracts	High antibacterial activity against all studied strains except *Escherichia coli*	–	–	–	–	Temerk et al. (2017)
	Methanol extract	Very effective against S. aureus and B. cereus	–	–	–	–	Motamedi et al. (2014)
	Methanolic, aqueous and ethanolic extracts	–	–	–	–	**Good antioxidant power with IC50 values of 21.4, 24.2 and 54.3 µg/mL for** methanolic, aqueous and ethanolic extracts, respectively	Khaleel et al. (2016)
	Ethanol extract	More effective to Proteus mirabilis	–	–	–	–	Motamedi et al. (2014)
	Ethanol extract	–	–	–	Positive on the MCF-7 (human breast adenocarcinoma) cell line: Low IC50 value (0.02 mg/mL)	–	Karimi et al. (2012)
	95% ethanol extract	–	–	High antifungal activity particularly against *D. biseptata*	–	–	Abu-Taleb et al. (2011)

(continued)

Table 15.5 (continued)

Part of plant used	Type of extract	Antibacterial activity	Anti-diabetic activity	Antifungal activity	Anti-proliferative activity	Antioxidant activity	Reference
Bark	Ethanol extract	Great activity against *Enterococcus faecalis*	–	–	–	–	Temerk et al. (2017)
	Methanol extract	Effective against *Klebsiella pneumonia*	–	–	–	–	Temerk et al. (2017)
	Methanol and 50% hydroethanol extract	Good activity against S. sobrinus	–	–	–	–	Mohieldin et al. (2017)
Roots	Extracts	–	A significant reduction of fasting serum glucose levels ($p < 0.001$) A marked increase in serum insulin level ($p < 0.001$)	–	–	–	Hussein et al. (2006)
	Ethanol extract	The largest clearing zone against *Enterococcus faecalis*	–	–	–	–	Temerk et al. (2017)

The methanolic extract of *Ziziphus spina Christi* showed greater effectiveness against S. aureus and B. cereus compared to the ethanolic extract even at a high concentration. While the ethanolic extract exhibited more activity against Proteus mirabilis. The minimum Inhibitory Concentration (MIC) and Minimum Bactericidal Concentration (MBC) indexes of the both extracts were found to be 8 mg/mL for S. aureus (Motamedi et al. 2014). Abdallah (2017) investigated the antibacterial activity of methanol fruit extract of Sudanese *Ziziphus spina-christi* against different ATCC strains of Gram-positive and Gram-negative bacteria. The fruit extract exhibited no antibacterial activity against the Gram-negative bacteria (*Escherichia coli* ATCC 35218 and *Klebsiella pneumoniae* ATCC 700603) and moderate antibacterial activities against the Gram-positive bacteria. Abdallah (2017) results are in agreement-in partial-with what was found by Ali et al. (2015) but they contradict results of Tom et al. (2009) who reported antibacterial activity against Escherichia coli, Pseudomonas aeruginosa (Gram-negatives) and antifungal potency against Candida albicans of the fruit of Ziziphus spina-christi.

Furthermore, antibacterial activity of *Ziziphus spina christi* methanolic and ethanolic extracts of bark, fruit, roots, and leaves were tested against seven pathogenic bacterial strains using the antibiotic erythromycin as positive control. The used strains include *Pseudomonas aeruginosa* (ATCC 278223), *Enterobacter cloaca*e (ATCC 13047), *Escherichia coli* (ATCC 25922), *Enterococcus faecali*s (ATCC 29212) *Enterobacter aerogenes* (ATCC 13084), *Klebsiella pneumonia*e (ATCC 13888), and *Methicillin-resistant Staphylococcus aureus* (MRSA, ATCC 43300) (Temerk et al. 2017). However, both gram negative and gram positive bacteria were sensitive to various plant extracts, bark extract was found to be the most active against all strains except for *Enterobacter aerogenes*. Ethanol bark extract showed great activity against *Enterococcus faecali*s, methanol bark extract was effective against *Klebsiella pneumonia*, leaves extracts showed high antibacterial activity against all strains except *Escherichia coli*, ethanol fruit extract exhibited high activity against *Pseudomonas aeruginosa* and ethanol roots showed the largest clearing zone against *Enterococcus faecali*s. Generally, the ethanolic extracts showed slightly better antibacterial (killing) activity compared to the methanolic extracts (Temerk et al. 2017). It could be concluded that ethanolic and methanolic extracts of the different parts of Z. *spina-christi* inhibited the growth of various species of Gram positive and Gram negative bacteria.

A 1000 μg/mL Z. *spina-christi* fruit extract showed good antibacterial activity against the growth of R. solanacerum and B. pumilus with an inhibition zone (IZ) values of 11.6 mm and 13.3 mm respectively. Furthermore, Z. *spina-christi* fruit oil has also exhibited a promising activity against P. carotovorum, with IZ value of 13 mm at 1000 μg/mL and 12.3 mm at 500 μg/mL. Furthermore, the fruit oil also showed activity against the growth of D. solani at the same concentrations with IZ values of 11.6 and 10.6 mm (EL-Hefny et al. 2018).

15.5 Antioxidant Activity

It was reported that medicinal plants possess many different natural antioxidants including tannins, phenolic acid, and flavonoids (Thirumalai et al. 2011). The flavonoids are considered to be the most potent nutritional antioxidants (Anderson 1980). Antioxidants are capable to lessen damaging of the tissues caused by the free radicals, and therefore, they can serve as cancer, arteriosclerosis, heart disease and many other diseases preventers (Bandyopodhyay et al. 2007). Furthermore, antioxidants can be used industrially to preserve foods and prolong their shelf life by preventing their deterioration caused by oxidation of lipids which results to formation of undesirable secondary lipid peroxidation.

EL-Hefny et al. (2018) reported that ZSC fruit extract exhibited antioxidant activity with IC50 values of 5.5 ± 0.1 and 4.1 ± 0.1 μg/mL, while, Ali et al. (2001) have reported an IC 50 value of 120 μg/mL for ethanolic extract of ZSC fruits against FL-cells, but no activity detected with both water and ethyl acetate extracts (Ali et al. 2001). Moreover, the antioxidant activity of *ziziphus* extracts were also investigated by Al-Jassabi and Abdullah (2013) and Singh et al. (2012) who expressed the activity as percentage of 2,2-diphenyl-1-picrylhydrazyl (DPPH) radical inhibition. Their reported values were ranged from 31.76% to 90.23% and 51% inhibition, respectively. These results indicate effective scavenging of DPPH radical and hence strong antioxidant capacity of *Zizyphus* spina Christi extract.

A work carried out by Youssif et al. (2011) to determine the antioxidant activity of *Zizyphus spina-christi* fruits against carbon tetrachloride (CCL_4) induced oxidative stress and hepatotoxicity in the Albino Wistar rats. They reported that ZSC fruits can provide protection to the liver against CCL_4-induced oxidative damage, as well as correlation of the hepatoprotective effect due to its antioxidant and free radical scavenger effects). The antioxidant activity of ZSC was also confirmed by others researchers (Othman et al. 2009). It was reported that *Ziziphus* kernel oils exhibited antioxidant activity ten times higher than BHT and Quercetin but comparable to that of pulp of the dry fruit (Azangueu Mafo et al. 2013; Biyanzi et al. 2012). This reported activity was attributed to the contents of polyphenols and tocopherols present in the plant.

15.6 Anti-Diabetic Activity

The main constituents of butanolic extract of *Zizyphus spina christi* fruits namely saponins glycoside, christinin-A have found to enhance glucose utilization in diabetic rats. A significant increase in serum insulin levels was noticed after 4 weeks of treatment with the extract. The butanolic extract has exhibited a pronounced antidiabetic effect than that of christinin-A (Glombitza et al. 1994; Adzu et al. 2002). Hussein and coworkers have carried clinical trials using 100 mg/kg root extracts of *Ziziphus spina christi* to treat diabetic rats orally. Their results revealed a significant reduction in fasting serum glucose levels ($p < 0.001$) as well as a noticeable enhancement of serum insulin levels ($p < 0.001$) of the tested diabetic rats, indicating anti-diabetic activity of ZSC extracts. The extracts also found to reduce some complications that associate diabetes such as hyperglycemia, hyperlipidemia and peroxidates and maintain liver and kidney function as well (Hussein et al. 2006).

In a previous other work a significant decrease in blood glucose levels was noticed when hydroalcoholic extract of *Z. spina-christi* had employed. The work also reported a concomitant significant increase in serum insulin in the diabetic dogs ($p < 0.05$). The hydroalcoholic extract also lowered the blood glucose. Therefore, a long term application of ZSC extract may be more preferable over chemical drugs for prevention or minimization of some chronic diseases and complications linked to diabetes (Avizeh et al. 2010). A test carried out by Nesseem et al. (2008) using ethanol 70% extract of ZSC leaves revealed a high anti hyperglycemic activity when compared to metformin ($P < 0.01$). The measured LD_{50} of the tested extract was 10 g/kg indicating safety of the extract, indicating no toxicity on cholesterol, triglycerides, urea, creatinine, and liver enzymes (Nesseem et al. 2008). The antidiabetic activity of ZSC was also proved by Othman and colleagues who reported a significant amelioration in glucose levels, increased body weight, as well as increased insulin and hemoglobin levels.

15.7 Antimicrobial Activity

About 26 different aromatic hydrocarbons were previously detected in ZSC fruits (EL-Hefny et al. 2018). Most likely, these aromatic hydrocarbons and their substitutions have a good antibacterial activity (Islam et al. 2008; Yanping et al. 2010; Jahangirian et al. 2013; Harami et al. 2010; Salem et al. 2014). The oil extract of *zizphus spina-christi* seed was found to be exhibited a good antimicrobial agent against four bacteria types (Escherichia coli, Staphylococcus aureus, Pseudomonas aeruginosa, and Bacillus subtilis) and two fungi (Aspergillus niger and Candida albicans) (Ahamed 2016). Among the previously different tested plants, an extract of *Ziziphus spina-christi* exhibited the best anti-*Candida* activity against *Candida albicans*. Extract with high concentrations (50–55 μL) were found to be significantly more effective against yeast (Pirbalouti et al. 2009).

15.8 Antihypertensive Activity

Zakaria et al. (1999) have investigated the antihypertensive activity of 10% ethanolic and they found that the extract did not have acute effect of systolic blood pressure, diastolic blood pressure and heart rate. Moreover, the aqueous extract was also found to exhibit the same activity.

15.9 Cytotoxicity

Cytotoxicity of *Z. spina-christi* L. Leaves was also investigated using cell lines (Brine shrimp) and the test revealed IC_{50} 300.0 μg/mL (McGaw et al. 2014; Gadir 2012). The anti-proliferative activity of methanol extract of ZSC leaves was also investigated by Abu-Raghif et al. (2017) who reported dose dependent inhibition of RD cell line, and their measured IC_{50} was 154.44 μg/mL. The MTT assay has also been used to investigate the cytotoxicity of different extracts (Hexane, chloroform, chloroform-methanol (9:1), methanol-water (7:1) methanol, butanol and water) of ZSC leaves against Hela and MDA-MB-468 tumor cells. Among the different extracts the chloroform-methanol extract was more potent one. From the review it can be concluded that *ZSC* is a good candidate for the development of anticancer drugs (Jafarian et al. 2014).

15.10 Food Uses

Based on the chemical constituents, *Ziziphus spina christi* fruits can be used as a tonic. The fruits contain a very high energy value and taste like a mixture of dates and can be eaten raw or dried. The seeds were found to contain high protein while leaves are rich in magnesium, calcium, and iron. In the western part of Sudan (Kordofan and Darfur) they used to produce a fine flour from the bitter-sweet dried pulp. The dried pulp-flour can be mixed with water and sesame to be shaped into small balls for either immediate or late consumption. The seed coats are sometimes being cracked to eparate the kernels which can be eaten raw (Glombitza et al. 1994; Zakaria et al. 1999). Therefore, ZSC represents an important naturally available cheap source of energy, minerals and protein. Summary of some bioactivities using leaves, stem, fruits and seeds/kernel oil of *Ziziphus spina-christi* are shown in Table 15.5.

15.11 Conclusion

The main objective of this chapter has been to review the description, bioactivity and chemical composition of ZSC as well as its potential as a nutritional and therapeutic agent. It is obvious that fruits of ZSC are rich in minerals, carbohydrates, vitamins indicating their nutritional value for both human and animals. From the information gathered, it is also clear that ZSC has multi-pharmacologicaly active agents including polyphenols, saponin glycosides (such as christinin-

A) and flavonoids. These active compounds can potentially serve as antibacterial, antioxidant, anti-diabetic, antimicrobial, antifungal, antihypertensive, antihyperglycemic, cytotoxic agents. Therefore, it can be concluded that ZSC can be considered as a potential natural source for nutritional uses as well as for many therapeutically active engredients from which new drugs can be developed successfully to cure various diseases.

References

Abalaka ME, Daniyan SY, Mann A (2010) Evaluation of the antimicrobial activities of two *Ziziphus* species (Ziziphus mauritiana L. and *Ziziphus spina-christi* L.) on some microbial pathogens. Afr J Pharm Pharmacol 4:135–139

Abdallah EM (2017) Antibacterial activity of fruit methanol extract of *Ziziphus spina-christi* from Sudan. Int J Curr Microbiol Appl Sci 6(5):38–44. https://doi.org/10.20546/ijcmas.2017.605.005

Abdel-Galil FM, El-Jissry MA (1991) Cyclopeptide alkaloids from *Zizyphus spina-christi*. Phytochemistry 30(4):1348–1349

Abdelmuti OMS (1991) Biochemical and nutritional evaluation of famine foods of the Sudan. Doctoral Dissertation in Biochemistry and Nutrition, University of Khartoum, Sudan

Abdel-Zaher AO, Salim SY, Assaf MH, Abdel-Hady RH (2005) Antidiabetic activity and toxicity of *Zizyphus spina-christi* leaves. J Ethnopharmacol 101:129–138

Abu-Raghif AR, Jasim GA, Hanoon MM (2017) Anti-proliferative activity of *Zizyphus spina christi* leaves methanol extract against rhabdomyosarcoma (rd) cell line. Int J Pharm Pharm Sci 9(2):279–282

Abu-Taleb AM, El-Deeb K, Al-Otibi FO (2011) Assessment of antifungal activity of *Rumex vesicarius* L. and *Ziziphus spina-christi* (L.) Willd. extracts against two phytopathogenic fungi. Afr J Microbiol Res 5(9):1001–1011

Adekunle AE, Adenike JO (2012) Comparative analysis of proximate, minerals and functional properties of Tamarindus indica pulp and Ziziphus spina christi fruit and seed. Greener J Agric Sci 2(1):21–25

Ads EN, Rajendrasozhan S, Hassan SI, Sharawy SMS, Humaidi JR (2017) Phytochemical, antimicrobial and cytotoxic evaluation of *Ziziphus spina christi* (L.) stem bark. Biomed Res 28(15):6646–6653

Adzu B, Amos S, Dzarma S, Wambebe C, Gamaniel K (2002) Effect of *Zizyphus spina-christi* wild aqueous extract on the central nervous system in mice. J Ethnopharmacol 79:13–16

Ahamed EMM (2016) Characterization and biological activity study of *Ziziphus spina-christi* seed oil. PhD thesis, Sudan University of Science and Technology

Ahmad N, Zuo Y, Lu X, Anwar F, Hameed S (2016) Characterization of free and conjugated phenolic compounds in fruits of selected wild plants. Food Chem 190:80–89

Ahmed FAM, Sati NME (2018) Chemotaxonomic study and botanical overview of some Ziziphus spp. in Sudan. Int Res J Biol Sci 7(5):32–39

Ahmed AI, Babeker AM, Elamin AMA, Ibrahim EA (2015) Physicochemical evaluation of seeds and oil of nontraditional oil seeds. Turk J Agric Food Sci Technol 3(8):687–691

Alhakmani F, Khan SA, Ahmad A (2014) Determination of total phenol, *in-vitro* antioxidant and anti-inflammatory activity of seeds and fruits of *Zizyphus spina-christi* grown in Oman. Asian Pac J Trop Biomed 4(2):S656–S660

Ali NAA, Julich WD, Kusnick C, Lindequist U (2001) Screening of Yemeni medicinal plants for antibacterial and cytotoxic activities. J Ethnopharmacol 74(2):173–179

Ali AB, Almagboul AZ, Mohammed OM (2015) Antimicrobial activity of fruits, leaves, seeds and stems extracts of *Ziziphus spina-christi*. Arab J Med Aromat Plants 1(2):94–107

Al-Jassabi S, Abdullah MS (2013) Extraction, purification and characterization of antioxidant fractions from *Zizyphus spina-christi* fruits. Am Eurasian J Toxicol Sci 5(3):66–71

Al-Wakeel HA (2008) Effect of topical cidir leaves extract in reducing skin sensitivity in atopic patients. Kufa Med J 11(2):7–11

Amellal M, Bronner C, Briancon F, Haag M, Anton R, Landry Y (1985) Inhibition of mast cell histamine release by flavonoids and biflavonoids. Planta Med 51:16–20

Anderson JW (1980) High-fibre diets for diabetic and hypertriglyceridemic patients. Can Med Assoc J 123:975–979

Anthony C, Dweck (2005) A review of Zizyphus spina-Christi. FLS FRSC FRSH – Technical Editor, PERSONAL CAR3, http://www.dweckdata.co.uk/Published_papers/Zizyphus_spina-Christi_paper.pdf

Asgarpanah J, Haghighat E (2012) Phytochemistry and pharmacologic properties of *Ziziphus spinachristi* (L.) Willd. Afr J Pharm Pharmacol 6(31):2332–2339. https://doi.org/10.5897/AJPP12.509

Avizeh R, Najafzaden H, Pourmahdi M, Mirazee M (2010) Effect of glibenclamide and fruit extract of *Ziziphus spinacristi* on alloxan induced diabetic dogs. Int J Appl Res Med 8(2):109–113

Aynehchi Y, Mahoodian M (1973) Chemical examination of *Zizyphus spina-christi* (L.) Willd. Acta Pharm Suec 10(6):515–519

Azangueu Mafo J, Kamga Gueobu D, Regonne R, Kamga R, Ndjouenkeu R (2013) Physico-chemical characterization of the oil of *ziziphus* kernels from the savannah area of Cameroon. In: 20th SAAFoST biennial conference Pretoria, 7–9 October

Bandyopodhyay M, Chakraborty R, Raychaudlhuri U (2007) A process for preparing a natural antioxidant enriched diary product. LWT-Food Sci Technol 40:842–851

Berry-Koch A, Moench R, Hakewill P, Dualeh M (1990) Alleviation of nutritional deficiency diseases in refugees. Food Nutr Bull 12:106–112

Biyanzi P, Ndjouenkeu R, Mbofung CMF (2012) Composition and antioxidant activity of four varieties of Ziziphus fruits (jujube) from savannah regions of Cameroon. Communication presented at the 2012 EFFoST annual meeting, Montpellier, France, 20–23 November 2012

Bors W, Saran M (1987) Radical scavenging by flavonoid antioxidants. Free Radic Res Commun 2:289–294

Brantner, A.H. and Males, Z. (1999) Quality assessment of Paliurus *spina-christi* extracts. J Ethnopharmacol 66(2):175–179

Cowan MM (1999) Plant products as antimicrobial agents. Clin Microbiol Rev 12:564–582

Duke JA (1985) Handbook of medicinal herbs. CRC Press, Boca Raton, p 677

Dweck AC (2005) A review of *Zizyphus spina-christi*. FLS FRSC FRSH-Technical Editor, Personal Care. http://www.dweckdata.co.uk/Published_papers/Zizyphus_spina-christi_paper.pdf

EL-Hefny M, Mohamed AA, Salem MZM, Abd El-Kareem MSM, Ali HM (2018) Chemical composition, antioxidant capacity and antibacterial activity against some potato bacterial pathogens of fruit extracts from Phytolacca dioica and *Ziziphus spina-christi* grown in Egypt. Sci Hortic 233:225–232

Farmani F, Moein M, Amanzadeh A, Kandelous HM, Ehsanpour Z, Salimi M (2016) Antiproliferative evaluation and apoptosis induction in MCF-7 cells by *Ziziphus spina christi* leaf extracts. Asian Pac J Cancer Prev 17(1):315–321

Gadir SA (2012) Assessment of bioactivity of some Sudanese medicinal plants using brine shrimp (Artemia salina) lethality assay. J Chem Pharm Res 4:5145–5148

Ghafoor AO, Qadir HK, Fakhri NA (2012) Analysis of phenolic compounds in extracts of *Ziziphus spina-christi* using RPHPLC method. J Chem Pharm Res 4(6):3158–3163

Gholamreza A, Mohamad J, Ehsan S (2012) Antimicrobial activity and chemical composition of essential oil from the seeds of Artemisia aucheri Boiss. Jundishapur J Nat Pharm Prod 7(1):11–15

Glombitza K, Mahran G, Mirhom Y, Michel K, Motawi T (1994) Hypoglycemic and antihyperglycemic effects of *Zizyphus spina christi* in rats. Planta Med 60:244–247

Godini A, Kazem M, Naseri G, Badavi M (2009) The effect of *Zizyphus spina-christi* leaf extract on the isolated rat aorta. J Pak Med Assoc 59:537–539

Goyal M, Nagori BP, Sasmal D (2012) Review on ethnomedicinal uses, pharmacological activity and phytochemical constituents of *Ziziphus mauritiana* (*Z. jujuba* Lam., non Mill). Spatula DD 2:107–116

Guo S, Duan JA, Tang Y, Qian Y, Zhao J, Qian D, Su S, Shang E (2011) Simultaneous qualitative and quantitative analysis of triterpenic acids, saponins and flavonoids in the leaves of two *Ziziphus* species by HPLC-PDA-MS/ELSD. J Pharm Biomed Anal 56:264–270

Hafiz TA, Mubaraki MA (2016) The potential role of *Ziziphus spina-christi* leaf extracts against Plasmodium berghei-induced liver and spleen injury. Biomed Res 27(4):1027–1032

Harami M, Adamu EO, Ekanem Suleiman B (2010) Identiffcation of essential oil components from Nigella sativa seed by gas chromatography-mass spectroscopy. Pak J Nutr 9(10):966–967

Hussein HM, Sayed EM, Said AA (2006) Antihyperglycemic, antihyperlipidemic and antioxidant effects of *Ziziphus spinacristi* and *Ziziphus jojoba* in alloxan diabetic rats. Int J Pharmacol 2(5):563–570

Ibrahim JA, Egharevba HO, Nnamdi RA, Kunle OF (2015) Comparative pharmacognostic and phytochemical analysis of Ziziphus spina-christi (L.) Desf. and Ziziphus abyssinica Hochst. Ex A. Rich. Int J Pharmacogn Phytochem Res 7(6):1160–1166

Ikram M, Tomlinson H (1976) Chemical constituents of *Zizyphus spina christi*. Planta Med 29(3):289–290

Islam M, Siddiqui AA, Rajesh R (2008) Synthesis, antitubercular, antifungal and antibacterial activities of 6-substituted phenyl-2-(3′-substituted phenyl pyridazin-6′-yl)-2,3,4,5-tetrahydropyridazin-3-one. Acta Pol Pharm 65(3):353–362

Ismail AE (1998) Effect of soil amendments with some hardwood barks on reproduction of rotylenchulus reniformis and growth of sunflower. Pak J Nematol 6(2):137–144

IUPAC (1992) Standard methods for the analysis of oils, fats and derivatives. Blackwell Scientific Publications, Oxford

Izuagie T, Hassan LG, Achor AUM, Sahabi DM, Sadiq IS (2012) Compaction properties of starch extracted from seeds of christ thorn (*Ziziphus spina-christi*). Int J Food Nutr Sci 1(2):10–15

Jafarian A, Zolfaghari B, Shirani K (2014) Cytotoxicity of different extracts of arial parts of Ziziphus spina-christi on Hela and MDA-MB-468 tumor cells. Adv Biomed Res 3:38

Jahangirian H, Haron MDJ, Ismail MHS, Raffee-Moghaddam R, Afsah-Hejri L, Abdollahi Y, Rezayi M, Vafai N (2013) Well diffusion method for evaluation of antibacterial activity of copper phenyl fatty hydroxamate synthesized from canola and palm kernel oils. Dig J Nanomater Biostruct 8(3):1263–1270

Karar MGE, Quiet L, Rezk A, Jaiswal R, Rehders M, Ullrich MS, Brix K, Kuhnert N (2016) Phenolic profile and *in vitro* assessment of cytotoxicity and antibacterial activity of *Ziziphus spina-christi* leaf extracts. Med Chem 6(3):143–156

Karimi E, Oskoueian E, Hendra R, Oskoueian A, Jaafar HZ (2012) Phenolic compounds characterization and biological activities of citrus aurantium bloom. Molecules 17:1203–1218

Kaur C, Kapoor HC (2001) Antioxidants in fruits and vegetables – the millennium's health. Int J Food Sci Technol 36:703–725

Khaleel SMJ, Jaran AS, Haddadin MSY (2016) Evaluation of total phenolic content and antioxidant activity of three leaf extracts of *Ziziphus spina-christi* (Sedr) grown in Jordan. Br J Med Med Res 14(6):1–8

Kris-Etherton PM, Hecker KD, Bonanome A, Coval SM, Binkoski AE, Griel AE, Etherton TD (2002) Bioactive compounds in foods: their role in the prevention of cardiovascular disease and cancer. Am J Med 113(9B):71S–88S

Liu J, Wang C, Wang Z, Zhang C, Lu S, Liu J (2011) The antioxidant and free-radical scavenging activities of extract and fractions from corn silk (Zea mays L.) and related flavone glycosides. Food Chem 126:261–269

Mahran GEDH, Glombitza KW, Mirhom YW, Hartmann R, Michel CG (1996) Novel saponins from *Zizyphus spina-christi* growing in Egypt. Planta Med 62(2):163–165

McGaw LJ, Elgorashi EE, Eloff JN (2014) Cytotoxicity of African medicinal plants against normal animal and human cells. Toxicol Surv Afr Med Plants:181–133. https://doi.org/10.1016/B978-0-12-800018-2.00008-X

Mishra T, Bhatia A (2014) Antiplasmodial effects of the aqueous ethanolic seed extract of *Ziziphus mauritiana* against Plasmodium berghei in Swiss albino mice. Int J Pharmacol Res 4(3):111–116

Mohammed GT, Abdulrahman FI, Khan IZ, Hussaini MI, Muazu J, Si Y, Wazis CH (2013) Antimicrobial efficacies of ethanolic extract and active column fractions of the stembark of *Zizyphus spina-christi* L. Int J Pharm Pharm Sci 5:455–460

Mohieldin EAM, Muddathir AM, Yamauchi K, Mitsunaga T (2017) Anti-caries activity of selected Sudanese medicinal plants with emphasis on Terminalia laxiflora. Rev Bras 27:611–618

Moroney MA, Alcaraz MJ, Forder RA, Carey F, Hoult JR (1988) Selectivity of neutrophil 5-lipoxygenase and cyclo-oxygenase inhibition by an anti-inflammatory flavonoid glycoside and related aglycone flavonoids. J Pharm Pharmacol 40:787–792

Motamedi M, Seyyednejad SM, Hasannejad Z, Dehghani F (2014) A comparative study on the effects of *Ziziphus spina-christi* alcoholic extracts on growth and structural integrity of bacterial pathogens. Iran J Pharm Sci 10(2):1–10

Nazif NM (2002) Phytoconstituents of Zizyphus spina-christi L. fruits and their antimicrobial activity. Food Chem 76:77–81

Nesseem DI, Michel CG, Saleem AA, El-Alfy TS (2008) Formulation and evaluation of antihyperglysimic leaf extracts of *Ziziphus spinacristi* (L.) wild. Pharmazie 64:104–109

Orwa C, Mutua A, Kindt R, Jamnadass R, Anthony S (2009) Agroforestree database: a tree reference and selection guide version 4.0. http://www.worldagroforestry.org/sites/treedbs/treedatabases.asp

Othman AI, Amer MA, Abdel-Mogib M, Samaha RF (2009) Effects of the methanolic extracts of *Zizyphus spina christi, Olea europaea* and *Morus alba* leaves in Streptozotocin- induced diabetic rats. Egypt J Hosp Med 37:759–771

Panduraju T, Raja Sridhar Rao P, Sateesh Kumar V (2009) A study on antimicrobial activity of Rumex vesicarius Linn. Int J Pharm Technol 1:21–25

Paroda RS, Mal B (1989) New plant sources for food and industry in India. In: Wickens GE, Haq N, Day P (eds) New crops for food and industry. Chapman and Hall, London, pp 135–149

Pawlowska AM, Camangi F, Bader A, Braca A (2008) Flavonoids of *Zizyphus jujuba* L. and Zizyphus spina-christi (L.) Willd (Rhamnaceae) fruits. Food Chem 112:858–862

Pirbalouti AG, Bahmani M, Avijgan M (2009) Anti-candida activity of some of the Iranian medicinal plants. Electron J Biol 5(4):85–88

Said A, Huefner A, Tabl E, Fawzy G (2006) Two new cyclic amino acids from the seeds and antiviral activity of methanolic extract of the roots of *Zizyphus spina-christi*. Paper presented at the 54th annual congress on medicinal plant research. Planta Medica 72

Saied AS, Gebauer J, Hammer K, Buerkert A (2008) Ziziphus spina-christi (L.) Willd.: a multipurpose fruit tree. Genetic Resources and Crop Evolution 55(7):929–937

Said A, Fawzy G, Abu tabl ESA, Tzakou O (2010) Volatile constituents of Zizyphus jujuba aerial parts and Zizyphus spina-christii fruits from Egypt. J Essent Oil Bear Plants 13(2):170–174

Salem MZM, Abdel-Megeed A, Ali HM (2014) Stem wood and bark extracts of Delonix regia (Boj Ex. Hook): chemical analysis, antibacterial, antifungal, and antioxidant properties. BioResource 9(2):2382–2395

Sameera NS, Mandakini BP (2015) Investigations into the antibacterial activity of *Ziziphus mauritiana* Lam. and *Ziziphus xylopyra* (Retz.) Willd. Int Food Res J 22(2):849–853

Shahat AA, Pieters L, Apers S, Nazeif NM, Abdel-Azim NS, Berghe DV, Vlietinck AJ (2001) Chemical and biological investigations on *Zizyphus spina-christi* L. Phytother Res 15:593–597

Shahidi F (2009) Nutraceuticals and functional foods: whole versus processed foods. Trends Food Sci Technol 20(9):376–387

Singh V, Guizani N, Essa MM, Rahman MS, Selvaraju S (2012) *In vitro* antioxidant activities of *Ziziphus spina-christi* fruits (red date) grown in Oman. Biotechnology 11:209–216

Suleiman MHA (2015) An ethnobotanical survey of medicinal plants used by communities of northern Kordofan region, Sudan. J Ethnopharmacol 176:232–242

Temerk HA, Salem WM, Sayed WF, Hassan FS (2017) Antibacterial effect of phytochemial extracts from *Ziziphus-spina christi* against some pathogenic bacteria. Egypt J Bot 57(3):595–604

Thirumalai T, Therasa SV, Elumalai EK, David E (2011) Hypolipidaemic and antioxidant effect of Enicostemmalittorale Blume. Asian Pac J Trop Biomed 1:381–385

Tom GM, Yesufu HB, Abdulrahman FI (2009) Antimicrobial screening and effect of the pulp extracts of *Zizyphus spina-christi* (Linnaeus Desf) on some biochemical parameters in rats. J Pharm Bioresour 6(2):58–64

Tripathi M, Pandey MB, Jha RN, Pandey VB, Tripathi PN, Singh JP (2001) Cyclopeptide alkaloids from *Zizyphus jujuba*. Fitoterapia 72:507–510

Tukan SK, Takruri HR, Al-Eisaw DM (1998) The use of wild edible plants in the Jordanian diet. Int J Food Sci Nutr 49:225–235

Van Wauwe J, Goossens J (1983) Effects of antioxidants on cyclooxygenase and lipoxygenase activities in intact human platelets: comparison with indomethacin and ETYA. Prostaglandins 26:725–730

Vinson A, Xuehui S, Ligia Z, Bose P (2001) Phenol antioxidant quantity and quality in foods: fruits. J Agric Food Chem 49:5315–5321

Yanping H, Suhua H, Jincheng WU, Shunquan L (2010) Identification of essential oils from the leaves of 11 species of Eriobotrya. Pak J Bot 42(6):4379–4386

Yossef HEE, Khedr AA, Mahran MZ (2011) Hepatoprotective activity and antioxidant effects of El Nabka *(Zizyphus spina-christi)* fruits on rats hepatotoxicity induced by carbon tetrachloride. Nat Sci 9(2):1–7

Younes ME, Amer MS, El-Messallami ADE (1996) Phytochemical examination of the leaves of the Egyptian *Zizyphus spina christi* "Nabc". Bull Natl Res Cent (Cairo) 21(1):35–40

Youssif HE, Khedr AA, Mahran MZ (2011) Hepatoprotective and antioxidant effects of *Zizyphus spina-christi* fruits on carbon tetrachloride induced hepatotoxicity in rats. Nat Sci 9(2):1–7. (ISSN: 1545-0740). http://www.sciencepub.net

Zakaria MNM, Islam MW, Radhakrishnan R, Ismail A, Habibullah M, Chan K (1999) Antidiabetic properties of *Zizyphus spina-christi* in STZ-diabetic mice. In: 2000 years of natural products research – past present and future, July 26–30, 1999, Amsterdam. Phytoconsult, Leiden

Chapter 16
Balanites aegyptiaca: Phytochemical Constituents, Bioactive Compounds, Traditional and Medicinal Uses

Essa E. M. Ahmad and Abdalbasit Adam Mariod

16.1 Introduction

Balanites aegyptiaca (desert date) belongs to the family of *Balanitaceae* and species of the genus *Balanites aegyptiaca (L) Delile* is spiny shrub alternately tree dependent upon 10 m tall, generally conveyed on dry area regions about Africa and South Asia. It is widespread in most arid, Semi-arid to sub-humid tropical savannas in Africa, everywhere throughout the Sahel extending starting with the Atlantic coastline from claiming Senegal to the red Sea, Indian Ocean, and the Arabian gulf (Von Maydell 1990). The tree is multibranched, spiny shrub. Crown spherical, in one or several distinct masses. Trunk short also regularly expanding from close those build. Bark with grey, profoundly fissured. Extensions armed with strong yellow alternately green thistles dependent upon 8 cm in length. Abandons with two separate leaflets; leaflets obovate, asymmetric, 2.5–6 cm long, brilliant green, leathery, with fine hairs. Blooms are in the leaf beet axils. Furthermore would fragrant, yellowish-green. Fruit is long, limited drupe, 2.5–7 cm long, 1.5–4 cm in diameter (Fig. 16.1). Junior fruits are green turning yellow and glabrous when full grown. Pulp is bitter-sweet and eatable. Seed is those pyrene (stone), 1.5 with 3 cm long, light brown, fibrous, Also greatly tricky. It makes up 50 with 60% of the apples and oranges. There need aid 500 on 1500 dry, clean seeds for every kilogram (Chothani and Vaghasiya 2011).

E. E. M. Ahmad (✉)
Department of Chemistry, Sudan University of Science and Technology, Khartoum, Sudan

A. A. Mariod
Indigenous Knowledge and Heritage Center, Ghibaish College of Science & Technology, Ghibaish, Sudan

A. A. Mariod (ed.), *Wild Fruits: Composition, Nutritional Value and Products*,
https://doi.org/10.1007/978-3-030-31885-7_16

Fig. 16.1 Balanites tree and fruits. (Source: https://commons.wikimedia.org)

16.2 Chemical Composition, Nutritional Properties of *B. aegyptiaca* Fruits

The chemical composition and the nutritional characteristics of *B. aegyptiaca* fruits were well known and have been investigated by a number of authors (Hardman and Sofowora 1972; Amadou and Le 2017; Nour et al. 1985; Wiesman and Chapagain 2006). *Balanites aegyptiaca* fruits were collected from two different locations in Sudan. The edible mesocarp contained 1.2–1.5% protein and 35–37% total sugars of which 81.3–91.1% is present as reducing sugars. In addition, it contains 2.4-2.9% ash, 0.14-0.43% oil and 14.4-21.5% alcohol insoluble solids. The kernels contained 45.0–46.1% oil. Palmitic, stearic, oleic and linoleic acids were the main fatty acids.

The ratio of saturated to unsaturated fatty acids was 10:26 in both samples. The protein was found to be low in aromatic amino acids (Nour et al. 1985). Moreover, the sample contain 86.9 mg/100 g Ca, 14.5 mg/100 g Fe, 43.4 and 71.3 mg/100g P. *B. aegyptiaca* fruits were collected from three African countries (Nigeria, Egypt and Tanzania) and one Asian country (India) and investigated as a commercially potential source of steroidal sapogenin in an industrial scale. Sapogenin content in the epicarp/mesocarp and the seeds was assayed using IR spectrophotometry. The results showed that the range of the total sapogenin in seeds was 0.74–1.74% (on dry weight basis of undefatted whole seeds) while in epicarp/mesocarp it was 1.01–4.08% (on dry weight basis). It was also estimated that 25β-sapogenin predominates in epicarp/mesocarp (56–71% proportion from the total sapogenin) while 25α-sapogenin prevails in the seeds (71–76% proportion from the total sapogenin). Moreover, thin layer chromatography of the crude sapogenin isolated from defatted seeds showed that diosgenin is the main constituent with the presence of yagomenin, 25α-spirosta-3:5 diene and β-sitosterol in trace amounts (Hardman and Sofowora 1972).

The level of diosgenin was determined in ten samples of defatted seeds kernel of *B. aegyptiaca* which were collected from five Israeli provenances, four Western African countries (Burkina Faso, Senegal, Mali and Niger) and one from India (Chapagain and Wiesman 2005). The results have shown that the level of diosgenin varies with the location and has a strong positive correlation with the percentage of the oil. The highest level was found for Israeli sample collected from Beat Shean Valley; 2.22% (on dry weight basis) and 50.22% oil percentage, while the lowest one was reported for the Indian sample; 1.09% (on dry weight basis) and 39.20% oil percentage.

Amadou and Le (2017) have assessed the nutritional properties and the sensory attributes of *B. aegyptiaca* fruit juice. Additionally, the chemical compositions of the fresh and the boiled juice samples were also evaluated. The results showed that the fresh juice comprises of 15.58 (°brix) soluble solids, 15.50 mg/mL reducing sugars, 3.85% polyphenols, 3.17% flavonoids, 88.79% water, 0.76% fat, 4.09% soluble proteins, 0.0966 mg/mL vitamin C and minor quantities of vitamins B1, B2, B6 and folic acid. On the other hand the boiling of the juice has significantly increased the concentration of soluble solids (°brix) and soluble protein and decreased the water and the vitamins (C, B1, B2 and B6). In addition, a number of 50 flavor volatile compounds were identified in fresh and boiled juices with cyclotrisiloxane, hexamethyl- and ethyl acetate (with more 95% probability) represent the highest probable flavor volatile compounds in the juice. It was noted that boiling has sharply increased these compounds. The results also revealed that the fresh juice contains 14.53% (13.12% for the boiled sample) of essential amino acids with an increase in the free amino acids for the boiled sample. Both fresh and boiled samples have exhibited high antioxidant activity of 85.28% and 81.82% with DPPH and a significant superoxide anion radical concentration of 5 mg/mL. Evaluation of sensory attributes, moreover, showed a slight variation between the fresh and the boiled samples on the overall acceptability of the product (2.97 and 2.7 respectively). The study concluded that the juice of *B. aegyptiaca* mesocarp fruit which is rich in vitamins and phenolic compounds has plentiful nutritional benefits besides its known medicinal attributes.

16.3 Medicinal Uses and Biological Activity of *B. aegyptiaca* Fruits

In one study (Morsy et al. 2010), the effect of different doses of *B. aegyptiaca* fruit mesocarp samples collected from some radioactive places in Wadi-El-Gemal (Egypt) on the levels of glucose, cholesterol, triglycerides, HDL-cholesterol, LDL-cholesterol and diabetes were examined on diabetic male rats. The results demonstrated that *B. aegyptiaca* fruit decreased the blood sugar content and could be considered as a high potential anti-diabetic product. Moreover, the fruit could be used to lower the levels of the total cholesterol (HDL- and LDL-cholesterol) and triglycerides.

Extracts of chloroform, ethylacetate, butanol and methanol of *B. aegyptiaca* fruit mesocarp were found to be effective against *Aedes aegypti* mosquito larvae (Wiesman and Chapagain 2006). Five fractions of the methanol extract were obtained using column chromatography and all have demonstrated larval mortality. Strong relationship was found between the amount of total saponins in the extracts and the fractions and their larvicidal effect. The amount of total saponins in these samples was found to be as follows: 22.5% (dry weight) fraction one, 15.6% fraction two, 12.3% fraction three, 11.2% methanol extract, 8.8% ethanol extract, 7.8% fraction four, 6.4% butanol fraction, 3.5% chloroform extract and 3% fraction five. Furthermore, a concentration of 0.0014% w/v from fraction one was found to be sufficient to inhibit the emergence of 50% of the larvae population compared to 0.0034% for the methanol extract.

Trigonelline which is scientifically well characterized hypoglycemic alkaloid was identified and quantified in the crude extract of the peel and pulp of *B. aegyptiaca* fruit (mesocarp) using NMR spectroscopy (Farag et al. 2015). It is concentration in the peel and pulp was found to be 8 and 13 mg/g respectively. Furthermore, ultra performance liquid chromatography-mass spectrometry (UPLC-MS) displayed the presence of quercetin/isorhamnetin conjugates and epicatechin as major flavonoids in the crude extract of peel and pulp of *B. aegyptiaca* fruit.

The molluscicidal and cercariacidal activities of aqueous extracts of whole fruit, mesocarp, endocarp and seeds of *B. aegyptiaca* were tested against Ethiopian *Biomphalaria pfeifferi (B. pfeifferi), Lymnaea natalensis (L. natalensis) and Schistosoma mansoni (S. mansoni) cercariae* (Molla et al. 2013). The results of molluscicidal properties revealed that the activities of these extracts are concentration dependent and they are potent towards *B. pfeifferi* more than *L. natalensis*. The lethal concentrations for aqueous extracts of *B. aegyptiaca* seeds, endocarp, mesocarp and the whole fruit that killed 50% (LC_{50}) of adult *B. pfeifferi* were 56.32, 77.53, 65.51 and 66.63 mg/L, respectively, while the respective LC_{90} values were 77.70, 120.04, 89.50 and 97.55 mg/L. The lethal concentrations for aqueous extracts of *B. aegyptiaca* seeds, endocarp, mesocarp and the whole fruit that killed 50% (LC_{50}) of adult *L. natalensis* were, 80.33, 92.61, 83.52 and 87.84 mg/L, respectively, while the respective LC90 values were 102.30, 138.21, 115.42 and 127.69 mg/L. Moreover, *in vivo* experiments demonstrated that lower concentrations (5 and 10 mg/L) of the aqueous extract of *B. aegyptiaca* seeds inhibit the infection of *S. mansoni cercariae* to mice and it is time dependent whereas complete inhibition was observed at 15 mg/L regardless of the time of exposure. Significant reduction in the mean egg load of tissue was observed after exposure of infected mice to 5 and 10 mg/L of the aqueous extract of *B. aegyptiaca* seeds.

Crude extract, butanol (polar fraction) and dichloromethane (nonpolar fraction) fractions of *B. aegyptiaca* fruits (fruit fleshes) were investigated for their antidiabetic potential in experimental diabetic rats (Hassanin et al. 2018). The results demonstrated that the extracts have hypoglycemic, hypolipidemic and insulinotropic effects. Ethanol and butanol extracts have shown similar activities compared to dichloromethane extract. Furthermore, liquid chromatography-high resolution mass spectrometry (LC-HRMS) revealed the presence of almost 23 components in these extracts and in particularly the presence of balanitin-2, hexadecenoic acid, methyl protodioscin and 26-(O-β-Dglucopyranosyl)-3-β-[4-O-(β-D-glucopyranosyl)-2-O-(α-L-rhamnopyranosyl)-β-D-glucopyranosyloxy]-22,26-dihydroxyfurost-5-ene that have reported hypoglycemic effect.

A mixture about steroidal saponins: balanitin-6 (28%) Furthermore balanitin-7 (72%), separated starting with b. Aegyptiaca kernels, exhibited calculable anticancer impacts in human cancer cell lines in vitro by utilizing against A549 non-small-cell lung malignancy (IC_{50}, 0.3 μM). Furthermore U373 glioblastoma (IC_{50}, 0.5 μM) cell lines. In vitro anticancer exercises are because of mostly exhaustion from claiming [ATP]i, heading thus should major disruption about actin. Furthermore it doesn't actuate a build clinched alongside intracellular sensitive oxygen species. Clinched alongside vivo, bal6/7 expanded those survival period for mice bearing murine L1210 leukemia grafts of the same degree accounted to vincristine (Gnoula et al. 2008; Pettit et al. 1991).

16.4 *Balanites aegyptiaca* Kernels and Oil

The proximate composition of *B. aegyptiaca* kernels have been investigated by a number of researchers (Hussain et al. 1949; Nour et al. 1985; Mohamed et al. 2002; Chapagain et al. 2009; Manji et al. 2013; Elbadawi et al. 2017). The findings of these studies have revealed considerable variations in the proximate composition based on various genotypes of *Balanites aegyptiaca* and the different origins of the analyzed samples. The lowest and the highest values reported for crude fat were 39–49.9%, crude protein 26.1–50.37%, ash 3.3–6.3%, and moisture 3.1–5.7%. It was also reported that the kernels contains calcium, iron, phosphorus, magnesium, potassium and zinc (Hussain et al. 1949; Nour et al. 1985; Mohamed et al. 2002). In addition, Mohamed et al. (2002) have reported that the kernels contain free saccharide (7.4 g/100 g DM), sapogenin (1.5 g/100 g DM), phytic acid (1.2 g/100 g DM), true protein (29.6 g/100 g DM), non-protein nitrogen (0.5 g/100 g DM) and crude fiber (1.4 g/100 g DM). The crude oil of *B. aegyptiaca* is a liquid which has a light yellow color. The values of the specific gravity (0.9013–0.927) and the refractive index (1.46–1.5142) vary somewhat and depend on genotype and locality (Hussain et al. 1949; Chapagain et al. 2009; Manji et al. 2013; Elbadawi et al. 2017). The melting point and the viscosity of the oil were found to be 3 to −10 and 49 cp, respectively (Chapagain et al. 2009). The chemical properties of *Balanites aegyptiaca* oil have been reported by a number of articles (Nour et al. 1985; Manji et al. 2013; Elbadawi et al. 2017). Their values were found to vary with genotype and origin of the sample. The lowest and the highest values reported for peroxide value were 2.7–13.34 mEq/kg, acid value 0.3–0.5 mg KOH/g and the saponification value 168.6–200.31 mg KOH/g.

The fatty acid composition of *B. aegyptiaca* oil revealed that the oil contains almost 13 fatty acids (Elbadawi et al. 2017). However, all the studies agree that the major fatty acids in the oil are palmitic, stearic, oleic and linoleic (Al Ashaal et al. 2010; Chapagain et al. 2009; Hussain et al. 1949; Mohamed et al. 2002; Nour et al. 1985; Elbadawi et al. 2017). Though, the percentages of these acids vary considerably with the genotype and the location. The noticeable variations in the percentages of the fatty acids of the oil were clarified by Chapagain et al. (2009) who has analyzed the fatty acid composition of six *Balanites aegyptiaca* genotypes. The percentage ranges of the four major fatty acids were found to be: palmitic acid 12.7–16.0%, stearic 10.2–12.1%, oleic acid 23.5–43.7%, and linoleic acid 31.50–51.6%. Moreover, the presence of cis-vaccenic (0.80–0.88%), margaric (0.11–0.12%), and eicosenoic (gondoic) (0.1%) acids in *B. aegyptiaca* oil were reported by Elbadawi et al. (2017).

References

Al Ashaal HA, Farghaly AA, Abd El Aziz MM, Ali MA (2010) Phytochemical investigation and medicinal evaluation of fixed oil of *Balanites aegyptiaca* fruits. J Ethnopharmacol 127:495–501

Amadou I, Le GW (2017) Nutritional and sensory attributes of desert date (Balanites aegyptiaca) juice. J Food Meas Charact 11(4):1978–1986

Chapagain B, Wiesman Z (2005) Variation in diosgenin level in seed kernels among different provenances of Balanites aegyptiaca Del (Zygophyllaceae) and its correlation with oil content. Afr J Biotechnol 4(11):1209–1213

Chapagain BP, Yehoshua Y, Wiesman Z (2009) Desert date (Balanites aegyptiaca) as an arid lands sustainable bioresource for biodiesel. Bioresour Technol 100(3):1221–1226

Chothani DL, Vaghasiya HU (2011) A review on *Balanites aegyptiaca* Del (desert date): phytochemical constituents, traditional uses, and pharmacological activity. Pharmacogn Rev 5(9):55–62

Elbadawi SMA, Ahmad EEM, Mariod AA, Mathäus B (2017) Effects of thermal processing on physicochemical properties and oxidative stability of Balanities aegyptiaca kernels and extracted oil. Grasas Aceites 68(1):184

Farag MA, Porzel A, Wessjohann LA (2015) Unraveling the active hypoglycemic agent trigonelline in Balanites aegyptiaca date fruit using metabolite fingerprinting by NMR. J Pharm Biomed Anal 115:383–387

Gnoula C, Mégalizzi V, De Nève N, Sauvage S, Ribaucour F, Guissou P (2008) Balanitin-6 and -7: Diosgenyl saponins isolated from *Balanites aegyptiaca* Del. display significant antitumor activity *in vitro*and *in vivo*. Int J Oncol 32:5–15

Hardman R, Sofowora EA (1972) A reinvestigation of *Balanites aegyptiaca* as a source of steroidal sapogenins. Econ Bot 26(2):169–173

Hassanin KM, Mahmoud MO, Hassan HM, Abdel-Razik ARH, Aziz LN, Rateb ME (2018) *Balanites aegyptiaca* ameliorates insulin secretion and decreases pancreatic apoptosis in diabetic rats: role of SAPK/JNK pathway. Biomed Pharmacother 102:1084–1091

Hussain SA, Dollear FG, O'Connor RT (1949) Oil from the kernels of lalob fruit, Balanites aegyptiaca. J Am Oil Chem Soc 26(12):730–732

Manji AJ, Sarah EE, Modibbo UU (2013) Studies on the potentials of Balanites aegyptiaca seed oil as raw material for the production of liquid cleansing agents. Int J Phys Sci 8(33):1655–1660

Mohamed AM, Wolf W, Spiess WEL (2002) Physical, morphological and chemical characteristics, oil recovery and fatty acid composition of Balanites aegyptiaca Del. kernels. Plant Foods Hum Nutr 57(2):179–189

Molla E, Giday M, Erko B (2013) Laboratory assessment of the molluscicidal and cercariacidal activities of *Balanites aegyptiaca*. Asian Pac J Trop Biomed 3(8):657–662

Morsy AMA, Ahmad IA, Kamel AM (2010) Some biomedical applications of Balanites aegyptiaca grown naturally in radioactive area, Southeastern Desert, Egypt. J Hazard Mater 178(1–3):725–728

Nour AAA, Ahmed AHR, Abdel-Gayoum AGA (1985) A chemical study of Balanites aegyptiaca L. (Lalob) fruits grown in Sudan. J Sci Food Agric 36(12):1254–1258

Pettit GR, Doubek DL, Herald DL (1991) Isolation and structure of cytostatic steroidal saponins from the African medicinal plant *Balanites aegyptiaca*. J Nat Prod 54:1491–1502

Von Maydell HJ (1990) Trees and shrubs of the Sahel: their characteristics and uses. Verlag Josef Margraf, Scientific Books, Weikersheim

Wiesman, Z. and Chapagain, B.P. 2006. Larvicidal activity of saponin containing extracts and fractions of fruit mesocarp of Balanites aegyptiaca. Fitoterapia 77(6):420–424

Chapter 17
Detarium microcarpum: Chemical Composition, Bioactivities and Uses

Abdalbasit Adam Mariod, Haroon Elrasheid Tahir, and Mahunu Gustav Komla

Abbreviations

ABTS	2, 2′-azino-bis-3-ethylbenzthiazoline-6-sulphonic acid
AR	Aldose reductase
CF	Chloroform fraction
CP	Cyclophosphamide
CRD	Completely randomized design
DM	*Detarium microcarpum*
DPPH	2,2-diphenyl-1-picrylhydrazyl
DSC	Differential scanning calorimeter
DSM	*Detarium microcarpum* seed meals
FCE	Fiber cement effluent
FI	Feed intake
FRAP	Ferric reducing antioxidant power assay
GAE	Gallic acid equivalent
GLC-CIMS	Gas-liquid chromatography-Chemical ionization mass spectrometry
HF	Hexane fraction
HIV	Human immunodeficiency virus
HPLC	High performance liquid chromatography
MF	Methanol fraction
NMR	Nuclear magnetic resonance
PER	Protein efficiency ratio
QE	Quercetin equivalent

A. A. Mariod (✉)
Indigenous Knowledge and Heritage Center, Ghibaish College of Science & Technology, Ghibaish, Sudan

H. E. Tahir
School of Food and Biological Engineering, Jiangsu University, Zhenjiang, China

M. G. Komla
Department of Food Science & Technology, Faculty of Agriculture, University for Development Studies, Tamale, Ghana

A. A. Mariod (ed.), *Wild Fruits: Composition, Nutritional Value and Products*,
https://doi.org/10.1007/978-3-030-31885-7_17

RVU	Rapid Visco Units
T2D	Type 2 diabetes
WHC	Water holding capacity

17.1 *Detarium microcarpum* Plant Description

Detarium microcarpum family Fabaceae is an African tree, with height of 15–25 m (Fig. 17.1); with distinguished green 8–12 cm leaves (FAO 1995). As far as development rate, the shoots of the storage compartment can achieve a tallness of 1.5–2 m in 1–2 years and are considerably more overwhelming than seedlings

Fig. 17.1 *Detarium microcarpum* small tree and fruits

which on proximate develop to 0.6 m following 3 years and may achieve 1.5 m in 4 years. It blossoms amid the rainy season (July to September/November), yet the principle blooming period just keeps going up to 8 days. It proves to be fruitful from September – January/May and in November; the tree sheds its leaves and delivers new leaves in March (Kouyaté and van Damme 2006). A tree can give 675 fruits about 7 kg. The fruit is sweet and usually eaten crisp, while the mash is utilized in the readiness of cakes and couscous. The fruit is about 4 cm long and 2.5 cm wide. The fruits can be stored for 1–3 years in jute bags (http://www.tropical.theferns.info).

17.2 *Detarium microcarpum* Distribution

D. microcarpum known in English as sweet dattock or tallow tree, it happens normally in the drier areas of West and Central Africa (Benin, Cameroon, Central African Republic, Chad, Gambia, Ghana, Guinea, Guinea Bissau, Côte d'Ivoire, Mali, Niger, Nigeria, Senegal, Sudan and Togo). The tree is broadly appropriated in dry Savannah zones of Africa, and in Sudan it is found in Darfour, Blue Nile and Kordofan States, where it is privately known as Abu-laili (Mariod et al. 2009).

This species is primarily found on shallow, stony and lateritic soils, regularly on slopes, and also in locales with a yearly rainfall of 600–1000 mm. It is most normal in lush savannahs or savannahs, semi-cleared dry forest regions and fallows, developing in sandy or hard soils with high iron substance (Kouyaté and van Damme 2006).

17.3 *Detarium microcarpum* Chemical Composition

The pulp has high extents of starch (40–42.0%) and protein (29.1–30.9%), which are viewed as dietary fundamental. DM tests have a high measure of major minerals and being the transcendent components (Mariod et al. 2009). A compositional investigation of this fruit showed that it is a rich wellspring of polysaccharide gum, despite the fact that this gum has not been up to this point used in food products. The dehulled seed flour contained 3.5% moisture, 3.5% ash, 2.9% rough fiber, 15% unrefined fat, 37.1% rough protein and 39% CHO (Akpata and Miachi 2001).

In 1999 Onweluzo and his co-workers evaluated *Detarium microcarpum* seed polysaccharide as a stabilizer, thickening and gelling operator in handled natural product items. The utilitarian characteristics of the Dm seed polysaccharide were examined concerning pectin in these items. The concentration of Dm polysaccharide in various handled organic product items (mango drink, orange refreshment, orange squash, tomato sauce and pineapple stick) was institutionalized. The shelf life of these items was assessed amid 2 months of capacity at temperature (26 °C). Mango and orange refreshments containing 1.5 g/L Dm, orange squash and tomato sauce containing 10 g/L Dm and pineapple jam containing 1.4 g/L Dm and 5.6 g/L pectin were profoundly adequate and had great storage dependability within 2 months at ambient storage (Onweluzo et al. 1999b).

The amino acids (g/100 g) of the essential of *D. microcarpum* was lysine (2.14), cystine (1.07), valine (2.13), methionine (0.74), isoleucine (2.35), leucine (2.35), phenylalanine (2.85) and threomine (2.17), respectively. Other non-essential amino acids were in various quantities in the seeds (Anhwange et al. 2004).

Aquino et al. (1992) made an extensive spectral study on the chloroform extract of *Detarium microcarpum* bark, their study afforded two tetranorditerpenes, 1-naphthalene acetic-5-carboxy-l,2,3,4,4a,7,8,8a-octahydro-1,2,4a-trimethyl acid and l-naphthalene acetic-7oxo l,2,3,4,4a,7,8,8a-octahydro-1,2,4a,stetramethyl acid, together with a clerodane diterpene, 2-oxo-kolavenic acid.

Adedeji et al. (2012) investigated the thermophysical characteristics of *Detarium microcarpum* seed flour. Thermophysical characteristics such as apparent density, bulk density, porosity and water activity of dry ground powder of the sample were evaluated and reported as 1.79 g/cm^3, 0.44 g/cm^3, 75.51% and 0.549, individually. The total soluble solids of the sample ranged from 1.45 to 4.00 °Bx. The heat capacity of the aqueous solutions of the sample ranged from 1.79 to 18.83 J/g °C, while dry sample heat capacity was from 0.81 to 1.79 J/g °C. There was a significant effect ($P < 0.05$) of heating temperature and sample concentration on the variation observed in heat capacity (Adedeji et al. 2012).

The characteristics of oil recovered from seeds of *Detarium microcarpum* have been investigated. The seed oil content was found low. The oils contained beta-carotene, plant sterols, phospholipid and glycolipids. Toxicological studies show absence of gossypol and no detectable mycotoxins. The gum content of the seeds was high. Linoleic acid was the predominant fatty acid (Njoku et al. 1999).

Detarium microcarpum seed contains about 7.0% oil. The extracted oil contains high level of saturated fatly acids with low iodine value (55 I_2 g 100 g^{-1} of oil). So, it is not suitable as alkyl resins for paint formulation but may, however, be used for soap production judging by it is saponification values (123.3 mg KOH g^{-1} of oil). DM seed oil showed refractive index of 1.465 (30–40 °C) (Kyari 2008).

Uhegbu et al. (2009) studied the proximate analysis of the dehulled and undehulled seed of *Detrarium microcarpum*. They reported protein content of the seed as 7.2 and 8.2 for undehulled and dehulled DM, respectively. Crude fiber and ash were 3% and 5%, respectively. Samples showed fat content of 18.5% and 15.5%, while carbohydrate was 52.0% and 57.0%, for the undehulled and dehulled samples respectively. The major minerals Na, K, Ca and Mg contents were less than 1% each. Alkaloids, tannins, saponins and flavonoids were found as the main phytochemicals.

Aviara et al. (2015) investigated the physical characteristics of DM seeds, they reported surface area, 1000-seed weight, and particle density, as 433.19 cm^2, 3.737 kg, and 1316 kg/m, respectively, whereas bulk density decreased from 652.5 to 617.2 kg/m^{-3} in the same moisture range.

17.4 *Detarium microcarpum* Main Bioactive Components

Polyphenols comprise the principle bioactive phytochemicals in the fruit, which have been demonstrated successful in the counteractive action of certain chronic diseases, for example, coronary heart illnesses, and diabetes, in light of their free radical-scavenging exercises (Asami et al. 2003).

Detarium microcarpum bark contain some terpene compounds and coumarin (Burkil 2004). Abreu and Relva (2002) studied the carbohydrate of DM bark extract, they isolated L-quino-1,5-lactone, D-(−)-bornesitol, D-pinitol, *myo*-inositol, sucrose, D-glucose, and D-fructose benzoates. The deep analysis of the root extract of DM showed the existence of proteins, carbohydrates and terpenoids in large amount while saponins, resins, glycosides and flavonoids were present in moderate amount (Okolo et al. 2012).

Lamien-Meda et al. (2008) investigated the fruit pulp of DM for their phenolic and flavonoid substance and their cancer prevention agent exercises utilizing the DPPH, FRAP and ABTS techniques. Their acquired information demonstrated that the aggregate phenolic and aggregate flavonoid levels were essentially higher in the $(CH_3)2CO$ than in the methanol extracts. DM fruit showed high phenolic 4946.67 and 5978.33 mg GAE/100 g of fruit and flavonoid content was found 116.05 and 155.90 (mg QE/100 g of fruit) for methanol and acetone extract, respectively. *Detarium microcarpum* fruit additionally demonstrated the most noteworthy cancer prevention agent movement utilizing the three antioxidant assays. These investigators found a solid connection between's total phenolic and flavonoid levels and antioxidant activities.

Pharmacognostic studies were carried out on DM stem bark. The anatomical areas and powdered examples of the plant parts were examined for their microscopical profiles. These uncovered the nearness of phloem tissues, xylem tissues, parenchyma cells, cork cells, calcium oxalate crystals, starch grains and secretary channels in the powdered bark. A preliminary screening revealed the presence of saponins, tannins, flavonoids, alkaloids, cardiac glycosides and carbohydrates as bioactive compounds. A co-chromatographic examination reported the existence of catechol, gallic corrosive, phloroglucinol, pyragallol and quercetin (Sani et al. 2014).

17.5 Food Uses

The fruit pulp is appropriate for concentrated squeeze and jam making including more an incentive than the fruit alone. Roasting or soaking seeds has healthful advantages as it builds the substance and properties of specific supplements. Functional properties of DM were investigated and found that it had the highest water absorption capacity of 18.03 g/g. The sample exhibited a comparable oil absorption capacity. The stability of the foam produced by *Detarium microcarpum* studied was relatively poor. *D. Microcarpum* exhibited significantly higher

($P < 0.05$) emulsifying activity of 95%, the legume flour exhibited significantly high ($P < 0.05$) emulsion capacity (Onweluzo et al. 1994).

Mariod et al. (2009) carried out study to produce concentrated juice and jam from the fruit pulp of *Detarium microcarpum*, and compared them with a commercial sample as control. They reported that the sensory evaluation of the products was essentially favored by the specialists for their color, smell, taste, texture and overall acceptability to the control.

Onweluzo et al. (1999a, b) studied incorporation of *Detarium Microcarpum* (Dm) water soluble gums at 0.0–0.5% levels in wheat flour to assess their impact on the rheological properties of wheat flour batter and white bread quality. Batter containing gums had higher blending resistance index than the control. Set back viscosities diminished by 4.0 and 9.0 Rapid Visco Units with expanded dimensions of DM gum consolidation, individually. Significantly ($p \leq 0.05$) higher oven spring occurred in all the gum substituted white bread when compared to the control. The 0.5% gum substituted breads had a significantly ($p \leq 0.05$) higher sensory score for crumb grain, texture but lower crumb firmness than the control as determined instrumentally. Textural analysis after 5 days storage acknowledged that Dm gums enhanced moisture reservation characteristics of the bread and decreased crumb firming tendency (Onweluzo et al. 1999a, b).

Beef burgers containing different levels of DM gums were produced. The studied samples were investigated for selected physicochemical and sensory properties so as to assess the stabilization potentials of polysaccharide gums relative to the conventional tragacanth gum. Beef burgers containing the polysaccharide gums showed lower shrinkage, higher water holding capacity (WHC), and better stability under ambient conditions (27 ± 1 °C and RH 90.6%) than the unstabilized burgers (Onweluzo et al. 2004).

Uzomah and Odusanya (2011) used hydrocolloids from *Detarium microcarpum* (DM) seeds together with their starches to study their pasting and thermal characteristic properties. Proximate investigation exhibited that DM and is a rich source of protein (27.01%), and fat (14.45%). The differential scanning calorimeter study for the defatted DM demonstrated that the gelatinization temperatures went from 40 to 100 °C, demonstrating a wide gelation temperature range ($Tc - To$) for the DM sample and therefore a wide enthalpy change was watched running from 229.2 to 775.8 J/g. All the samples demonstrated a solitary endotherm amid gelation. This investigation exhibited the pasting and thermal properties of a starch-hydrocolloid blend of DM seed flour and the conceivable utilization of this mixture in frozen food applications is apparent (Uzomah and Odusanya 2011).

Umaru et al. (2007) investigated DM chemically for the presence of four antinutritional factors, they reported oxalate as 13.50%, phytate as 2.13%, saponin as 12.10%, and tannin as 3.54%. These authors concluded that, values obtained for this fruit is lower than the established toxic level. Thus DM fruit can be devoured with no limitation. In any case, eating of too much fruits with more elevated amounts of antinutrients ought to be maintained and averted.

Igbabul et al. (2012) studied the effect of fermentation on the proximate composition and functional properties of DM seed flour. Their results showed that fermentation for 72 h increased protein and crude fiber whereas there was no change in the moisture and oil content, while ash and carbohydrates were decreased. The study showed a significant decrease in the functional properties (emulsification, foaming, and water absorption capacities), while no change observed on the viscosity of fermented DM flour (Igbabul et al. 2012).

Obun et al. (2011) investigated the effects of feeding DM meals on growth performance, nutrient digestibility values and carcass characteristics of broiler chicks. They announced that, the organ loads of the heart, lungs, kidney and proventriculus were comparative over the treatments, however had logically higher qualities with an expanded in the dietary amounts of DSM. The consequences of this examination propose that crude DSM can be incorporated chickens diet up to 5% with no antagonistic impacts on the birds (Obun et al. 2011).

17.6 Medicinal Uses

Detarium microcarpum is considered as one of important African medicinal plant. The bark extract of DM is utilized to cure diuretic, anti-inflammatory and antiparasitic properties, while the fruits and leaves are utilized in the recovering of dysentery and syphilis (Abreu and Relva 2002).

Okolo et al. (2012) investigated the antidiabetic activity of root extract of *Detarium microcapum* in rat model. They prepared a methanol root extract by Soxhlet extraction and they separated it into fraction using chloroform, n-hexane and methanol to yield chloroform fraction (CF), n-hexane fraction (HF) and methanol fraction (MF). Their outcomes revealed that infusion of ME, MF, CF and HF turned around the impact of alloxan in rats by various degrees. Their outcomes legitimize the utilization of root concentrate of DM as conventional treatment for diabetes mellitus. These authors also indicated that *D. microcarpum* root extract and fraction reduced significantly the blood sugar level in alloxan diabetic rats. The methanol extract exhibited significant ($p < 0.05$) antihyperglycemic effect without causing hypoglycaemia. The methanol fraction produced similar effect like glibenclamide. Glibenclamide directly improves insulin action and is effective only in the presence of insulin (Bailey 1992). It is not evident whether the antidiabetic effect of *D. microcarpum* may be due to increased insulin secretion as occurred with glibenclamide. The antidiabetic effect of the extract may be partly due to the presence of flavonoids (Okolo et al. 2012).

Clerodane diterpenes molecules isolated from the pulp of *Detarium microcarpum* showed antifungal activity and inhibition of the enzyme acetylcholinesterase implicated in Alzheimer's disease (Cavin et al. 2006).

Irondi et al. (2015) evaluated the inhibitory impact of methanol extract of DM seed flour on type 2 diabetes enzymes; and their antioxidant properties. These authors found the concentrate of the flour repressed the enzymes in a portion subordinate way. This inhibitory impact might be credited to its joined activity of polyphenols and aggregate saponins, and this might be a conceivable component of activity offering help for their utilization in overseeing hyperglycemia and the confusions of type 2 (Irondi et al. 2015).

DM gum is obtained from the seeds of the DM tree. Generally, the seeds are utilized as sauces in dry shape (Akpata and Miachi 2001). It is utilized chiefly as soup thickener (Odoh et al. 2004). Other utilization of DM gum incorporate enhancing the moisture maintenance of bread, diminishing crumb-firming inclination (Onweluzo et al. 1999a, b), and as binder in tablet producing (Chukwu 1992). DM is additionally utilized as fumigant against insects (Audu 1989). Silver nano particles were synthesized from the leaf extract of DM and were characterized using spectroscopic techniques which indicated hydroxyl and carbonyl useful groups to be in charge for their synthesis. The synthesized nano-particles noticed to be crystalline and spherical. Their normal size, dictated by transmission electron microscopy (TEM) was 17.05 nm, with moderate to good antioxidant activity (Labulo et al. 2016).

Most of plant parts are broadly utilized for their diuretic and astringent properties. Modern research has verified the presence of medicinally active compounds. The bark has been found to contain different diterpenes and coumarin. An ethanol concentrate of the bark has shown antimicrobial activity against *Pseudomonas aeruginosa, Citrobacter freundii, Klebsiella pneumoniae, Staphylococcus aureus, Streptococcus pyogenes* and *Listeria monocytogenes*. The extract likewise indicated moderate antitumor action against breast cancer cells. The flavones present in a methanol extract of the plant indicated solid inhibitory impacts on HIV-1 or HIV-2 disease. A bark extract demonstrated huge molluscicidal action against *Lymnaea natalensis* (Burkil 2004). The bark, leaves and roots are set up as decoctions to treat an extensive variety of diseases including rheumatism; venereal diseases and urogenital infections; haemorrhoids; caries; problems of the digestive system such as biliousness, stomach-ache, intestinal worms, diarrhoea and dysentery. A decoction of the leaves is taken to treat fainting and convulsions. The leaves, combined with the leaves of *Sclerocarya birrea* and *Acacia macrostachya*, and disolved in milk, are used for snakebites. Eatening of fruit cure meningitis and malaria. A preparation of the fruits is taken to treat dizziness. Externally, the fruit mash is utilized for treating skin contaminations (http://www.prota.org).

A methanol extract of DM leaves containing four clerodane diterpenes showed strong feeding deterrent activity against termites (Lajide et al. 1995). Kela and Bowen (1995) investigated the molluscicidal impact of DM extract. The molluscicide treatment results in a particular restraint of ATPase action which gives off an impression of being a trademark highlight of prompted cell death.

17.7 Other Uses

D. microcarpum produces a fragrant resin. The seeds are used as frankincense. The boiled roots are sweetly scented and are utilized as a perfume and as a mosquito-repellent. The seeds are utilized to decore women necks. In southern Mali the leaves are used as roofing material. The leaves are used to make masks. The dark brown wood is used for carpentry, fence poles and joinery. The wood is utilized as firewood, as it burns slowly and gives off a pleasant aroma. The tree has great potential for poverty elevation where growing conditions are poor. The leaves are used as an organic fertilizer (http://www.prota.org).

Rouamba et al. (2017) evidenced the capability of an ethanol fruit extract from DM to preserve DNA integrity across oxidative genomic loss. The genoprotective activity was assessed ex vivo by comet assay, on liver cells of NMRI female mice using cyclophosphamide (CP) as genotoxic agent. They found that, the ethanol extract from *D. microcarpum* fruit pulp exhibited interesting antioxidant activity in 2,2-diphenyl-1-picrylhydrazyl (DPPH), deoxyribose degradation and lipid peroxidation assays. These authors noticed that, the extract did not present any genotoxic effect, but protected DNA against cyclophosphamide CP-induced damages with a portion subordinate way. The genoprotective effect observed was related to the antioxidant molecules of the fruit that scavenged the hydroxyl radical as well as the peroxyl and alkoxyl radicals issued from lipid peroxidation. They concluded that, the wild edible fruit from *D. microcarpum* could be beneficial on consumer's health by its antioxidant and genoprotective effects, particularly during chemotherapies exhibiting genotoxic effects like CP in cancer treatment (Rouamba et al. 2017).

The extracted gum from DM showed viscosity of 115.05 cps measured by Haake rotovisko viscometer. The extracted gum from *Detarium microcarpum* contains some important monsacchrides with better emulsion property.

Ani et al. (2012) used turbidimetry to study the coagulation performance of DM in a fiber cement effluent (FCE). They found that, the presence of proteins favors DM coag-flocculation performance in different FCE pH conditions (Ani et al. 2012). *Detarium microcarpum* seed powder exhibited an excellent potential coagulation–flocculation performance (Okolo et al. 2016).

References

Abreu P, Relva A (2002) Carbohydrates from *Detarium microcarpum* bark extract. Carbohydr Res 337:1663–1666

Adedeji AA, Alakali J, Adewale PO, Ngadi MO (2012) Thermophysical properties of *Detarium microcarpum* seed flour. LWT-Food Sci Technol 47:233–237

Akpata MI, Miachi OE (2001) Proximate composition and selected functional properties of *Detarium microcarpum*. Plant Foods Hum Nutr 56:297–302

Anhwange BA, Ajibola VO, Oniye SJ (2004) Amino acid composition of the seeds of *Moringa oleifera* (Lam), *Detarium microcarpum* (Guill & Sperr) and *Bauhinia monandra* (Linn.). ChemClass J:9–13

Ani JU, Nnaji NJN, Onukwuli OD, Okoye COB (2012) Nephelometric and functional parameters response of coagulation for the purification of industrial wastewater using Detarium microcarpum. J Hazard Mater 243:59–66

Aquino R, Ciavatta ML, Tommasi ND, Gacs-baitz E (1992) Tetranorditerpenes From *Detarium microcarpum*. Phytochemistry 31(5):1823–1825

Asami DK, Hong Y-J, Barret DM, Mitchell AE (2003) Comparison of the total phenolic and ascorbic acid content of freeze-dried and air-dried marionberry, strawberry, and corn grown using conventional, organic, and sustainable agricultural practices. J Agric Food Chem 51:1237–1241

Audu JA (1989) Medicinal herbs and their uses in Bauchi state. Niger Field 54:157–168

Aviara NA, Onaji ME, Lawal AA (2015) Moisture-dependent physical properties of detarium microcarpum seeds. J Biosyst Eng 40(3):212–223

Bailey CJ (1992) Biguanides and NIDDM. Diabetes Care 15:755–775

Burkil HM (2004) The useful plants of west tropical Africa. Royal Botanic Gardens, Richmond

Cavin A, Hay A, Marston A, Stoeckli-Evans H, Scopelliti R, Diallo D (2006) Bioactive diterpenes from the fruits of *Detarium microcarpum*. J Nat Prod 69(5):768–773

Chukwu A (1992) Studies on detarium microcarpium gum. I: Comparative evaluation as a binder in tablets containing tartrazine dye. STO Pharma Sci 2(6):463–468

FAO (1995) Fruit and vegetable processing. FAO agricultural services bulletin, vol 119. FAO, Rome

http://www.prota.org

http://www.tropical.theferns.info

Igbabul BD, Idikwu HO, Inyang CU (2012) Effect of fermentation on some functional properties of *Munca sloanei* and *Detarium microcarpum*. J Food Technol 10(3):83–86

Irondi EA, Oboh G, Akindahunsi AA (2015) Methanol extracts of *Brachystegia eurycoma* and *Detarium microcarpum* seeds flours inhibit some key enzymes linked to the pathology and complications of type 2 diabetes *in vitro*. Food Sci Hum Well 4:162–168

Kela SL, Bowen ID (1995) The histopathological effect of *Detarium microcarpum* extract, a naturally occurring plant molluscicide on the mid-gut and digestive gland of *Lymnaea stagnalis*. Cell Biol Int 19(3):175–181

Kouyaté AM, van Damme P (2006) Medicinal plants/plantes médicinales: *Detarium microcarpum* Guill. & Perr. Prota 11, no. 1. http://database.prota.org/PROTAhtml/Detarium%20microcarpum_En.htm. Accessed 24 Sept 2018

Kyari MZ (2008) Extraction and characterization of seed oils. Int Agrophysics 22:139–142

Labulo AH, Adesuji ET, Dedeke OA, Bodede OS, Oseghale CO, Moodley R, Nyamori VO, Dare EO, Adegoke OA (2016) A dual-purpose silver nano-particles biosynthesized using aqueous leaf extract of *Detarium microcarpum*: Anunder-utilized species. Talanta 160:735–744

Lajide L, Escoubas P, Mizutani J (1995) Termite antifeedant activity in *Detarium microcarpum*. Phytochemistry 40(4):1101–1104

Lamien-Meda A, Lamien EC, Compaore MM, Meda NR, Kiendrebeogo M, Zeba B (2008) Polyphenol content and antioxidant activity of fourteen wild edible fruits from Burkina Faso. Molecules 13(3):581–594

Mariod AA, Mirghani MES, Abdul AB, Abdelwahab SI (2009) *Detarium microcarpum* Guill and Perr fruit proximate chemical analysis and sensory characteristics of concentrated juice and jam. Afr J Biotechnol 8(17):4217–4221

Njoku OU, Obioma U, Frank EU (1999) Investigation on some nutritional and toxicological properties of *Afzelia africana* and *Detarium microcarpum* seed oil. Boll Chim Farm 138(4):165–168

Obun CO, Yahaya MS, Kibon A, Ukim C (2011) Effect of dietary inclusion of raw *Detarium microcarpum* seed meal on the performance and carcass and organ weights of broiler chicks. Am J Food Nutr 1(3):128–135

Odoh UE, Ezugwu CO, Nwodo JN, Ajali U (2004) Antidiabetic property of *Detarium microcarpium* Guill and Perr. gum (family: Caesalpinaceae). Afr J Pharm Res Dev 1(1):39–41

Okolo CE, Akah PA, Uzodinma SU (2012) Antidiabetic activity of the root extracts of *Detarium microcarpum* (Fabacaee) Guill and Perr. Phytopharmacology 3(1):12–18

Okolo BI, Nnaji PC, Onukwuli OD (2016) Nephelometric approach to study coagulation–flocculation of brewery effluent medium using *Detarium microcarpum* seed powder by response surface methodology. J Environ Chem Eng 4:992–1001

Onweluzo JC, Obanu ZA, Onuoha KC (1994) Functional properties of some lesser known tropical legumes. J Food Sci Technol 31:302–306

Onweluzo JC, Leelavathi K, Haridas Rao P (1999a) Effect of *Detarium microcarpum* (DM) and *Mucuna flagellipes* (MF) gums on the quality of white bread. Plant Foods Hum Nutr 54(2):173–182

Onweluzo JC, Vijayalakshmi MR, Vijayanand P, Eipeson WE (1999b) Detarium microcarpum polysaccharide as a stabilizer in processed fruit products. LWT-Food Sci Technol 32:521–526

Onweluzo JC, Obanu ZA, Okwandu MC (2004) Potentials of gum from *Detarium microcarpum* (DM) and *Mucuna flagellipes* (MF) seeds as raw beef burger stabilizers. Plant Foods Hum Nutr 59(4):137–141

Rouamba A, Ouedraogo M, Kiendrebeogo M (2017) Antioxidant capacity and genoprotective effect of ethanol fruit extract from Detarium microcarpum Guill. and Perr. (Caesalpiniaceae). Asian Pac J Trop Biomed 7(1):32–36

Sani A, Agunu A, Danmalam UH, Hajara I (2014) Pharmacognostic studies of the stem bark of *Detarium Microcarpum* – Guill. and Perr. (Fabaceae). Nat Prod Chem Res S1:004:1–8. https://doi.org/10.4172/2329-6836.S1-004

Uhegbu FO, Onwuchekwa CC, Iweala EEJ, Kanu I (2009) Effect of processing methods on nutritive and antinutritive properties of seeds of *Brachystegia eurycoma* and *Detarium microcarpum* from Nigeria. Pak J Nutr 8(4):316–320

Umaru HA, Adamu R, Dahiru D, Nadro MS (2007) Levels of antinutritional factors in some wild edible fruits of northern Nigeria. Afr J Biotechnol 6(16):1935–1938

Uzomah A, Odusanya OS (2011) *Mucuna sloanei, Detarium microcarpum* and *Brachystegia eurycoma* seeds: a preliminary study of their starch-hydrocolloids system. Afr J Food Sci 5(13):733–740

Chapter 18
Sclerocarya birrea Chemical Composition, Bioactivities and Uses

Abdalbasit Adam Mariod, Haroon Elrasheid Tahir, and Mahunu Gustav Komla

Abbreviations

CLA	Conjugated linoleic acid
MIC	Minimum inhibitory concentration
SCO	*Sclerocarya birrea* oil

18.1 Introduction

The historical backdrop of the Marula tree returns a large number of years. Archeological proof demonstrates the marula tree was a wellspring of nutrition as long as prior as 10,000 years B.C. *Sclerocarya birrea* is a savannah tree, belonging to the family Anacardiaceae. The common English name is Marula or cider tree, and is commonly known in the Sudan as Homeid. The tree requires sandy or alluvial soils and propagation is with seeds or cuttings (Voget 1995). The nonexclusive name Sclerocarya is gotten from the Ancient Greek words 'skleros' signifying 'hard' and 'karyon' signifying 'nut'. This alludes to the hard pit of the fruit. The explicit appellation 'birrea' originates from the regular name 'birr', for this sort of tree in Senegal. The marula has a place with indistinguishable family Anacardiaceae as the mango, cashew, pistachio and sumac, and is firmly identified with the variety Poupartia from Madagascar. Normal names incorporate jelly plum, cat thorn,

A. A. Mariod (✉)
Indigenous Knowledge and Heritage Center, Ghibaish College of Science & Technology, Ghibaish, Sudan

H. E. Tahir
School of Food and Biological Engineering, Jiangsu University, Zhenjiang, China

M. G. Komla
Department of Food Science & Technology, Faculty of Agriculture, University for Development Studies, Tamale, Ghana
e-mail: gmahunu@uds.edu.gh

A. A. Mariod (ed.), *Wild Fruits: Composition, Nutritional Value and Products*, https://doi.org/10.1007/978-3-030-31885-7_18

morula, cider tree, marula, maroola nut/plum, and in Afrikaans, Maroela (https://en.wikipedia.org/wiki/Sclerocarya_birrea). In Western Sudan, the dry seeds are crushed; the kernels obtained are an item of local trade for edible purposes (Mariod et al. 2005). Daldoum et al. (2012) reported that *S. birrea* populations established portion of the absolute tree density in South Kordofan province of Sudan, the tree had 1:1 male to female proportion. A female tree yielded 31,350 organic fruits on the normal (940,500 fruits ha), which relate to a yield of 14.7 tons of fruit products ha yearly.

18.1.1 Plant Description

Marula is a tree with large branches and a light, rounded crown. It usually grows to about 13 m tall, with exceptional specimens up to 20 m (Fig. 18.1). The plant has a pale yellow fruit, which is plum-like, 3–4 cm in diameter with a plain tough skin and a juicy mucilaginous flesh. The fruit is edible and contains a hard brown seed. The seed encloses a soft white kernel 1.0–1.5 cm in length and 0.5–0.75 cm in

Fig. 18.1 *Sclerocarya birrea* tree and fruits. (Source: https://commons.wikimedia.org)

width. The fruit is yellow with leathery rind like mango covering the seed. The taste is acid but of a pleasant flavour when fully ripe. The tree produces very large amounts of fruit each year (FAO 1988; Mariod et al. 2005). The tree has rough mottled stem-bark, with pale brown patches. The dark-green leaves are shown in small leaflets, each about 60 mm long. The plant has small flowers, with red sepals and yellow petals, the male and female are separated, usually but not always on separate trees. The fruit is flattened, 30 mm in diameter and pale yellow when ripe. The fruit cycle is from summer to mid-winter. The fruit weight is range from 15 to 25 g (Mariod and Abdelwahab 2012).

18.1.2 Plant Distribution

The tree is found in many African countries: South Africa, Malawi, Namibia, Niger, Botswana, Gambia, Zambia, Zimbabwe, Sudan, Swaziland, the Democratic Republic of Congo, Ethiopia, Kenya, Tanzania, Angola and Uganda. Marula is also grow in Madagascar and has been introduced into Mauritius and Reunion. It has been grown as an experimental crop in Israel and has been introduced into Australia and India. The tree is grown worldwide and has been valuable to millions of people, particularly in Central and East Africa, for thousands of years. It is also grown in parts of Asia, Europe and America and has been introduced into Australia (https://www.daff.gov.za). The tree grows in a wide variety of soils but prefers well-drained soil.

It exists at altitudes varying from sea level to 1800 m and an annual rainfall range of 200–1500 mm. Its major habitat limitation is probably its sensitivity to frost (Mariod and Abdelwahab 2012).

18.2 *Sclerocarya birrea* Chemical Composition

18.2.1 **Sclerocarya birrea** *Fruits Chemical Composition*

The fruit is rich in sugars and vitamin C, contains 2–3 edible kernels which are rich in oil and protein (FAO 1988). High levels (403.3 mg/100 g) of ascorbic acid were found in fruits. *Sclerocarya birrea* fruit contains 31.9 Mg, 36.2 Ca, 0.34 Zn, 0.11 Mn, 0.10 Cu, 0.13 Co, 1.12 Fe, and 18.0 Pb mg/100 g (Glew et al. 2004). The fruit contains 7.0, 0.5 and 0.1 of 100 g edible portion as carbohydrate, protein and fat, respectively (Wehmeyer 1966). Marula juice contains sesquiterpene hydrocarbons. The total protein content of Marula fruit was found as 36 mg/g dry weight, while the amino acid content of the fruit was reported as 3.8 ASP, 4.5 GLU, 1.9 SER, 1.9 GLY, 0.8 HIS, 2.1 ARG, 1.5 THR, 2.7 ALA, 3. PRO 3 (mg/g dry weight) (Glew et al. 1997).

18.2.2 Sclerocarya birrea *Seed Kernel Chemical Composition*

The fruit contains 2–3 edible kernels, which contain 53.0%, 28.0% and 8.0% of oil, protein and carbohydrates, respectively. In Western Sudan, the dry seeds are crushed, and the kernels obtained are an item of the local trade for edible purposes. Although the tree represents a potential new oil source, it has never been commercially exploited for the production of oil or protein (Mariod et al. 2005).

Although, the tree represents a potential new oil source, has never been commercially exploited for its oil or protein. Moreover, these products have been partially characterized. Reports in literature are variable as far as oil is concerned. Salama (1973) working on Sudanese Sclerocarya, reported an oil fatty acid composition markedly different from that of (Ogbobe 1992).

Who studied Nigerian *Sclerocarya* contained 50.7% stearic, 22.5% palmitic, 8.4% arachidonic, with 100.2 iodine value and 3.06 unsaponifiable matter. Mariod et al. (2005) investigated *Sclerocarya birrea* kernel oil samples originating from Western Sudan, they reported the physico-chemical properties of the oil and found that, the specific gravity was 0.9224, oil color was described as pale yellow, the refractive index was 1.4685, the saponification value was 193.54, while the unsaponifiable matter was 0.72, the acid value was 05.16, the iodine value was 64.20, and the phosphorus content was found as 0.110. They concluded that *Sclerocarya birrea* kernel oil could serve as a source of edible stable oil, and of fatty acids of technical grade, mainly conjugated linoleic acid (CLA) as an antioxidant, anticarcino-genic and antiatherogenic.

The seed kernel contained 24.8 mg/g (dry wt) of copper, 4210 mg/g of magnesium, and 62.4 mg/g of zinc. 36.4% of the kernel dry weight a protein content, however, the protein fraction contained relatively low proportions of leucine, phenylalanine, lysine, and threonine. Fatty acids accounted for 47 mg/g dry wt of the pit, two-thirds of which was due to oleic acid, followed by linoleic acid which was accounted as 24.5 mg/g dry wt (Glew et al. 2004).

Tocopherols are minor parts in charge of oil stability. Changes in the substance of these segments amid capacity for 2 years at 30 °C were considered in stored *S. birrea oil.* α, γ, and δ-tocopherols were distinguished in *S. birrea* oil where γ-tocopherol was the most overwhelming. After the storage time the level of total tocopherols fell by 25.0% (Mariod et al. 2008). *S. birrea* oil contained 13.7 mg/100 g of tocopherols where gamma-tocopherol spoke to 95% of the absolute tocopherols. The measure of sterols in the oil was 286.6 mg/100 g, with β-sitosterol as the prevailing one. 16% of the absolute sterols was Δ5-avenasterol, which referred to go about as a cell reinforcement and as an antipolymerization operator in frying oils (Mariod et al. 2004).

The protein contained great dimension of sulfur-containing amino acids (methionine and cystine), when contrasted with that of four diverse food materials. The in-vitro protein absorbability of *Sclerocarya birrea* was practically like that of soy bean protein and lupine. With a chemical score of 33.0%, in light of the fundamental amino acids design prerequisites for kids, the restricting amino acid in *Sclerocarya* protein was lysine (Mariod et al. 2005).

Phenolic mixes were gotten from *Sclerocarya birrea* seedcake utilizing two techniques for extraction. Progressive ultrasonic extraction brings about expanding measures of total phenolic mixes. Successive ultrasonic treatment presumably results in a simpler infiltration of the samples by the solvent. The antioxidative impacts of portions gotten by overnight extract were superior with the impacts of parts from ultra-sonic extract. The decreasing intensity of each fraction related well with the all out substance of phenolics present (Mariod et al. 2006b).

18.2.3 Sclerocarya birrea Leaves Chemical Composition

The leaves methanol extract of *Sclerocarya birrea* contains flavonol glycoside, quercetin 3-*O*-α-L-(5-galloyl)-arabinofuranoside, and eight known phenolic compounds; two epicatechin derivatives were also isolated from the same extract. Basic oils from leaves of *Sclerocarya birrea* were extricated by steam distillation. The oil yield from plant gathered was in the range 0.10–0.24%. The oils contained about 96% sesquiterpenes among which 7-epi-α-selinene, α-muurolene, valencene, β-selinene, β-caryophyllene, allo-aromadendrene-epoxide, and 14-hydrox-α-humulene (Kpoviessi et al. 2011). the leaves of *S. birrea* is considered a good source of tannins and flavonoids. The total phenolic substance of *S. birrea* leaf extract was 13.95 ± 0.05 mg GAE/g. Despite the fact that phenolic compounds are in all plant parts, their quantitative circulation changes between various organs in a plant. Plants with large amounts of phenolic compounds have been appeared to display high antioxidant capacities. A large portion of the antioxidant capability of medicinal plants is because of the redox properties of phenolic compounds, which empower them to go about as reducing agents, hydrogen donors and singlet oxygen scavengers (Moyo et al. 2011).

18.2.4 Sclerocarya birrea Bark Chemical Composition

The antidiarrhoeic activity of the bark of *Sclerocarya birrea* was related to an inhibition of intestinal transit rather than to inhibition of net secretion of fluid and electrolytes provoked by the laxative agents. A condensed tannin was isolated from the crude drug which produced inhibition in intestinal motility, and the monomer of which was identified as procyanidin (a condensed tannin), which found to be responsible for the antimotility activity of the *Sclerocarya birrea* bark (Galvez et al. 1991).

18.3 *Sclerocarya birrea* Bioactives

Numerous bioactive compounds have been distinguished from the stem bark and pulp of *S. birrea* including tannins, flavonoids, alkaloids, steroids, sesquiterpene hydrocarbons, ascorbic acid, oleic, myristic, stearic acids, coumarins, catechins and epicatechins. Almost certainly, a portion of these mixes might be independently or synergistically in charge of the counter *H. pylori* activity (Njume et al. 2011). Phytochemical analysis of the methanol extract of wild and cultivated leaves of *Sclerocarya birrea* led to the isolation of one new flavonol glycoside, quercetin 3-*O*-α-L-(5′-galloyl)-arabinofuranoside, and two epicatechin derivatives were also isolated from the same extract of the species (Kpoviessi et al. 2011).

18.4 *Sclerocarya birrea* Uses

It is a multipurpose tree giving numerous items like wood, natural products, fodder and rendering numerous administrations like luxury, agroforestry medications, shade. In South Africa, the reception taming and use of items got from this tree have continued to full commercialization status both locally and universally (Leakey 2005). The skin of the fruit can be bubbled to make a beverage or consumed to be utilized as a substitute for coffee. The wood is delicate and utilized for carving; the internal bark can be utilized to make rope. Archeological locales have indicated Marula fruit to be utilized as a nourishment source since antiquated occasions by Africa's clans. The bark can likewise be utilized to make a light darker color. Inside the substance are a couple of little delectable nuts which are wealthy in protein. Oil is utilized as a skin corrective. Their green leaves are eaten to alleviate acid reflux. The bark contains antihistamines and is likewise utilized for purifying by soaking in boiling water and breathing in the steam. A bit of bark is squashed into a pulp, blended with cold water and drink in the treatment of dysentery and diarrhoea. The bark likewise is utilized as a malaria prophylactic (http://www.krugerpark.co.za/africa_marula.html).

18.4.1 Sclerocarya birrea Food Uses

S. birrea has a wide assortment of utilization including the consumption of the new fruit, utilization of the fresh fruit to make juice and jam, utilization of the kernels as a sustenance source and as an enhancing and additive for different foods (Shackleton, Shackleton et al. 2002).

All parts of the plant are utilized; its plum-like organic fruits are generally prepared into a beer, yet additionally loan themselves extremely well to being utilized in the drink industry. Marula flavor is like grapefruit on account of the non-unstable substances that reason bitterness; its smell is reminiscent of pineapple. The organic fruit suggests a taste like a cross between an apple, a litchi, and a guava. Numerous items containing Marula are as of now available on the market in juices and alcohols, and a lot more are certain to pursue (Gruenwald 2009).

18.4.2 Sclerocarya birrea Medicinal Uses

Sclerocarya birrea is a plant utilized generally to treat numerous ailments in various African nations (Belemtougri et al. 2001) examined the impacts of *Sclerocarya birrea* leaf extricates on calcium signaling in rat cultured skeletal muscle cells. They found that the diverse concentrates (crude decoction, watery, ethanolic and chloroformic extracts have significant opposing impact on caffeine-incited calcium discharge from sarcoplasmic reticulum. Crude decoction is the most dynamic pursued by ethanolic, watery and chloroformic extracts in portion subordinate way and can somewhat legitimize the utilization of the plant in traditional drug.

The plant was found to contain saponins, tannins, flavonoids, alkaloids, steroids, heart glycosides, terpenoids and reducing sugars. Antimicrobial action of methanol, acetone and ethyl acetic acid derivation concentrates of the stem bark and roots of *Sclerocarya birrea* against clinical isolates of *Salmonella spp*, *Staphylococcus aureus*, *Escherichia coli*, *Aspergillus niger* and *Candida albicans* was researched. Results got uncovered that every one of the concentrates displayed inhibitory impact on the development of the tested microorganisms (Mohammed et al. 2015).

In Mali the tree is used in traditional Malian medicine for the treatment of several diseases. The leaves and the pulp of fruit are used for hypertension, and the leaves are used against diabetes, dysentery, snake and scorpion bites, malaria, and inflammations Further on, the plant is also utilized as a tonic, and the fruits are often fermented to give a refreshing drink (Braca et al. 2003). Chemical and pharmacological studies on the bark have reported antidiarrheal activity of the decoction and the isolation of phenolic compounds such as procyanidins and (−) epicatechin-3-galloylester (Mariod 2005).

The minimum inhibitory concentrations (MIC) estimations of bark and leaves acetone concentrate of Sclerocarya birrea were resolved utilizing a microplate sequential dilution system with Staphylococcus aureus, Pseudomonas aeruginosa, Escherichia coli and Enterococcus faecalis as test microbes. All concentrates were dynamic with MIC values from 0.15 to 3 mg/ml and internal bark extracts would in general be the most powerful pursued by external bark and leaf extracts. There were two noteworthy bioactive compounds of leaf extract. Utilizing bark might be inconvenient to the plant, however leaf material can likewise be utilized for antibacterial application (Eloff 2001).

The anti-inflammatory impacts of *Sclerocarya birrea* aqueous and methanolic stem-bark concentrates of the plant were inspected on rat paw oedema. Both the aqueous and methanolic stem-bark concentrates of S. birrea continuously and time-conditionally decreased rat paw oedema. Albeit both the aqueous and methanolic concentrates of S. birrea stem-bark are less strong than anti-inflammatory agent

(Ojewole 2003a). The aftereffects of hypoglycemic impact of Sclerocarya birrea stem-bark aqueous extract in ordinary and in streptozotocin (STZ)-treated, diabetic rats had hypoglycemic activity, and in this manner loan belief to the recommended folkloric utilization of the plant in the administration or potentially control of type-2 diabetes mellitus in some African people group (Ojewole 2003a, b).

Bark and leaves of Marula were extracted with acetone and minimum inhibitory concentrations (MIC) values were resolved utilizing a microplate sequential dilution procedure with Staphylococcus aureus, Pseudomonas aeruginosa, Escherichia coli and Enterococcus faecalis as test organisms. In light of least inhibitory focus esteems, internal bark extracts would in general be the most intense pursued by external bark and leaf extracts, yet the distinctions were not statistically significant. There were two noteworthy bioactive compounds obvious after bioautography of leaf extracts one unequivocally polar and the other exceptionally non-polar. The bioactive compounds could be isolated from 92% of the non-dynamic dry issue by solvent – dissolvable fractionation into the carbon tetrachloride, chloroform and n-butanol portions; these portions, be that as it may, in any case contained a wide range of compounds. Utilizing bark might be hindering to the plant, yet leaf material can likewise be utilized for antibacterial application (Eloff 2001).

18.4.3 Sclerocarya birrea Industrial Uses

Mariod et al. (2006b, c) contemplated the conduct of crude *Sclerocarya birrea* bit oil amid deep-frying of potatoes, as to compound, chemical, physical, and sensory parameters. The oil did not surpass the limits for the substance of polar mixes and oligomer triglycerides amid the frying test. Just the measure of free fatty acids was surpassed; this was on the grounds that the measure of free fatty acids toward the start of the examination was higher than for refined oils. The outcomes demonstrated that the oil was appropriate for deep-frying of potatoes, however amazing contrasts in the time amid which the oil delivered palatable items were found.

The stable oil of *Sclerocarya birrea* was transesterified utilizing methanol or ethanol within the sight of sulfuric acid; the produced biodiesel attributes were considered as per the DIN EN 14214 specifications for biodiesel. The greater part of the biodiesel attributes met the DIN details. The kinematic consistency estimations of the example was higher than those of biodiesel standard points of confinement. Concerning the oxidative stability, *Sclerocarya birrea* oil has an induction period higher than as far as possible. It was conceivable to set up the methyl and ethyl esters catalyzed by H2SO4 from the oil (Mariod et al. 2006a).

Marula seed husk biomass might be utilized as an elective minimal effort, eco-accommodating and powerful biosorbent to expel Pb(II) and Cu(II) from aqueous solution and waste waters. The biosorption of Pb(II) and Cu(II) onto marula seed husk biosorbent relies upon elements, for example, pH, biosorbent measurements, contact time, convergence of metal particles and temperature. The balance sorption limits of the seed husk were 20 mg g^{-1} for Pb(II) and 10.20 mg g^{-1} for Cu(II).

The biosorption information adjusts best with Langmuir model for both Pb(II) and Cu(II) with correlation factors of 0.999 and 0.998, separately. Kinetic information were appropriately fitted with the pseudo-second order Kinetic model. The thermodynamic parameters (ΔG0, ΔS0 and ΔH0) demonstrated that the biosorption was feasible, unconstrained and endothermic (Moyo et al. 2015).

Stripped and non-stripped *Sclerocarya birrea* oil (SCO), generally utilized for healthful applications in Sudan, was explored for its fatty acids and tocopherols, and their oxidative stability. Three stripping techniques were utilized, phenolic mixes extraction, silicic acid column, and aluminum oxide column. The stripping strategies did not influence the fatty acid composition. Non-stripped SCO contained oleic, palmitic, stearic and linoleic acids, which were not significantly ($P < 0.05$) not quite the same as stripped SCO. The stripping strategies' impact on the tocopherol composition of the investigated oil, the sum of tocopherol in non-stripped oil diminished by extraction of phenolic mixes, imply that part of the tocopherols was extracted with the phenolic mixes. No hints of tocopherols were found in oil stripped utilizing silicic and aluminum column and the tocopherols were wiped out amid the stripping forms. The stability of SCO was 43, it diminished by 22.0%, after extraction of phenolic mixes. This stability diminished by 96.9%, when stripped utilizing the aluminum column and diminished by 92.6% when stripped by the silicic acid column. It is conceivable to accept that the tocopherols and phenolic mixes assume a progressively dynamic role in the oxidative stability of the oil than the fatty acid and phytosterols composition (Mariod et al. 2011).

References

Belemtougri RG, Constantin B, Cognard C, Raymond G, Sawadogo L (2001) Effects of Sclerocarya birrea (A. Rich) hochst (anacardiaceae) leaf extracts on calcium signalling in cultured rat skeletal muscle cells. J Ethnopharmacol 76(3):247–252

Braca A, Politi M, Sanogo R, Sanou H, Morelli I, Pizza C, De Tommasi N (2003) Chemical composition and antioxidant activity of phenolic compounds from wild and cultivated Sclerocarya birrea (Anacardiaceae) leaves. J Agric Food Chem 51(23):6689–6695

Daldoum D, Massaud M, Adam YO (2012) Distribution, fruit production and natural regeneration of Sclerocarya birrea (A. Rich.) Hochst. subsp. birrea in the Nuba mountains, Sudan. World J Agric Sci 8(3):234–239

Eloff JN (2001) Antibacterial activity of Marula (Sclerocarya birrea (A. Rich.) Hochst. subsp. caffra (Sond.) Kokwaro) (Anacardiaceae) bark and leaves. J Ethnopharmacol 76(3):305–308

FAO (1988) Traditional food plants. Food and nutrition, paper no. 42. U.N. Rome, Italy

Galvez J, Zarzuelo A, Crespo ME, Utrilla MP, Jiménez J, Spiessens C, de Witte P (1991) Antidiarrhoeic activity of Sclerocarya birrea bark extract and its active tannin constituent in rats. Phototherapy Research 5(6):276–278 https://doi.org/10.1002/ptr.2650050611

Glew RH, VanderJagt DJ, Lockett C, Grivetti LE, Smith GC, Pastuszyn A, Millson M (1997) Amino acid, fatty acid, and mineral composition of 24 indigenous plants of Burkina Faso. J Food Compos Anal 10(3):205–217

Glew RS, VanderJagt DJ, Huang YS, Chuang LT, Bosse R, Glew RH (2004) Nutritional analysis of the edible pit of Sclerocarya birrea in the Republic of Niger (daniya, Hausa). J Food Compos Anal 17(1):99–111

Gruenwald J (2009) Novel botanical ingredients for beverages. Clin Dermatol 27(2):210–216
http://www.krugerpark.co.za/africa_marula.html
https://en.wikipedia.org/wiki/Sclerocarya_birrea
https://www.daff.gov.za
Kpoviessi DSS, Gbaguidi A, Kossouoh C, Agbani P, Yayi-Ladekan E, Sinsin B, Moudachirou M, Accrombessi GC (2011) Chemical composition and seasonal variation of essential oil of Sclerocarya birrea (A. Rich.) Hochst subsp birrea leaves from Benin. J Med Plants Res 5(18):4640–4646
Leakey R (2005) Domestication potential of Marula (Sclerocarya birrea subsp. caffra) in South Africa and Namibia: 3. Multiple trait selection. Agrofor Syst 64(1):51–59
Mariod AA (2005) Investigations on the oxidative stability of some unconventional Sudanese oils, traditionally used in human nutrition, Verlag nicht ermittelbar
Mariod AA, Abdelwahab SI (2012) Sclerocarya birrea (Marula), an African tree of nutritional and medicinal uses: a review. Food Rev Intl 28(4):375–388
Mariod A, Matthäus B, Eichner K (2004) Fatty acid, tocopherol and sterol composition as well as oxidative stability of three unusual Sudanese oils. J Food Lipids 11(3):179–189
Mariod A, Ali AO, Elhussein S, Hussien I (2005) Quality of proteins and products based on sclerocarya birrea (marula) seed. J Sci Technol 6(1):0–0
Mariod A, Klupsch S, Hussein I, Ondruschka B (2006a) Synthesis of alkyl esters from three unconventional Sudanese oils for their use as biodiesel. Energy Fuel 20(5):2249–2252
Mariod AA, Matthäus B, Eichner K, Hussein IH (2006b) Antioxidant activity of extracts from Sclerocarya birrea kernel oil cake. Grasas Aceites 57(4):361–366
Mariod A, Matthäus B, Eichner K, Hussein IH (2006c) Frying quality and oxidative stability of two unconventional oils. J Am Oil Chem Soc 83(6):529–538
Mariod A, Matthäus B, Eichner K, Hussein IH (2008) Long-term storage of three unconventional oils. Grasas Aceites 59(1):16–22
Mariod A, Matthäus B, Hussein IH (2011) Effect of stripping methods on the oxidative stability of three unconventional oils. J Am Oil Chem Soc 88(5):603–609
Mohammed M, Adamu H, Bello K (2015) Phytochemical analysis of bioactive extracts of Sclerocarya birrea. In: 2nd International conference on chemical, biological, and environmental sciences (ICCBES'15), May
Moyo M, Finnie JF, Van Staden J (2011) Antimicrobial and cyclooxygenase enzyme inhibitory activities of Sclerocarya birrea and Harpephyllum caffrum (Anacardiaceae) plant extracts. S Afr J Bot 77(3):592–597
Moyo M, Guyo U, Mawenyiyo G, Zinyama NP, Nyamunda BC (2015) Marula seed husk (Sclerocarya birrea) biomass as a low cost biosorbent for removal of Pb(II) and Cu(II) from aqueous solution. J Ind Eng Chem 27:126–132
Njume C, Afolayan AJ, Ndip RN (2011) Preliminary phytochemical screening and in vitro anti-helicobacter pylori activity of acetone and aqueous extracts of the stem bark of Sclerocarya birrea (Anacardiaceae). Arch Med Res 42(3):252–257
Ogbobe O (1992) Physico-chemical composition and characterisation of the seed and seed oil of Sclerocarya birrea. Plant Foods Hum Nutr 42(3):201–206
Ojewole JA (2003a) Evaluation of the anti-inflammatory properties of Sclerocarya birrea (A. Rich.) Hochst.(family: Anacardiaceae) stem-bark extracts in rats. J Ethnopharmacol 85(2–3):217–220
Ojewole JAO (2003b) Hypoglycemic effect of Sclerocarya birrea {(A. Rich.) Hochst.} [Anacardiaceae] stem-bark aqueous extract in rats. Phytomedicine 10(8):675–681
Salama R (1973) The seed oil of Sclerocarya birrea. Sudan J Food Sci Technol 5:4–6
Shackleton SE, Shackleton CM, Cunningham T, Lombard C, Sullivan CA, Netshiluvhi TR (2002) Knowledge on Sclerocarya Birrea subsp. caffra with emphasis on its importance as a non-timber forest product in south and southern Mrica: a summary: part 1: taxonomy, ecology and role in rural livelihoods. South Afr For J 194(1):27–41
Voget K (1995) Common trees and shrubs of dryland Sudan. SOS Sahel International, London
Wehmeyer A (1966) Nutrient composition of some edible wild fruits found in the Transvaal. S Afr Med J 40(45):1102–1104

Chapter 19
Tamarindus indica: Phytochemical Constituents, Bioactive Compounds and Traditional and Medicinal Uses

Ghada Hashim Abdelrahman and Abdalbasit Adam Mariod

19.1 Introduction

19.1.1 Tamarind Plant Origin

The tamarind tree (*Tamarindus indica* L.) is occupant to the dry savannas of tropical Africa (Fig. 19.1). The tree was acquainted with Asia by Arab dealers with its pleasant, acidic tasting natural product. Later on the tamarind achieved the new world from West Africa. In Caribbean and Latin America the plant is abundantly valued, as in Africa and Asia, for the succulent, sweet-acrid mash that fills its units. In any case, India remains the main nation misused the tamarind broadly, more than 250,000 tons are reaped there every year, 3000 tons of which are sent out to Europe and North America for use in meat sauces and refreshments (Gunasena and Hughes 2000). The tamarind tree is chiefly developed as a decorative and for shade however has been utilized both as a starvation sustenance and for restorative purposes.

The tamarind tree is a wonderful, fine finished tree and it makes a fantastic shade tree in extensive scenes. Usually planted out in the open parks and as a road tree in tropical. For all intents and purposes all aspects of the tree (wood, root, leaves, bark and organic products) has some an incentive in trade and especially in the subsistence of rural individuals. The name tamarind gets from the Arabic name tamarhind which implies date of India (Morton 1987).

G. H. Abdelrahman (✉)
National Food Research Center, Khartoum, Sudan

A. A. Mariod
Indigenous Knowledge and Heritage Center, Ghibaish College of Science & Technology, Ghibaish, Sudan

A. A. Mariod (ed.), *Wild Fruits: Composition, Nutritional Value and Products*,
https://doi.org/10.1007/978-3-030-31885-7_19

Fig. 19.1 Tamarind tree. (Source: https://commons.wikimedia.org)

19.1.2 Tamarind Plant Description

Tamarindus indica is an evergreen tree, up to 24 m in stature. It is an extensive evergreen tree with an incredibly lovely spreading crown. Leaves alternate, compound, with 10–18 sets of inverse leaflets; petiole and rachis finely haired, midrib and net veining pretty much clear on both surface. Blossoms appealing light yellow or pinkish, in little, lax spikes about 2.5 cm in width. Flour buds totally encased by two bracteoles, which fall early. Natural fruit is a pod, indehiscent, subcylindrical, 10–18 × 4 cm, straight or bended, smooth, corroded darker; the shell of the pod is brittle and the seeds are inserted in a sticky palatable pulp. Seeds are 3–10, around 1.6 cm long, sporadically molded, and testa hard, gleaming and smooth (Bhadoriya et al. 2011; El-Siddig et al. 2006).

19.1.3 Distribution of Tamarind Plant

Tamarind plant exists broadly all through tropical Africa, where it is as often as possible planted as a shade tree. The tamarind tree is presently generally spread all through semi-arid South and Southeast Asia. It is as of now developed in home nurseries, farmlands, on roadsides, on basic grounds and on a constrained plantation scale in India and Thailand. While tamarind tree is acquainted with the American mainland from Asia over the pacific with the Spanish. It is now produced commercially in different states of Mexico. Also tamarind grows throughout the Caribbean islands such as Jamaica and Cuba. Likewise there are business plantation in Brazil and some Latin American nations (El-Siddig et al. 2006).

19.1.4 Adaptation

The tamarind trees are well adapted to semiarid tropical conditions, although they do well in many humid tropical areas of the world with seasonally high rainfall. Young trees are very susceptible to frost, but mature trees will withstand brief periods of 28 °F without serious injury. Dry weather is important during period of fruits development (Gunasena and Hughes 2000).

19.1.5 Pests Infestation

Tamarind trees are at risk to assault by an expansive number of creepy crawly insect. The significant pest which assault tamarind incorporate shot gap borers, toy beetles, leaf nourishing caterpillars, bag worms, mealy bugs and scale insects. A portion of these bugs assault the flower buds and the growing fruits and seeds, while others harm the fruits amid storage (Coronel 1991).

19.2 Chemical Compositions of Tamarind Fruits

The most extraordinary attributes of tamarind fruit is its acidity which is because of for the most part tartaric acid (2,3-dihydroxybutanedioic acid, C4H6O6) extending from 12.3% to 23.8% which is unprecedented in other plant tissues (Ulrich 1970). Albeit tartaric acid happens in other sour fruits, for example, grapes, grapefruit, and raspberries, yet it is absent in such high sums as in tamarind.

Sweet tamarind cultivars are grown in Thailand, where tartaric acid content varied from 2.0% to 3.2% and the sugar content was as high as 39.1–47.7% (Feungchan et al. 1996). Total sugars and tartaric acid content were reported to be 41.2% and 8–18% respectively in sour tamarind cultivars (Duke 1981; Ishola et al. 1990; Shankaracharya 1998).The brown edible pulp is high in sugar (about 20–30%), where the reducing sugars are the main sugars (glucose and fructose) and the sucrose level is low in comparison to the level of reducing sugars, it is pleasant acid taste due to the presence of tartaric acid (Siliha and Askar 2000). The reducing sugars content are in the range of 25–45% (Ishola et al. 1990; Duke 1981). The moisture content of the tamarind fruit is in the range of 17–35% (Shankaracharya 1998;

Feungchan et al. 1996; Coronel 1991). Protein content is in the range of 2–8.79% (Shankaracharya 1998; Feungchan et al. 1996; Coronel 1991; Ishola et al. 1990).

Fiber content is in the range of 2.2–18.3% (Shankaracharya 1998, Feungchan et al. 1996, Coronel 1991; Ishola et al. 1990). Small amount of ascorbic acid as reported that varies from 2–20 mg/100 and no any detectable amounts of phytic acid (Ishola et al. 1990). Tamarind pulp is rich in potassium, phosphorus, calcium and less iron (Saka and Msonthi 1994; Ishola et al. 1990; Marangoni et al. 1987; Bhattacharyya et al. 1983). The total ash content is in the range of 2–3.9% (Shankaracharya 1998, Feungchan et al. 1996, Coronel 1991; Ishola et al. 1990).

19.3 Industrial Uses of Tamarind

19.3.1 Fruit Pulp

Tamarind is grown basically for its pulp (Fig. 19.2) which is utilized to set up a refreshment and in flavor sweets, curries and sauces, and is made into jelly and syrups. Jugo or Fresco de Tamarindo is a most loved drink in numerous Latin Americans nations and is produced industrially in some of them (James 1981).

Tamarind extract is used as one of the ingredients in preparing a traditional Sudanese fermented drink (Hulu Mur) to provide acidity and enhance the perception of flavors (Agab 1985). Tamarind powder is one of the convenience foods developed from tamarind by modern food processing methods. It is conveniently used as a souring agent in the culinary preparation of rasam, sambar, puliogare mix, sauces and chutney (Manjunath et al. 1991; Ahmed 2009).

Fig. 19.2 Tamarind fruits

Tamarind extract from the fruit is utilized as alternative for phosphoric acid, citrus extract and different acids that are added to soft drinks. Drinks containing the concentrate have an enhanced shelf life, of realistic usability, because of low pH, likewise flavor profile comparable to or superior to refreshments improved with aspartame (Linda 1995; Zoblocki and Pecore 1996; Mustafa 2007). Rich aroma tamarind extract has been used widely as beverage drink in Egypt and India (Askar et al. 1987).

Tamarind juice concentrate (TJC) is a convenience product easy to disperse and reconstitute well in hot water. The concentrate is hygienic and can be stored well for long periods. It was made by extracting the tamarind pulp in boiling water, filtered and concentrated under vacuum, to a jam-like consistency and then filled in bottles or cans. Also pickles and pastes can be made from tamarind fruit (Shankaracharya 1998).

The extraction and processing techniques for the preparation of canned tamarind pulp and the manufacture of tamarind soft drinks have been reported. Wine and vinegar production from tamarind fruit was also reported by. A process for making drinks, syrup, juice, liquor and solid extracts based on tamarind were developed. Good quality ready-to-serve (RTS) beverage, syrup, and concentrate were prepared from tamarind fruits. All these products have been satisfactory preserved and stored at 33 °C for over 180 days without affecting their quality (Kotecha and Kadam 2003).

19.3.2 Tamarind Seeds

The major modern use for tamarind seeds (Fig. 19.3) is the maker of tamarind bit powder which is an essential material for the jute and textile. The seeds additionally are picking up significance as a rich wellspring of protein and amino acids

Fig. 19.3 Tamarind seeds

(Anon 1984). Likewise seed bits have been utilized as foods in scarcity periods either alone or blended with grain flours. Also seed kernels have been used as foods in times of scarcity either alone or mixed with cereal flours. Polysaccharides obtained by extraction from the kernel of tamarind seeds form mucilaginous dispersions with water and possess the characteristics property of forming gels with sugar concentrates, as do fruit pectin's. However, the tamarind polysaccharides fruit pectin are capable of forming gels over a wide pH range, including neutral and basic pH conditions (Bhattacharyya et al. 1983). Several products made from tamarind fruits pulp and seeds e.g. tamarind juice concentrate, pulp powder, kernel powder, tartaric acid, and pectin.

19.4 Phytochemical and Bioactive Constituents of Tamarind

Many Foods are rich in phytochemical compounds which play an important role in the prevention of diseases. The antibacterial activities from plant origin have been linked to the presence of phytochemicals, these compounds contain secondary metabolites such as alkaloids, saponin, tannin, terpenoids and phenolic compounds which have been associated with antimicrobial, antioxidants and anti-inflammatory properties (Daniel 2006). Researchers have investigated for possible alternative of synthetic antibiotics by natural ones, many plants could be usefully for this purpose for example black pepper, curly leaf, coriander, onion, ginger, bay leaf and tumaric (Bag and Chattopadhyay 2015).

Tamarind plant contains many bioactive compounds in the leaves, seeds, bark, pulp and flowers with the useful effects to the human health and the opportunity of application in the pharmaceutical industry. Several studies have been carried out about tamarind compounds in its different parts, which have a significant role as antimicrobial, antidiabetic, anti-inflammatory, control of satiety, prevention of obesity and other chronic diseases (Menezes et al. 2016).

Bark of tamarind is a good source of tannin compounds e.g. proanthocyanidin. So the bark has the properties of antiallgeric, antimicrobial, antibiotic, antityrosiase, antioxidant and analgesic (Menezes et al. 2016).

Tamarind seed could be considering as a good source of protein and starch, sulfur amino acids and phenolic antioxidants. It has antiflammmatory activity, also it has impact on the control of satiety. It has a great potential for treatment or prevention of obesity and gastro protective effects (Menezes et al. 2016). Extracts from seeds were found to be enriched in xyloglycans which is an active agent in a cosmetic and/or pharmaceutical product for tropical usage for the skin and/or other exposed parts of the body (Pauly 1999).

Leaves of tamarind plant being consider as source of protein, lipid, fiber and vitamins like thiamine, riboflavin, niacin, and ascorbic acid and B- carotene. Also composed 13 essential oils like benzoate, benzyle, pentadecaol and hexadecaol. Leaves have an antiemetic activity and also have the property of protection the liver (Leng et al. 2017; Menezes et al. 2016). Recent research findings showed that, the tamarind leaves warm water extracts have higher zone inhibition against *Pseudomonas putida.*

Stem bark of tamarind plant contains flavonoids, cardiac glycosides, alkaloids, saponins and tannins. Tea made from stem bark used for sore throat. The stem bark has many bioactive activities such as spasmogenic, analgestic, antimicrobial and hypoglycemic.

Tamarind fruit pulp contains vitamins, minerals, many acids, amino acids, invert sugar (25–30%), pectin, protein, fat, some pyranzines, and some thiazoles. Also it has alkaloids, flavonoids, saponins and tannins. Tamarind pulp has many bioactivities such as hypolipidemic activity, antioxidant, antifluorose, analgesic, hepatoregenerativa and antispasmodic (Menezes et al. 2016).

Early research studied the bacteriostatic effect of tamarind. Ethanol extraction from tamarind fruit was the most effective inhibitor against all tested organisms in soft drinks (Alian et al. 1983). A recent research was carried out on antimicrobial activity of ethanol and water extracts of tamarind fruit pulp against some pathogenic bacteria. The research findings revealed that, the ethanol extracts produce strong antibacterial activity against *E.coli, Klebsiella pneum* A and *Pseudomona aeruginosa* (Daniyany and Muhammad 2008). While tamarind pulp hot water extracts have good antimicrobial activity against some bacteria such as *Aeromonas hydrophila* and *Hafnia alvei* (Adeniyi et al. 2017). The antibacterial activity of tamarind fruit pulp against some bacteria propose that, there is a scientific basis for their utilization in traditional medicines for the treatment of some bacterial infections.

Tamarind wastes such as tamarind fruits shells and tamarind seed coat extracts also have antioxidant like (−)-epicatechine (Ganesapilla et al. 2017).

19.5 Traditional Medicinal Uses of Tamarind Fruits Pulp

Numerous therapeutic employments of the tamarind were accounted for, where the pulp has been utilized by numerous individuals in Africa, Asia and America (Kuru 2014). Tamarind uses are all around perceived as refrigerants in fevers and as diuretic and carminatives. Alone or in a blend with lime juice, nectar, milk, dates flavors or camphor, the pulp is viewed as powerful as a digestive even for elephants and as a solution for biliousness and bile issue.

For quite a long time in numerous nations, the pulp of the fruit has been used as an antiscorbutic, in Brazil as diaphoretic, emollient, urgative and for hemorrhoids. In Eritrea, the pulp is sold for looseness of the bowels and jungle fever; in Indonesia for hair nourishment; in Madagascar for worms and stomach issue; in Mauritius as a liniment for stiffness; in Tanganyika for snakebite; in Sri Lanka for jaundice, eye illnesses and ulcer and in Cambodia for conjunctivitis (Duke 1981). The pulp is said to enhance loss of hunger, it is accessible economically in tablet shape in Thailand for the decrease of overabundance weight. In Southeast Asia, the pulp is recommended to neutralize the impacts of overdoses of chaulmoogra (Hydnocarpus anthelmintica Pierre), which treat leprosy (Gunasena and Hughes 2000). In Colombia, the tamarind pulp is powerful in destruction of vermin in residential animals, through the use of pulp with butter and different compounds (Morton 1987).

19.6 Future Prospects for the Utilization of *Tamarindus indica*

19.6.1 Medical Field

The aphrodisiac potential and reproductive safety of profile of aqueous extract of *Tamarindus indica* pulp in male Wistar rates was evaluated. The results showed that, the aqueous extract of *Tamarindus indica* possessed aphrodisiac together with spermatogenic potential (Rai et al. 2018). Research results showed that, the benefits of tamarind fruit extract in colon carcinanogensis might be related to it is hepatic protection against lipid peroxidation (Martinello et al. 2017).

19.6.2 Food Industry Field

The fermentation process of tamarind seeds substantially decreased phytic acid content, tannin content and trypsin inhibitor activity. While the phosphorus content which is the major mineral in the seed was increased (Olagunju et al. 2018). Mucilage has been isolated from tamarind seed (Reyes et al. 2017). The mucilage has great functional properties such as solubility, water holding capacity and oil holding capacity which increased with the temperature. Additionally, the swelling index. Emulsifying capacity of tamarind mucilage expanded and emulsifying soundness diminished when the heaviness of powder of tamarind seed mucilage/oil volume expanded. In this manner mucilage of tamarind possibly can be utilized in food processing as practical, cheep and ecological benevolent hydrocolloids rather than business hydrocolloids. Now globally there is a great interest for natural antioxidant sources like seeds, leaves and fruits to be utilized as food preservatives instead of using synthetic preservatives. The phenolic content of tamarind seed is a promising source of natural antioxidant in the food industry (Reis et al. 2016). Pectin from tamarind pulp was found to have significantly higher antioxidant activity as compare to apple pomace pectin, citrus peel pectin and commercial pectin. So tamarind pectin could be utilize as an excipient for food and pharmaceutical industry (Sharma et al. 2015).

References

Adeniyi VO, Olaifa EF, Emikpe O, Band TS (2017) Phytochemical components and antibacterial activity of *Tamarindius indica* Linn. Extracts against some pathogens. Biotechnol J Int 17(2):1–9

Agab MA (1985) Fermented food products: 'Hulu' Mur drink made from Sorghum bicolor. Food Microbiol J 2(2):147–155

Ahmed SA (2009) Development of spray dried tamarind powder. M.Sc. thesis, Sudan Academy of Sciences, Sudan

Alian A, El Ashwah E, Eid N (1983) Antimicrobial properties of some Egyptian non-alcoholic beverages with special reference to tamarind. Egypt J Food Sci 11:109–114

Anon (1984) Tamarind kernal powder- a brief technical note profodil Bulletin 19:24–26

Askar A, El Nemr SE, Siliha H (1987) Aroma constituents of Egyptian tamarind pulp. Deu Leb-Run J 83(4):108–110

Bag A, Chattopadhyay RR (2015) Evaluation of synergistic antibacterial and antioxidant efficacy of essential oils of spices and herbs in combination. PLoS One 10:e0131321. https://doi.org/10.1371/journal.pone.013132.

Bhadoriya SS, Ganeshpurkar A, Narwaria J, Rai G, Jain AP (2011) Tamarindus indica: extent of explored potential. Pharmacogn Rev J 5(9):73–81. https://doi.org/10.4103/0973-7847.79102

Bhattacharyya S, Gangopadhyay H, Chaudhuri DR (1983) Tamarind kernel powder as pectin substitute in food industries. In: Proceedings of National Conference on Fruits and Vegetables Proceeding Industries, Association of Food Scientists and Technologists, India, pp 61–64

Coronel RE (1991) *Tamarindus indica* L. In: Verheji EWM, Cornel RE (eds) Plant resources of South East Asia, Wageningen; Pudoc. No.2. Edible fruits and nuts. PROSEA Foundation, Bogor, Indonesia, pp 298–301. Cited from Fruits for the Future 1

Daniel M (2006) Chemistry and properties, Chapter 1. In: Medicinal plants. CRC Press, Boca Raton

Daniyany S, Muhammad BH (2008) Evaluation of the antimicrobial activities and phytochemical properties of extracts of *Tamarindus indica* against some diseases causing bacteria. Afr J Biotechnol 7(14):2451–2453

Duke JA (1981) Hand book of legumes of world economic importance. Plenum Press, New York, pp 228–230

El-Siddig K, Gunasena HP, Prasad BA, Pushpakumare DK, Ramana KV (2006) Tamarind (*Tamarindus indica* L), Fruits for the future 1, Revised edn. RPM Print and Design, West Sussex

Feungchan S, Yimsawat T, Chindapraseat S, Kitpowsong P (1996) Tamarind (*Tamarindus indica* L.) Plant genetic resources in Thailand. Thai J Agric Sci (Special Issue No. 1): 1–11. Cited from Fruits for the Future 1

Ganesapilla M, Venugopal A, Simha P (2017) Preliminary isolation, recovery and characterization of polyphenols from waste Tamarindus indica L. Mater Today Proc J 4:10658–10661

Gunasena HP, Hughes A (2000) Fruits for the future 1. Tamarind. International Centre for Underutilized Crops, Southampton

Ishola MM, Agbaaji EB, Agbaji AS (1990) A chemical study of *Tamarindus indica* (Tsamiya) fruits grown in Nigeria. J Food Sci Agric 51:141–143. Cited from Fruits for the Future 1

James AD (1981) Handbook of legumes of world economic importance. Plenurn Press, New York, p 228

Kotecha PM, Kadam SS (2003) Preparation of ready-to-serve beverage, syrup, and concentrate from tamarind. J Food Technol 40(1):76–79

Kuru P (2014) *Tamarindus indica* and its health related effects. Asian Pac J Trop Biomed 4(9):676–681

Leng LY, Nadzri N, Shaari A, Abdul Razak N, Yee KC (2017) Antioxidant capacity and total phenolic content of fresh, dried and stir fried tamarind leaves. Curr Res Nutr Food Sci J 5(3):282–287. https://doi.org/10.12944/CRNFSJ.5.3.13

Linda Z (1995) Improved beverages using tamarind extracts. World Intellectual Property Organization – International Bureau, No. WO 95/10949

Manjunath MN, Sattggeri VD, Rama SN, Usha MR, Nagaraja KV (1991) Physico-chemical composition of commercial tamarind powder. Indian Food Pac J 45:39–42

Marangoni A, Alli I, Kermasha S, Morton JF (1987) Fruits of warm climates. J.F. Morton, Miami, pp 115–121

Martinello F, Kannen V, Franco JJ, Gasparotto B, Sakita JY, Sugohara A, Garcia SB, Uyemura SA (2017) Chemopreventive effects of a *Tamarindus indica* fruit extract against colon carcinogenesis depends on the dietary cholesterol level in hamsters. Food Chem Toxicol J 107:261–269. https://doi.org/10.1016/j.fct.2017.07.005

Morton J (1987) Fruits of warm climates. Miami Fl: 115

Mustafa GA (2007) Preparation of carbonated beverage from locally grown tamarind fruit Aradaib. M.Sc. thesis, Sudan Academy of Sciences, Sudan

Olagunju OF, Ezekiel OO, Ogunsh AO, Oyeyinka SA, Ijabadeniyi OA (2018) Effects of fermentation on proximate composition, mineral profile and antinutrients of tamarind (*Tamarindus indica* L) seed in the production of daddawa-type condiment. LWT-Food Sci Technol J 90:455–459. https://doi.org/10.1016/j.lwt.2017.12.064

Menezes AP, Cerini SC, Barbalho SM, Guiguer EL (2016) *Tamarindus indica* L. a plant with multiple medicinal purposes. J Pharmacogn Phytochem 5(3):50–54

Pauly G (1999) Use of extracts of tamarind seeds enriched in xyloglycans and cosmetic or pharmaceutical product containing such extracts. United States Patent, No. 5,876,729

Rai A, Das S, Chamallamudi MR, Nandakumar K, Shetty R, Gill M, Susmalatha S, Devkar R, Gourishetti K, Kumar N (2018) Evaluation of the aphrodisiac potential of a chemically characterized aqueous extract *Tamarindus indica* pulp. J Ethnopharmacol 210:118–124. https://doi.org/10.1016/j.jep.2017.08.016

Reis MP, Dariva C, Vieira GA, Hense H (2016) Extraction and evaluation of antioxidant potential of the extracts obtained from tamarind seeds (*Tamarindus indica)* sweet variety. J Food Eng 173:116–123. https://doi.org/10.1016/j.jfoodeng.2015.11.001

Reyes EA, Navas HC, Rivera RG, Guerrero VV, Ramirez JA, Alonso CP (2017) Functional properties and physicochemical characteristics of tamarind (*Tamarindus indica* L) seed mucilage powder as a novel hydrocolloid. J Food Eng 209:68–75. https://doi.org/10.1016/j.jfoofeng.2017.04.02

Saka JD, Msonthi JD (1994) Nutritional value of edible fruits of indigenous wild tree in Malawi. For Ecol Manage J 64:245–248

Shankaracharya NB (1998) Tamarind – chemistry, technology and uses – a critical appraisal review. J Food Sci Technol 35(3):93–208

Sharma R, Kamboj S, Khurana R, Singh G, Rana V (2015) Physicochemical and functional properties performance of pectin extracted by Qbd from *Tamarindus indica* L pulp. Carbohydr Polym J 134:364–374. https://doi.org/10.1016/j.carbpol.2015.07.073

Siliha H, Askar A (2000) Determination of sugars in carob, roselle and tamarind. Alimenta 5:127–129

Ulrich H (1970) Organic acids, In: The biochemistry of fruits and their products. II. Academic press, London and New York

Zoblocki L, Pecore S (1996) Tamarind extract. United States Patent US 5,474,791

Chapter 20
Dacryodes edulis: Composition and Physico-chemical Properties

Ngozika Chioma Onuegbu

20.1 Introduction

Dacryodes edulis is a tropical fruit tree popularly known as African pear, African black pear, African plum, ube (Ibo), elemi (Yoruba), oruru (Benin), safou, safoutier, bush butter (Cameroun), Native pear (Ghana). It can be found in the Central and West African regions (Emebiri and Nwufo 1990; Leaky 1999; Awono et al. 2002; Anegbeh et al. 2005). The trees are usually domesticated and grown around homesteads. They grow in humid and sub humid climates of the tropics and can grow on all soil types (Kengue et al. 2002). According to Awono et al. (2002), the amount of *Dacryodes edulis* commercialised in Cameroun in 1997 has been estimated at 11,000 tonnes (which amounts to about US$7.5 million). Also the quantity of *Dacryodes edulis* shipped into the United Kingdom and Belgium from Nigeria and Central Africa was worth over US$2 million. It has been estimated that each grower or collector makes an income of up to US$161 from the African pear annually (Ayuk et al. 1999).

The African pear tree belongs to the family Burseraceae. The fruit is described by most authors as an ellipsoidal drupe. The mature fruits are purplish/bluish black on ripening. Flowering takes place from December to April, while the fruits ripen from April to September (Aiyelaagbe et al. 1998; Anegbeh et al. 2005).

N. C. Onuegbu (✉)
Department of Food Science and Technology, Federal University of Technology, Owerri, Nigeria
e-mail: ngozika.onuegbu@futo.edu.ng

A. A. Mariod (ed.), *Wild Fruits: Composition, Nutritional Value and Products*,
https://doi.org/10.1007/978-3-030-31885-7_20

20.2 Botanical Descriptions and Characteristics of the African Pear Tree

Dacryodes edulis is an evergreen tree. It grows up to the height of 18–40 m in forests and about 12 m in plantations and homesteads. The tree is usually well branched with a deep, dense crown. The trunk is 50–175 cm in diameter (Aiyelaagbe et al. 1998). The tree has compound leaves that are generally glossy and the flowers are usually yellow. The tree bark is grey in colour.

20.2.1 Physical Properties of the Fruit

Dacryodes edulis fruits (Table.20.1) obtained from different trees usually vary in several properties. Understanding the physical properties of the fruit is important for proper industrial utilization. It is crucial for equipment design and for providing adequate procedures for the handling, packaging and distribution of fresh produce as well as processing.

The single fruit weight of *Dacryodes edulis* (Fig. 20.1), ranged from 16.1 to 31 g (Adedokun and Onuegbu 2011), 22.6 to 24.7 g (Onuegbu et al. 2011b) and 15.57 to 39.39 g (Onuegbu and Ihediohanma 2008). These values were lower than some of the values obtained by Mbofung et al. (2002) and Fonteh et al. (2005) which ranged from 37.4 to 48.8 g and 53.28 to 95.82 g respectively. Varietal difference, soil fertility and growth conditions could be responsible (Waruhiu et al. 2004; Kinkela et al. 2006). Other physical characteristics investigated include fruit length, width, volume, density, pulp/seed ratio, etc. as shown on Table 20.1.

In the research by Onuegbu and Ihediohanma (2008), the fruit length and width were 3.98–8.07 cm and 2.55–3.34 cm respectively, while Adedokun and Onuegbu (2011) reported a fruit length and width of 4.26–6.24 cm and 2.60–3.20 cm respectively. A study in Cameroun by Mbofung et al. (2002), identified two varieties (the small and the large fruit types). The small fruit length was 4.32–4.0 cm and the large fruit length was 6.2–6.25 cm as reported by Kadji et al. (2016). Other researchers such as Ondo-Azi et al. (2009), have also identified the smaller variety as *Dacryodes edulis* var. *parvicarpa* and larger variety as *Dacryodes edulis* var *edulis*. The report by Onuegbu and Ihediohanma (2008) revealed that the African pear fruit is not perfectly cylindrical as generally reported by most authors (Omoti and Okiy 1987), but is made of four sides with rounded ends. In the research by Onuegbu and Ihediohanma (2008), it was discovered that the transverse section of the fruit revealed a rectangle shape with curved edges (Fig. 20.2). The longer sides were reported as width 1 and the shorter side as width 2. This was consistent in all the fruits gathered. These trends were also confirmed by Adedokun and Onuegbu (2011) and Onuegbu et al. (2011b). Ondo-Azi et al. (2009) also described the fruit in terms of its length and width instead of diameter implying that it is not cylindrical. A negative correlation

Table 20.1 Mean values of some physical properties of some African pear samples

Pear samples	Length (cm)	Width 1 (cm)	Width 2 (cm)	Fruit weight (g)	Fruit volume (cm^3)	Fruit density (g/cm^3)	Pulp thickness (cm)	Pulp weight (g)	Seed weight (g)	% pulp	% seed	Pulp seed ratio
1	3.08	2.66	2.56	15.97	18.33	0.86	0.33	10.02	5.96	62.74	37.31	1.66
2	4.34	2.60	2.39	17.85	18.33	0.98	0.38	11.27	6.60	63.13	36.92	1.72
3	4.89	2.73	2.63	20.98	20.00	1.09	0.33	12.23	8.88	58.29	42.32	1.38
4	5.48	2.97	2.68	24.24	26.00	0.93	0.39	15.05	9.18	62.08	37.97	1.66
5	4.73	3.26	2.91	28.00	27.66	1.01	0.56	6.19	11.55	57.82	41.25	1.43
6	5.48	2.84	2.60	22.01	21.66	1.02	0.40	13.76	8.28	62.51	37.61	1.66
7	5.28	3.40	3.35	31.54	32.66	0.96	0.46	17.78	13.73	56.37	43.53	1.30
8	6.05	2.98	2.83	25.88	33.33	0.78	0.39	17.99	7.86	69.59	30.37	2.34
9	8.76	3.36	3.34	39.36	50.00	0.78	0.54	31.52	8.84	80.08	20.45	3.60

Source: Onuegbu and Ihediohanma (2008)

Fig. 20.1 Different sizes of ripe fruits

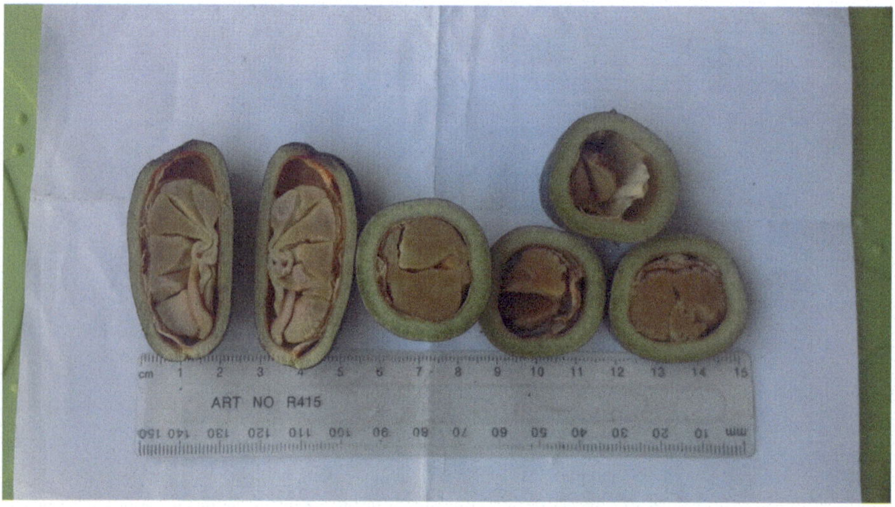

Fig. 20.2 Cut fruits showing the seed structure

(r = −0.86) was found between fruit density and pulp/seed ratio (Onuegbu and Ihediohanma 2008). This was because fruits with high pulp/seed ratios, usually have small seeds which resulted in an air space between the pulp and the seed. This results in low fruit density.

20.2.2 Physical Changes During Fruit Development and Ripening

The changes that occur in the physical properties of the fruit during development and ripening have been studied by Onuegbu et al. (2011b), as shown in Table 20.2. Determining the properties of the fruit during development will enable producers to ascertain the proper harvest time for optimum yield and reduction of losses. In this research, three separate trees were used. The fruiting period (fruit set to senescence) was between 20 and 21 weeks. However the measurements started from the fifth week and was done fortnightly.

There was a rapid increase in fruit weight, length and width up to the ninth week, but the rate of increase reduced after the ninth week. The mean values of the pulp and seed mass steadily increased from 5.48 to 17.7 g and 0.22 to 7.0 g respectively, from the 5th week to the 21st week. However the percentage pulp, reduced from 95.4% to 72.86%, while percentage seed increased from 3.74% to 27.12% within the period. This is due to the rapid rate of increase in seed mass. In accordance with that, the pulp/seed ratio gradually decreased from the 5th week value of 24.91 to 2.53 in the 21st week. The colour also changes from pink during the early weeks to light yellow at about the 11th week. The fruit then starts to have blue/violet streaks or spots on top of the pale yellow background at the 15–17th week. Indicating maturity and onset of ripening (Fig. 20.3). The darkening intensifies during the weeks that follow, until a dark blue/violet colour is achieved in the 20–21st week.

20.3 Chemical Composition and Properties of the Fruit and Seed of African Pear

20.3.1 Fruit Pulp

The chemical properties, composition of *Dacryodes edulis* fruits vary from tree to tree based on fruit type, climatic conditions and soil type. For the fruit pulp, the range of values reported by Onuegbu and Ihediohanma (2008) was 11.00–15.75%, 18.99–38.10%, 4.65–10.68% and 36.56–52.62% for protein, fat, carbohydrate and moisture respectively (Table 20.3). Ash content ranged from 0.26% to 3.88%, while pulp acidity ranged from 0.92% to 1.69%. The authors further noted that consumer acceptability scores which ranged from 5.5 to 7.9 on a nine-point hedonic scale did not correlate with any of the parameters studied. This suggests that the acceptability of the pulp by consumers may be dependent on an interplay of several factors such as fat content, acidity, moisture content, pulp thickness, etc.

Some differences were also discovered between the two varieties of *Dacryodes edulis* in some parameters. Mbofung et al. (2002), noted that the two varieties varied

Table 20.2 Mean values of physical properties of African pear during fruit development

Stage (weeks)	Fruit length (cm)	Width 1 cm	Width 2 cm	Fruit weight (g)	Fruit volume (cm^3)	Fruit density (g/cm^3)	Pulp thickness (cm)	Pulp weight (g)	Seed weight (g)	% pulp	% seed	Pulp/seed ratio
5	2.16	0.48	0.42	5.77	4.33	1.36	0.14	5.48	0.22	95.47	3.74	24.91
7	3.66	0.92	0.86	9.68	8.56	1.15	0.27	9.02	0.52	92.22	5.22	17.34
9	5,05	2.42	2.37	15.87	17.22	0.96	0.46	13.72	2.11	86.47	13.10	6.50
11	4.96	2.57	2.52	18.22	21.11	0.86	0.36	14.98	3.21	84.56	15.09	4.65
13	5.24	2.82	2.77	18.86	24.44	0.81	0.45	16.46	3.45	85.30	14.44	4.77
15	5.29	3.00	2.93	22.60	27.56	0.89	0/40	16.98	5.56	77.07	22.59	3.05
17	5,13	2.91	2.84	22.70	27.00	0.95	0.43	17.42	5.28	81.25	18.82	3.29
19	5.43	3.04	2.99	23.43	27.72	0.89	0.37	17.10	6.30	74.90	24.78	2.71
21	5.43	3.01	2.82	24.75	27.67	0.94	0.38	17.70	7.00	72.86	27.12	2.53

Source: Onuegbu et al. (2011b)

Fig. 20.3 Mature and Immature fruits on the tree

Table 20.3 Mean values for proximate composition, titratable acidity and acceptability of nine samples African pear pulp

Sample	Protein (%)	Oil (%)	Carbohydrate (%)	Dry matter	Moisture (%)	Ash (%)	Titratable acidity (%)	Consumer acceptability (%)
1	8.56	16.62	29.13	54.71	45.29	0.39	0.50	6.3
2	7.98	17.44	27.51	53.46	46.54	0.53	0.56	7.9
3	5.17	13.38	28.02	46.68	52.32	0.65	0.54	6.6
4	9.99	16.19	29.46	55.78	44.22	0.14	0.94	6.4
5	6.61	8.81	29.97	47.18	52.82	1.79	0.57	5.5
6	10.31	13,72	21.61	53.73	46.27	0.63	0.71	5.8
7	8.98	19.17	28.70	57.49	42.51	0.64	0.75	6.3
8	8.33	20.44	23.81	52.90	47.10	0.32	0.76	7.8
9	9.14	24.51	8.41	63.44	36.56	0.47	0.93	6.2

Source: Onuegbu and Ihediohanma (2008)

in their chemical composition. This variation based on fruit size or variety, was not clearly observed by some other authors (Onuegbu and Ihediohanma 2008; Ene-Obong et al. 2017) whose reports show variations in chemical composition within the species, which was not found to be related to fruit size. However, in the research by Kadji et al. (2016), the larger variety (*D. edulis var edulis*) had significantly higher ($p \leq 0.05$) values for Calcium (1307.66–1337.02 mg/kg), Sodium (104.42–108.01 mg/kg), Mg (71.95–73.18 mg/1 kg) and Fe (7.71–8.67 mg/kg) as against the values of 531.31–533.41, 51.54–53.69, 23.13–26.70 and 4.91–6.06 mg/kg recorded for Ca, Na, Mg and Fe for the *D. edulis var parvicarpa*. But no significant difference ($p \geq 0.05$) was found between the two varieties in the potassium (542.39–646.05 mg/kg), Manganese (25.96–27.38 mg/kg) and Zinc (130.48–136.18 mg/kg) as well as the proximate composition.

The mean values of the mineral, vitamin and amino acid profile reported by Onuegbu et al. (2011b). Sodium had the highest value (3756.34 mg/kg) among the mineral elements determined, followed by Manganese (89.72 mg/kg) and Copper (63.69 mg/kg). The lowest mean value (2.21 mg/kg) was given by Calcium. The ascorbic acid, niacin and pyridoxine values were 164.34 mg/100 g, 17.11 mg/100 g and 33.90 mg/100 g respectively. Methionine was found to be the limiting amino acid with a chemical score of 32.12% Leucine had the highest chemical score of 179.33%.

A study was done by Adedokun and Onuegbu (2011) on the characteristics of oil extracted from fruits of three randomly selected trees from the three senatorial zones of Imo State Nigeria. No significant difference was found in the chemical properties of the extracted oils from these different fruits. The yield was 41–47%. The range of values for acid, saponification and iodine values was 0.34–0.46 mg KOH/g, 180.89–189.77 mg KOH/g, 32.46–37.11mL/g respectively. The Refractive index (at 27.3 °C), melting point, smoke point, flash point were 1.4674–4.4687, 75–78 °C, 184–185 °C, 274–278 °C respectively.

The percentage unsaturated fatty acid in the pulp oil is 69.88% which were mainly oleic acid and linoleic acid as reported by Ajayi and Adesanwo (2009). Udo et al. (2012) reported that the fatty acids present in Dacryodes edulis pulp oil to be Oleic acid (45–60%), palmitic (30–35%), linoleic (15–20%) and stearic acid (2%).

In another study by Onyeka et al. (2005), the oil yield of fruit pulp was 29.5%, 24.24% and 38.53% for raw, steamed and roasted samples. The specific gravity value of the oil ranged 0.904 to 0.91 and there was no significant difference ($p \geq 0.05$) between the values obtained for raw, steamed and roasted samples. However the saponification value and peroxide values were reduced significantly ($p \leq 0.05$) by the steaming and roasting pre-treatments.

Further research reveals the presence of two major phenolic compounds, caffeic acid which accounts for 82.09% and ellagic acid which accounts for 8.62% of all phenolics in the African pear (Onwuka et al. 2013). Ene-Obong et al. (2017) reported the presence of other phytochemicals such beta-carotene (1670.77–2927.80 µg/100 g), phytosterols, squalene and caffeine.

The chemical properties of the African pear pulp were also investigated at different stages of maturity (Nwosuagwu et al. 2009). Results showed that percent mean values for fat, crude fibre, protein, and ash all increased from their 5th week values of 8.89%, 3.53%, 3.22% and 0.49% to their 21st week values of 18.04%, 7.46%, 7.47%, and 1.46% respectively. On the contrary the percent moisture content decreased from their 5th week values of 66.73% to the 21st week values 48.52%. The carbohydrate values ranged from 16.38% to 18.24% with no observed trend of increase or decrease during the period of fruit development. In a similar study by Missang et al. (2001), on the behaviour of cell wall polysaccharides during ripening, it was shown that the hemicelluloses and sugar composition of an alkali extract of the pulp did not vary during ripening.

20.3.2 African Pear Seed Composition

The proximate composition of African pear seed shows a high lipid content (19.47%), thus indicating that the *Dacryodes edulis* seed can find usefulness in oil production (Onuegbu et al. 2016). Antinutrients present in the seed includes the trypsin inhibitors (7.33%), saponins (1.14%), tannins (1.05%), polyphenols (0.35%), oxalates (0.64%) and phytates (0.77%).

The oil yield for both industrial and food uses is high. The fruit and the seed oils are of the same composition, therefore extraction of the pulp before oil extraction will not be necessary (Onocha et al. 1999). The residue (cake) after extraction has found useful application in the livestock industry. In the research by Onuegbu et al. (2016), the seed cake was utilized in the poultry feed formulation. The Protein Efficiency Ratio (PER) of the substituted feed was not significantly different ($P \geq 0.05$) from that of the commercial feed samples. The seeds also contain up to 1.5% essential oil content. These include mycrene (45%), α-pinene (9%), α-terpineol (8%), germacrene – D (4%) and to lesser extent E-α-cadinol, δ-cadinol and β-eudesmol (Onocha et al. 1999). Some of the properties of the seed oil reported by Onuegbu et al. (2016), include melting point (32.35 °C), smoke point (176.13 °C), and flash point (210.44 °C), iodine value (48.16 mL/g) and saponification value (213.54 mg KOH/g). All the values falling into the range of cooking oil as well as oil for soap making. In a similar research, the oil yield of the African pear seed was 2.74%, 11.08%, and 13.21% for raw, steamed and roasted samples, showing that the oil yield was improved by the pretreatment operations (Onyeka et al. 2005). The saponification and peroxide values ranged from 191.72 to 195.66 mg KOH/g and 4.49 to 7.22 mEq/kg respectively. The values decreased with the two pre-treatment operations (steaming and roasting).

The African pear seed contains two major phenolic compounds (Onuegbu et al. 2013). One was identified as gallic acid while the other peak could not be identified from the standards available at that time. The gallic acid formed 74% of all phenolics of the African pear seed. A comparative study of the properties of the *Dacryodes edulis* pulp and seed has been done by Ajayi and Adesanwo (2009), which revealed several similarities in their proximate composition, oil characteristics but showed some differences in the mineral profile.

20.4 Uses of *Dacryodes edulis*

Several uses and benefits have been identified for African pear pulp, seed, root, stem bark, trunk and leaves. They range from food uses, animal feed, medicines to wood and timber production. (http://docsdrive.com; Ajibesin 2011; Omonhinmin and Agbara 2013).

20.4.1 Food Uses

The fruit pulp is the focus for the use of African pear as food. It contains cell wall degrading enzymes that slowly soften the pulp within a period of 7–10 days at room temperatures. However, warmer temperatures (60–85 °C) tend to hasten the rate of softening as it increases enzyme activity. It takes about 2–3 minutes to soften under such warm conditions. But if the fruit is boiled, hardening of the pulp takes place and the enzyme is also denatured (Okolie and Obasi 1992).

African pear pulp is generally consumed after softening in hot water, hot ash or near a fireplace. It is commonly eaten in South East Nigeria as an accompaniment to boiled/roasted fresh maize during the months of April to September when both products are in season. The pulp is sometimes preserved traditionally by removing the seed, boiling the deseeded pulp and drying under the sun (Ayuk et al. 1999). The dried pulps are eaten dry as snacks or rehydrated by addition of small amounts of warm water or by steaming. Several attempts have been made to utilize this fruit (pulp, seed oil and cake) in other ways apart from this. The pulp, and cake have been used in the formulation of high fibre, high protein biscuits (Onuegbu et al. 2011c; Mbofung et al. 2002). The biscuits produced by 100% substitution of bakery fat with African pear pulp, compared favourably with the control in terms of consumer acceptability. In other research by Onuegbu et al. (2007), the African pear pulp and oil were evaluated as ingredients in bread-making, the samples produced with pear pulp had significantly ($p \geq 0.05$) lower loaf volume and lower acceptability scores. However, the samples produced with a 25% and 50% substitution of bakery fat using African pear pulp oil were not significantly different ($p \leq 0.05$) from those produced by the control (100% bakery fat).

African pear seed is usually fed to domestic animals. In a research by Onuegbu et al. (2016), the performance of the African pear seed (APS) cake in the poultry feed formulation was investigated. No significant difference ($p \leq 0.05$) was observed in the protein efficiency ratio (PER) for the test diet (containing 40:25:15 maize:APScake:soybean) and the control (commercial poultry feed product).

In apiculture, honey produced from African pear nectar is highly rated in quality and acceptability (Ayuk et al. 1999). The high oil content of the fresh pulp and seed yielding up to 7.2 t/ha of oil is an unexplored source of raw material for the oil processing and food industry (Bejeng 2016).

20.4.2 Medicinal Uses

The leaves, bark, stem and root of *D.edulis* tree are used as a local medicine against diseases (Duru et al. 2012). Different parts of the plant and extract obtained from them have been applied in different ways by different people groups in disease prevention and treatment. Resins from the bark has been used in traditional medicine in Nigeria to treat skin infections, The leaf decoction is used for the treatment of diarrhea, dysentery and anaemia and fevers, tooth ache and earache. The root and bark decoctions are also used in the treatment of fevers, headaches, leprosy, and skin infections (Ayuk et al. 1999; Aiyelaagbe et al. 1998). Attempts have been made by several authors to identify, quantify and characterize the compounds necessary for these beneficial health attributes.

In the research by Duru et al. (2012), the phytochemical contents of the fruit pulp from the mature undarkened fruit to the darkened fruit were analysed. The highest value was obtained for the tannins (3.10–5.78 mg/100 g), followed by oxalates (1.34–4.97 mg/100 g), flavonoids (0.67–2.47 mg/100 g), alkaloids (0.42–1.50 mg/100 g), saponins (0.21–1.29 mg/100 g) and cyanogenic glycosides (0.03–0.05 mg/100 g). The work by Erukainure et al. (2017), showed that ethanol extract of the fruit inhibited DNA fragmentation and showed free radical scavenging ability. Other antioxidants and bioactive compounds found in the fruit pulp include β-carotene, α-tocopherol, squalene, γ and β sitosterol and caffeine.

Researchers (Anowu et al. 2015; Tee et al. 2014) also analysed the seed extracts of *Dacryodes edulis* and several phytochemicals were found. They include anthraquinone, steroid/triterpenes, tannin, and saponin. Due to the presence of these secondary plant metabolites, the extracts exhibited some antimicrobial activity towards the test organisms (*Staphylococcus aureus, Eschrichia coli, Salminella typhi, Shigella senteriae, Psuedomonas aeruginosa, Klebsiella pneumonia, Candida albican, Microsporum* sp.) The research by Udo et al. (2016), revealed the presence of alkaloids, saponins, carbohydrates and some resins. Some mineral elements such as Sodium 90.25 μg/g, Potassium 16.54 μg/g, Calcium 500.56 μg/g, Magnesium 40.60 μg/g.

The seed extracts (using ethanol and hexane) were able to give a mortality of 70.6% and 49.3% respectively at 48 hours against the larvae of *Aedes vittatus* (Akande-Grillo and Nzelibe 2017) while the ethanol and hexane extracts of the leaf gave a mortality of 40% and 70.67% at 48 hours against *Aedes vittatus.*

The broad spectrum antimicrobial potential of the *Dacryodes edulis* leaf extract as reported by Tee et al. (2014) and Oyetunji and Opeyemi (2017) has been attributed to the presence of saponins, alkaloids, flavonoids andtanins contained in the leaf.

The tree bark contains alkaloids (28.23 mg/kg), phenolic compounds (22.01 mg/kg), flavonoids (60.91 mg/kg), tannins (18.16 mg/kg), saponins (3.16 mg/kg), anthraquinones (12.16 mg/kg), Cyanogenic glycoside (0.81 mg/kg), and steroids (0.91 mg/kg). In the report by Udo et al. (2016), tree bark exudate contained calcium (500.56 μg/g), potassium (16.54 μg/g) and magnesium (40.60 μg/g). It also had a very high content of saponins (2.08 mg/100 g) followed by tannins (0.47 mg/100 g), and alkaloids (0.28 mg/100 g). These plant metabolites and their synthetic derivatives are essential medicinal ingredients for wound healing, analgesic, antispasmodic, antimicrobial activities. This is why the tree bark has been used for medicinal purposes for several years in many cultures.

In the research by Okolo et al. (2016), it was shown that the hexane extract of Dacryodes edulis fruit administered to alloxan-induced diabetic rats and sub toxic levels (400–1600 mg/kg), for 15 days compared favourably with glibenclamide (2.5 mg/kg). The results showed decreased blood glucose levels. It down-regulated hyperglycemia, total cholesterol, triglyceride, LDL-C but HDL-cholesterol increased significantly.

The cholesterol lowering ability of the ethanolic seed extract was reported by Iyawe et al. (2007). The serum cholesterol level of the rats were reduced from 70.60 to 50.10 mg/L. In the work by Zofou et al. (2013) the anti-plasmodic effect of *Dacryodes edulis* stem bark hexane-extract was reported, revealing its potential as raw material in production of preventive and curative medicine for malaria.

20.4.3 Other Uses

The Dacryodes edulis tree is used as a shade crop in coffee, cocoa and cocoyam plantations. The wood is useful as firewood (Ayuk et al. 1999). It is also useful for tool handles because they are strong and useful for carpentry work. The trees are also used for shade in homesteads and gardens and as ornamental plants especially at the early stages.

References

Adedokun II, Onuegbu NC (2011) The physical properties of pulp and chemical characteristics of edible oil extracted from the pulp of African pear *(Dacyodes edulis)*. Pak J Nutr 10(6):558–560

Aiyelaagbe IOO, Adeola AO, Popoola L, Obisesan KO (1998) Agroforestry potential of *Dacryodes edulis* in the oil palm cassava belt of southeastern Nigeria. Agrofor Syst 40:263–274

Ajayi IA, Adesanwo O (2009) Comparative study of the mineral element and fatty acid composition of *Dacyodes edulis* pulp and seed. World J Agric Sci 5(3):279–283

Ajibesin KK (2011) *Dacryodes edulis* (G. Don) H.J. Lam: a review on its medicinal, phytochemical and economical properties. Res J Med Plant 5(1):32–41

Akande-Grillo HT, Nzelibe HCC (2017) Larvicidal activities of *Dacryodes edulis* (G. Don) H J Lam on Aedes bigot and Culex quinquefasciatus Say (Diptera: Culicidaae). Int J Mosq Res 4(4):32–36

Anegbeh PO, Ukafor V, Usoro C, Tchoundjeu Z, Leakey RRB, Schreckenberg K (2005) Domestication of Dacryodes edulis: 1. Phenotypic variation of fruit traits from 100 trees in southeast Nigeria. New For 29(2):149–160

Anowu JN, Tor-Anyiin TA, Igoli JO (2015) Studies on *Dacryodes edulis* I: phytochemical and medicinal principles of raw seeds. J Nat Prod Plant Res 5(2):13–19

Awono A, Ndoye O, Schreckenberg K, Tabuna H, Isseri F, Temple L (2002) Production and marketing of safou (*Dacryodes edulis*) in Cameroon and internationally: market development issue. For Trees Live 12(1):125–147

Ayuk ET, Duguma B, Franzel S, Kengue J, Moiiet M, Tiki-Manga T, Zekeng P (1999) Uses management and economic potential of *Dacryodes edulis* (Burseracae) in the humid lowlands of Cameroon. Econ Bot 53(3):292–301

Bejeng MT (2016) Dacryodes edulis (The African plum) Journal of African Medicine https://www.journalofafricanmedicine.com/index.php/en/

Duru M, Amadi C, Ugbogu A, Eze A, Benjamin A (2012) Phytochemical, vitamin and proximate composition of *Dacryodes edulis* at different stages of maturation. Asian J Plant Sci Res 2(4):437–441

Emebiri IC, Nwufo MI (1990) Effects of fruit types and storage treatments on the biodeterioration of African pear (*Dacryodes edulis*) G. Don, H. J. Lam. Int Biodeterior 26(1):43–50

Ene-Obong H, Igile G, Egbung E, Ekpo A, Agbo M (2017) Varietal in the chemical composition of different lines of West African pear (*Dacryodes edulis*). Paper presented at the 12th IFDC, Buenos Aires Argentina. 11th – 13th Oct 2017

Erukainure OL, Mopuri R, Oyebode OA, Koorbanally NA, Islam MS (2017) *Dacryodes edulis* enhances antioxidant activities, suppresses DNA fragmentation in oxidative pancreatic and hepatic injuries; and inhibits carbohydrate digestive enzymes linked to type 2 diabetes. Biomed Pharmacother 96:37–47

Fonteh FA, Niba TA, Tchoumboue J (2005) Some physicochemical properties of African pear and Black olive from different agro-ecological zones of Cameroun. Fruits 60(2):107–113

http://docsdrive.com/pdfs/academicjournals/rjmp/2011/32-41.pdf

Iyawe HOT, Osagie VE, Esekhigbe A, Azih MC (2007) Nutrient composition and effect of seed extract of African black pear (*Dacryodes edulis*) on rats. Asian J Plant Sci 6(5):878–880

Kadji BRL, Kone FMT, Sika AE, Dabonne S (2016) Physico-chemical properties of safou (*Dacryodes edulis*) grown in Côte d'Ivoire. J Appl Biosci 105:10103–10110

Kengue J, Chuenguem T, Fabuo FN, Adewusi HG (2002) Towards improvement of safou (*Dacrudes edulis*), population and rproductive biology. For Trees Liv 12:73–84

Kinkela T, Niamayoua RK, Mampouya D, Silou T (2006) Variations in morphological characterisyics, lipid content and chemical composition of safou (*Dacryodes edulis*) G. Don, H. J. Lam. According to fruit distribution: a case study. Afr J Biotechnol 5(12):1233–1238

Leaky RRB (1999) Potential for novel food products from agroforestry tres. A review. Food Chem 66:1–14

Mbofung CMF, Silou T, Mouragadja I (2002) Chemical properties of safou (*Dacryodes edulis*) and evaluation of its potentials as an ingredient in nutritious biscuits. For Trees Liv 2:105–117

Missang CE, Renardc MGC, Baron A, Drilleau JF (2001) Cell wall polysaccharides of bush butter (*Dacryodes edulis* (G Don) H. J. Lam) fruit pulp and their evolution during ripening. J Sci Food Agric 81(8):773–780

Nwosuagwu UH, Onuegbu NC, Nkwoala C (2009) The chemical properties of African pear pulp at different stages of development. Int NGO J 9:380–385

Okolie PN, Obasi BN (1992) Implication of cell wall degrading enzymes in the heat-induced softening of the African pear (*Dacryodes edulis* (G. Don) H.J. Lam). J Sci Food Agric 59:59–63

Okolo CA, Ejere VC, Chukwuka CO, Ezeigbo II, Nwibo DD, Okorie AN (2016) Hexane extract of Dacryodes edulis fruits possesses anti-diabetic and hypolipidaemic potentials in alloxan diabetes of rats. Afr J Tradit Complement Altern Med 13(4):132–144

Omonhinmin AC, Agbara I (2013) Assessment of in vivo antioxidant properties of *Dacryodes edulis* and *Ficus exasperata* as anti-malaria plants. Asian Pac J Trop Dis 3(4):294–300

Omoti U, Okiy DA (1987) Characteristics and composition of pulp and cake of cake of African pear, *Dacryodes edulis*. J Sci Food Agric 38(1):805–809

Ondo-Azi AS, Missang E, Silou T (2009) Classification of *Dacryodes edulis* (G. Don) H. J. Lam by using morphological and physical characteristics of the fruit: a statistical approach. For Trees Liv 19(2):99–109

Onocha PA, Ekundayo O, Oyelola O, Laakso I (1999) Essential oils of *Dacryodes edulis* (G.Don) H.J. Lam (African pear). Flavour Fragr J 14:135–139

Onuegbu NC, Ihediohanma NC (2008) Some proximate analysis of African pear *(Dacryodes edulis)*. J Appl Sci Environ Manag 12(1):83–85

Onuegbu NC, Onyeka EU, Agbayi D (2007) Evaluation for the potentials of African pear pulp and pulp oil in bread manufacture. Niger Food J 25(2):171–175

Onuegbu NC, Adedokun II, Kabuo NO, Nwosu JN (2011a) Amino acid profile and micronutrient composition of African pear (*Dacrodesedulis*) pulp. Pak J Nutr 10(6):555–557

Onuegbu NC, Nwosuagwu UC, Kabuo NO, Nwosu JN, Ihediohanma NC (2011b) The physical properties of Ube (*Dacryodes eclulis*) at different stages of fruit development. Nat Sci 9(9):71–75

Onuegbu NC, Ihediohanma NC, Okorafor UC, Nwankwere AC (2011c) Evaluation of the potentials of African pear (*Dacryodes edulis)* pulp as an ingredient in biscuit-making. Niger Food J 29(1):78–82

Onuegbu NC, Onwuka GI, Clifford H (2013) Development of extraction protocol for phenolic compounds in African pear (*Dacryodes edulis*) seed. Int J Biol Pharm Allied Sci 2(5):1044–1066

Onuegbu NC, Nwuka MU, Ojukwu MO, Kabuo NO (2016) Nutritional properties of African pear seed and performance of defatted cake in poultry feed formulations. Int J Life Sci 5(3):149–151

Onwuka GI, Clifford H, Onuegbu NC (2013) The evaluation of extraction efficiencies of different solvent mixtures and extraction procedures on the phenolics in African pear (*Dacryodes edulis*) pulp. Int J Biol Pharm Allied Sci 2(5):1010–1023

Onyeka EU, Onuegbu NC, Onuoha NU, Ochonogor F (2005) Effect of extraction pretreatment on the composition and characteristics of seed and pulp oil of African black pear (*Dacryodes edulis)*. Niger Food J 23:13–20

Oyetunji OO, Opeyemi PO (2017) Antibacterial and antioxidant activities of *Dacyodes edulis* methanolic leaf extract. J Adv Med Pharm Sci 14(1):1–11

Tee HL, Yang B, Negendra KP, Ramanan RN, Sun J (2014) Nutritional compositionand bioactives of Dacryodes species. A review. Food Chem 165:247–255

Udo IE, Akpabio UD, Akpakpan AE (2012) Physicochemical studies on the hydrolysis and saponification product of *Dacryodes edulis* exudates. Adv Appl Sci Res 4(4):1741–1747

Udo IE, Obuzor G, Horsefall M (2016) Proximate and phytochemical studies of exudate of Dacryodes edulis. Res J Chem Sci 6(3):1–9

Waruhiu AN, Kengue J, Atangana AR, Tchoundjeu Z, Leakey RRB (2004) Domestication of *Dacryodes edulis*: phenotypic variation of fruit traits in 200 trees in Cameroon. J Food Agric Environ 2(1):340–346

Zofou D, Tematio EL, Ntie-kang F, Tene M, Ngemenya MN, Tane P, Titanji VPK (2013) New antimalaria hits from *Dacyodes edulis* (Burseraceae) – Part 1: Isolatio, in vitro activity, in silico 'drug-likeness' and pharmacokinetic profiles. PLoS One 8(11):e79544. https://doi.org/10.1371/journal.pone.0079544

Chapter 21
Irvingia gabonensis: Phytochemical Constituents, Bioactive Compounds, Traditional and Medicinal Uses

Gustav Komla Mahunu, Lydia Quansah, Haroon Elrasheid Tahir, and Abdalbasit Adam Mariod

Abbreviations

NTFP	Non-timber forest product
OAC	Oil Absorption Capacity
WAC	Water Absorption Capacity

21.1 Introduction

The nutritional status of huge proportion of the African population remains alarming notwithstanding the encouraging socio-economic progress that is ongoing. The worrying nutritional problems include undernutrition, and deficiency in iron and vitamin A. Nutritional deficiencies and malnutrition aggravate numerous diseases and health situations. In addition, the African continent is battling with undernutrition and the prevalence of over nutrition and obesity are escalating, alongside the

---

The original version of this chapter was revised. The correction to this chapter is available at https://doi.org/10.1007/978-3-030-31885-7_42

G. K. Mahunu (✉)
Department of Food Science & Technology, Faculty of Agriculture, University for Development Studies, Tamale, Ghana
e-mail: gmahunu@uds.edu.gh

L. Quansah
Department of Biotechnology, Faculty of Agriculture, University for Development Studies, Tamale, Ghana

H. E. Tahir
School of Food and Biological Engineering, Jiangsu University, Zhenjiang, Jiangsu, China

A. A. Mariod
Indigenous Knowledge and Heritage Center, Ghibaish College of Science & Technology, Ghibaish, Sudan

A. A. Mariod (ed.), *Wild Fruits: Composition, Nutritional Value and Products*,
https://doi.org/10.1007/978-3-030-31885-7_21

related health conditions including diabetes and coronary heart diseases (Ogungbenle and Omaejalile, 2010). This has called for the search and use of indigenous fruits with health benefit to curb these health problems.

Fruits, particularly indigenous edible fruits characterize one of such sources representing dietary antioxidants (Kolar et al. 2011); thus, compelling the demand for their enhanced intake in recent times (Hall et al. 2009). The overall antioxidant (free radical scavenging) potential of fruits is mostly ascribed to the added and synergistic properties of innate phytochemicals most especially those belonging to phenolic compounds (Cartea et al. 2010; Liu 2003). Phytochemicals do confer other beneficial toxicological or pharmacological properties in humans in addition to their antioxidant activity. Consequently, analysis of the free radical scavenging properties and phytochemical content of fruits provides a clue to the potential health importance to be derived from their frequent intake (Pisoschi et al. 2009). It is widely known and has been reported that antioxidants such as fruits and vegetables of natural sources are more beneficial to health compared to the synthetic counterparts/supplements (Liu 2003). Therefore, one of such important indigenous fruit tree found in West and Central Africa is *Irvingia gabonensis*, which is being recognized as an important source of traditional food, medicine and economic benefits.

To the best of our knowledge, this book chapter on *Irvingia gabonensis* is the first to compile extensive research outcomes on its phytochemical constituents, bioactive properties, traditional and medicinal uses as well as possible role in contributing to the improvement of malnutrition and related health problems.

21.2 Botanical Background and Distribution

Irvingia gabonensis, commonly known as bush mango is a non-timber forest product, comprising of tree trunk (stem), leaves, roots and fruits. It belongs to the family, Irvingiaceae (Ekundayo et al. 2013). There are two varieties that have been recognized in Nigeria; Var *gabonensis* and Var *excelsa* (Okafor and Ujor 1994). *Irvingia* encompasses seven species, out of these, six of them occurs in tropical Africa and one found in south-east Asia. The species are *Irvingia gabonensis, I. excels, I. wombolu, I. orbor, I. smithil* and *I. midbr*. However, two common species of the tree; *I. gabonensis* has a sweet edible pulp while *I. wombulu* has a bitter edible pulp though both are fit for human consumption and used as such in Africa (Atangana et al. 2001, 2002) and forms part of the traditional diets of several people especially in the rural areas. Furthermore, *I. gabonensis* seeds is prominent in international trade in West Africa. The major destinations of the products are Nigeria, Gabon, Liberia, and Sierra Leone, while the humid lowlands countries of Cameroon, Nigeria, and Côte d'Ivoire are the main sources of the products (Ainge and Brown 2004a). Actually, it ranked first among the ten priority species by farmers ranged and from 60 in Cameroon to 172 in Nigeria, where it is valued for its cotyledons (Leakey and Ajayi 2008). Although, it was not ranked among the top ten species in Ghana, the fruits are being widely harvested for export to Nigeria. The fruits are varied with respect to size, shape and weight which is revealed by fruit mass, flesh nut mass, shell mass, kernel mass, length, width, flesh depth, extent of variability (Table 21.1).

Irvingia gabonensis has other common names: English (wild mango, native mango, dika nut, bush mango, bread tree, African mango tree); French (manguier sauvage, bobo); Hausa (goron, biri); Kuwing (Agoi); Igbo (obono); Trade name (dika nut); Yoruba (oro); Ewe (atɔ). It is named differently in other countries. It is named oba in Gabon, sioko in Ivory Coast, wilder mango baum in Germany, Afurika mango no ki in Japan, and arbol chocolate in Spain (Brands 1989).

Generally, *I. gabonensis* based-products have high nutritive value. However, inadequate knowledge of their food value has led to their wastage in terms of economic revenues or/and postharvest losses. A major problem associated with handling both fresh fruit and seeds of *I. gabonensis* is storage. Usually, they have very short shelf life and perhaps there is no available information on cold storage currently. It is common practice to find processing of seeds in fresh state by pounding and fermentation; to inhibit spoilage and improve its nutritive value. Coupled with the poor research on extended storage life, there is the need to improve upon varying methods of processing in addition to utilization of the kernels to enhance its domestication potentials and commercialization initiatives. *I. gabonensis* has been found to be top among the list of NTFP being promoted for domestication in West Africa (Agbogidi et al. 2005; Leakey et al. 2004; Ndoye et al. 1997), and rapidly becoming the tree of choice in agroforestry practices (Koyejo and Omokhua 2001). It is much valuable in Mvila compared to other vital species (such as *Dacryodes edulis*, *Garcinia kola*, *Ricinodendron heudelotii*) in the region (Ayuk et al. 1999).

The tree is native to Angola, Cameroon, Central African Republic, Congo, Cote d'Ivoire, Democratic Republic of Congo, Equatorial Guinea, Gabon, Ghana, Guinea-Bissau, Liberia, Nigeria, Senegal, Sierra Leone, Sudan and Uganda. While it is existing as exotic tree in Benin, Sao Tome et Principe (Fig. 21.1).

Table 21.1 Characteristics of *Irvingia gabonensis* fruit

Parameters	Range of values
Fruit weight	69.0–419.8 g
Flesh/pulp weight	59.5–388.8 g
Nut weight	9.5–40.6 g
Shell weight	4.9–30.9 g
Kernel weight	0.41–7.58 g
Length	49.2–89.3 mm
Width	46.2–100.5 mm
Flesh/pulp depth	12.9–31.4 mm
Fruit share	
Particular	Share (%)
Mesocarp	89
Endocarp	8
Seed	3

Fig. 21.1 Natural geographical distribution of *I. gabonensis* (https://epo.wikitrans.net/Irvingia_gabonensis)

Characteristically, *I. gabonensis* is a moist forest tree species; mainly found in the tropical evergreen rain forest or in the moist semi-deciduous forest, within the altitude 0–500 m above sea level (CAB 2000). But then, for example in Nigeria, it may occur in deciduous forests, derived forests and the Guinea savannah forest (Onyekwelu and Stimm 2004). However, the tree doesn't thrive in swampy areas. As a gregarious tree species, about 2 or 3 individuals usually may be found growing together in the wild in low forest areas (Onyekwelu and Stimm 2004). Particularly, human settlements or cultivation are known to have contributed to the high density of the trees, which may look like natural forest. In addition, the trees are often found in home gardens, school compounds, farms, outskirts of villages or cities, though their population with time may decline as may be affected by human settlements and poor natural regeneration. At maturity, the tree can attain the height of 30–40 m and may take 6 to 10-year time scale to reach fruiting maturity. However, there are variations with respect to early fruiting observed in young trees of the species (Ladipo and Anegbe 1995).

As a highly, economically important tree, the fruit and seeds of *I. gabonensis* play a relevant role in food security of rural and urban dwellers. Many rural inhabitants living across the zone of its natural distribution are engaged in the collection, processing and sale of the plant products as a means of livelihood through rural employment. *I. gabonensis* ranks among the five major fruit trees occurring on traditional farms, out of 171 indigenous woody plants of economic importance identified within the forest zone of Nigeria (Onyekwelu and Stimm 2004). However, the prices of the products fluctuate with season, availability of the resource, dryness and appearance of the kernels, method of processing and bargaining power of the collectors and traders. In addition, the products are also processed into cubes and packaged for export.

21.3 Phytochemical Constituents, and Bioactive Compounds

21.3.1 Oil Seeds

The seeds of *I. gabonensis* is among the best oil seeds because it contains high amount of crude fat. The fat content is 65.46% (Egbekun and Ehieze 1997), which is higher than those of African breadfruit, 41.6% (Okoye et al. 1999), soybeans, 19.10% (Onimawo et al. 2003), rubber seeds 44.63% (Ikwuagwu et al. 2000), *Annona muricata*, 41.5%, *Pencrodethera macrophylla*, 50.8% and *Tetracarpidium conaphorous*, 41.5% (Okoye et al. 1999). Therefore, it is a very important contributor of oil with alternative uses in the industry aside providing food energy.

In addition, *I. gabonensis* seeds contained crude protein (7.76%), relatively high fiber content (10.23%) and low moisture (3.36%). However, they are reported not to be a very good source of plant protein compared to bambara groundnut (17–21%) (Onimawo et al. 1998), pigeon peas (23.77%) and cowpea seeds (24.14%) (Ifon and Bassir 1980). Essentially, the low moisture content implies that *I. gabonensis* seeds will keep well at room temperature. As a matter of fact any seed that has a moisture level greater than 15% is liable to rapid deterioration caused by mold growth, heat, insect damage and sprouting (Onimawo et al. 2003). Besides, the fiber content may provide bulk; thus, enhance bowel function.

Irvingia gabonensis seed oil is light yellow in colour, with yield of 67.33 ± 1.2, a pH of 6.32 ± 0.07 and a specific gravity of 0.85 ± 0.03 kg/dm^3 (Ekpe et al. 2018). In addition, aflatoxins were not detected, representing toxin free *I. gabonensis* seed oil.

The relative density in *I. gabonensis* seed oil is 0.82 ± 0.01, and refractive index (RI) of 1.37 ± 0.01. Compared to most oils, *I. gabonensis* seed oil has low RI and therefore, less thick. Fat refractive index is measured as the ratio of speed of light at a defined wavelength to its speed in the fat itself. The values may vary per the wavelength and temperature, the degree and type of unsaturation, the type of substitutions of component fatty acids and with accompanying substances. Usually, refractive index is utilized in quality control to check for the purity of ingredients and to follow hydrogenation and isomerization. However, refractive index (1.37 ± 0.01) recorded in *I. gabonensis* seed oil falls within the acceptable range of 1.4677–1.4707 for virgin, refined and refined-pomace oils per the Codex Standards for fats and oils from vegetable origin (Alimentarius 1999).

When fatty materials are heated in the presence of air, the smoke point is determined by its thermal stability. Glycerides are much stable than fatty acids, hence, in ordinary oils, the presence of smoke points depend mainly in their amount of free fatty acids. The increase in the degree of unsaturation triggers melting point to decrease. Therefore, oils with lower melting point are probably preferred and important in the manufacturing of soft and easy-to- digest margarine while those with higher melting point are used in the manufacture of oil creams. Free fatty acid content was 7.37 ± 0.12%.

Irvingia gabonensis seed oil has a low iodine value of 13.40 ± 0.17 which is highly saturated with a high saponification value of 230.95 ± 1.62 mgkOH/g. Hence, the low iodine and high saponification value is an indication that the oil is suitable for the manufacturing of soap, cosmetics, pharmaceuticals, margarine and cooking oil (Okoronkwo et al. 2014). In addition, *I. gabonensis* oil has a peroxide value of 2.67 ± 0.35 meq/kg which is lower than the peroxide value of other oils such as groundnut oil and refined olive oil with reported amount of 9 and 11 meq/kg, respectively (Matos et al. 2009).

The percentage fatty acid composition in *Irvingia gabonensis* seed oil are caproic acid (0.55 ± 0.05), caprylic acid (0.07 ± 0.00), capric acid (1.6 ± 0.1), lauric acid (40.5 ± 0.2), myristic acid (42.75 ± 0.25), palmitic acid (4.7 ± 0.1), stearic acid (0.85 ± 0.05), oleic acid (4.5 ± 0.1), erucic acid (0.06 ± 0.01), linoleic acid (0.65 ± 0.05), alpha-linoleic acid (4.2 ± 0.1) and arachidonic acid (0.14 ± 0.00). According to these findings, myristic acid has been the most abundant followed by lauric acid; which implies that *I. gabonensis* seed oil can be considered as myristic-lauric oil (Ekpe et al. 2018).

The changes in oil stability from 2.67 ± 0.35 at day 0 to 5.18 ± 0.02 at 28 days indicate very slow changes, largely depicted a stable *I. gabonensis* oil (Table 21.2). Therefore, it implied that the oil was stable (longer shelflife) if it is properly handled and kept in a favorable environment (Okoronkwo et al. 2014).

The raw and defatted *I. gabonensis* seeds have been reported to contain significant amount of the minerals: magnesium, potassium, sodium, phosphorus, calcium, zinc, copper, manganese plus exceptionally low lead level of 0.03 mg/g (Fig. 21.2). The highest mineral was magnesium with the value of 48.2 mg/g followed by potassium (43.1 mg/g) and sodium (30.1 mg/g) respectively (Niyi 2014). The Na/K and Ca/P ratios are nutritionally essential. According to (Shils and Shike 2006), the modern diets rich in animal proteins and phosphorus may enhance the absence of calcium in the urine. This formed the concept of the Ca/P ratio. Thus, if the Ca/P is low (low calcium, high phosphorus) more than the normal levels of calcium may be absent in the urine in that way calcium level will be decreased in bones (Adeyeye 2013). Hence, food is considered 'good' if the proportion is beyond one and 'poor' if the ratio is less than 0.5 (David and Aderibigbe 2010). The Ca/P ratio of *I. gabonensis* ranged between 0.60 and 0.91 an indication that Ca/P ratio of defatted sample was good but that of raw was fair. The ratios of Na/K in raw and defatted

Table 21.2 Storage stability of *Irvingia gabonensis* seed oil

Parameters %	Value (%)
Oil stability (0 day)	2.67 ± 0.35
Oil stability (7 days)	3.18 ± 0.28
Oil stability (14 days)	3.91 ± 0.03
Oil stability (21 days)	4.37 ± 0.03
Oil stability (28 days)	5.18 ± 0.02

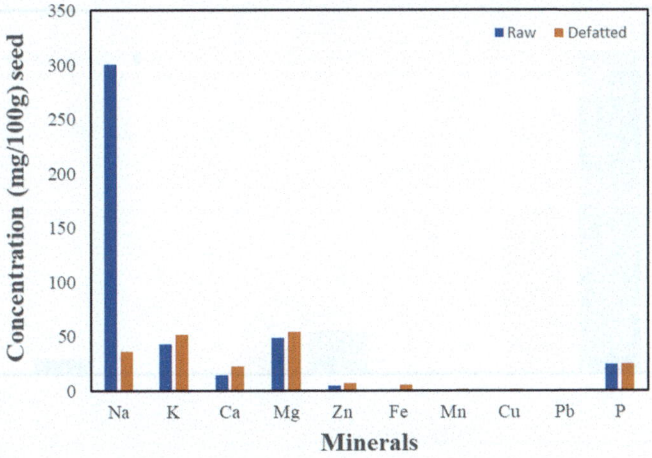

Fig. 21.2 Mineral composition (mg/100 g) of raw and defatted *Irvingia gabonensis* seeds

was 0.70 and 0.71, respectively. A Na/K ratio of 0.60 is suggested (David and Aderibigbe 2010), but the reported values were slightly above 0.60 indicating that they would not promote high blood pressure diseases. Generally, the iron levels were in the range of 2.25–5.15%. The iron content of defatted *I. gabonensis* seeds was comparable with that of *Luffa cylindrical* kernel (5.3%) (Olaofe et al. 2008), but higher than that of kidney bean (Olaofe et al. 2010), quinoa (Ogungbenle 2003) and date palm fruit (Ogungbenle 2011) recording 4.58%, 2.6% and 3.68%, respectively. Iron is very important in blood formation (Niyi 2014) and defatted *I. gabonensis* contained a good amount of iron (Fig. 21.2).

Though *Irvingia gabonensis* seeds contains anti-nutritional factors but the levels are lower compared to other crops. For instance, the phytate value shown in Fig. 21.3 is lower than the values found in kidney bean (40.8 mg/g) (Olaofe et al. 2010), soy bean (40.5%) and cowpea (20.4 mg/g) (Aletor and Omodara 1994). High phytate levels can negatively affect digestibility by chelating with calcium or bind to substrate or proteolytic enzyme (Niyi 2014). In addition, phytate increases time of cooking in legumes (Nwokolo and Bragg 1977). The oxalate in defatted seeds was lower than values of other crop seeds such as 3.65 mg/g in kidney bean, 2.54 mg/g in soy bean, 7.34 mg/g in cowpea (Aletor and Omodara 1994), 6.01 mg/g in date palm fruit (Ogungbenle 2008), 5.22 mg/g in sorghum (Oberleas and Harland 1981), and 4.06 mg/g millet (Mills 1980). Tannin level in raw and defatted *I. gabonensis* seeds were 3.75% and 3.22%, respectively which were higher than those found in cooked walnut (2.33%) (Ogungbenle 2009), *Afzelia Africana* 0.43% (Ogungbenle and Omaejalile 2010) and kidney bean 0.77% (Olaofe et al. 2010). Therefore, the defatting process contribute to the decreasing anti-nutritional factors in *I. gabonensis* seeds.

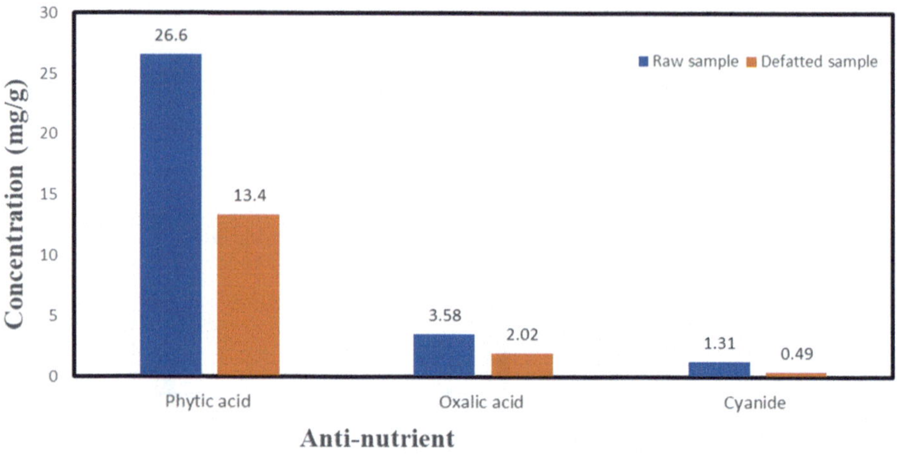

Fig. 21.3 Anti-nutritional factors in raw and defatted seeds of *Ivengia gabonensis*

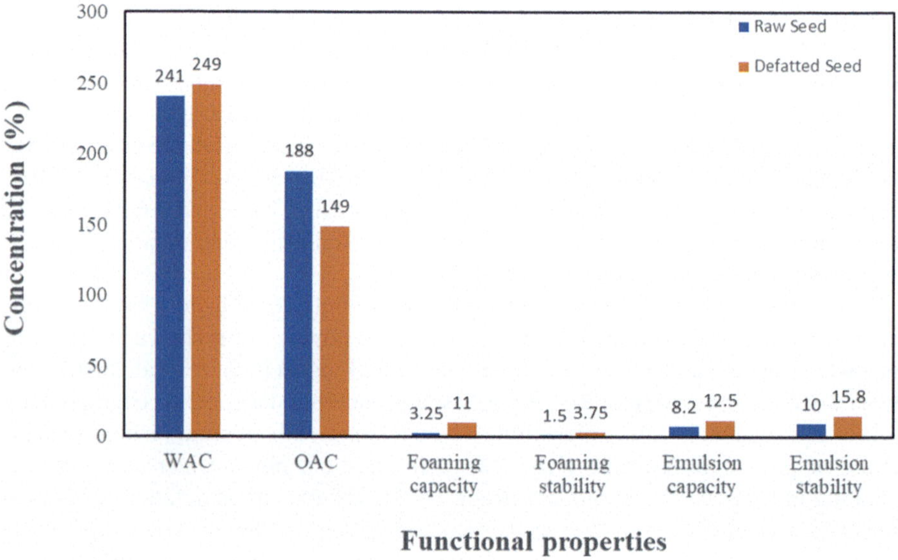

Fig. 21.4 Functional characteristics of raw and defatted *I. gabonensis* seeds. WAC; water absorption capacity, OAC; oil absorption capacity

Raw and defatted *I. gabonensis* seeds showed varying levels of functional characteristics (Fig. 21.4). Water absorption capacity (WAC) is an important factor considered in the production of viscous foods (Niyi 2014). However, the WAC was 241% in raw seeds and 249% in the defatted seeds. Clearly, the values are higher than 165%, 182%, 115% and 147% recorded in kidney bean (Olaofe et al. 2010), beniseed, pearl millet and quinoa, respectively (Oshodi et al. 1999) and the range of 100–200% for some legumes (Ogungbenle 2006). Hence, the WAC values present in *I. gabonensis* seeds makes it more hydrophilic. Therefore, the favorable WAC of

I. gabonensis seed presents it as useful raw material in the production of soups, gravies and dough. Oil absorption capacity (OAC) in *I. gabonensis* was 148.9–188.3% which is higher than soy flour (142.5%) (Lin et al. 1974), pigeon pea (89.7%) (Oshodi and Ekperigin 1989) and fluted pumpkin seed flour (142.5%) (Fagbemi and Oshodi 1991), however, it was found to be lower than those found in cowpea flour (281–321%) (Olaofe et al. 1994) and African nutmeg (256%) (Ogungbenle and Adu 2012). This functional property of OAC plays an important role in increasing the mouth perception of foods and acts to retain flavour(Tolstoguzov 1998). The gelation concentration for raw and defatted *I. gabonensis* seeds was 12% w/v and 14%w/v, respectively; probably this values might have been induced by the high value of carbohydrate content recorded in the seed (Niyi 2014). Consequently, the seed may be considered important in the curd production or as an additive to other materials for forming gel in food products. More so, the foaming capacity will enrich the functionality of *I. gabonensis* seeds in the production of cakes, whipping and toppings in bakery (Tolstoguzov 1998).

The amino acid composition of *I. gabonensis* protein after acid hydrolysis showed varying concentration of amino acid in the seed (Fig. 21.5) Methionine, which is needed for the synthesis of other important compounds including choline was present and had an amount of 0.86 g/100 g. Methionine intake may prevent fatty liver and enhances development and growth (Adeyeye 2009). Also, *I. gabonensis* seeds contain 1.16 g/100 g phenylalanine, whereas Isoleucine (2.32 g/100 g) is the major abundant amino acid present in the seed. However, the total essential amino acid of12.43 g/100 g in the seeds appear to be lower than African locust bean (81.2 g/100 g) (Adeyeye 2006), pumpkin seeds (38.3 g/100 g) (Asiegbu 1987), soybean (44.4 g/100 g) (Holland et al. 1989), scarlet runner bean (Aremu et al. 2006), cowpea (42.6% g/100 g) (Olaofe et al. 1994), and pigeon pea (43.60 g/100 g)

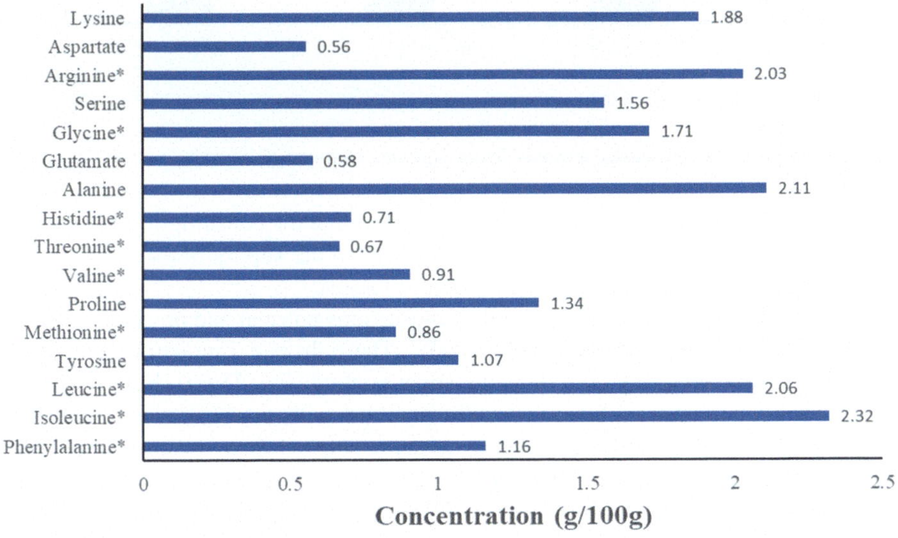

Fig. 21.5 Amino acid profile of *I. gabonensis* seeds. ∗ denotes essential amino acids

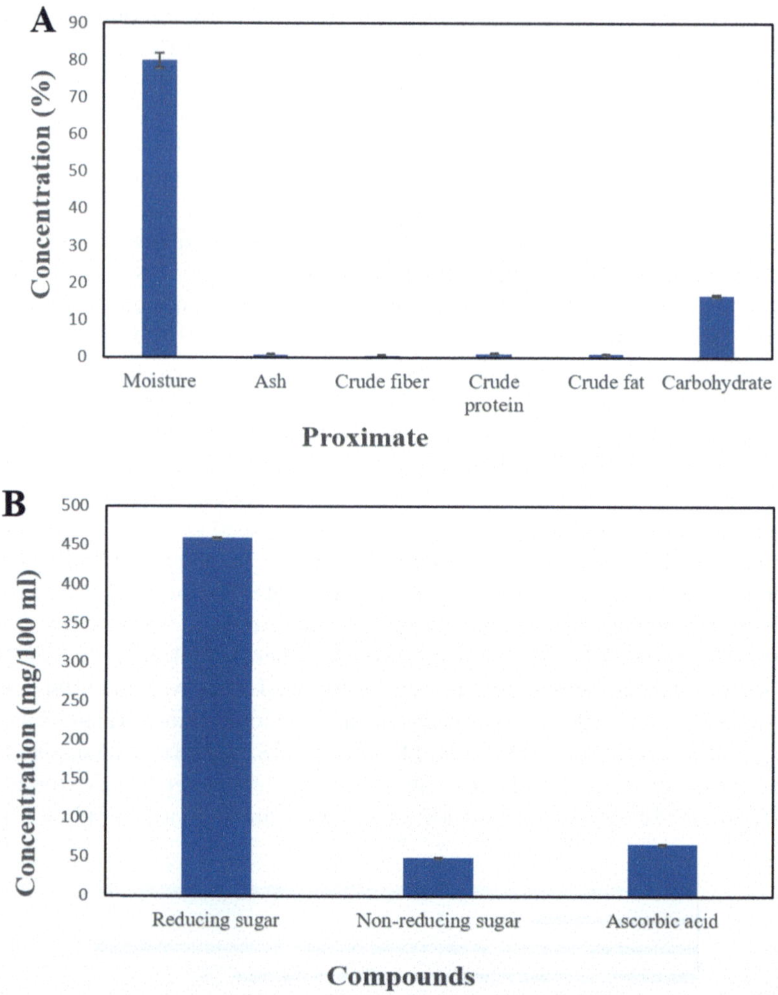

Fig. 21.6 Proximate and chemical properties of pulp of *I. gabonensis*

(Oshodi et al. 1993). Therefore, the results suggest that the consumption of *I. gabonensis* may contribute little or significant amount to the supply of essential amino acid in the diet as compared to other legume-oil seeds.

Fruit pulp of *I. gabonensis* is palatable and various nutrient and physicochemical properties have been reported (Fig. 21.6a/b). The pulp has a moisture content of 80% (Fig. 21.6a) and very high reducing sugar content (Fig. 21.6b) (Egbekun et al. 1996), showing that the edible pulp has the potential to be used for juice production (Onimawo et al. 2003). The mineral content in the pulp is low, a reflection of the low ash content (0.8%), out of which 0.12% is made up of water soluble ash (as potassium trioxo carbonate (iv), K_2CO_3). Whiles, the calcium content of the ash is relatively high (262.3 mg/100 g). Other values for crude fiber, crude fat and crude protein of the pulp are indicated in Fig. 21.6 (Egbekun et al. 1996). The pulp of *I. gabonensis* contains 10.0 (Brix°) total soluble solids (TSS). The low value of TSS may be an indication that osmophilic organisms like yeast may find it difficult to grow in the pulp; consequently they are not likely to contribute to its spoilage (Enujiugha and Ayodele-Oni 2003).

The pulp has slightly bitter taste due to its slightly acidic value (pH of 5.8). The ascorbic acid content (66.7 mg/100 ml) in the pulp is relatively high, signifying a good source of vitamin C. This also implies that processing of the pulp into juice will supply a good amount of vitamin C, simple sugars (459.7 mg/100 ml) and some protein (1.09%). The pulp also has the following physical characteristics: viscosity, 1.2 poiseuille (Nsm − 2), refractive index (1.3355), and specific gravity (1.012).

21.4 Traditional Uses

21.4.1 Fresh Fruits/Pulp

The fruit is composed of a fleshy part and nut (Fig. 21.7). The nut consists of a hard shell and kernel/seed. The seeds have an outer brown testa (hull) and two white cotyledons (Ndoye et al. 1997). Fruits are eaten fresh although the taste varies between sweet and bitter depending on the ripening stage. The juice extracted from fruit is recommended for production of wine (Akubor 1996). Also, the juicy pulp can be further processed into jelly and jam. The fruit juice of *I. gabonensis* was also produced into ancient wines and liquors, which were left to ferment for 28 days to obtain its alcohol content.

According to the sensory assessment of the alcoholic drinks produced from the fruit pulp dika, there was no significant difference from a German reference wine in terms of colour, mouth feel, sweetness, flavour and general acceptability (Leakey et al. 2003). The fruit pulp can also be used to produce black dyes for cloth coloration.

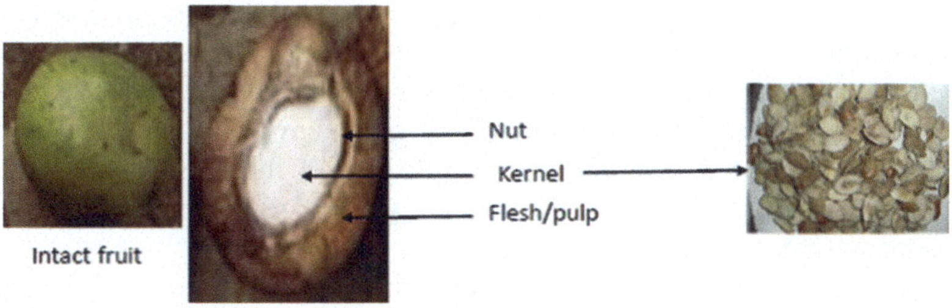

Fig. 21.7 Section of *Irvingia gabonensis* fruit

21.4.2 Seeds and Kernels

Apart from the fresh intact fruits that are eaten fresh, the kernel is used to make a variety of products and they are widely marketed in countries where the tree occurs (Ndoye et al. 1997). The kernel forms a significant part of diet, providing carbohydrate, oil, and protein, to enhance health and nutrition. Whereas, the sun-dried seeds are processed into flour and used as soup thickeners, thus, the white cotyledons are roasted and eaten; besides the roasted seeds enhances flavour and aroma on foods especially vegetables (Iponga et al. 2018). The food gum component of the seeds can be utilized as a thickening agent in water (especially hot water) (Ndjouenkeu et al. 1996). Also, the kernels are processed manually by traditional techniques and used as thickener in soups or as paste (Adamson et al. 1990). Similarly, in Cameroon, the kernels are used as condiment and are extremely treasured for their food thickening properties to prepare "ndo'o" or "draw soup" as well as groundnut or njansang (*Ricinodendron heudelotii*) (Tchoundjeu 2006).

Ofoefule et al. (1997) reported that dika fat, a vegetable oil obtained from the kernel is also used in the formulation of sustained released frusemide granules and a highly gross energy is obtained from it compared to other tropical seeds; this is because of its high fat content. Recently, Ekpe et al. (2018) indicated that seed endosperm is removed by cracking the seed coat and it this be eaten raw or roasted. Often, they are pounded to butter-like or chocolate-like block. Also, the seeds can be used to produce edible oil (solid at ambient temperature) or margarine, and can be used for cooking. The extracted oil can be used produce soap or cosmetics. The by-product (press cake) after pressing oil out of the seeds can be used as cattle feed or as thickening agent for soup as mentioned earlier. According to Onwuka et al. (1984), solvent extraction of dika nut seed yielded between 68% and 75% fat. Similarly, other authors (Ogunsina et al. 2012; Okolo 1987) indicated that dika nut seed nutrients consist of 8.65% protein, 14.1% carbohydrate, 2.1% moisture, 1.4% crude fibre, 16.8% ash and 38.9% dietary fibre.

21.4.3 Bark and Stem

The bark of *I. gabonensis* is used for treating dysentery, scabies, toothache and skin diseases (Okolo et al. 1995). It is used internally as a purgative, anti-poison agent and in the treatment of hernias and yellow fever (Ayuk et al. 1999) and considered to be a powerful aphrodisiac. The stems of *I. gabonensis* tree have been used as chewing sticks to clean the teeth (Ainge and Brown 2004b). In Senegal, the decoction of the stem bark is used in the treatment of gonorrhea, hepatic and gastrointestinal disorders (Donfack et al. 2010).

The wood (non-timber forest product) obtained from the stem is pale brown, very hard and fine grained, but not easy to cut that limits its usefulness. Though the weight limits it from all construction works, it is most rugged for other works such as railway ties. Also, the wood is suitable for making pestles, yam mortars, also suitable for making canoes, boards, planking, ship decking and paving blocks.

21.4.4 Leaves

The *I. gabonensis* leaves, when combined with palm oil are used to stop haemorrhage in pregnant women (Gbadegesin et al. 2014). The leaf extracts are known to have diuretic and hypotensive effects (Nosiri et al. 2009). The leaf extract has also been reported to have hepatoprotective, cytoprotective and anti-ulcer effects on aspirin-induced ulcer (Nosiri et al. 2018). It has also been reported to demonstrate hepatoprotective and anti-clastogenic effects on sodium arsenite-induced toxicity in Wistar rats (Gbadegesin et al. 2014). The leaf extract also has hepato-protective and nephronprotective effect on cadmium chloride-induced toxicity in Wistar rats (Ewere et al. 2016). The juice has been reported to have renal and hepato-protective effects on sodium fluoride-induced toxicity in Wistar rats (Emejulu et al. 2016).

Apparently, drink extract can be prepared from leaves of *I.* gabonensis, which has an appreciable phytonutrients content. Phytonutrients represents an important nutritional indicator of the usefulness of *I. gabonensis* leaves in producing vegetable drink extract. Among phytonutrients, total phenol, flavonoids and alkaloids were most enhanced by shade drying before extraction, whereas carotenoids, saponins and steroids were decreased by same (Mbaeyi-Nwaoha and Akobueze 2018). The concentrations of anti-nutrients in the leaves drink extracts were significantly low and unable to interfere with nutrients utilization. The presence of anti-nutritional factors in many tropical plants constitute a major factor limiting the wider food use as anti-nutritional factors hinder the bio-availability of nutrients of. In addition, the leaves drink also had considerable contents of vitamins A, E and B_1. Vitamin A is essential for several bodily function, including the production pigments responsible for vision, aids in resistance to infectious agents and maintains and ensure that many epithelial cells are in good of health. Vitamin E is important for absorption of iron, slowing of the ageing process, and fertility. It is a powerful antioxidant which helps protect cells from damage by free radicals. Vitamin B complexes are necessary for converting blood sugar to energy, keeping the nervous system healthy; B_1 promotes growth.

21.5 Medicinal Uses

Every part of a typical *I. gabonensis* tree such as the bark, leaves, seeds and seed kernels are used for medicinal purposes. In recent times, consideration has been given to the use of medicinal plants in the treatment of several diseases or disorders. According to Ngondi et al. (2005), the seed has been used solely or as supplement in treating type II diabetics and in reducing obesity incidence. Obese patients given

I. gabonensis seed extract had a sufficient decrease in fasting blood glucose, triglycerides, total cholesterol, LDL-cholesterol, with an increase in HDL-cholesterol. However, the placebo showed no changes in blood glucose and lipid components (Ngondi et al. 2005). Also, it has been proven by research that *I. gabonensis* seeds delay stomach emptying, which eventually lead to a more gradual absorption of dietary sugar. The consequence of the above phenomenon can reduce the elevation of blood sugar levels, which is typical after a meal (Vuksan et al. 1999).

The sticky wax (mucilage) in the seeds is suitable for making medicinal tablets. The wax acts as a binding agent during the production of tablets. Tablets manufactured with *I. gabonensis* seeds have displayed increased brittleness and reduced tensile strength, compared with gelatin tablets. The powdered kernels of seeds can be used as an astringent applied to the skin to soothe burns and reduce bleeding from minor abrasions (Irvine 1961), and treat wounds (Okwu and Josiah 2006). Aqueous and ethanol extracts of *I. gabonensis* seeds have been used to improve renal and hepatic functions (Bassey et al. 2018; Obianime and Uche 2010). In addition, the bark of *I. gabonensis* contains antibiotic properties, and it's very effective for treating skin bruises, the boiled bark is used to cure toothaches, whereas the outer portion of stem bark and leaves are ingested orally for treating dysentery and hernia (Ayuk et al. 1999; Okolo et al. 1995).

21.6 Services

Irvingia gabonensis, when interplanted with other species is able to control soil erosion (Orwa et al. 2009). The entire tree can also be cultivated to provide shade to cocoa and coffee plants. In addition, rural household could be encouraged to cultivate and establish plantation which could go a long way to help household food security. Furthermore, the many uses that the plant has could enable rural households earn extra income through the sale of the fruits and the products.

21.7 Conclusions

The available data show that *I. gabonensis* is often neglected, however, the findings from this work present to us the essential information on *I. gabonensis* and its derivable potential benefits. The various parts of *I. gabonensis* are rich in chemical composition; thus, serve as good sources of edible oil, protein and essential minerals. Also, these qualities enable it to be suitable in formulations and fortifications in the food and pharmaceutical industries. The seed contains appreciable amount of carbohydrate, protein and high viscosity; hence, it serves as both soup thickener and food condiment. The other parts including the fruits, leaves, bark and root could be effectively used as ingredients in health and functional food to improve certain disease conditions. Presently, fruits of *I. gabonensis* are regarded underutilized and the status warrants a transformation to offer substantial scope for enhancing the nutritional and economic security of rural livelihood particularly in the tropical areas.

References

Adamson I, Okafor C, Abu-Bakare A (1990) A supplement of Dikanut (*Irvingia gabonesis*) improves treatment of type II diabetics. West Afr J Med 9:108–115

Adeyeye E (2006) Nutritional composition and food properties of whole and defatted kernel flours of *Gossypium barbadense* (L). J Technoscience 6:22–30

Adeyeye E (2009) Amino acid composition of some aquaculture fauna resources in Nigeria. In: Haugen S, Meijer S (eds) Handbook of nutritional biochemistry: genomics, metabolomics and food supply, pp 225–254

Adeyeye, E. I. 2013. Proximate, mineral and antinutrient composition of dika nut (Irvingia gabonensis) kernel. Elixir. Food Sci. 58: 14902–14906.

Agbogidi O, Okonta B, Dolor D (2005) Participation of rural women in sustainable forest management and development. In: Environmental sustainability and conservation in Nigeria. Jubee-Niyi Publishers, Akure

Ainge L, Brown N (2004a) Bush Mango (*Irvingia gabonensis* and *I. wombolu*). The Key Non-Timber Forest Products of Central Africa: State of the Knowledge, 15

Ainge L, Brown N (2004b) *Irvingia gabonensis* and *Irvingia wombolu*: a state of knowledge report. Clark, L., Sunderland, T., The key non-timber forest products of Central Africa: state of the knowledge. United States Agency for International Development (USAID), Technical Paper 122, 15–36

Akubor P (1996) The suitability of African bush mango juice for wine production. Plant Foods Hum Nutr 49:213–219

Aletor V, Omodara O (1994) Studies on some leguminous browse plants, with particular reference to their proximate, mineral and some endogenous anti-nutritional constituents. Anim Feed Sci Technol 46:343–348

Alimentarius C (1999) Codex standard for named vegetable oils. Codex Stan 210:1–13

Aremu MO, Olaofe O, Akintayo TE (2006) A comparative study on the chemical and amino acid composition of some Nigerian under-utilized legume flours. Pak J Nutr 5:34–38

Asiegbu JE (1987) Some biochemical evaluation of fluted pumpkin seed. J Sci Food Agric 40:151–155

Atangana A, Tchoundjeu Z, Fondoun J, Asaah E, Ndoumbe M, Leakey R (2001) Domestication of Irvingia gabonensis: 1. Phenotypic variation in fruits and kernels in two populations from Cameroon. Agrofor Syst 53:55–64

Atangana A, Ukafor V, Anegbeh P, Asaah E, Tchoundjeu Z, Fondoun J-M, Ndoumbe M, Leakey R (2002) Domestication of *Irvingia gabonensis*: 2. The selection of multiple traits for potential cultivars from Cameroon and Nigeria. Agrofor Syst 55:221–229

Ayuk ET, Duguma B, Franzel S, Kengue J, Mollet M, Tiki-Manga T, Zenkeng P (1999) Uses, management and economic potential of *Irvingia gabonensis* in the humid lowlands of Cameroon. For Ecol Manag 113:1–9

Bassey SO, Udefa A, Ekpe O, Essien NM, Eteng MU (2018) Ethanol extract of *Irvingia gabonensis* (bush mango) seed improves renal and hepatic functions in wistar rats. Int J Pharm Bio Sci 13:46–50

Brands S (1989) The taxonomicon. Universal Taxonomic Services, Zwaag. Accessed January 15, 2012

CAB (2000) Forestry Compendium Global Module 1

Cartea ME, Francisco M, Soengas P, Velasco P (2010) Phenolic compounds in Brassica vegetables. Molecules 16:251–280

David OM, Aderibigbe EY (2010) Microbiology and proximate composition of ogiri, a pastry produced from different melon seeds. New York Sci J 3:18–27

Donfack JH, Fotso GW, Ngameni B, Tsofack FN, Tchoukoua A, Ambassa P, Abia W, Tchana A, Giardina S, Buonocore D (2010) In vitro hepatoprotective and antioxidant activities of the crude extract and isolated compounds from Irvingia gabonensis. Asian J Trad Med 5:79–88

Egbekun M, Ehieze M (1997) Proximate composition and functional properties of fullfat and defatted beniseed (*Sesamum indicum* L.) flour. Plant Foods Hum Nutr 51:35–41

Egbekun MK, Akowe JI, Ede RJ (1996) Physico-chemical and sensory properties of formulated syrup from black plum (*Vitex doniana*) fruit. Plant Foods Hum Nutr 49:301–306

Ekpe OO, Bassey S, Udefa A, Essien NM (2018) Physicochemical properties and fatty acid profile of Irvingia gabonensis (Kuwing) seed oil. Int J Food Sci Nutr 3:153

Ekundayo F, Oladipupo O, Ekundayo E (2013) Studies on the effects of microbial fermentation on bush mango (*Irvingia gabonensis*) seed cotyledons. AJMR 7:4363–4367

Emejulu A, Alisi C, Asiwe E, Igwe C, Nwogu L, Onwuliri V (2016) Renal and hepato-protective effects of Irvingia gabonensis juice on sodium fluoride-induced toxicity in Wistar rats. J Clin Toxicol 6:296. https://doi.org/10.4172/2161-0495.1000296

Enujiugha VN, Ayodele-Oni O (2003) Evaluation of nutrients and some anti-nutrients in lesser-known, underutilized oilseeds. Int J Food Sci Technol 38:525–528

Ewere EG, Oyebadejo SA, Peter VC (2016) Ethanolic leaf extract of Irvingia gabonensis (O'Rorke) Baill protects against nephrotoxicity and hepatotoxocity in cadmium-induced Wistar albino rats. Int J Pharm Tox 4:105–110

Fagbemi T, Oshodi A (1991) Chemical composition and functional properties of full fat fluted pumpkin seed flour. Nig Food J 9:26–32

Gbadegesin M, Adegoke A, Ewere E, Odunola O (2014) Hepatoprotective and anticlastogenic effects of ethanol extract of *Irvingia gabonensis* (IG) leaves in sodium arsenite-induced toxicity in male Wistar rats. Niger J Physiol Sci 29:29–36

Hall JN, Moore S, Harper SB, Lynch JW (2009) Global variability in fruit and vegetable consumption. Am J Prev Med 36:402–409. e405

Holland B, Unwin I, Buss D (1989) Milk products and eggs. Fourth supplement to McCance and Widdowson's the composition of foods, 4th edn. Royal Society of Chemistry, Cambridge

Ifon E, Bassir O (1980) The nutritive value of some Nigerian leafy green vegetables—part 2: the distribution of protein, carbohydrates (including ethanol-soluble simple sugars), crude fat, fibre and ash. Food Chem 5:231–235

Ikwuagwu O, Ononogbu I, Njoku O (2000) Production of biodiesel using rubber [Hevea brasiliensis (Kunth. Muell.)] seed oil. Ind Crop Prod 12:57–62

Iponga D, Yobo CM, Ingram V, Bengone NN, Ngoye A (2018) Livelihoods, economic contribution and sustainability of the bush mango (*Irvingia gabonensis*) value chain from three provinces of Gabon. Int For Rev 20:115–129

Irvine FR (1961) Woody plants of Ghana. Woody plants of Ghana

Kolar FR, Kamble VS, Dixit GB (2011) Phytochemical constituents and antioxidant potential of some underused fruits. Afr J Pharm Pharmacol 5:2067–2072

Koyejo A, Omokhua G (2001) Propagation studies on *Irvingia wombulu* (syn. Excelsa) Ex. Okafor and agroforestry species. Proceedings of FAN Conference held in Abuja, 17th, pp. 183–187

Ladipo D, Anegbe P (1995) Flowering and fruiting in field grown bush mango (*I. gabonensis* and *I. wombolu*) in southeast Nigeria. 23rd Annual, Conference of the Genetics Society of Nigeria

Leakey RRB, Akinnifesi FK (2008) Towards a domestication strategy for indigenous fruit trees in the tropics. In: Akinnifesi FK, Leakey RRB, Ajayi OC, Sileshi G, Tchoundjeu Z, Matakala P, Kwesiga F (eds) Indigenous fruit trees in the tropics: domestication, utilization and commercialization. CAB International, Wallingford, pp 28–49.

Leakey R, Schreckenberg K, Tchoundjeu Z (2003) The participatory domestication of West African indigenous fruits. Int For Rev 5:338–347

Leakey RR, Tchoundjeu Z, Smith RI, Munro RC, Fondoun J-M, Kengue J, Anegbeh PO, Atangana AR, Waruhiu AN, Asaah E (2004) Evidence that subsistence farmers have domesticated indigenous fruits (*Dacryodes edulis* and *Irvingia gabonensis*) in Cameroon and Nigeria. Agrofor Syst 60:101–111

Lin MJ-Y, Humbert E, Sosulski F (1974) Certain functional properties of sunflower meal products. J Food Sci 39:368–370

Liu RH (2003) Health benefits of fruit and vegetables are from additive and synergistic combinations of phytochemicals. Am J Clin Nutr 78:517S–520S

Matos L, Nzikou J, Matouba E, Pandzou-Yembe V, Mapepoulou TG, Linder M, Desobry S (2009) Studies of *Irvingia gabonensis* seeds kernels: oil technological applications. Pak J Nutr 8:151–157

Mbaeyi-Nwaoha I, Akobueze E (2018) Effect of processing methods on the chemical composition and microbiological quality of vegetable drink extract of african bush mango (*Irvingia gabonensis*). Afr J Microbiol Res 12:433–447

Mills B (1980) Recommended dietary allowances. National Academy of Sciences, 2101 Constitution Avenue, Washington, DC 20418, USA (1980), 185, $6.00. Elsevier

Ndjouenkeu R, Goycoolea F, Morrisa E, Akingbala J (1996) Rheology of okra (*Hibiscus esculentus* L.) and dika nut (*Irvingia gabonensis*) polysaccharides. Carbohydr Polym 29:263–269

Ndoye O, Pérez MR, Eyebe A (1997) The markets of non-timber forest products in the humid forest zone of Cameroon. Overseas Development Institute, London, UK

Ngondi JL, Oben JE, Minka SR (2005) The effect of *Irvingia gabonensis* seeds on body weight and blood lipids of obese subjects in Cameroon. Lipids Health Dis 4:12

Niyi OH (2014) Chemical and amino acid composition of Raw and Defatted African Mango (*Irvingia gabonensis*) kernel. Br Biotechnol J 4:244–253

Nosiri I, Abdu-Aguye I, Hussaini M, Abdurahaman E (2009) Leaf extracts of Irvingia Gabonensis increase urine output and electrolytes in rats. Internet J Altern Med 8

Nosiri C, Atasie O, Idume J, Barney D (2018) Cytoprotective and anti ulcer activities of the ethanolic leaf extract of *Irvingia gabonensis* on aspirin-induced ulcer. IOSR J Pharm Biol Sci 13(2):16–20

Nwokolo E, Bragg D (1977) Influence of phytic acid and crude fibre on the availability of minerals from four protein supplements in growing chicks. Can J Anim Sci 57:475–477

Oberleas D, Harland BF (1981) Phytate content of foods: effect on dietary zinc bioavailability. J Am Diet Assoc 79:433–436

Obianime A, Uche F (2010) The phytoconstituents and the comparative effects of aqueous extract of *Irvingia gabonensis* seeds and proviron on the biochemical parameters of male Guinea pigs. Asian Pac J Trop Med 3:101–104

Ofoefule S, Chukwu A, Okore V, Ugwah M (1997) Use of dika fat in the formulation of sustained release frusemide encapsulated granules. Boll Chim Farm 136:646–650

Ogungbenle H (2003) Nutritional evaluation and functional properties of quinoa (*Chenopodium quinoa*) flour. Int J Food Sci Nutr 54:153–158

Ogungbenle H (2006) Chemical composition, functional properties and amino acid composition of some edible seeds. Rivista Italiana Delle Sostanze Grasse 83:81–86

Ogungbenle H (2008) Nutritional quality and amino acid composition of date palm fruit (Phoenix doctylifera L.) flour. Rivista Italiana Delle Sostanze Grasse 85:54–59

Ogungbenle HN (2009) Chemical and amino acid composition of cooked walnut (*Juglans regia*) flour. Pak J Sci Ind Res 52:130–133

Ogungbenle H (2011) Chemical, *in vitro* digestibility and fatty acids composition of African nutmeg. Annals Sci Biotechnol 2:46–51

Ogungbenle HN, Adu TO (2012) Proximate composition and functional properties of dehulled African nutmeg (*Monodora myristica*). Series A: Physical Sciences 80

Ogungbenle H, Omaejalile M (2010) Functional and anti-nutritional properties, in-vitro protein digestibility and amino acid composition of dehulled *Afzelia africana* seeds. Pak J Sci Ind Res 53:265–270

Ogunsina B, Bhatnagar A, Indira T, Radha C (2012) The proximate composition of African Bush mango kernels (*Irvingia gabonensis*) and characteristics of its oil. IFE J Sci 14:177–183

Okafor J, Ujor G (1994) Varietal differences in Irvingia gabonensis. Bush mango and close relatives. Proceedings of a West African Collection Workshop held in Ibadan, Nigeria, pp. 10–11

Okolo H (1987) Indigenous Manufacturing of selected Nigerian foods in packaged form for defence usage. First Ministry of Defence TRADOC Seminar Benin City

Okolo CO, Johnson PB, Abdurahman EM, Abdu-Aguye I, Hussaini IM (1995) Analgesic effect of Irvingia gabonensis stem bark extract. J Ethnopharmacol 45:125–129

Okoronkwo C, Agoha E, Ogodo A, Nwachukwu N (2014) Physical and chemical characteristics of the African bush mango (*Irvingia gabonensis* var *garbonensis*) seed oil. Int J Adv Eng Manag 1:28–31

Okoye W, Okobi O, Okonkwo E (1999) Extraction and characterization of oils from nine lesser known Nigerian oil seeds. Proceedings of 23rd Annual NIFST conference, 25th, pp. 232–234

Okwu DE, Josiah C (2006) Evaluation of the chemical composition of two Nigerian medicinal plants. Afr J Biotechnol 5:357–361

Olaofe O, Umar Y, Adediran G (1994) The effect of nematicides on the nutritive value and functional properties of cowpea seeds. Vigna unguiculata. Food Chem 46:337–341

Olaofe O, Okiribiti B, Aremu M (2008) Chemical evaluation of the nutritive value of smooth luffa (Luffa cylindrica) seed's kernel. Elec J Env Agricult Food Chem Title 7:3444–3452

Olaofe O, Famurewa J, Ekuagbere A (2010) Chemical and functional properties of kidney bean seed (*Phaseolus vulgaris* L.) flour. Int J Chem Sci 3:51–69

Onimawo I, Momoh A, Usman A (1998) Proximate composition and functional properties of four cultivars of bambara groundnut (*Voandezeia subterranea*). Plant Foods Hum Nutr 53:153–158

Onimawo I, Oteno F, Orokpo G, Akubor P (2003) Physicochemical and nutrient evaluation of African bush mango (*Irvingia gabonensis*) seeds and pulp. Plant Foods Hum Nutr 58:1–6

Onwuka N, Peczak R, Babuchowski A (1984) The characteristics indexes and the major fatty acid components of oils from the less common oilseeds. 1: Pentaclethra macrophylla (PM) and *Irvingia gabonensis* (IG). Rivista Italiana delle Sostanze Grasse (Italy) 61:569

Onyekwelu JC, Stimm B (2004) Irvingia gabonensis. In: Enzyklopädie der Holzgewächse: Handbuch und Atlas der Dendrologie, 1–14

Orwa C, Mutua A, Kindt R, Jamnadass R, Simons A (2009) Agroforestree database: a tree species reference and selection guide version 4.0. World Agroforestry Centre ICRAF, Nairobi

Oshodi A, Ekperigin M (1989) Functional properties of pigeon pea (*Cajanus cajan*) flour. Food Chem 34:187–191

Oshodi A, Olaofe O, Hall G (1993) Amino acid, fatty acid and mineral composition of pigeon pea (*Cajanus cajan*). Int J Food Sci Nutr 43:187–191

Oshodi A, Ogungbenle H, Oladimeji M (1999) Chemical composition, functional properties and nutritionally valuable minerals of benniseed, pearl millet, quinoa seed flour. Int J Food Sci Nutr 50:325–331

Pisoschi AM, Cheregi MC, Danet AF (2009) Total antioxidant capacity of some commercial fruit juices: electrochemical and spectrophotometrical approaches. Molecules 14:480–493

Shils ME, Shike M (2006) Modern nutrition in health and disease. Lippincott Williams & Wilkins, Philadelphia

Tchoundjeu Z (2006) Ndjanssang: Ricinodendron Hendelotii (Baill.). In: Crops for the future. Southampton Centre for Underutilized Crops, University of Southampton, Southampton, UK

Tolstoguzov V (1998) Functional properties of protein-polysaccharide. In: Functional properties of food macromolecules. Elservier, London, p 252

Vuksan V, Jenkins D, Spadafora P, Sievenpiper JL, Owen R, Vidgen E, Brighenti F, Josse R, Leiter LA, Bruce-Thompson C (1999) Konjac-mannan (glucomannan) improves glycemia and other associated risk factors for coronary heart disease in type 2 diabetes. A randomized controlled metabolic trial. Diabetes Care 22:913–919

Chapter 22
Parkia biglobosa: Phytochemical Constituents, Bioactive Compounds, Traditional and Medicinal Uses

Lydia Quansah, Gustav Komla Mahunu, Haroon Elrasheid Tahir, and Abdalbasit Adam Mariod

Abbreviations

FAO Food and Agriculture Organization
NFE Nitrogen free extract

22.1 Introduction

Parkia biglobosa Benth (African Locust bean) is an underutilized and under-exploited tropical legume which is found to have both health benefits and functional properties essential in fighting malnutrition (Ajayi et al. 2018). As a perennial tree legume, *P. biglobosa* Benth belongs to the sub-family Mimosoideae and family Leguminosae (now family Fabaceae) (Ak et al. 2010; Sambe et al. 2010). Similar names for *P. biglobosa* are *P. africana* R. Br., *P. intermedia* Oliver, *P. oliveri* J. F. Macbr., *P. clappertoniana* Keay. Again, other common names aside African

L. Quansah
Department of Biotechnology, Faculty of Agriculture, University for Development Studies, Tamale, Ghana

G. K. Mahunu (✉)
Department of Food Science & Technology, Faculty of Agriculture, University for Development Studies, Tamale, Ghana
e-mail: gmahunu@uds.edu.gh

H. E. Tahir
School of Food and Biological Engineering, Jiangsu University, Zhenjiang, Jiangsu, China

A. A. Mariod
Indigenous Knowledge and Heritage Center, Ghibaish College of Science & Technology, Ghibaish, Sudan

A. A. Mariod (ed.), *Wild Fruits: Composition, Nutritional Value and Products*,
https://doi.org/10.1007/978-3-030-31885-7_22

locust bean are fern leaf, arbre à farine, monkey cutlass tree, two-ball nitta-tree. In Francophone West African countries, it is usually known as *néré, netto*, *ulele*, *séou*, and *ouli* (Booth and Wickens 1988).

Historically, the genus *Parkia* was first described by Robert Brown and he named it after Mungo Park, who explored West Africa in 1795–1797 and in 1805 (Orwa et al. 2009). At present the species described within the genus *Parkia* are 24, with ten species occurring in tropical South America, tropical Asia also has ten species, and Africa has four species (Hall 1997).

22.2 Plant Distribution

Generally, the *P. biglobosa* trees are established through natural regeneration and the ecological range is widely distributed in African savannahs, stretching through the Guinea savannah and Sudan vegetation zones (Campbell-Platt 1980). The geographical frontiers of the tree distribution overlap with climatic zones (Fig. 22.1).

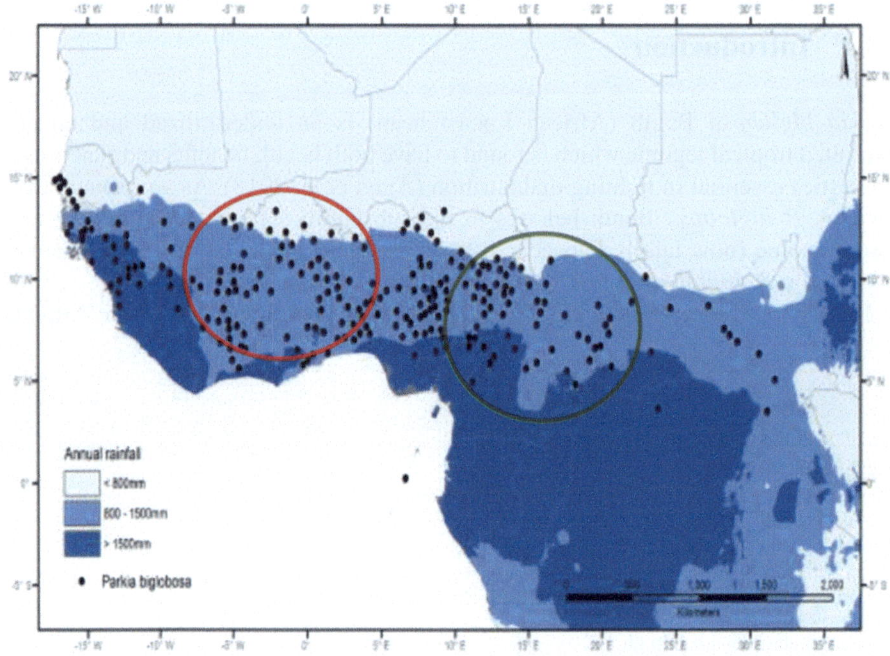

Fig. 22.1 Ecological distribution of *P. Biglobosa* in relation to annual rainfall with slight modification (Hall, 1997). The dark dotes show the general distribution belt of *P. Biglobosa* trees in Africa. The red colour circle (LEFT) shows area of high genetic diversity in the middle part of West Africa (Lompo et al. 2017). While, green colour circle (RIGHT) shows high density of tree population in Central Africa Republic (Depommier and Fernandes 1985)

P. biglobosa occurs in over 19 African countries north of the equator including Senegal, The Gambia, Guinea Bissau, Guinea, Sierra Leone, Ghana, Togo, Benin, Mali, Côte d'Ivoire, Burkina Faso, Niger, Nigeria, Cameroon, Chad, Central African Republic, Sudan, Zaire, and Uganda) (Hall 1997).

Parkia biglobosa is adapted to a varied habitats and commonly grows in deep loamy and sandy soils. It's found within an altitude (0–600 m), annual rainfall between 700 and 2600 mm with a mean annual temperature of 26 °C (Sacande et al. 2006). Its seen in intense seasonal climates with the dry season extending from 4 to 8 months (Gbadamosi et al. 2005). The adaptation of trees to varied conditions translates into considerable differences in tree population density. Tree population densities of up to 40 trees per hectare have been found in the Central African Republic, (Depommier and Fernandes 1985). Burkina Faso is located in the centre of species' range, where density varies from less than one tree/hectare in the sub-Sahelian zone to up to 25 trees/hectare in the southern part of the Sudanian zone (Ouédraogo 1995). In Ghana, *P. biglobosa* tree population is concentrated mostly in the Northern sector, and the tree distribution also follow common ecological factors in its range. Figure 22.1 shows the existence of a high genetic diversity in the middle part of West Africa (Benin, Burkina Faso, Ghana, La Cote d'ivoire, Mali, Togo) (Lompo et al. 2017).

22.3 Plant Description

The *Parkia biglobosa* tree can grow in height between 7 and 20 m but can reach 30 m under favorable conditions (Fig. 22.2) (Ojewumi et al. 2017a, b). Flowers are reddish and immature fruits are green in appearance (Fig. 22.3). The alternate leaves have dark green bipinnate forms up to 30 cm long, pinnae up to 17 pairs of leaflets (Booth and Wickens 1988) and the hermaphrodite flower balls are red or orange in colour, which normally flower in January (Irvine 1961). Seeds are surrounded by mealy, yellowish, edible pulp of sweet taste (Aliero 2004). Also the testa of the seed is hard with mean weight (0.26 g/seed) with large cotyledons that constitute about 70% of the weight (Booth and Wickens 1988). As a drought tolerant tree, it develops deep tap root system and its also able to restrict transpiration (Okunlola et al. 2011). The tree can start to producing fruits between 5 and 7 years after planting and, a matured tree of between 20 and 30 years can produce approximately a tone and above of harvested fruits (Musa 1991).

Parkia biglobosa tree is monoecious with more than 95% outcrossing. The long distance pollination is carried out by megabats and the insects undertake the short distance pollination. Polyads perhaps offer a discriminating advantage for this animal-mediated pollination. Each infructescence has up to 25 pods, each with a full subfamily inside. Seeds are mainly dispersed by primates, rodents, and birds (Lompo et al. 2017).

Fig. 22.2 *Parkia biglobosa* tree. (Source: https://commons.wikimedia.org)

Fig. 22.3 Left reddish flower; right mature dry pods fruit of *P. biglobosa* tree. (Source: https://commons.wikimedia.org)

22.4 Nutritional Components of *Parkia biglobosa* Benth

Quality nutrition is an essential component of life as it is key in the intellectual and physical development of humans especially children (Boadi and Kobina 2017), but it is reported that Africa is one of the major continent challenged with the issue of malnutrition as one in every four people (representing a total of 23.2 percent of the populace in sub-Saharan Africa) is estimated to be undernourished in 2014–2016 (FAO 2015). According to FAO (2012), about two million people in the world today are deficient in key vitamins and minerals (such as vitamin A, iron, zinc and iodine). In solving the problem of malnutrition/undernourishment to internationally acceptable levels, underutilized leguminous crops/plants such as *P. biglobosa* which is rich in nutrients (plant protein and vitamin substitutes) can be utilized especially in diet of rural people.

Parkia biglobosa Benth (African Locust bean) has industrial potential because of its nutrition (Ogunwusi and Ibrahim 2016), it is said to be rich in nutrients and hence have a wide range of application. Proximate and fibre quantification conducted by Elemo et al. (2011) revealed that the seeds of *P. biglobosa* are rich in nutrients (Fig. 22.4), minerals such as potassium (1101.5 mg/100 g) and phosphorus (170.0 mg/100 g) are also reported to be high. The ratio of unsaturated-saturated fatty acid was reported to be high, which has a beneficial effect on human with linoleic acid having the highest level (90.4 mg/100 g) (Elemo et al. 2011). According to Sadiku (2010), the oil content is good for consumption as it contains low acid and iodine. Besides bean oil has high saponification value and would be useful in soap industry. *P. biglobosa* has a high protein content (27.9%) with low quantities of methionine and cysteine which therefore limits the availability of sulphur making this legume ideal for fortification applications (Elemo et al. 2011). The high protein

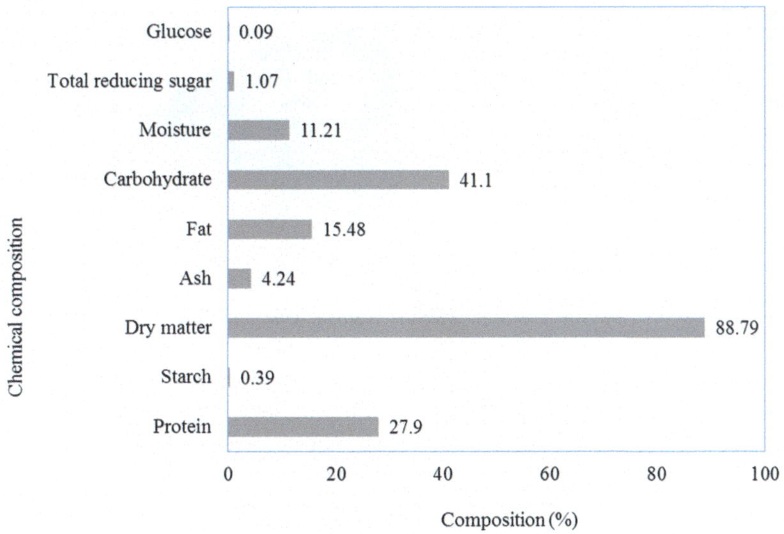

Fig. 22.4 Proximate composition of *Parkia biglobosa*

content of this legume was also reported by Borquaye et al. (2017). The seeds have been shown to have appreciable fibre content (Fig. 22.5).

Gernah et al. (2007) also indicated that fruit pulp of *P. biglobosa* contains moisture content of 8.41%, protein (6.56%), fat (1.80%), crude fibre (11.75%), ash (4.18%) and carbohydrate (67.30%). Again, the bean has sugar content of 9.00 °Brix, total carotenoids (49,175 μg/100 g) which are normally converted to large quantities of retinol/vitamin A, and ascorbic acid/vitamin C of 191.20 mg/100 g. These same proximate parameters were also identified and quantified in locust bean-based powder, as well as vitamins (A and C) and minerals (potassium, zinc, sodium, iron, calcium and copper) (Murtala et al. 2016). A study by Olujobi et al. (2012) on the vitamins content of both locust bean seed and pulp revealed the presence of vitamin B_1 (thiamine) and B_2 (riboflavin) in low quantities with high levels of vitamin C, D, E and ß-carotene.

In comparing the nutritional composition of raw to fermented *P. biglobosa* bean seeds; it was revealed that, there is a significant difference in their nutrient composition after conducting proximate analysis. Crude protein, carbohydrate and crude fibre were found to have reduced in the fermented seeds whiles fat, moisture, phosphorus and energy rather increased in the fermented seeds (Ndukwe and Solomon 2017). But according to Ojewuni et al. (2018), fermentation of African Locust bean seeds/seed flour increases the protein content; this was additionally confirmed by Ezugwu et al. (2018). However, a study conducted by Ojewuni et al. (2018), revealed that reduction in protein content of fermented African Locust bean seeds can occur when the seeds are excessively heated during processing which, causes denaturation of protein structures.

Nitrogen free extract (NFE) is the total carbohydrate contents (Ndukwe and Solomon 2017) consisting of carbohydrates, sugars, starches and a major portion of materials classed as hemicellulose in feeds. Carbohydrates normally serve as the

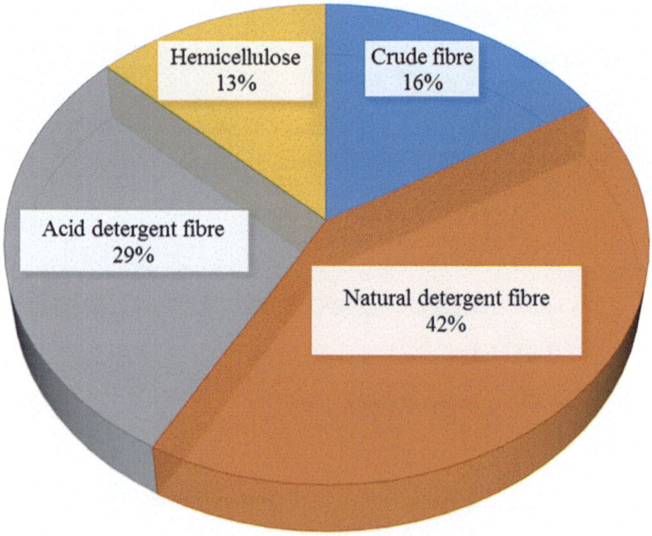

Fig. 22.5 Various sections indicate the fibre composition of *Parkia biglobosa* seed

most important source of bodily energy; this portion is called the NFE. However, the reduction in carbohydrate and crude fibre in fermented African Locust bean seeds/seed flour could be due to degradation and utilization of these components by microbes used in the fermentation process (Tamburawa et al. 2017).

Also, vitamins are generally reported to have reduced in processed *P. biglobosa* bean seed powder (Ezugwu et al. 2018), this was attributed to oxidation and leaching of these vitamins during boiling (Oboh et al. 2008). Some mineral composition in raw and fermented seeds of *P. biglobosa* were studied and the results indicated that aside copper, all other minerals present were significantly reduced after fermentation (Abdulrahman et al. 2016).

22.5 Phytochemical and Bioactive Components of *Parkia biglobosa* Benth

Phytochemicals are plant chemicals which are non-nutritive but contain protective and/or disease preventive properties. Flavonoids are natural phenolic compounds with antioxidant properties and can easily pick-up free radicals; preventing oxidative cell damage which could lead to cancers. Flavonoids also lower the potential risk of heart diseases when present in the intestine (Yao et al. 2004).

Total phenols are bioactive compounds which protect humans from oxidative damage just as in plants, these compounds also block inflammatory causing enzymes and enhance the human body immune system as well as destroy cancer cells (Yao et al. 2004). Total phenols again have antitumor properties.

Fermented *P. biglobosa* bean seeds is given the attribute as a functional food as it have high levels of medicinal and nutritional values (Oluwaniyi and Bazambo 2014). In a research conducted, it was revealed that fermentation led to an acceptable loss in the levels of anti-nutritional compounds such as oxalate, phytate and tannins (Ndukwe and Solomon 2017; Abdulrahman et al. 2016; Oluwaniyi and Bazambo 2014) whiles increasing the availability of phytochemicals such as flavonoids and total phenolics (Oluwaniyi and Bazambo 2014) in *P. biglobosa* bean seeds. A study by Abdulrahman et al. (2016) confirmed that fermentation of *P. biglobosa* seeds leads to significant reduction in anti-nutritional compounds as it causes approximately 29%, 85% and 61% reductions in levels of tannin, phytic acid and oxalate, respectively.

22.6 Traditional and Medicinal Uses for *Parkia biglobosa*

22.6.1 *Traditional Uses*

Parkia biglobosa is considered an important multipurpose tree in the West African savannah parkland, and one of the commonest species of the grassland agroforestry system. This tree and its products are critically important for rural livelihood. The traditional uses for *P. biglobosa* include wood energy (fuelwood and charcoal), food, medicine, glazes, animal fodder, soil amendments, charcoal, and firewood (Chandrasekharan 1993). Food is the most important product from *P. biglobosa*. Similar to other indigenous fruits trees, the food products obtained from *P. biglobosa* are unique as a result of the seasonality of fruit maturation and their time of availability during the year. For instance, the fruit trees ripen at different times of the year, therefore several rural households depend on these indigenous fruits as a coping strategy during severe seasonal hunger period, which usually lasts 4 months and more per year (Akinnifesi et al. 2007). Recently, research works have shown that *P. biglobosa* powder can be used as partial substitute for wheat flour in bread making. The incorporation of the powder in bread suggest more sources of nutritious ingredient to improve the health of consumers. Furthermore, the inclusion of *P. biglobosa* seed powder improved most of the mineral composition and storability of the frankfurter sausages. In all the tests, the crude protein, crude fat and moisture content of the sausages were not adversely affected by and *P. biglobosa* seed powder (Teye et al. 2013)

22.6.1.1 Fruit (Pod and Seeds)

All through West Africa the entire pod is edible and it's eaten when fresh although when dried, a large proportion remains edible. Often the pods were carried during long journeys as a staple ration (Fig. 22.6). In Ghana, between February and March (peak of the dry season), the men roast and eat the young green whole pods. Then, between March and April (the beginning of "hunger" season) when other foods are scarce, mature pods are collected for food (Shao 2002). Dried seeds are processed into dawadawa, a protein and fat rich food. In Ghana, dawadawa (local condiment obtained from *P. biglobosa* seeds) or the powder can also be mixed with water to produce a drink called "dozim" by the Dagombas (a major tribe in northern Ghana) and "*bololo*" in Hausa (Campbell-Platt 1980). The processed fermented seeds (dawadawa) are used extensively as flavouring and nutritious condiments in soups and stews, which contains crude fat (26.85%) and crude protein (49.69%) in addition to vitamins and minerals (Appiah et al. 2012). In addition, "Dobulong" as it's called by Dagombas (a major tribe in northern Ghana), is the yellow starchy pulp that surrounds the seed that is used as an important food supplement rich in Vitamin C and carbohydrates (Fig. 22.6). The pod husk can also use to poison fish and when burnt to ash, it can be used to make soap (Irvine 1961).

A series of fermentation process (Fig. 22.7) of *P. biglobosa* seeds transforms the indigestible seeds into the nutritious and flavoured food, dawadawa (Figs. 22.6 and 22.7). The hard seed coat is removed to make the soft cotyledons of the beans edible, while subsequent fermentation increases nutritional value, digestibility, and flavour. Traditionally, products of the *P. biglobosa* seed are available food rich in protein and fat recommended for households with poor balanced diet (nutrition and calories) in West Africa. Usually, when most of the stored grains in the household have been used, the women harvest *P. biglobosa* pods to sell fresh and processed beans to purchase food. This practice over the years has transformed *P. biglobosa*

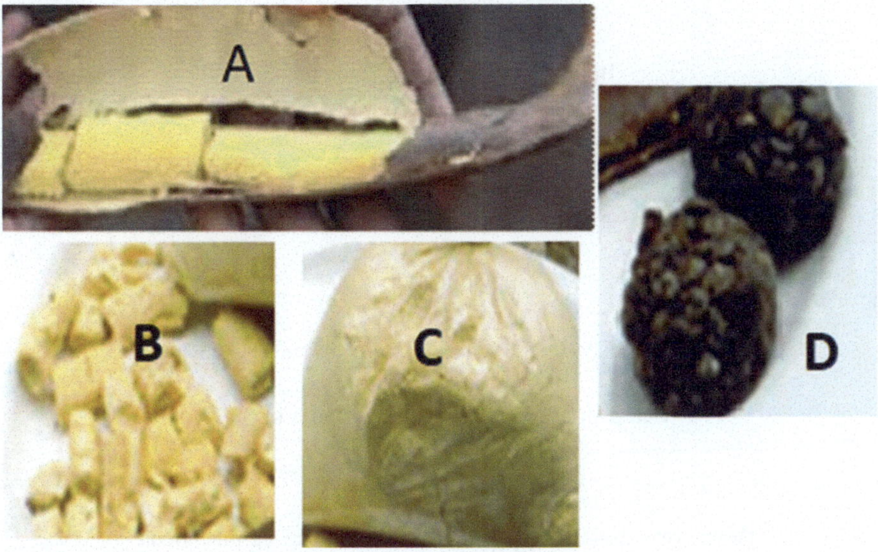

Fig. 22.6 The *P. biglobosa* different fruit components and products (**a**: pod, **b**: pulp, **c**: Powdered pulp, **d**: Dawadawa)

into considerable desired income by most rural households West Africa. Locally, *P. biglobosa* plant is used to lessen the effects of two types of snake venoms (*Naja nigricollis*, and *Echis ocellatus*) Asuzu and Harvey (2003).

In addition, a plaster used to paint the walls of houses which is produced from steeping and boiling the husks of *P. biglobosa* and then mixed with mud. Similarly, the plaster can be used to paint compressed earthen floors. The polymeric nature of tannins existing in the husk is able to hold together the soil and make the surface impermeable to water. Pot makers in Burkina Faso splash their pots with a vegetal gum solution produced from the pods and husks; this actually acts as a sealant and creates a dark, mottled surface. Also, leathers are dyed and cured with tannins found in the bark and husks of the *P. biglobosa* pods (Campbell-Platt 1980). The bright red-brown colour effect of leather is typical paintings in the Upper East Region of Ghana. The tannin content in the bark of *P. biglobosa* is 12–14% and the 27–44% in the husk (Hall 1997).

22.6.1.2 Leaves

The green looking leaves of *P. biglobosa* often remain on the tree during the dry season, and they form important source of animal fodder. Normally, farmers will prune the lower of the tree branches to feed their livestock. Compared to other savannah tree species, *P. biglobosa* has high amount of crude protein and high-energy value, which make the tree important quality leguminous fodder (Sabiiti and Cobbina 1992). The broad umbrella shaped canopy as well as leaf retention of the *P. biglobosa* during the dry season provide shade against the hot and harsh climate

Seeds
↓
Boil in water for 14 hours or more
↓
Cool
↓
Pound boiled seeds with wood ash in mortar to remove testa
↓
Lay out to dry
↓
Wash vigorously several times to remove wood ash and testa
↓
Select out seeds which still have testa intact
↓
Boil cotyledons for 2 hours
↓
Drain cotyledons and place in loose weave bag, eg jute cocoa bag
↓
Put full bag in a large basket put heavy stones to compress cotyledons
↓
Leave and let ferment for 72 hours
↓
Take dawadawa to mortar and pound to a paste or use the whole beans then form into desired shape
↓
Air dry for 24 hours
↓
dawadawa

Fig. 22.7 Flow chart to illustrate steps in processing *P. biglobosa* seeds into dawadawa (Shao 2002)

in the Savannah and Sahel ecological zones, particularly. In cases where livestock take shelter under the tree to escape the heat, their faecal matter deposit under the trees enriches the soil. In addition, leaf fall contributes organic matter to the soil beneath and around the trees. In Gambia, farmers gather the decomposed leaves as fertilizers. While in Burkina Faso, as part of byproduct utilization, the testa from processing is used as fertilizers in crop production (Hall 1997).

22.6.1.3 Bark and Stem

The rich tannin bark is used to tan leather. The branches are sometimes cut for firewood, whereas the roots and pods are used as sponges and as springs for local musical instruments (Okunlola et al. 2011). In agroforestry systems, the *P. biglobosa* can be used to fix atmospheric nitrogen in soil (Leopold and Kriedemann 1975).

22.6.2 Medicinal Uses

Various medicinal products are derived from *P. biglobosa*, which stands as a great value to rural communities where the tree are found, where "modern medicine" is expensive or not accessible (Tijani et al. 2009). Different parts of the tree (bark, pulp, leaves, flowers, root) have medicinal properties used to cure many disease including wounds, ulcers, rickets, toothache, fever, diarrhoea, orchitis and many more. Although, *P. biglobosa* is used for food but the primary usage is medicine. The boiled bark is used to make tea to relieve diarrhoea, as well as bloody diarrhoea (Dalziel 1937). In addition, the boiled bark is applied topically to treat infections, wounds and fever (Shao 2002). It's important to note that *P. biglobosa* stem bark has been found to have anti-diarrhoea action against microorganisms that cause diarrhoea (Tijani et al. 2009). The use of *P. biglobosa* to treat leprosy, hypertension and diarrhoea is increasing its value (Assane et al. 1993). The seeds have been revealed to contain cardiac glycosides and alkaloids (Ajaiyeoba 2002) as well as high protein and amino acid content (Fetuga et al. 1974).

Extracts from leaves, flowers and seeds are also used to treat various ailments such as toothaches, lumbago and hemorrhoids to diabetes mellitus (Abbiw 1990; Abo et al. 2008; Hall 1997). It was confirmed that traditionally, the powdered stem bark of *P. biglobosa* mixed with hot pap ingested or as infusion treats diarrhoea and dysentery. In other words, the hot water extract has potentials to control toxin-induced diarrhoea (for instance, cholera, shigellosis, and traveler's diarrhoea), diarrhoea due to intestinal malabsorption (such as radiation enteritis, lactose intolerance) or dysentery due to microbial infections (Tijani et al. 2009). Various parts of *P. biglobosa* are used in respective preparations and medicinal uses (Abbiw 1990; Booth and Wickens 1988; Hall 1997).

22.7 Conclusion

Parkia biglobosa represents a highly economically important tree to the people of West and Central Africa. The various parts of the tree are highly valued for their health, medicinal, nutritional and economic benefits. Generally, the tree shows high potential as source of income generation and food security to the rural folks. However, *P. biglobosa* like other wild fruit trees is still limited to traditional market with much focus on just the supply of raw materials. There is little or no value addition in some cases. Besides, the trees are threatened as a result of the following overexploitation, bush fires, and a progressive habitat degradation leading to disintegration of tree populations, domestic animals overgrazing leading to lack of natural regeneration and an over-aging of tree individuals in savannah parklands. Other anticipated potential threats is the absence or declining number of pollinators. In view of the wide range of *P. biglobosa*, the development of inter-country strategies for the conservation and sustainable use of the tree and its resources, would require harmonization through well-organized international partnership. It will also ensure economic viability of products, well-adjusted to the socio-cultural practices of the people.

References

Abbiw DK (1990) Useful plants of Ghana. Intermediate Technology Publication Ltd. Southampton Row, London, UK

Abdulrahman BO, Osibemhe M, Idoko AS (2016) The status of mineral and anti-nutritional composition of raw and fermented seeds of African locust bean (*Parkia biglobosa*). Int J Curr Res Biosci Plant Bio 3(2):1–4

Abo K, Fred-Jaiyesimi A, Jaiyesimi A (2008) Ethnobotanical studies of medicinal plants used in the management of diabetes mellitus in South Western Nigeria. J Ethnopharmacol 115:67–71

Ajaiyeoba EO (2002) Phytochemical and antibacterial properties of Parkia biglobosa and Parkia bicolor leaf extracts. Afr J Biomed Res 5

Ajayi K, Adepoju OT, Taiwo OM, Omojola ST, Aladetuyi ME (2018) Nutritional potential of underutilized gum arabic tree seeds (*Acacia nilotica*) and locust bean seeds (*Parkia biglobosa*). Afr J Food Sci 12(8):196–203

Ak F, Adejumo O, Adamade C, Bodunde J (2010) Processing of locust bean fruits: challenges and prospects. Afr J Agric Res 5:2268–2271

Akinnifesi F, Ajayi O, Sileshi G, Kadzere I, Akinnifesi A (2007) Domesticating and commercializing indigenous fruit and nut tree crops for food security and income generation in Sub-Saharan Africa. Paper presented at the New Crops International Symposium, 3–4 September, 2007, Southampton, UK

Aliero B (2004) Effects of sulphuric acid, mechanical scarification and wet heat treatments on germination of seeds of African locust bean tree, Parkia biglobosa. Afr J Biotech 3:179–181

Appiah F, Oduro I, Ellis W, Adu G (2012) Comparative assessment of the performance of *Parkia biglobosa*, *Glycine max* and *Treculia africana* in the production of a local condiment (dawadawa) in Ghana. Afr J Food Sci 6:111–116

Assane M, Baba RM, Bassene E, Sere A (1993) Antihypertensive action of *Parkia biglobosa*+ (Jacq) Benth seeds in the rat. Dakar Med 38:49–54

Asuzu IU, Harvey AL (2003) The antisnake venom activities of *Parkia biglobosa* (Mimosaceae) stem bark extract. Toxicon 42:763–768

Boadi R, Kobina S (2017) Dietary diversity and child malnutrition in Ghana. Heliyon 3(5):e00298. https://doi.org/10.1016/j.heliyon.2017.e00298

Booth FE, Wickens GE (1988) Non-timber uses of selected arid zone trees and shrubs in Africa, FAO, 19. Food and Agriculture Organization of the United Nations, Rome

Borquaye LS, Darko G, Laryea MK, Gasu EN, Afia N, Amponsah A, Appiah EN (2017) Nutritional and anti-nutrient profiles of some Ghanaian spices nutritional and anti-nutrient profiles of some Ghanaian spices. Cogent Food Agric 3:1–12. https://doi.org/10.1080/23311932.2017.1348185

Campbell-Platt G (1980) African locust bean (*Parkia* species) and its West African fermented food product, dawadawa. Ecol Food Nutr 9:123–132

Chandrasekharan C (1993) Issues involved in the sustainable development of non-wood forest products. In CSC/FAO, Regional Expert Consultation on Non-Wood Forest Products for English-Speaking African Countries, Arusha, Tanzania 17-22 October 1993, pp III, 1-24, Commonwealth Science Council/FAO, London.

Dalziel JM (1937) The useful plants of west tropical Africa, 2nd edn. pp. 648 pp, UK

Depommier D, Fernandes E (1985) Aspects du parc à karités-nérés (*Butyrospermum parkii, Parkia biglobosa*) dans la région de l'Ouham (République Centrafricaine). Nairobi, ICRAF 27

Elemo GN, Elemo BO, Oladunmoye OO, Erukainure OL (2011) Comprehensive investigation into the nutritional composition of dehulled and defatted African locust bean seed (*Parkia biglobosa*). African J Plant Sci 5(May):291–295

Ezugwu EN, Okoye JI, Ene GI (2018) Effects of processing on the chemical composition and functional properties of African locust bean (*Parkia Biglobosa*). IOSR JESTFT 12(7):9–14

FAO (2015) The state of food insecurity in the world 2015. Meeting the 2015 international hunger targets: taking stock of uneven progress. Rome, FAO

Fetuga B, Babatunde G, Oyenuga V (1974) Protein quality of some unusual protein foodstuffs. Studies on the African locust-bean seed (*Parkia filicoidea* Welw.). Br J Nutr 32:27–36

Food and Agricultural Organization of the United Nations (2012) FAO statistical yearbook, 2013-World food and agriculture. FAO. Rome, Italy

Gbadamosi A, Faboye O, Oni O (2005) Seed morphological traits, germination and early seedling growth of *Parkia biglobosa* (Jacq.) R. Br. Ex. G. don from four provenances in Nigeria. Niger J For 35:129–137

Gernah DI, Atolagbe MO, Echegwo CC (2007) Nutritional composition of the African locust bean (*Parkia biglobosa*) fruit pulp. Niger Food J 25(1):190–196

Hall J (1997) Parkia biglobosa: a monograph. School of Agricultural and Forest Sciences, University of Wales, Bangor. 107p. ISBN 0 904567, 4

Irvine FR (1961) Woody plants of Ghana with special reference to their uses. Q Rev Biol 39:304, 868pp

Leopold AC, Kriedemann PE (1975) Plant growth and development, 2nd edn. 545 pp. McGraw-Hill, New York, London

Lompo D, Vinceti B, Gaisberger H, Konrad H, Duminil J, Ouedraogo M, Sina S, Geburek T (2017) Genetic conservation in *Parkia biglobosa* (Fabaceae: Mimosoideae)—what do we know? Silvae Genet 66:1–8

Murtala Y, Babandi A, Ja M, Hm U, Ibrahim S, Ibrahim A, Aj A (2016) Proximate, vitamins and mineral compositions of locust and soya beans-based Daddawa sold in Kano Metropolis, Kano, Nigeria. Bayero J Biomed Sci 1(1):1–7

Musa H (1991) Ginger and locust bean tree. History, growth, uses and potentials.'Tuk-Ham' Symposium, Kurmin-Musa, pp. 1–16

Ndukwe MN, Solomon MD (2017) Proximate and antinutrient composition of some local food condiments in their raw and fermented forms. Int J Biochem Res Rev 20(1):1–8. https://doi.org/10.9734/IJBCRR/2017/37727

Oboh G, Alabi K, Akindahunsi A (2008) Fermentation changes the nutritive value, polyphenol distribution, and antioxidant properties of *Parkia biglobosa* seeds (African locust beans). Food Biotechnol 22:363–376

Ogunwusi AA, Ibrahim HD (2016) Industrial potential of underutilized plants in Nigeria. J Nat Sci Res 6(6):20–28

Ojewumi ME, Eluagwule B, Ayoola AA, Ogunbiyi AT, Adeoye J, Emetere M, Joseph OO (2017a) Termiticidal effects of African locust bean (*Parkia biglobosa*) seed oil extracts. Int J Curr Res:53929–53934

Ojewumi ME, Eluagwule B, Ayoola AA, Ogunbiyi AT, Adeoye JB, Emetere ME, Joseph OO (2017b) Termiticidal effects of African locust bean (*Parkia biglobosa*) seed oil extracts. Int J Curr Res 9(7):53929–53934

Ojewuni ME, Omoleye JA, Ayoola AA, Ajayi AA, Adekeye BT, Adeyemi AO (2018) Effects of salting and drying on the deterioration rate of fermented *Parkia biglobosa* seed. J Nutr Health Food Eng 8(1):1–5

Okunlola A, Adebayo R, Orimogunje A (2011) Methods of braking seed dormancy on germination and early seedling growth of African locust bean (*Parkia biglobosa*)(JACQ.) Benth. J Hortic For 3:1–6

Olujobi OJ, Management F, Ekiti A (2012) Comparative evaluation of nutritional composition of African locust bean (*Parkia biglobosa*) fruits from two locations. Niger J Basic Appl Sci 20(3):195–198

Oluwaniyi OO, Bazambo IO (2014) Antinutritional and phytochemical evaluation of raw and fermented African locust bean (*Parkia Biglobosa*) seeds. Global J Pure Appl Sci 20(1989):105–109

Orwa C, Mutua A, Kindt R, Jamnadass R, Simons A (2009) Agroforestry tree database: a tree species reference and selection guide version 4.0. World Agroforestry Centre ICRAF, Nairobi

Ouédraogo A-S (1995) *Parkia biglobosa* (Leguminosae) en Afrique de l'Ouest: Biosystématique et amélioration. Ouedraogo

Sabiiti EN, Cobbina J (1992) Initial agronomic evaluation of Parkia biglobosa in the humid zone of Nigeria. Agroforestry Systems, 17 (3):271–279. https://link.springer.com/journal/10457

Sacande M, Rønne C, Sanon M, Jøker D (2006) *Adansonia digitata* L. Seed leaflet (106)

Sadiku OA (2010) Processing methods influence the quality of fermented African locust bean (Iru/ogiri/dadawa) *Parkia biglobosa*. J Appl Sci Res 6(11):1656–1661

Sambe M, Sagna M, Sy M (2010) Seed germination and *in vitro* plant regeneration of *Parkia biglobosa* (Jacq.) Benth. Afr J Biotech 9:3099–3108

Shao M (2002) *Parkia biglobosa*: changes in resource allocation in Kandiga. Michigan Technolocical University, Ghana

Tamburawa M a S a, Ogundipe SO, BabatundeTegbe TS, Olugbemi TS, Hassan MR (2017) Effect of soaked and fermented African locust bean seeds meal on the performance, organs and carcass characteristics of broiler chickens. Anim Vet Sci 5(1):8–14

Teye GA, Taame F, Bonsu KO, Teye M (2013) Effect of Dawadawa (Parkia biglobosa) as a spice on sensory and nutritional qualities of meat products: a preliminary study. Int J Curr Res 5(02):110–112

Tijani A, Okhale S, Salawu T, Onigbanjo H, Obianodo L, Akingbasote J et al (2009) Antidiarrhoeal and antibacterial properties of crude aqueous stem bark extract and fractions of *Parkia biglobosa* (Jacq.) R. Br. Ex G. Don. Afr J Pharm Pharmacol 3:347–353

Yao LH, Jiang Y, Shi J, Tomas-Barberan F, Datta N, Singanusong R, Chen S (2004) Flavonoids in food and their health benefits. Plant Food Hum Nutr 59:113–122

Chapter 23
Garcinia Kola (Bitter Kola): Chemical Composition

Abdel Moneim and Elhadi Sulieman

23.1 Introduction

Garcinia kola, also called bitter kola originated in western and central Africa, belongs to the species of a tropical flowering plant in the *Clusiaceae* or *Guttiferae* family. It is found in *Benin, Cameroon, Democratic Republic of the Congo, Ivory Coast, Gabon, Ghana, Liberia, Nigeria, Senegal and Sierra Leone.* The natural *habitat* of the tree is subtropical or tropical moist lowland *forests* (https://en.wikipedia.org/wiki/Garcinia_kola).

It is a large species of spreading forest trees commonly found in the rain forest region of West Africa (https://mavcure.com/garcinia-kola-health-efits/ben).

Garcinia kola is scientifically classified as follows:

Kingdom:	*Plantae*
Division:	*Magnoliophyta*
Class:	*Magnoliopsida*
Order:	*Theales*
Family:	*Clusiaceae or guttiferae*
Genus:	*Garcinia*
Species:	*G. kola*
Binomial name:	*Garcinial kola, Heckel.*

About 400 species of *G. kola* are found in tropical regions especially in Asia and in South Africa (Mabberley 1987). Other species of *Garcinia* found in Nigeria as well as generally across the humid low land plains of West Africa extending from

A. Moneim (✉) · E. Sulieman (✉)
Department of Biology, Faculty of Science, University of Hail,
Hail, Kingdom of Saudi Arabia
e-mail: am.sulieman@uoh.edu.sa

A. A. Mariod (ed.), *Wild Fruits: Composition, Nutritional Value and Products*,
https://doi.org/10.1007/978-3-030-31885-7_23

Sierra Leone to Zaire according to Vivien and Faure (1996) and Angola (Keay 1989) include; *Garcinia livingstonei, gnetoides, standtii, smeath emanii, ovalivolia, brevipediellata* and *manni.*

Garcinia kola is an evergreen Tree developing to 12 m (39 ft) by 10 m (32 ft) at a medium rate. The trunk is straight with brown bark. The leaves are rough. The flowers are greenish-white. The seeds can be eaten raw. The rosy yellow fruits are extremely sour however is eatable. In Nigeria, these are eaten as a cure for body torments. The tree is additionally planted as shade tree in cocoa plantations. The bark yields a resinous gum that has water-proofing capacity. The wood is sturdy and to some degree impervious to termite assaults. The roots are utilized as bite sticks (Mabberley 1987).

23.1.1 Plant Distribution

The kola tree develops in a somewhat constrained area, comprising that portion of western Africa between Sierra Leone and the Congo and Lower Guinea. It flourishes at about or somewhat higher than the sea level, in hot and moist circumstances. At the point when states of soil and atmosphere are ideal, it grows inland from the points mentioned 500 or 600 miles. It has been found elsewhere, however without a doubt presented, as in Jamaica, where it was distributed by slave dealers. The English and French have brought it into a significant number of their belonging, and the gulf and Pacific coast districts of the United States are said to have the required climate and soil for its development in this country. The movements of Leo Africanus (in the sixteenth century) alluded to this tree as the Gora or Guru, and he wrote concerning its bitter nut. Clusius (1591) depicted and outlined the seeds. J. Bauhin first outlined to its restorative utilize, noticing its employment by the natives in fevers (Adesina et al. 1995).

Kola seeds have been utilized by the African natives from time immemorial as a need and an extravagance. It figured as a fundamental need in numerous ceremonials—social, political, and religious. It was utilized as an assertion of war (red nut), and as an image of peace (white nut). It figured in romance and marriage, compacts of friendship, as a sign of friendliness, and was put into the graves of the dead to nourish them on their long journey.

In civilized countries kola was referred to predominantly as an oddity until as of late. In 1883, Heckel and Schlagdenhauffen published a detailed monograph concerning it, entitled "Sur les Kolas Africains." It currently has a broad commerce in this country (Adesina et al. 1995).

23.2 Botanical Description

This is a tree from 40 to 60 ft high, to some degree looking like the normal chestnut tree. The trunk is erect, smooth, and cylindrical. The bark is green and thick (Fig. 23.1). The leaves are alternate, entire, slightly revolute, smooth, green, and

Fig. 23.1 *Garcinia kola* tree. (Source: https://commons.wikimedia.org/wiki/File:Garcinia_kola_-_Plant.jpg)

elongated taper, from 8 to 6 inches long by 1–2 inches broad (Fig. 23.2). They are borne on petioles from 1 to 3 inches long. The younger leaves are pubescent. The flowers are polygamous, and borne in both terminal and axillary cymose panicles, beset with stellate hairs. The flowers are greenish-yellow or white and purple at the margins of the petals. The fruit is made out of follicles, containing from 1 to 10 elongated coldhearted seeds, with a cartilaginous, purplish testa. The cotyledons are generally 2 in number (may be 3–5), red or greenish-yellow, flatly ovate, compressed, and thick. Kola nuts, alleged, are the cotyledons of the seeds, denied of their purplish, cartilaginous testa. The irregular seeds (nuts), attributable to close settling in the follicle, have a packed, to some degree triangular and subtetragonal shape, and look to some extent like the stallion chestnut (Fig. 23.3). The cotyledons, which may be from 2 to 5, are fleshy and thick, and around 1 inch long. They have, when fresh, a bitterish, somewhat astringent taste. Whenever dried, however, they possess a mild and faintly aromatic taste, and an odor that has been compared to that of nutmegs(Fig. 23.4). There are two varieties of kola nut—the *white kola*, which is more nearly "a pale greenish-yellow," and the *red kola*—both being yielded by the similar species, and often occurring in a similar pod (Adesina et al. 1995).

Fig. 23.2 Garcinia kola leaves. (Source: https://commons.wikimedia.org/w/index.php?search=bitter+kola)

Fig. 23.3 *Garcinia kola* dry nut. (Source: https://commons.wikimedia.org/w/index.php?search=bitter+kola)

23.3 Cultivation, Growth and Development

Garcinia kola is grown either by seeds or by cuttings. Seed cultivation includes setting up an appropriate seed bed estimating 3 × 4 m (12 m^2) on a flat ground inside a shade house to protect the little plants from direct radiation of the sun and strong

Fig. 23.4 *Garcinia kola tree with fruits*. (Source: https://commons.wikimedia.org/w/index.php?search=bitter+kola)

Fig. 23.5 *Garcinia kola* mature fruit. (Source: https://commons.wikimedia.org/w/index.php?search=bitter+kola)

rains. The shade is usually built from local materials like bamboo, stakes cut in forest or palm tree branches. Seeds that germinated effortlessly are generally those from developed ripe fruits that tumbled to the ground before the seeds are expel (Fig. 23.5).

To accomplish pre-nursery germination, a few banana or plantain tree trunks are cut and extensive openings made on the level of their bases to take into consideration demolition of the focal bud (meristem) of the cut plant. Seeds are thereafter inserted into the trunks of the banana tree and the two ends of each trunk attached solidly. Trunks are then arranged hermetically closed under a hanger. Following 3 months the coupling wires are withdrawn to recoup the seeds which had already germinated (Esiegwu and Udedibie 2009).

The germinated seeds are precisely sown inside polyethylene sacks with 2/3 of it loaded with a blend of dark soil and sand. The pots of young seedlings are then laid out in the seed bed, and kept up by watering each 2 days, weeding, use of 3 g of N.P.K 20:10:10 manure like fertilizer every 3 months and fighting off fungal and insect attacks with the appropriate fungicides and insecticides. Following a year, the seedlings are planted in the field, usually at the beginning of the rainy season with a standard spacing of 10 × 10 m. In the field, further agronomic practices are observed, and production of fruit will start approximately following 7 years (Esiegwu and Udedibie 2009).

Cultivation by cutting is typically done utilizing bitter kola cuttings acquired from extremely delicate branches and stems with young solid leaves and vertical branches looking upwards. Ideally, cuttings are best cut very early in the morning and just after rain to abstain drying. Moved into a wet plastic pack, which is solidly closed and held under shade from where they are traded explicitly to the propagator. Cutting are commonly around 12 cm. The youthful seedlings that sprout are purposely inserted into polyethylene sacks stacked with mix of dim soil and sand up to 2/3 level. After expansion of the youthful seedlings, the polyethylene sacks are stacked with soil to the overflow and sprinkled daintily with water. The pots of seedling are from that point exchanged to the seed informal lodging to the field as in spread by seeds. *Garcinia kola* usually produces fruits between the months of July and October (Keute 2014).

Bitter (unpleasant) kola is characterized by moderate rate of development. Difficulties were constantly experienced in endeavoring to raise its seedlings in nurseries, and the tree has a long incubation period before flowering and fruiting (Adebisi 2004). This has disheartened farmers from cultivating it. However, a considerable lot of the germination troubles have been overwhelmed by methods developed by Okafor (1998), and intrigue is developing in cultivating the tree in plantations. Because of this, Ladipo (1995) has developed projected production figures indicating that a mature fruit tree produces 85–1717 fruits, with 208–6112 seeds annually. Taking the mean of these values at 834 fruits and 2627 nuts per tree, he has anticipated a fruit production of 26 tons per ha per year, with 278 trees per ha planted at 6 m × 6 m spacing. Fruiting initiates in July and ends in October. Fruit harvests proceed discontinuously as the ready fruit falls and is then gathered for the extraction of seeds. The fruit is rosy yellow, about 6 cm in diameter, and each fruit contains two to four brown seeds embedded in an orange-colored pulp (Ladipo 1995). A few open doors for enhanced rustic advancement are connected to non-timber forest products, one of which is Bitter kola. In numerous territories, rural populations are traditionally dependent on local forest resources to provide additional income through collection and marketing. Where employment opportunities from traditional industries are declining, laborers searching for elective sources of income often turn to the collection of these products from the nearby forest (Adepoju and Salau 2007) (Fig. 23.6).

Fig. 23.6 *Garcinia_kola*_Germination. (Source: https://commons.wikimedia.org/wiki/File:Garcinia_Kola_Germination.jpg)

23.4 Chemical Composition of *Garcinia kola*

Chemical analysis of *Garcinia kola* seed has been studied by some investigators. *Garcinia kola* contains nutrients such as proteins, carbohydrates, fiber, minerals, fat and oils. Ibekwe et al. (2007) revealed that *Garcinia kola* seed has poor nutrient composition however profoundly esteemed in trado-medicine because of its helpful active phytochemical composition.

Incontrast to Ibekwe et al. (2007) as appeared in Table 23.1, Esiegwu and Udedibie (2009) found that the nutrient compositions of *Garcinia kola* is as appeared in Table 23.2. Odebunmi et al. (2009) detailed the moisture content of the seeds to be 60.48%, dry matter of 39.52%, crude fat of 4.51%, crude protein of 2.48%, ash content of 0.79%, crude fiber of 5.23% and total carbohydrates (+ fiber) of 35.64%. These values were not quite the same as what had beforehand been accounted for bitter kola. Eleyinmi et al. (2006) reported a protein content of 3.95%, lipid of 4.33%, ash of 1.14% and a crude fiber content of 11.4% in the seed.

Asaolu (2003) reported that the fresh seeds of bitter kola (wet weight) contains high moisture content of 75.50% and dry weight of 24.50, while the ash content was 5.90%, crude fat was 14.50%, carbohydrate was 10.85%, crude fat was 14.50% and crude protein was observed to be very low (4.25%). In contrast Dosunmu and Johnson (1995) reported the nutritive value of the fresh fruit from Nigeria demonstrated that crude protein was higher in the mesocarp (7.8%) than in the pericarp (3.9%), while the pericap was more extravagant in crude fiber (13.9–16.5%). The mesocarp was additionally more extravagant in crude lipid (6.9–8.7%). Unsaturated fatty acids (linoleic acid, 40.5%, oleic acid, 30.8%) are the fundamental components of the lipids (4.5%) found in the seeds of this species (Essien et al. 1995; Omode et al. 1995).

Table 23.1 Nutrient composition of *Garcinia kola* (Dry matter basis)

Component (%)	Ibekwe et al. (2007)	Esiegwu and Udedibie (2009)
Moisture	14.6	7.30
Crude protein	0.58	2.64
Crude fibre	0.10	20.51
Ether extract	3.00	9.47
Ash	5.00	1.07
Nitrogen-free extract	91.32	57.54

Table 23.2 Vitamin and mineral composition of *Garcinia kola* seeds (on dry weight basis) please redesigned your table as Table 23.1

Vitamin	Quantity (mg/100 g)
Thiamin	0.5
Riboflavin	0.22
Niacin	0.16
Ascorbic acid	23.10
Mineral	
Magnesium	0.42
Calcium	0.80
Potassium	2.50
Phosphorous	0.33
Sodium	0.72
Iron	17.75
Zinc	2.3
Copper	0.78
Manganese	2.01
Chromium	0.00
Cobalt	0.55
Cadmium	0.29

Source: Okwu (2005)

The micronutrients content of bitter kola have also been reported. In Nigeria, Okwu (2005) demonstrated that it contains an extensive range of vitamins and minerals as shown in Table 23.2. As indicated by Odebunmi et al. (2009), *Garcinia kola* has 722.10 mg/100 g of potassium (K), 67.07 ± 0.12 mg/kgDM of calcium (Ca), 114.83 ± 3.47 mg/kgDM of magnesium (Mg), 6.10 ± 0.43 mg/kgDM of iron (Fe), 2.30 ± 0.08 mg/kgDM of zinc (Zn), and 188.57 ± 0.37 mg/kgDM of phosphorus (P).

23.5 Phytochemical Components of *Garcinia kola*

The role of phytochemicals in enhancing body cell immunity against diseases in the body cannot be overemphasized. The active constituents adding to the defensive impact of *Garcinia kola* on animals is credited to the presence of phytochemicals, vitamins and minerals (Okwu and Ekeke 2003). Phytochemicals exhibit an extensive

Table 23.3 Phytochemical components of Garcinia kola seeds (Dry matter basis)

Components	Quantity (mg/100 g)
Phenols	0.11
Cyanogenic glycosides	0.54
Alkaloids	0.36–4.93
Tannins	0.26–0.34
Flavonoids	1.98
Saponins	10.06

Source: Okwu (2005) and Esiegwu and Udedibie (2009)

variety of biological activities because of the anti-oxidant properties of some of these chemicals. Several types of polyphenols (phenolic acid, hydrolysable tannins and flavonoids) indicate anti-carcinogenic and mutagenic impacts (Uruquiaga and Leighton 2000). Okwu (2005) and Esiegwu and Udedibie (2009) reported the phytochemical values as shown in Table 23.3.

Garcinia kola stem has been shown to contain a complex blend of phenolic for example, biflavonoids, xanthones and benzophenone (Iwu and Igboko 1982) which have anti-microbial activity as kolanone (Hussain et al. 1982), kola flavonone and garcinia flavonone (Iwu 1993). Phytochemical contemplates have demonstrated that the seeds constituents include biflavonoids, xanthones and benzophenones. Therefore, the seeds of *Garcinia kola* are known to have a general counteractant impact in traditional medicine in Africa. These conceivably clarify its revealed aphrodisiac properties and in the treatment of catarrh and abdominal colicky torment. Moreover, their utilization is believed to improve singing voice and relieve cough (Irvine 1961).

23.5.1 Amino Acid Composition of Garcinia kola

Glutamic acid was the most prominent acid in bitter kola as determined by Adeyeye et al. (2007). The total amino acids was 112.90 mg/g protein. The percentage of total essential amino acids was 47.05%. The calculated isoelectric points were 0.7, showing they can all be precipitated at acidic pH. valine was the limiting amino acid in *Garcinia kola*. The percentage of cystine in the total sulphur amino acids was 37.75%. So consumption of bitter kola be avoided by ulcer patients because of their high levels of acidic amino acids (Table 23.4).

23.6 Traditional Uses

Garcinia kola is cultivated all through West Africa for its consumable fruit and seeds which are utilized as restoring agent for masticatory purposes and as a general cure (Ibiblio 1983). Among the Igbos of Nigeria it is presented to guests as an

Table 23.4 Amino acid composition of *Garcinia kola* nuts (mg/g crude protein) dry weight

Amino acid	Content (mg/g crude protein)
Lysine (Lys)[a]	6.11
Histidine (His)[a]	5.32
Arginine (Arg)[a]	9.55
Aspartic acid (Asp)	15.51
Threonine (Thr)[a]	5.85
Serine (Ser)	6.21
Glutamic acid (Glu)	19.05
Proline (Pro)	3.69
Glycine (Gly)	5.44
Alanine (Ala)	3.55
Cystine (Cys)	2.45
Valine (Val)[a]	3.74
Methionine (Met)[a]	4.04
Isoleucine (Ile)[a]	3.24
Leucine (Leu)[a]	8.07
Tyrosine (Tyr)	3.88
Phenylalanine (Phe)[a]	7.20

[a]Essential amino acid. Source: Adeyeye et al. (2007)

indication of peace and welcome. It is likewise used to engage visitors during functions and celebrations. Once more, it is prominently utilized among other Nigerian groups for anxious readiness and induction of insomnia (sleeping disorder) when chewed.

The natives chewed kola to relieve hunger, avert thirst, promote digestion, and maintain strength. Like the supposed Indian clearing nut, it was licensed with the property of cleansing and sweetening water. The natives prefer it over tea and coffee, and incalculable are the spectacular ideals credited to it (Tcheghebe et al. 2016).

Traditionally, the nuts of *Garcinia kola* have utilized as sialegogue to empower the flow of saliva (Leakey 2001). The kernels of the nuts are extensively exchanged and eaten as a stimulant (Omode et al. 1995; Atawodi et al. 1995; Leakey 2001). It is accepted to clean the stomach, without reactions, for example, stomach issues, notwithstanding when a lot of it is eaten (Onochie and Stanfield 1960). In customary drug, the dried nut is ground and blended with nectar to make a conventional hack blend. The ground nut may likewise be blended with water and given to new conceived infants to alleviate stomach relieve stomach spasms.

Experimentations utilizing *Garcinia kola* kernels as bounce substitutes a few indigenous mixed beverages and in addition enhance enhancer in the beverage industry also exist (FDA 1999). Ofor et al. (2004) perceived a few ethno-herbal uses to which the indigenes of Imo state in South–eastern Nigeria put the *Garcinia kola* seeds. These include as a remedy to snake bites, poison and overdose, for cough, vomiting and as a snake repellent. The seeds which serve as a bitter stimulant likewise fill in as a snake repellent when they are put round the compound (Nair 1990). The seed is utilized as a part of the treatment of the diarrhoea (Braide 1991), bronchitis and throat infections (Orie and Ekon 1993; Adesina et al. 1995), liver disorders (Iwu et al. 1990) and a poison antitoxin (Kabangu et al. 1987). As per Farombi et al. (2005), the seeds of *Garcinia kola* have pharmacological uses in treating hacks, throat disease, bronchitis, hepatitis and liver disorders.

23.6.1 Processing of Bitter Kola

For bitter kola to meet export standard, it might require some processing which can accomplished without the utilization of any apparatus or gear s as this should be possible normally. Planned exporters can enter the business either in a little or enormous way. It relies upon the current money requirements of the purchasers. Small scale exporters can begin the business appropriate from their room with only a useful email address. It is neither a short-lived great nor is it fragile. Exporters are guaranteed that the products will get to the purchaser securely. But there is the requirement for an effective strategy for quality control with a specific end goal to keep it new and healthy looking. The essential method of exporting bitter kola is by means of air cargo. It is expected that it will get to the destination in less than 4 days. This guarantees that the product doesn't spend much travel subsequently bringing about the loss of quality and content. The packaging method used in bitter kola export business is easy and can be learned by anybody (https://www.vanguardngr.com/2016/09/exportable-commodities-bitter-kola/).

23.7 Medicinal Uses of Bitter Kola

Due to its complex chemical composition, *Garcinia kola* plant has an anti-bacterial properties in mitigating fever, sore throat and symptoms of cold and cough.

23.7.1 Antimicrobial Properties

Garcinia. kola seeds extract exhibited strong antibacterial activity against bacterial isolates including: *Staphylococcus aureus, Escherichia coli, Klebsiella pneumoniae, Bacillus subtilis* and *Pseudomonas aeruginosa*. And fungal isolates including *Candida albicans* and *Aspergillus niger* at the different treatment regimens i.e. 10 mg/ml, 20 mg/ml and 30 mg/ml concentration of the extracts (Arekemase et al. 2012). Extracts from the bark, stem and seed of *Garcinia kola* have been reported to inhibit the growth of *plasmodium falciparum* by well over

60% *in vitro* at a concentration of 6 mg/ml (Tona et al. 1999). A study by Akoachere et al. (2002) to investigate the anti-bacterial activity of bitter kola on four respiratory tract pathogens, viz. *Staphylococcus aureus, Streptococcus pyogens, Streptococcus pneumoniae and Haemophilus influenza* revealed that the extract exhibited anti-bacterial activity against the pathogens. A study by Elekwa et al. (2003) on the impact of fluid concentrates of *Garcinia kola* seeds on membrane stability of human erythrocytes showed the possible use of the extract for the management of sickle cell crisis.

The antimicrobial activity of bitter kola might be attributed to the presence of these phytochemical compounds.

Iwu et al. (2002) indicated that kolanone, one of the major constituents of the petroleum spirit extract of *G. kola,* has critical antimicrobial properties against both Gram-positive and Gram negative organisms. Its immune-boosting activity accepted to be because of the presence of biflavonoids, GB1, GB2, kolaflavonone, and garciniflavonone. Other activities ascribed to the biflavonoids include anti-inflammatory, antimicrobial, antiviral and antidiabetic impacts. They have demonstrated a noteworthy action against an assortment of viruses including Punta Toro, Pichinde, Sandfly fever, Influenza A, Venezuelan Equine Encephalomyelitis and Ebola.

23.7.2 Antioxidant Properties

Forest fruits are good nutrient and antioxidant sources that contribute significantly to human diets. Garcinia kola fruits is a promising source of essential nutrients, thus its consumption can make significant contributions to rural diets by ameliorating nutrient deficiencies. The high antioxidant content in bitter kola seed kernel proposes that they could be potential sources of natural antioxidant, thus they could be utilized as supplements in food production and the development of nutritional drinks (Onyekwelu et al. 2015).

23.7.3 Arthritis, Healthy Lungs and Malaria Treatment

The consumption of bitter kola assists in the decreasing the inflammation, pain, and immovable joints. The considerate amount of regular consumption of the bitter kola seed helps in strengthening the fibers and the lung tissue, stabilizing any counter effects, and further assists in maintaining a good respiratory track and treats chest colds. *Garcinia kola* hostile to anti-malarial properties. It contains Kolaviron which is a biflavonoid inspired antimalarial activity in P. berghei infected mice. What's more, kolaviron at the administered doses enhanced the parasite-induced anaemia and body weight modifications, perhaps through interfering with lipid peroxidation process and saving endogenous primary antioxidant enzymes reserves (Oluwatosin et al. 2014).

23.7.4 Anti-cancerous

Bitter kola can be utilized in treating breast cancer as it has an anti-cancerous effect. Phytochemicals with anti-oxidative and anti-inflammatory properties are known to restrain tumour inception, promotion and progression. Thus, there is an increasingly-convincing reason for utilizing cures containing those phytochemicals in the treatment of cancers and furthermore as pain relieving and mitigating adjuvants in treatment. The bitter kola stem bark and root, have been reported to be part of different indigenous anti-cancer regimens (Popoolaa et al. 2016).

23.7.5 Additional Medicinal Uses

Plant parts, for example, bark, fruit, seeds and nuts have been utilized as a part of conventional prescription for the treatment of different conditions like coughs, fever, gonorrhea, wounds, malignant tumors, chronic urethral discharge, stomach pains, pulmonary and gastro-intestinal conditions, and general body torments (Rajash and Rattan 2008). *Garcinia kola* helps to prevent glaucoma eye problem which results from an increase in eye pressure (Rajash and Rattan 2008). *Garcinia kola* plant sometimes cures impotency (male fertility) by boosting up a man's performance. Bitter kola has been known as a characteristic craving suppressant, accordingly, expanding the inclination to drink water (Tcheghebe et al. 2016). Bitter kola plant is a natural immunity booster against various deadly viruses like HIV, AIDS, ebola and flu (Tcheghebe et al. 2016). The amazing properties of bitter kola treat diabetes naturally on regular and considerate consumption (Iwu et al. 2002).

The bark and the seed when eaten together assists in detoxification of the human body during food poisoning (Tcheghebe et al. 2016). Saponin found in bitter kola is utilized as a liver tonic, assisting in the gallbladder capacities because of its purifying properties (Iwu 1999). Certain African cultures and scientific studies trust that bitter kola prolongs life by breaking down glycogen in the liver (Iwu et al. 2002; Okigbo and Mmeka 2008).

References

Adebisi AA (2004) A case study of *Garcinia kola* nut production-to-consumption system in J4 area of Omo forest reserve, south-West Nigeria. In: Sundenland T, Ndoye O (eds) Forest products, livelihoods and conservation: case studies on non-timber forest systems, vol 2- Africa. CIFOR, Bogor, pp 1–24

Adepoju AA, Salau AS (2007) Economic valuation of non-timber forest products. MPRA paper no. 2689. Online at http://mpra.ub.uni-muenchen.de/2689/. Accessed 8 Aug 2007

Adesina SK, Gbile ZO, Odukoya OA, Akinwusi DD, lloh HC, Jayeola AA (1995) Survey of indigenous useful plants of West Africa with special emphasis on medicinal plants and issues associated with their management. The United Nations University programmed on natural resources in Africa. 2nd edn. 84–85

Adeyeye EI, Asaolu SS, Aluko AO (2007) Amino acid composition of two masticatory nuts (*Cola acuminata* and *Garcinia kola*) and a snack nut (*Anacardium occidentale*). Int J Food Sci Nutr 58(4):241–249

Akoachere JF, Ndip RN, Chenwi EB, Ndip LM, Njock TE, Anong DN (2002) Antibacterial effect of Zingiber officinale and *Garcinia kola* on respiratory tract pathogens. East Afr Med J 79(11):588–592

Arekemase MO, Aliyu MB, Kayode RM, Ajiboye AE, Ajijolakewu AK (2012) Antimicrobial effects of Garcinia Kola (bitter kola) on some selected pathogens from University of Ilorin Teaching Hospital Ilorin, Nigeria. J Asian Sci Res 2(4):159–169

Asaolu MF (2003) Chemical composition and phytochemical screening of the seeds of *Garcinia kola*. Pak J Sci Ind Res 46:145–147

Atawodi S, Mende P, Pfundstein B, Preussmann R, Spiegelhalder B (1995) Nitrosatable amines and nitrosamide formation in natural stimulants, cola acuminate, C. nitida and *Garcinia kola*. Food Chem Toxicol 33(8):625–630

Braide VP (1991) Antihepatotoxic biochemical effects of kolaviron, a biflavonoid of *Garcinia kola* seeds. Phytother Res 5:35–37

Dosunmu MI, Johnson EC (1995) Chemical evaluation of the nutritive value and changes in ascorbic acid content during storage of the fruit of bitter kola (*Garcinia kola*). Food Chem 54:67–71

Elekwa I, Monanu MO, Anosike EO (2003) Effects of Aqeous extracts of *Garcinia kola* seeds on membrane stability of HbAA, HbAS and Hbss human erythrocytes. Glob J Med Sci 2(2): 97–101

Eleyinmi AF, Bressler DC, Amoo IA, Sporns P, Oshodi AA (2006) Chemical composition of bitter kola (*Garcinia kola*) seed and hulls. Polish J Food Nutr Sci 15(4):395–400

Esiegwu AC, Udedibie ABI (2009) Growth performance of and anti-microbial activities in broilers fed supplementary bitter kola (*Garcinia kola*). Anim Production Res Adv 5(1):20–24

Essien EU, Esenowo GJ, Akpanabiatu MI (1995) Lipid composition of lesser known tropical seeds. Plant Foods Hum Nutr 48:135–140

Farombi EO, Adepoju BF, Oladavies OE, Emerole GO (2005) Chemoprevention of aflatoxin B1-induced genotoxicity and hepatic oxidative damage in rats by kolaviron, a natural biflavonoid of *Garcinia kola* seeds. Eur J Cancer Prev 14(3):207–214

FDA (1999). FDA/CFSAN/OPA: Agency Response Letter: GRAS Notice No. GRN000025 on the use of Garcinia kola seed in distillation @ http://vm.cfsan.fda.gov/~rdb/opag025.html

https://commons.wikimedia.org/wiki/File:Garcinia_Kola_Germination.jpg

https://mavcure.com/garcinia-kola-health-benefits/

https://www.vanguardngr.com/2016/09/exportable-commodities-bitter-kola

Hussain RA, Owe-Gby AG, Arimoo PP, Waterman PG (1982) Kolanone, a novel polyisoprenylated benzophenone with antimicrobial properties from the fruit of *Garcinia kola*. Planta Med 44:78–81

Ibekwe AM, Grieve CM, Yang C-H (2007) Survival of Escherichia coli O157:H7 in soil and on lettuce after soil fumigation. Can J Microbiol 53:623–635. https://doi.org/10.1139/W07-003

Ibiblio JO (1983) Some medicinal plants of Nigeria. Ibadan publishers Nig. Ltd., pp 132–133

Irvine FR (1961) Woody plants of Ghana. Oxford University press, Oxford, pp 146–147

Iwu MM (1993) Handbook of African medicinal plants. CRC Press, London, pp 183–184

Iwu M (1999) *Garcinia kola*: a new adaptogen with remarkable immunostimulant, antiinfective and anti-inflammatory properties. A colloquium on *Garcinia kola* presented in International Conference on Ethnomedicine and Drug Discovery, Maryland, USA, PL-26

Iwu MM, Igboko O (1982) Flavonoid of *Garcinia kola* seeds. J Nat Prod 45:650–651

Iwu MM, Ogboko OA, Okunji CO, Tempesta MS (1990) Antidiabetic and aldose reductase activities of biflavanones of *Garcinia kola*. J Pharm Pharmacol 42:290–292

Iwu M, Diop A, Meserole L, Okunij C (2002) "Garcinia kola" a new look at an old adaptogenic agent. Ethnomedicine and Drug Discovery_9 2002 Elsevier Science B.V

Kabangu L, Gleffi C, Aonzo E, Nicoletti M, Messana I (1987) New biflavone from the bark of *Garcinia kola*. Planta Med 11:275–277

Keay RWJ (1989) Trees of Nigeria. Int J Bot Mycol 11(3) Clarendon Press Oxford:322–322
Keute V (2014) Toxicological survey of African medicinal plants. Pages: i–iii, Copyright © 2014 Elsevier Inc. All rights reserved. Elsevier
Ladipo DO (1995) Physiological/morphological growth rate and fruit/nut yields in *Garcinia kola* trees on acid soil of Onne, Port-Harcourt. ICRAF In-House Report
Leakey R (2001) Potential for novel food production from agroforestry trees: a review. http://www.wanatca.org.au/acotanc/papers/leakey_1
Mabberley DJ (1987) The plant book. Cambridge University Press. 707 pp
Nair PKR (1990) The prospects for agroforestry in the tropics. Technical paper 131, Nairobi, ICRAF
Odebunmi EO, Oluwaniyi OO, Awolola GV, Adediji OD (2009) Proximate and nutritional composition of kola nut (*Cola nitida*), bitter cola (*Garcinia kola*) and alligator pepper (Afromomum melegueta). Afr J Biotechnol 8(2):308–310
Ofor MO, Ngobili CA, Nwufo MI (2004) Ethno-botanical uses and trade characteristics of *Garcinia kola* in Imo State, Nigeria. Int J Agric Rural Dev 5:140–144
Okafor JC (1998) Mass propagation of species for immediate utilization. Paper presented at the meeting of underutilized crops of Nigeria. 4–8 May 1998. NACGRAB Moor Plantation, Ibadan
Okigbo RN, Mmeka EC (2008) Antimicrobial effect of three tropical plant extracts on Staphylococcus aureus, *Escheria coli and* Candida albicans. Afr J Tradit Complement Altern Med 5(3):226–229
Okwu DE (2005) Phytochemical, vitamin and mineral contents of two Nigerian medicinal plants. Int J Mol Med Adv Sci 1(4):375–381
Okwu DE, Ekeke O (2003) Phytochemical screening and mineral composition of chewing sticks in south-eastern Nigeria. Global J Pure Appl Sci 9:235–238
Oluwatosin A, Tolulope A, Ayokulehin K, Patricia O, Aderemi K, Catherine F, Olusegun A (2014) Antimalarial potential of kolaviron, a biflavonoid from *Garcinia* kola seeds, against Plasmodium berghei infection in Swiss albino mice. Asian Pac J Trop Med 7(2014):97–104
Omode AA, Fatoki OS, Olaogun KA (1995) Physicochemical properties of some underexploited and non-conventional oilseeds. J Agric Food Chem 43:2850–2853
Onochie CFA, Stanfield DF (1960) Nigerian trees. Gov. Printer, Lagos, pp 5–10
Onyekwelu JC, Oyewale O, Stimm B, Mosand R (2015) Antioxidant, nutritional and anti-nutritional composition of *Garcinia kola* and Chrysophyllum albidum from rainforest ecosystem of Ondo State, Nigeria. J For Res 26(2):417–424
Orie NN, Ekon EUA (1993) The bronchodilator effect of Garcinia kola. East African Med J 70:143–145
Popoolaa TD, Awodelea O, Omisanyaa A, Obia N, Umezinwaa C, Fatokun A (2016) Three indigenous plants used in anti-cancer remedies, *Garcinia kola* Heckel (stem bark), Uvaria chamae P. Beauv. (root) and Olax subscorpioidea Oliv. (root) show analgesic and anti-inflammatory activities in animal models. J Ethnopharmacol 194(2016):440–449
Rajash B, Rattan LI (2008) Essentials of medical microbiology, 4th edn. Jaypee Brothers Medical Publishers Ltd., New Delhi. 500pp
Tcheghebe OT, Signe M, Seukepe JM, Ngouafong F (2016) Ethnobotanical uses, phytochemical and pharmacological profiles, and cultural value of Momordica chlarantia Linn. : an overview. Merit Res J Med Med Sci 4(11):480–489
Tona L, Ngimbi NP, Tsakala M, Mesiak K, Cimanga K, Apers S, Bruyne TDE, Peters L, Totte J, Vlietinck AJ (1999) Antimalarial activity of 20 crude extracts from nine African medicinal plants used in Kinshasa, Congo. J Ethnopharmacol 15(68):193–203
Uruquiaga I, Leighton F (2000) Plant polyphenol antioxidants and oxidative stress. Biol Res 33:159–16522
Vivien J, Faure JJ (1996) Fruitiers samvages d'Afrique: especes du cameroun. Ministere de la cooperation/CTA, Paris/Wageningen, p 416

Chapter 24
Ricinoden dronheudelotii (Njangsa): Composition, Nutritional Values and Product

Norazalina Saad, Nur Khatijah Mohd Zin, Shafinah Ahmad Suhaimi, Muhammad Ehsan Fitri Rusli, Norsharina Ismail, Siti Nurulhuda Mastuki, and Rozita Rosli

Abbreviations

NTFP non-timber forest product
PUFA polyunsaturated fatty acid

24.1 Introduction

*R. heudelotii*or Njangsa is a deciduous, dioecic plant native to the coastal countries of West Africa. The common name of 'Njangsa' is established from the Greek words with the meaning of tick and tree as the seeds were believed to look like ticks.

N. Saad (✉) · N. K. Mohd Zin · S. Ahmad Suhaimi · M. E. F. Rusli
Laboratory of Research UPM-MAKNA (CANRES), Institute of Bioscience, Universiti Putra Malaysia, Serdang, Selangor, Malaysia
e-mail: norazalina@upm.edu.my

N. Ismail
Laboratory of Molecular Biomedicine, Institute of Bioscience, Universiti Putra Malaysia, Serdang, Selangor, Malaysia
e-mail: norsharina@upm.edu.my

S. N. Mastuki
Laboratory of Natural Products, Institute of Bioscience, Universiti Putra Malaysia, Serdang, Selangor, Malaysia
e-mail: ctnurulhuda@upm.edu.my

R. Rosli
Laboratory of Research UPM-MAKNA (CANRES), Institute of Bioscience, Universiti Putra Malaysia, Serdang, Selangor, Malaysia

Department of Biomedical Sciences, Faculty of Medicine and Health Sciences, Universiti Putra Malaysia, Serdang, Selangor, Malaysia
e-mail: rozita@upm.edu.my

A. A. Mariod (ed.), *Wild Fruits: Composition, Nutritional Value and Products*,
https://doi.org/10.1007/978-3-030-31885-7_24

Aside from Njangsa, different countries refer to the tree with alternative names such as Manguella (Angola); Esseenang (Cameroon); Bofeko (Zaire); Wama (Ghana); Okheun (Nigeria); Kishingo (Uganda); Akpi (Cote d'Ivoire); Groundnut tree, Corkwood, African wood-oil nut tree, African wood, African nut tree (English); Bois Jasanga (France); Afrikanisches Mahagoni (Germany); Muawa (Swahili); and erimado (Yoruba) (Orwa et al., 2009;Yirankinyuki et al. 2018). Until now, two varieties of the plant have been identified, which are *R. heudelotii*var. *heudelotii*in Ghana, and *R. heudelotii*var. *africanum*in Nigeria and westwards (Leone 2009).

24.2 Botanic and Biological Description

*R. heudelotii*is a rapid growing tree with an average of 20–30 m in height and may grow up to 50 m (Plenderleith 1997). It has a straight trunk that can reach up to 2.7 m in diameter with short buttresses. The trunk is straight with a smooth grey bark that fissures with age with a red slash bark where it is densely mottled with orange stone-cell granules and scattered pits. The inner bark, however, ranges from pink to red. The heartwood is soft, white to pale yellow and not differentiated from the sap. While the crown is broad with candelabra-like branches, the root is thick and often heavy superficial roots (Vivien and Faure 1985). *Ricinodendron heudelotii* achieves maturity at around 4–5 years.

The flowers, the inflorescences appear hairy, yellowish white, and approximately 5 mm long with the ability to develop a long terminal panicle that ranges from 15 to 40 cm. In particular, the male panicles are longer and more slender compared to the female panicles and can reach up to 41 cm in length. The flowers of Njangsa start to bloom between April and May whereas the fruits are abundantly produced and matured during September and October. When the fruits fall, the tree appears leafless for several weeks, and *R. heudelotii* will only produce fruits between the 7th to 10th years if grown in open spaces with maximal light intensity (Plenderleith 1997) (Fig. 24.1).

The fruit is a 2 to 3-lobed yellow-greenish drupe which is indehiscent as it does not open spontaneously when fully mature. It has a spherical shape, with a length of 2–5 cm, a width of 2.5–4 cm and weight ranges from 19 to 47 g (Ogbuagu et al. 2018). Moreover, the smell of the fruits is akin to an overripe apple and reaching at maturity around August to September. The fruit, which contains around 2–3 seeds, has a fleshy mesocarp and a woody endocarp. These red, brown and black seeds are flat, round, approximately 1 cm in diameter and have an oily texture. Unlike the fruits which smell like overripe apples when mature, the seeds have an odour similar to oily chocolate. However, the seeds have a unique taste which is slightly aromatic and leaves a mildly bitter aftertaste; hence, they are commonly used as a food source and supplement. The seeds, which are the edible part of the fruits, are extracted from the hard endocarp via a crude procedure involving boiling of the kernels for several hours at a high temperature of more than 100 °C. Following seeds extraction, the fruit pulp and endocarp are usually discarded (Ezekwe et al. 2014). In contrast, the extracted oil is light yellow, which tastes similar to groundnut oil.

Fig. 24.1 (**a**) Tree; (**b**) fruits; (**c**) seeds; and (**d**) oil of *Ricinodendronheudelotii*(Njangsa). (Source: https://commons.wikimedia.org)

As for the leaves, they grow in an alternate order as in Fig. 24.2 (a), and the leaflets grow palmately as in Fig. 24.3 (b). The leaflets are fan-shaped with 6– 20 cm long and 2.5–12 cm broad united at the base with 3–8 leaflets each leaf. Generally, there are 10–16 pairs of veins per leaflet, and the margins have small glands. Young leaflets range from glabrous to densely stellate hairy which can disappear when adult. The leaflet base is lengthily attenuated at the top of the petiole, and the tips are acute and acuminate. The petioles are 5– 20 cm long with 2 lateral glands at the top. The stipules are toothed, leafy and persistent, and clasp the stem.

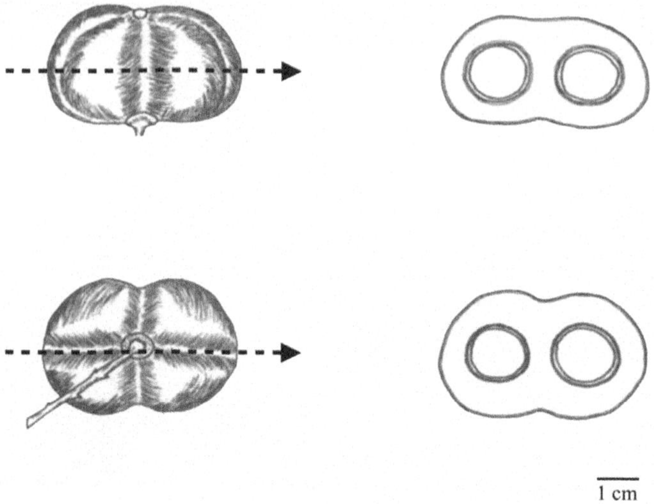

Fig. 24.2 Illustration of *R. heudelotii* fruit and its cross-section. (Adapted from: Thiselton-Dyer (1913))

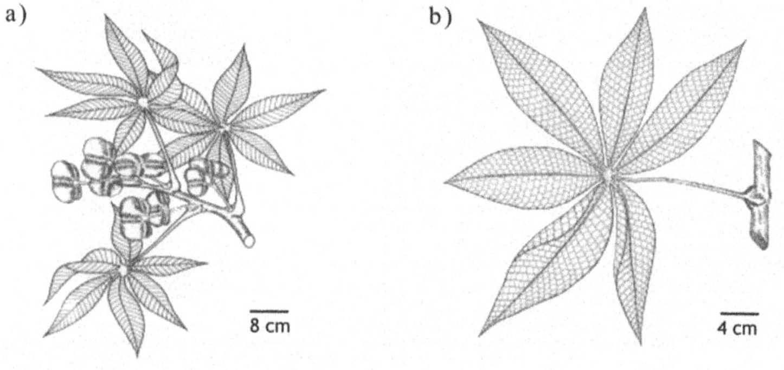

Fig. 24.3 Illustration of *R. heudelotii* leaf arrangements (**a**) and its leaflet (**b**). (Adapted from: Thiselton-Dyer (1913))

24.3 Phytochemical Composition

24.3.1 R. heudelotii *Leaves and Bark*

In recent years, the various phytochemical composition of the leaves and bark has been done on *R. heudelotii*. A study by Yatindo Yves et al. (2017) reported that the leaves of *R.heudelotii* contain numerous families of chemical groups which include alkaloids, flavonoids, tannins, reducing compounds and mucilage. The presence of alkaloids in the leaves supports the traditional use of Njangsa leaves decoction in treating malaria. Likewise, a study conducted by Uzoekwe and Hamilton-Amachree

Table 24.1 Qualitative phytochemical analysis of leaf and bark

Test	Leaf (ethanol)	Bark (ethanol)
Flavonoids	+	+
Tannins	+	+ +
Cardiac glycosides	+ + +	+
Saponins	−	+
Steroids	+	+ +
Terpenoids	+ +	+ +
Anthraquinones	−	−
Alkaloids (Hager's)	−	−
Alkaloids (Wagner's)	−	+ +
Reducing sugar	+ +	+ + +

Table 24.2 Quantitative phytochemical analysis of leaf and bark

Parameters	Leaves	Bark
Calcium	5932	4409
Magnesium	1994	2431.20
Sodium	0.00	0.00
Potassium	0.00	0.00
Iron	32.30	15.60
Zinc	18.60	11.50
Copper	22.70	19
Manganese	41.30	27.70
Phosphorous	688	249

(2016) also reported the phytochemical composition of leaves and bark of R. heudelotii consists of flavonoids, tannins, cardiac glycosides, steroids, terpenoids, saponins and reducing sugars. Therefore, the presence of these active compounds may contribute to the existing therapeutic properties of the tree. For example, the bark of RicinodendronHeudelotii is conventionally utilised for managing pain and treating, rheumatism, diarrhoea, stomach problems, oedema in children, infertility rashes, mouth ulcers, chest complications as well as serves as a galactagogue.

As presented in Table 24.1, both the ethanol extracts of leaves and bark contains flavonoids, tannins, cardiac glycosides, terpenoids, and reducing sugars. However, the leaves extract lack saponins, anthraquinone and alkaloids, as were the presence of anthraquinone in the bark extract. Nonetheless, although the alkaloid was moderately present using Wagner's test, the presence of alkaloid by using Hager's test was negative.

Estimation of mineral elements of both the leaves and the bark as of Table 24.2 shows that both leaves and bark extract lacks sodium and potassium. It also shows that although the bark contains higher magnesium compared to the leaves extract, it has lower calcium, iron, zinc, copper, manganese and phosphorus content compared to the leaves extract.

24.3.2 **R. heudelotii** *Seeds and its Oil*

Qualitative (Table 24.3) and quantitative (Table 24.4) analyses for the seed extract of *R. heudelotii* were carried out by Tamuno-Boma (2016) and demonstrated substantial amounts of phytochemicals, such as tannins, saponins, flavonoids, alkaloids, phenols, carotenoids and anthraquinones. Tannins are said to be a natural defence against infections, and it is also known to have effects such as lowering blood pressure, assisting in the clotting of blood as well as modulates the immune response. Flavonoids were also part of the chemical found in *R. Heudelotii*, and it is known that phenolic compounds have antioxidant, anti-inflammatory and anti-carcinogenic properties. Besides, small amounts of phytate and oxalates were also observed. In addition, there were also unidentified residues of an alkaloid and resin in the seeds (Okwu and Josiah 2002). Other than the phytochemicals, the seeds were found to contain 31.4% crude protein and 44.7% lipid (Ezekwe et al. 2014).

As for the seeds oil, approximately 47% (w/w) of its oil comprises the fatty acids, particularly 44% eleostearic acid, 16% oleic acid, and 10% each of palmitic, stearic, linoleic and linolenic acids (Orwa et al. 2009). The geographical origin of *R. heudelotii* kernels being harvested is also important to be considered as different locations in Cameroon yields fat contents that range between 50.0 and 65.2% (Manga et al. 2000). Regardless, the total fat content of the oil may be generally accepted to be within the range, as supported by Ogbuagu et al. (2018) which reported that the yield of extracted oil from the Njangsa kernels was found to be in the range of 49–63%. Typically, the presence of fatty acid indicates a high content of polyunsaturated fatty acids (PUFA) (C18–3) and essential amino acids (Tchiegang et al. 1997). Indeed, a food composition analysis of *R. heudelotii* showed the presence of distinct long chain omega-3 fatty acids that was not commonly found in plants. Specifically, the identified lipid comprises of approximately 73% of polyunsaturated fatty acids (PUFA), with nearly all the lipid was eicosapentaenoic acid and around 18% of them was found to be oleic acid (Ezekwe et al. 2014).

Table 24.3 Qualitative phytochemical analysis of seed oil

Component	Result
Tannin	+
Phlobatannins	–
Flavonoid	+
Alkaloid	+
Cardiac glycoside	+
Terpenoid	+
Saponin frothing	+
Saponin emulsifying	+
Combined anthraquinone	+
Free Anthraquinone	–
Carotenoid	+
Reducing compound	–

Table 24.4 Quantitave phytochemical analysis of seed oil

Component	Sub-class	Concentration µg/g
Phenol		2.2357
Tannin		39.9463
Oxalate		1.1581
Phytate		0.2946
Alkaloid	Lumamarine	7.9079
	Ribalidine	5.608
	Spartein	0.0002
Saponin	Sapogenin	14.2632
Flavonoid	Catechin	11.5693
	Epicatechin	1.9145
	Rutin	5.9573
	Kaempferol	3.1418
	Naringenin	18.7939
	Anthocyanin	2.4412

24.4 Nutritional Values and Products

24.4.1 Leaves

In traditional practice, herbal formulations of the leaves of *R. heudelotii* are consumed for numerous medicinal reasons such as a disinfectant and as treatments of tonsillitis, ophthalmic, stomach and back pains. Also, several reports on medicinal virtues have mentioned the application of *R. heudelotii* leaves for treating abscesses, boils, fever and fungal infections (Yatindo Yves et al. 2017). In Cameroon, the crushed leaves were also applied daily to the skin to combat Varicella infection. The treatments are may be effective due to the existence of several active compounds such as flavonoids, tannins, cardiac glycosides, saponins, steroids, terpenoids, and alkaloids in the leaves. Additionally, the leaves of *R. heudelotii* are also high in fibre, minerals such as ash as well as proteins. However, *R. heudelotii* leaves are not a suitable source of energy due to the small amounts of carbohydrate and fat (Ezekwe, et al. 2014).

R. heudelotii leaves are frequently used as animals feed and treatments. In particular, the leaves are an essential source of high-quality fodder for sheep and goats during the dry season and also utilised to treat male infertility as well as malaria (Yatindo Yves et al. 2017). The leaves provide an important supply of high quality fodder for sheep and goat in dry season, the values of crude protein identified in the study by Anigbogu was 5.0% for the leave and 9.9% for the bark which is comparatively lesser than 16.0% achieved for green foliage of *R. heudelotii* (Anigbogu 1996). The broad uses of the leaves on humans and animals are supported by a toxicity analysis in which the ethanolic and aqueous leaves extracts of *R. heudelotii* were non-toxic with LC50 values greater than 0.1 mg/mL (LC = limit of toxicity) Hence, the leaves of *R. heudelotii* are safe for consumption by both humans and animals (Yatindo Yves et al. 2017).

24.4.2 Bark, Wood and Root

For Ricinodendronheudelotii, the bark was revealed as the main part used for the treatment of several diseases. The decoctions of the bark are used to treat ovarian cysts, injuries due to childbirth such as obstetric fistulas, haemorrhoids, kidney problems, sores and muscle pain as it has a purgative effect (Yatindo Yves et al. 2017). It has also been used to prevent abortion and facilitate childbirth in combination with Ehretiacymosa (Yakubu et al. 2018). In some regions of Cameroon, a mixture of the bark of *R. heudelotii* has been described to enhance sexual desires and improve urination. Additionally, the addition of the dried bark to a soap locally known as "Coto" was used to treat a chronic fatigue syndrome known as asthenia and also against venomous bites (Yatindo Yves et al. 2017).

Although the *R. heudelotii* wood is not feasible as a fuel source since it burns very rapidly, it is, however, useful for the production of paper pulp, rough planks, coffins, fishing net floats and rafts due to its timber properties which are fibrous, soft, light and perishable. Its timber properties also allow it to be conveniently shaped into products such as fetish masks, spoons, ladles, plates, platters, bowls, dippers and stools (Yakubu et al. 2018). As mentioned earlier, the applications of *R. heudelotii* as musical instruments differ in each country. For example, in the Democratic Republic of Congo, the wood of the tree is used to produce drums since the wood appears to be highly resonant. However, in southern Nigeria, Gabon and Angola, it is carved to create the entire or the sonorous features of musical instruments. In Ghana, instead of as a musical instrument, the wood is beneficial for insulation whereas the sawdust is utilised in sun helmets production.

As for the medicinal properties of *R. heudelotii*, in Nigeria, the root is used to ease constipation when powdered and mixed with pepper, salt and the bark. On the other hand, the Temne which is the largest ethnic group in Sierra Leone believes that tying its people to the beaten with warmed *R. heudelotii* bark was able to cure elephantiasis or lymphatic filariasis due to the parasitic worms, while ingestion of the root-bark decoction in Cote d'Ivoire works against dysentery, an infection of the intestines (Plenderleith 1997). For women, bark liquor of the plant is consumed by pregnant Liberian women to reduce pains and avoid miscarriage. Also, in Gabon, the bark decoction is taken for blennorrhoea, painful menstruation and to neutralise poison (Tchoundjeu et al. 2006).

24.4.3 Seed and its Oil

Different countries in Africa utilise Njangsa seeds for distinct purposes. For instance, in West and Central Africa, the seeds are valued for its distinctive flavour. Hence, they are mainly used as an ingredient in soups, the flavour in grilled meats and fish, as well as seasoning and thickening agent in food (Plenderleith 1997). Moreover, they are also used when roasting chicken and vegetables, made into a paste and can even be consumed on its own. In particular, in Cote d'Ivoire, the kernels can be eaten after boiling in water, in sauce, or mixed with fish, meats or vegetables. In Gabon, the seeds are used as a relish for small, white mushrooms known as dibindi (https://www.appropedia.org/Ndjanssang). On the other hand, in several countries, the seeds are used in musical instruments as well as for local games. For example, in Sierra Leone, they are employed in rattles for bundu dances, used in 'songo' music in Cameroon and 'okwe' game in Nigeria (Ogbuagu et al. 2018).

Since seed oil is an inexpensive NTFP, it is commercially valuable especially in Cameroon, particularly for industrial, pharmaceutical and nutraceutical purposes (Sagara et al. 2011; Mori et al. 1999). The seed oil, which is light, yellow, dry, and has a sweet flavour, it is applicable in varnishes and soft soaps and used industrially in waterproofing supplies (Southampton Center for Underutilised Crops (SCUC 2006)). Due to the abundance of α-tocopherol, the oil is very stable when sealed and transformed gradually into rancid when exposed. In addition, Yirankinyuki et al. (2018) revealed that the *R. heudelotii* seed oil could be a suitable binder for making the formulation of oil paints for surface coating and could serve as raw materials for the production of resins.

On the other hand, for nutraceutical purposes, the nature of the seed oil which is low in carbohydrate, cholesterol and triglyceride levels is beneficial for diabetics as well as against cardiovascular disease and atherosclerosis (Yeboah et al. 2011, Matig et al. 2006)). Interestingly, as mentioned earlier, a high percentage of polyunsaturated fatty acids makes the seed oil promising in health improving management, particularly for the heart. Not only that, but the seeds oil is also effective against smallpox via skin massage (YatindoYves et al. 2017). Moreover, it has been proposed that high yield of the seed oil together with a high proportion of polyunsaturated fats in the oil make it profitable for production of cooking oil and margarine, soaps as well as pharmaceutical products (Ogbuagu et al. 2018). The presence of elongated chain omega-3 fatty acids in Ricinodendronheudelotii seeds oil would also generate a higher demand for manipulation of this plant for human use and animal nutrition. Indeed, there is a growing trend towards obtaining leaner pork via dietary or genetic modifications, specifically by altering the fatty acid profile as healthier market products. This trend is promising since the patents for improving lean meat using omega-3 fatty acids already existed.

To sum up, *R. heudelotii* is extensively used in traditional medicine among Africans. Although there are an increasing number of scientific literatures to support the use of the plant not only as medicine but also as food and raw materials in the industries, there is a dire need to continuously attempt to relate the chemical components and physicochemical characteristics of *R. heudelotii* with its broad applications.

References

Anigbogu NM (1996). Nature's gifts improving trees and shrubs around the world: R. heudelolti in Nigeria. Agroforestry Today 8(2):18

Ezekwe MO, Besong SA, Johnson R (2014) Nutritive composition of omega-3 fatty acids-rich Ricinodendron heudelotii and its potential for nutrition. Int J Nutr Metab 6(6):56–62. http://www.academicjournals.org/IJNAM

Leone, IS (2009) RicinodendronHeudelotii(Baill.) Pierre EuphorbiaceaeRicinodendronHeudelotii (Baill.) Pierre. 0: 1–5

Manga TT, Foundoun JM, Kengue J, Thiengang C (2000) Chemical composition of Rheudelotii an Indigenous Fruit Tree in Southern Cameroon. Afr Crop Sci J 8:195–201

Matig OE, Ndoye O, Kengue J, Awono A (2006) Les fruitiers forestiers comestibles du Cameroun. Bioversity International, Yaounde

Mori TA, Bao DQ, Burke V, Puddey IB, Beilin LJ (1999) Docosahexaenoic acid but eicosapentaenoic acid lowers ambulatory blood pressure and heart rate in humans. Hypertension 34(2):253–260

Ogbuagu AS, Eric SN, Obumselu FO, Ogbuagu JO (2018) Chemical & Pharmaceutical Research Qualitative and Quantiative Phytochemical and Physicochemical Analyses on The Oil and Extracts from RicinodendronHeudelotii. Chem Pharm Res 1(1):1–7

Okwu DE, Josiah C (2002) Afr J Biotechnol 5(4):357–361. https://www.ajol.info/index.php/ajb/article/view/137808

Orwa C, Mutua A, Kindt R, Jamnadass R, Anthony S (2009) Agroforestree Database: a tree reference and selection guide version 4.0. World Agroforestry Centre, Kenya

Plenderleith K (1997) RicinodendronHeudelotii. Oxford Forestry Institute Department of Plant Sciences University of Oxford United Kingdom

Sagara M et al (2011) Effects of Docosahexaenoic Acid Supplementation on Blood Pressure, Heart Rate, and Serum Lipids in Scottish Men with Hypertension and Hypercholesterolemia. Int J Hypertens 2011(Cvd):1–7

Southampton Centre for Underutilized Crops, SCUC (2006) Njansa, Ricinodendronheudelottii, Field Manual for Extension Workers and Farmers. pp. 143–156

Tamuno-Boma O, Worlu-Wodo QE, Deekae S (2016) Bioprospective Screening of Ricinodendron Heudelotti Seeds. J Anal Pharm Res 3(7):00084

Tchiegang C, Kapseu C, Ndjouenken R, Ngassoum MB (1997) Ricinodendron heudelotii (bail) kernels: a novel ingredient for tropical agro. J Food Eng 32:1–10

Tchoundjeu Z et al (2006) In: Williams JT, Smith RW, Haq N, Dunsiger Z (eds) *NdjanssangRicinodendronHeudelotii (Baill.)*. Southampton Centre for Underutilised Crops, University of Southampton, Southampton

Thiselton-Dyer WT (1913) Flora of tropical Africa, vol VI, Section I. Lovell Reeve and Co., London

Uzoekwe NM, Hamilton-Amachree A (2016) Phytochemicals and nutritional characteristics of ethanol extract of the leaf and bark of Njangsa (*RicinodendronHeudelotii*) plant. J Appl Sci Environ Manag 20(3):522

Vivien J, Faure JJ (1985) Arbres des Forêts Denses d'Afrique Centrale. Ministère des Relations Extérieures, Coopération et Développement, ACCT, Paris

Yakubu OF et al (2018) Antimicrobial and toxicological studies of ricinodendron heudelotii (Baill.). Asian J Pharm Clin Res 11(1):299

Yatindo Yves BOKO-HAYA et al (2017) Traditional uses, phytochemistry and in vitro evaluation of toxicity of ricinodendron heudelotii (Baill Pierre ex Heckel) leaves in benin. Int J Recent Sci Res 8(10):21227–21236

Yeboah SO, Mitei YC, Ngila JC, Wessjohann L, Schmidt J (2011). Compositional and Structural Studies of the Major and Minor Components in three Cameroonian seed oils by GC–MS, ESI-FTICR-MS and HPLC. J Am Oil Chem Soc 88:1539–1549.

Yirankinyuki FF, Lamayi DW, Muhammad UA, Musa B (2018) Assessing the suitability of RicinodendronHeudelotii seed oil for paint formulation. IOSR J Appl Chem (IOSR-JAC) 11(7):37–42

Chapter 25
Artocarpus heterophyllus (Jackfruit): Composition, Nutritional Value and Products

Raquel P. F. Guiné and Sofia de Guiné e Florença

25.1 Introduction

Jackfruit (*Artocarpus heterophyll*us Lam.) is a fruit commonly found in tropical areas. The fruit itself is protected by a case that can be of large dimensions, 30–50 cm long for the greater fruits, and which contains inside up to 500 individual fruits, with a golden colouration. The aroma can be quite strong, and therefore it may displease some people. However, when the fruit is dried this drawback is eliminated. Because of the high perishability of the jackfruit, a high index of post-harvest loss is observed, and therefore dehydration is frequently an alternative method for its preservation. This method reduces moisture content, decreasing the growth of microorganisms and preventing biochemical reactions that depend on moisture, and in addition reduces the costs of storage and transportation (Oliveira et al. 2011).

The edible part of jackfruit includes fruits and seeds. The fruit can be consumed either raw or cooked in multiple ways. The pulp is consumed in its almost totality *in natura* by the most diverse layers of the population in diverse countries. When the fruit is young, and due to its considerable amount of carbohydrates, it is frequently cooked as a vegetable. Along maturation the fruit becomes sweeter, being for that reason eaten raw when fully ripe, although it can also be eaten after cooking. The rind can be used to produce a jelly and the seeds can be consumed cooked, either boiled or roasted, having flavour and textural characteristics comparable to chestnuts. Alternatively, the seeds can be ground into a powder and incorporated into biscuits or other bakery products' formulations, due to their high starch and protein.

R. P. F. Guiné (✉)
CI&DETS and CERNAS Research Centre, Polytechnic Institute of Viseu, Viseu, Portugal
e-mail: raquelguine@esav.ipv.pt

S. de Guiné e Florença
Faculty of Nutrition Sciences, University of Porto, Porto, Portugal

A. A. Mariod (ed.), *Wild Fruits: Composition, Nutritional Value and Products*,
https://doi.org/10.1007/978-3-030-31885-7_25

When the fruits are still very young they can be cooked in soups and stews together with leaf shoots. Also young male flowers are edible, usually consumed mixed with chillies, sugar, salt or other spices. Finally many drinks are obtained from jackfruit, like juices, liquors or spirits (Madruga et al. 2014; Fonseca 2016).

Jackfruit is composed of a significant number of berries with yellow pulp and brown seeds enclosed in a hard case. The ripe buds are rich in nutritive substances, such as carbohydrates, proteins, minerals (calcium and phosphorus) and vitamins (complex B), as well as other bioactive molecules like phenolic compounds, thus providing nutritional and bioactive properties (Oliveira et al. 2011; Madruga et al. 2014; Fonseca 2016; Jayus and Giyarto 2016). The pulp contains several classes of compounds such as carotenoids, flavonoids, volatile acids, sterols and terpenes, contributing also for its distinctive flavour (Hu et al. 2016). Besides, jackfruit provides components for pharmaceutical use, owing to the presence of several phytochemicals that provide antioxidant, anti-inflammatory, antibacterial, anti-carcinogenic or anti-fungal activities (Baliga et al. 2011).

25.2 Botanical Characteristics

Jackfruit belongs to the family Moraceae, gender *Artocarpus* and includes four synonyms: *Artocarpus integra* Merr., *Artocarpus integrifolia* Linn. F., *Artocarpus brasiliensis* Lam. and *Artocarpus heterophyllus* Lam. It is a fruit original from Asia (India, Malaysia, Philippines), having been spread to other parts of the globe (Fig. 25.1), including Brazil where was introduced by the Portuguese. In India, despite the existence of dozens of varieties, two assume particular importance for cultivation: wareka, with firm rind and Vela, with soft rind. There are also some distinct sub-varieties like kuru-wareka (small rounded fruits), peniwareka (sweet pulp, being for that also called honey jackfruit) or johore (with small oblong fruits with a very pleasant aroma) (Oliveira et al. 2011; Fonseca 2016).

The jackfruit tree is monoecious giving male and female inflorescences in the same plant (Fig. 25.2). The time necessary for complete fruit development is highly variable (taking form 3 to 7 months), depending on the climatic conditions in each country. (Falcão et al. 2001; Baliga et al. 2011).

The fruit is oblong with a cylindrical form measuring from 30–40 cm long, but in some cases it can go to 80 cm by 30 cm diameter. Its weight is also variable form 10 to 25 kg, but in extreme cases it can achieve 50 kg. The fruit is composed of different parts (Fig. 25.3). The axis is the core of the fruit, being rich in latex owing to the presence of laticiferous cells. This part is not edible and its mission is to hold the fruits together. The most important part of the fruit is the perianth, constituting the bulk. It comprises three zones: the bulb, which is the lower fleshy edible part; the fused part in the middle that forms the rind of the syncarp, and the upper free and thorny non-edible part. The ripe fruit, arils and flesh, contains yellow good flavoured and sweet bulbs and seeds (Fig. 25.4) (Falcão et al. 2001; Baliga et al. 2011).

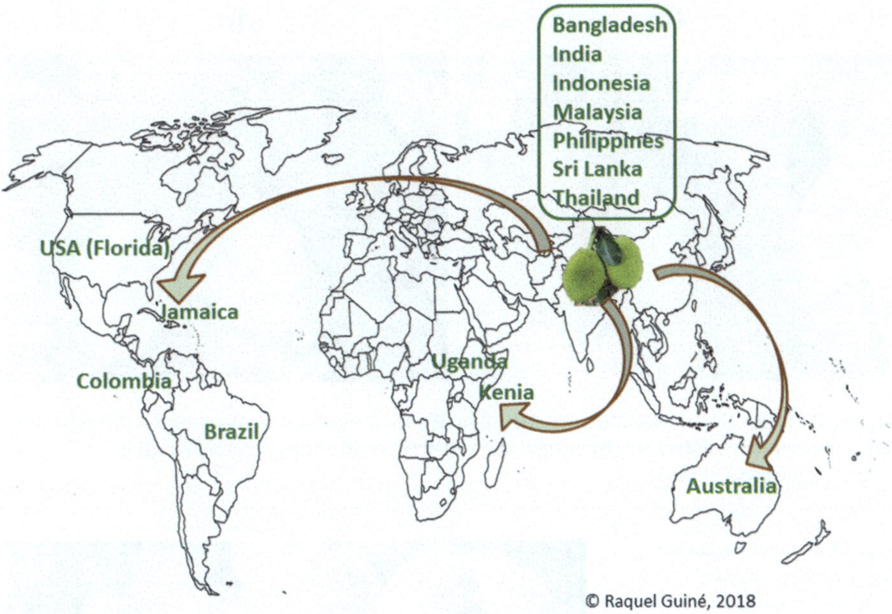

Fig. 25.1 World distribution of jackfruit production

Fig. 25.2 Jackfruit Tree. (Source https://commons.wikimedia.org/w/index.php?curid=63866076)

Fig. 25.3 Jackfruit: (**a**) outside view. (By Len Worthington https://commons.wikimedia.org/w/index.php?curid=57560922); (**b**) inside view. (By Wiel46, https://commons.wikimedia.org/w/index.php?curid=2143863)

Fig. 25.4 Jackfruit seeds. (By Gharouni, https://commons.wikimedia.org/w/index.php?curid=66850649)

25.3 Production and Transformation of Jackfruit

25.3.1 Production

The jackfruit is an evergreen tree, typical of advanced successional stages and grows naturally in sub-areas of moist tropical rain forest until 1300 m of altitude. It is tolerant to shade, but needs a bit of light and space for its development during the early stages of life, germinating better in clearings. In the original habitat (in Asia), its fruits mature between July and August, and the seeds vary in size and mass (from 4 to 14 g) (Abreu and Rodrigues 2005, 2010).

Young seedlings should be fertilized with generous amounts of nitrogen while bearing trees need regular applications of phosphorous and potassium. It is advisable to apply organic fertilizer around the trees. Although the water requirements are not critical in jackfruit production, when the climatic conditions are unfavourable, like during extreme drought, irrigation should be provided. Periodic ring weeding and underbrush is advisable every 3 months. The trees must be pruned when they achieve 2 years of age. Pruning comprises the removal of small unproductive branches as well as diseased and insect-damaged ones. Because fruits are usually produced on the trunk and large branches, the removal of small unwanted branches shall provide more favourable conditions for the development and growth of the fruits. Jackfruit plantations provide high yields, up to 26 ton/ha, with a mature tree producing around 700 fruits annually (Jagadeesh et al. 2007; Foo and Hameed 2012).

25.3.2 Harvesting and Handling

The indicators of fruit ripeness commonly used to harvest the fruits include: the last leaf on the stalk turns yellow; the fruit produces dull, hollow sounds when tapped; the well-developed and widely spaced spines yield to moderate pressure. When harvesting the peduncle is cut (normally using a sharp knife), and the fruit is hold, normally wearing gloves to protect from the spines. Harvesting should be carried out at mid-morning to late afternoon, in order to minimize latex flow, since during these periods the latex cells are less turgid (Pua et al. 2010).

When handling the fruit, it should be left with the stalk faced down to let the latex flow and coagulate. To minimize injuries and quality loss, it is preferable to transport the fruits in single layers, using leaves in between them and on the container to prevent the fruits from getting bruises, scars, and breaks (Pua et al. 2010).

Post-harvest includes the removal of immature, over-ripe, damaged and misshapen fruits. The remaining fruits are classified according to categories (Large: weighing 16 kg and above or Medium: weighing 8–16 kg) and then washed with chlorinated water (at a concentration of 100 ppm) to remove dirt, latex stains and any field contamination. Finally, the fruits must be drained in order to remove the excess of moisture from the surface (Pua et al. 2010).

25.3.3 Packaging and Storage

Graded and washed fruits are packed into plastic containers or bamboo baskets for storage. The freshly harvested ripe fruits can be stored for 4 to 5 days at a temperature ranging from 25 to 35 °C. For longer storage, from 2 to 6 weeks, they must be kept as lower temperatures, 11–13 °C, and relative humidity of 85–95%, depending on cultivar and maturity stage. Jackfruit is a climacteric fruit possessing a short shelf life, and therefore it rapidly becomes unacceptable for the consumer (Pua et al. 2010).

Jackfruits are susceptible to chill injury when stored at temperatures below 12 °C before transfer to higher temperatures. This results in dark-brown discoloration of the skin, pulp browning, flavour deterioration and increased susceptibility to degradation (Vargas-Torres et al. 2017).

For the minimally processed (pre-cut) jackfruit storage, Vargas-Torres et al. (2017) recommend manual peeling, conservation at 5 °C and use of edible coatings to ensure a good quality and sensory acceptability while prolonging shelf life.

The microbial quality of minimally processed jackfruit was studied by Adiani et al. (2014) during storage at 4 and 10 °C, for products packed in polystyrene trays enclosed with a cling film. They used GC spectral data and total mass spectral data to find that ethanol, ethyl acetate and 3-methyl-1-butanol were the major compounds responsible for the associations between the spectral data and the variables that indicate microbial deterioration: total viable count (TVC) and yeast and mold count (Y&M) (Adiani et al. 2014).

25.3.4 Processing

Because jackfruit is perishable and has a short shelf life, it is processed into different products, like jams for example, or alternatively is dried so that its moisture is drastically reduced thus providing an appreciable extension of the shelf life. Drying is an essential operation in food processing industry aimed at preserving food quality and food stability by minimizing the water activity owing to a reduction in moisture content. Nevertheless, although providing preservation, it causes many physical, chemical and biochemical changes in the tissues, some that may confer different and appreciated characteristics to the product and others, compromising the organoleptic or nutritional properties (Kaushal & Sharma 2016).

Drying is a popular method used both for fruits and vegetables, which results in a high degree of water loss thus resulting in an appreciable reduction in bulk volume, as well as low packaging and transportation costs, besides providing microbial safety and shelf stability. However, due to the intensive reduction in water mass and a high degree of shrinking, the quality of the dried products can be somewhat affected in terms of texture, colour, nutritive value and rehydration characteristics (Saxena et al. 2015).

The influence of pre-treatments for a combination of freeze drying and hot air-drying were investigated by Saxena et al. (2015) aiming at the optimization of processing conditions to produce jackfruit bulb crisps (Saxena et al. 2015). The obtained results showed that the optimized combined drying procedure allowed obtaining a product with similar quality to that of the freeze-dried product, thus providing potential for commercialization. In order to produce a stable product with an improved shelf life, Pua et al. (2010) investigated the drum drying process for the obtaining of jackfruit powder, using response surface methodology (RSM). They investigated the influence of some drying conditions (steam pressure and drum rotation speed) on the physical properties (colour parameters), chemical properties (moisture content, water activity and solubility index) and sensorial properties (by quantitative descriptive analysis – QDA and hedonic test) of the obtained powder. Their results showed that optimum quality of the jackfruit powder could be obtained with drum drying accomplished with a rotating speed of 1.2 rpm at a steam pressure of 336 kPa (Pua et al. 2010). Also the quality of the powder products was investigated by Pua et al. (2008), namely the total colour difference, the rate of moisture absorption as well as sensory attributes, for the product under storage for 3 months The storage conditions consisted in temperatures of 28 and 38 °C and variable relative humidity (50%, 75% and 90%). The containers were plastic pouches, in two variations: aluminium laminated polyethylene and metallized co-extruded biaxially oriented polypropylene. The final results evidenced that aluminium laminated polyethylene packaging and storage conditions at the lowers tested temperature (28 °C) and relative humidity lower than 75% showed better suitability for preservation of the jackfruit powder (Pua et al. 2008).

Osmotic dehydration consists in dipping the product into a hypertonic aqueous solution, originating a loss of water owing to the osmotic pressure. The water molecules pass through the cell membranes of the product and then flow in the intercellular spaces until reaching the product surface and finally diffuse into the osmotic solution. The osmotic solution is prepared with a high concentration of salts or sugars. Osmotic dehydration constitutes one of the most significant pre-treatments prior to the drying or dehydration of food products, because it presents some benefits like reducing the damage caused by high temperature exposure, namely in terms of flavour and colour, inhibiting enzymatic browning, and at the same time decreasing the global energy costs (Kaushal and Sharma 2016). Osmotic pre-treatments with a 15% NaCl solution were used for jackfruit samples which were lately submitted to convective drying with hot air at 50, 60 and 70 °C at constant velocity of 1.5 m/s, in experiments performed by Kaushal and Sharma (2016). The results obtained showed that the temperature of 60 °C was more beneficial in terms of product's final colour when compared to 50 and 70 °C.

Winemaking is one of the oldest technologies allowing to obtain commercially successful biotechnological products. Because many tropical fruits have been used as substrates for the production of wines, Jagtap et al. (2011) tried using this fermentation technology in ripe fruits or their juices as a means of adding value to surplus or over-ripe jackfruits. Furthermore, the obtained fermented drink showed good characteristics, including important antioxidant and DNA damage protecting properties, providing health benefits for those who consume it, thus constituting a source of antioxidant rich neutraceuticals.

25.4 Chemical Composition

Tables 25.1 and 25.2 present the chemical composition of raw jackfruit, including macronutrients and micronutrients lime minerals and vitamins. According to Kaushal and Sharma (2016) the raw jackfruit is a good source of minerals like Potassium, Magnesium, Manganese, and Iron.

Table 25.1 Proximate composition of raw jackfruit (USDA 2018)

Property	g/100 g of edible portion
Water	73.46
Protein	1.72
Fat	0.64
Saturated fatty acids	0.20
Monounsaturated fatty acids	0.16
Polyunsaturated fatty acids	0.10
Trans fatty acids	0.00
Cholesterol	0.00
Carbohydrates	23.25
Fiber	1.50
Sugars	19.08
Energy (Kcal/100 g)	**95.00**

Table 25.2 Micronutrients in raw jackfruit (USDA 2018)

Minerals	**mg/100 g of edible portion**
Potassium [K]	448.00
Magnesium [Mg]	29.00
Calcium [Ca]	24.00
Phosphorus [P]	21.00
Sodium [Na]	2.00
Iron [Fe]	0.23
Zinc [Zn]	0.13
Vitamins	**μg/100 g of edible portion**
Vitamin C (ascorbic acid)	13,700
Vitamin B_3 (niacin)	920
Vitamin E (alpha-tocopherol)	340
Vitamin B_6 (pyridoxine)	329
Vitamin B_1 (thiamine)	105
Vitamin B_2 (riboflavin)	55
Vitamin B_9 (folate, DFE[a])	24
Vitamin A (retinol, RAE[b])	5

[a]*DFE* dietary folate equivalent
[b]*RAE* retinol activity equivalent

Jagadeesh et al. (2007) studied the chemical composition of the edible bulbs in 24 jackfruit types cultivated in India. Total soluble solids (TSS) was found to vary from 19.87 to 35.00 °Brix and acidity from 0.190% to 0.595%, while the ratio TSS/acidity varied in the range 54–134. This variation is very important having in mind that the samples studied were supposedly all in the edible ripe stage.

Regarding the sugar composition, the work by Jagadeesh et al. (2007) highlighted that jackfruit bulbs are rich in sugars and starch. The amount of total sugars varied between 19.1% and 32.1%, while the reducing sugars accounted for 8.6–14.6%. The quantity of starch was also highly variable, varying from 0.63% to 5.13%, which was explained as resulting from differences in the intrinsic capacity of the jackfruit tissues to accumulate starch and amylase activity, originating different rates of starch hydrolysis associated with ripening (Rahman et al. 1995; Jagadeesh et al. 2007). Similarly to what was observed previously for other chemical properties in the work of Jagadeesh et al. (2007), the carotenoids content was also highly variable among the clones evaluated, covering a range from 0.363 to 0.879 mg/100 g edible portion, with the highest values corresponding to types of jackfruit possessing bulbs with deeper yellow-orange colouration. According to Kaushal and Sharma (2016), the fresh fruit contains vitamin-A and flavonoid pigments (beta-carotene, xanthin, lutein and beta-cryptoxanthin). The ripened fruit is normally fibrous and contains sugars like glucose, fructose, xylose, rhamnose, arabinose and galactose (Kaushal and Sharma 2016). Xu et al. (2018) extracted pectins from jackfruit peel and the physicochemical analysis indicated that these pectins were rich in galactose (1.0 molar ratio), rhamnose (0.4–0.5 molar ratio), arabinose (0.5–0.8 molar ratio), glucose (3.3–3.7 molar ratio), and galacturonic acid (2.8–3.1 molar ratio), besides possessing antioxidant activity.

Begum et al. (2014) also studied pectins extracted from jackfruit, whose fractions of Calcium pectate varied in the ranges: 1.57–2.28%,1.00–1.92% and 1.14–1.60%, respectively for the rind, core and bulb. The evaluated pectins presented moisture content varying from 7.72% to 14.73% and ash varying from 3.71% to 8.15%. A study by Gupta et al. (2011) focused on the study of the composition of the jackfruit seeds, which are also edible, and for that reason constitute a means of ingestion of important nutrients. They calculated the proximate composition and found that seeds are rich in protein, 11.85%, besides other components: 61.8% moisture, 26.20% carbohydrates, 1.00% fat and 0.15% ash.

In the study by Fernandes et al. (2017), which evaluated the jackfruits seeds, it was found that they contain organic acids in the amounts 13.92 and 3.26 g/kg d.m. (dry matter) for the seed kernel and the seed coating, respectively. Specifically, the organic acids present in the kernel were: aconitic, citric, fumaric, malic, oxalic and quinic, while in the coating, besides those present in the core, were also found the acetic, pyruvic and shikimic acids. Furthermore, their study allowed the quantification of the amino acids (AA), separated as essential AA and nonessential AA. The essential AA found were: isoleucine, histidine, leucine, lysine, phenylalanine, threonine, tryptophan and valine, in amounts of 1884 and 268 mg/kg d.m., respectively for the kernel and coating. The nonessential AA present in the kernel and coating were 11.73 and 1.34 g/kg d.m., respectively, and accounted for alanine, arginine, asparagine, aspartic acid, cysteine, glutamic acid, glutamine, glycine, ornitine, proline, serine and tyrosine. Table 25.3 shows the fatty acid composition of jackfruit seed kernel and coating. The fatty acids found by Fernandes et al. (2017) included dodecanoic (C12:0), tridecanoic (C13:0), tetradecanoic (C14:0), cis-10-pentadecenoic (C15:1n-5c), pentadecanoic (C15:0), cis-9-hexadecenoic (C16:1n-7c), hexadecanoic (C16:0), cis-10-heptadecenoic (C17:1 n-7c), heptadecanoic (C17:0), cis-9,12-octadecadienoic (C18:2n-6c), cis-9-octadecenoic (C18:1n-9c), trans-9-octadecenoic (C18:1n-9 t), octadecanoic (C18:0), cis-9,12,15-octadecatrienoic (C18:3n-3c), eicosanoic (C20:0), heneicosanoic (C21:0), docosanoic (C22:0), tricosanoic (C23:0), tetracosanoic (C24:0).

Table 25.3 Fatty acid composition of jackfruit seeds (Fernandes et al. 2017)

Fatty acids	Seed kernel (mg/kg d.m.)	Seed coating (mg/kg d.m.)
Saturated fatty acids	1383	1545
Monounsaturated fatty acids	162	264
Polyunsaturated fatty acids	803	152
Omega-3	2	4
Omega-6	801	148
Omega-9	127	213
Total	**2348**	**1961**

Table 25.4 Chemical composition of jackfruit starches (Zhang et al. 2016)

Component	Content (g/100 g d.m.)
Moisture	10.31–11.68
Starch	99.06–99.17
Protein	0.41–0.44
Fat	0.12–0.18
Ash	0.29–0.34
Amylose	26.41–38.24

In the work by Fernandes et al. (2017) are also reported 18 phenolic compounds, including phenolic acids, in extracts obtained from jackfruit seeds with two solvents, water and methanol: 3-caffeoylquinic acid, feruloylglucaric acid isom., feruloylglucaric acid isom., feruloylglucaric acid isom., feruloylglucaric acid isom., 5-caffeoylquinic acid, caffeoylquinic acid isom., feruloylglucaric acid isom., feruloylsinapic acid isom., feruloylsinapic acid isom., 4-p-coumaroylquinic acid, 5-p-coumaroylquinic acid, 4-feruloylquinic acid, 5-feruloylquinic acid, p-coumaroyl derivative, feruloyl derivative, sinapoyl derivative and feruloyl derivative.

Jackfruit seed cotyledons are fairly rich in starch and protein (Kaushal & Sharma 2016). This starch from jackfruit has been object of diverse studies aimed at its characterization and potential applications. Zhang et al. (2016) undertook a study aimed at characterizing high purity starches obtained from seeds of several jackfruit varieties. The chemical and physical properties varied considerably, with particle size varying from medium (12.59 μm) to small (10.12–6.62 μm) and very small (4.28 μm). Chemical composition of the isolated starches is indicated in Table 25.4. Because the isolated seed starches presented protein, fat and ash contents lower than 0.50% these jackfruit seeds constitute a good source of high purity starches. Also Madruga et al. (2014) studied the chemical properties of jackfruit seeds starch, and found moisture contents varying from 2.75% to 2.86% (depending whether it was the soft or hard varieties), fat equal to 0.37% (for both varieties), protein in the range 1.53–0.62% and ash varying from 0.16% to 0.07% (for soft and hard varieties, respectively).

The work by Zhu et al. (2018) focused on the use of jackfruit starch as a coating on microcapsules containing flavourings. Also the work by Suryadevara et al. (2017), evaluated the applicability of jackfruit starch to facilitate disintegration of medicinal tablets. The study by Resendiz-Vazquez et al. (2017) focused on the protein isolate obtained from jackfruit seeds, evaluating the technological and functional properties as well as structure, concluding that these protein isolates could be successfully applied in the manufacture of food products. Nagala et al. (2013) evaluated the lipid profiles of different varieties of seed oils obtained from jackfruit by gas chromatography (GC), and identified some fatty acids, fatty acid methyl esters and fatty acid esters, namely tetradecane, methyl decanoate, methyl undecanoate, methyl dodecanoate and 2-octanone, thus concluding that the analysed oils were rich in unsaturated fatty acids.

25.5 Biological Effects and Health Properties

Jackfruit does not contain saturated fats or cholesterol, being thus considered as a healthy fruit. Furthermore, Jackfruit and jackfruit residues constitute a source of compounds that may be used to produce functional foods, with important benefits aimed at meeting the present demands on the food market (Feili et al. 2013).

Many compounds are present in jackfruit, including a variety of phenolic compounds like flavonoids, arylbenzofurans, and stilbenoids, which are known for having major bioactive properties (Inoue et al. 2018).

25.5.1 Traditional Usages

The richness in bioactive compounds such as antioxidants, carotenoids, phenolics and other phytonutrients potentiate the medicinal value of jackfruit and its seeds (Sarkar et al. 2015). Some of the health effects attributed to jackfruit and for which it is used in traditional medicine include anti-inflammatory, antibacterial, antioxidant and antidiabetic properties (Navarro-García et al. 2012). Besides, it has hypoglycemic activity being for that useful for the treatment of diabetes (Beidokhti & Jäger 2017).

According to Sivasankari et al. (2014) jackfruit medicinal usages include laxative, aphrodisiac and sedative effects as well as ability to treat skin ophthalmologic diseases (Sivasankari et al. 2014). In Mauritius island jackfruit is used for gastrointestinal disorders, particularly for stomach ache and constipation (Sreekeesoon & Mahomoodally 2014; Mahomoodally & Ramalingum 2015). In India, the seeds from jackfruit have been used in traditional medicine as ingredient for the preparation of an antidote for excessive drinkers (Kaushal & Sharma 2016). Jackfruit, as other members of the genus *Artocarpus*, has for long been employed in South-East Asia as traditional folk medicine for the cure of inflammation, malarial fever as well as for the treatment of ulcers, abscess and diarrhoea (Jagtap et al. 2011). Table 25.5 shows the usage of different parts of the *Artocarpus* fruit and tree.

Table 25.5 Traditional usages of some parts of *jackfruit* plant (Jagtap et al. 2011)

Component	Usage
Edible parts (pulp and seeds)	Cooling tonic Pectorial
Roots	Diarrhea Fever
Leaves/stems	Activate milk production (in women and animals) Antisyphilic Vermifuge Ulcers wounds Anemia Asthma Dermatitis Diarrhea Cough Anti-expectorant
Latex	Abscesses Snakebites Glandular swellings

25.5.2 Antioxidant Properties

Fresh jackfruit contains carotenoids, which play important roles as antioxidants and for improvement of vision functions (Kaushal & Sharma 2016). The kernel of jackfruit is rich in beta-carotene, alpha-carotene, beta-zeacarotene, alpha-zeacarotene, beta-carotene-5,6alpha-epoxide, a dicarboxylic carotenoid and crocetin (Baliga et al. 2011). In the work by Faria et al. (2009) some key carotenoids from jackfruit were quantified: all-trans-lutein (24–44%), all-trans-beta-carotene (24–30%), all-trans-neoxanthin (4–19%), 9-cis-neoxanthin (4–9%) and 9-cis-vio-laxanthin (4–10%).

Jackfruit pulp also contains phenolic compounds, which are well established for acting like antioxidants through various mechanisms (Gupta et al. 2011). In the study by Ruiz-Montañez et al. (2015) the antioxidant activity of jackfruit pulp extract was evaluated by two methodologies, using the ABTS[1] and DPPH[2] radicals. The ranges of antioxidant activity were 5.01–42.18 μg TE/mL and 6.27–33.42 μmol TE/mL, respectively for DPPH and ABTS determinations. Gupta et al. (2011) evaluated Jackfruit seeds and found that the total phenolics content varied from 1.45 to 2.12 μg GAE[3]/mg and the total flavonoid content varied from 290.6 to 457.1 μg RE[4]/mg, depending on the type of extract (acetone or dichloromethane:methanol). The DPPH free-radical scavenging ability in the extracts was IC_{50} = 0.64–0.79 mg/mL, while the ABTS scavenging assay gave IC_{50} = 0.05–0.06 mg/mL, varying

[1] ABTS: 2,2′-azino-bis(3-ethylbenzothiazoline-6-sulphonic acid)

[2] DPPH: 2,2-diphenyl-1-picrylhydrazyl

[3] GAE: galic acid equivalent

[4] RE: rutin equivalent

according to the type of extract. The FRAP[5] reducing activity was 2.20–3.22 μg TE[6]/mg, RPA[7] was 14.03–16.67 μg GAE/mg and the Fe^{2+} chelating activity gave values IC_{50} = 0.06–0.08 mg/mL (Gupta et al. 2011). Nagala et al. (2013) evaluated the antioxidant activities of seed oils obtained from jackfruit of different varieties, and found that these oils exhibited extremely high antioxidant capacity: DPPH assay gave 51.6–98.4% of inhibition/50 μL, FRAP antioxidant power was 136–189 μmol AAE[8]/mL, RPA reducing activity was 40.3–137.0 μg AScAE[9]/mL and finally HRSA[10] scavenging activity was 63.0–82.2% of inhibition/0.1 μL (Nagala et al. 2013).

25.5.3 Antifungal and Antimicrobial Activities

Although infections caused by fungi are very frequent, they are very dangerous to human health and can even lead to death. One such example is the paracoccidioidomycosis (PCM), which is caused by fungi of the gender *Paracoccidioides*, and which can be fatal for immunocompetent people. Because the conventional treatment has undesirable side effects, alternative or complementary therapies to combat this pathogen are investigated. Lectin, a compound obtained from jackfruit seeds, has shown to exhibit immunomodulatory activity against many pathogens, including *Paracoccidioides* species. Hence, Ruas et al.(2018) evaluated the influence of lectin on patients with PCM and found that it stimulated the secretion of pro-inflammatory cytokines, which have a protective effect against infection by *Paracoccidioides*. In conclusion, they verified the immunostimulatory action of lectin and its potential use for the immunotherapy of PCM.

According to Septama et al. (2017) artocarpanone isolated from *Artocarpus* species exhibits antibacterial activity. Hence, the authors investigated synergistic activity of artocarpanone when coupled with tetracycline, ampicillin, and norfloxacin against methicillin-resistant *Staphylococcus aureus* (MRSA), *Pseudomonas aeruginosa*, and *Escherichia coli*. Their results evidenced that, although the antibacterial activity of artocarpone was weak against MRSA and *P. aeruginosa*, it was strong against *E. coli*.

Furthermore, they found that this compound presented a synergistic effect on the antibacterial activity of norfloxacin. Therefore, artocarpanone could successfully be used to improve the antibacterial activity of norfloxacin against MRSA (Septama et al. 2017).

---

[5] FRAP: ferric reducing antioxidant power

[6] TE: trolox equivalent

[7] RPA: reducing power assay

[8] AAE: ascorbic acid equivalent

[9] AscAE: ascorbic acid equivalent

[10] HRSA: hydroxyl radical scavenging activity

25.5.4 Anti-Inflammatory Activity

Septama et al. (2018) investigated the effects of some flavonoids extracted from jackfruit on the chemotaxis, phagocytosis, reactive oxygen species (ROS) production and myeloperoxidase (MPO) activity of human phagocytes. They found that two compounds (artocarpanone and artocarpin) were potent inhibitors of all steps of phagocytosis due to their structure, and in particular to the presence of hydroxyl and prenyl groups. Hence, because these flavonoids could powerfully suppress phagocytosis, they showed potential to produce potent anti-inflammatory medicines.

Artocarpin is a prenylated flavonoid that was isolated from *Artocarpus* species and studied for its anti-inflammatory power by Qu and Liu (2018) using recombinant human UGTs (UDP-glucuronosyltransferasesupersomes) and human liver microsomes, and they confirmed its inhibitory effect. The study by Meera et al. (2017) aimed at evaluating the anti-inflammatory activity of ethanolic extracts obtained from different parts of jackfruit. Their results evidenced that the anti-inflammatory activity of all extracts was similar to that of a commercial pharmacological drug (diclofenac) tested both *in vitro* and *in vivo*.

Because infection by *Toxoplasma gondii* is very frequent around the world (affecting more than 30% of world population) it is essential to investigate more effective and non-toxic substances for the treatment of patients with toxoplasmosis. The work by Souza et al. (2016) focused on the evaluation of immunostimulatory effect of lectins obtained from jackfruit for the treatment of *T. gondii* infection during acute phase, using an animal model. They observed a high capacity of these compounds to induce pro-inflammatory and anti-inflammatory cytokines production. Furthermore, the treated mice presented a significantly lower parasite burden and a higher survival rate when compared to control mice, thus allowing concluding about the immunotherapeutic potential of these jackfruit lectins (Souza et al. 2016).

25.5.5 Diabetes Treatment

Diabetes mellitus is undoubtedly one of the principal endocrine syndromes, characterized by reduced insulin response. Ajiboye et al. (2018) evaluated the effect of an ethanol extract obtained from jackfruit in alloxan-induced diabetic rats, and found that after treatment, the diabetic rats showed significant improvement in the health status as compared with non-diabetic control rats, namely in terms of weight reduction, better haematological parameters, more adequate serum lipids concentrations (except HDL[11]), as well as better creatinine, bilirubin, urea and albumin levels. In

[11] HDL: high density lipotrotein

view of their findings, the authors concluded that jackfruit may be useful in improving the general health status of patients with diabetes mellitus (Ajiboye et al. 2018).

The work by Indrowati et al. (2017) evaluated the potential of *Artocarpus* active compounds on serum glucose and insulin response of beta-cells in diabetic rats. Their results indicated that the extracts studied can be used as an antidiabetic, and particularly for type 1 diabetes.

25.5.6 Skin Regeneration

After skin damage, wound healing is essential to restore regular tissue functioning. Artocarpin, which is present in jackfruit, was evaluated by Yeh et al. (2017) for its wound healing possibilities, through *in vitro* and *in vivo* (a mice model) essays. Their results showed that artocarpin improved skin wound healing, and the authors suggested the mechanics through which skin regeneration occurred. In conclusion, their work indicated that this compound can present useful capacities to be used as a therapeutic agent for the treatment of skin wounds (Yeh et al. 2017).

Prostaglandins have a short life *in vivo* due to rapid metabolization caused by NAD^+-dependent 15-hydroxyprostaglandin dehydrogenase (15-PGDH). Hence, some potent inhibitors of 15-PGDH were studied *in vitro* by Karna (2017) for their wound healing effect on a keratinocyte cell line scratch model. The ethanol extract obtained from jackfruit showed the highest inhibitory capacity with lowest cytotoxicity, as well as high intracellular and extracellular prostaglandins levels, as compared with other extracts, thus confirming its applicability to treat dermal wounds (Karna 2017).

Artocarpin possesses melanogenesis inhibitory activity and was tested for its ability to decrease skin pigmentation, thus allowing a significant improvement in people suffering from hyperpigmentation, without any observed adverse effects. Hence, the results indicate that an artocarpin based formulation could be used for depigmenting action in cosmetic preparations (Kwankaew et al. 2017).

25.5.7 Cancer Prevention and Management

Glioblastoma multiforme (GBM) is a malignant tumour affecting the central nervous system with devastating and many times fatal aggressiveness. Artocarpin is a prenylated flavonoid found in jackfruit, which demonstrated several anti-inflammatory and anti-tumour properties. The study by Lee et al. (2018) with two different cell lines demonstrated that treatment with artocarpin induced apoptosis, causing the death of the tumour cells. Besides the *in vitro* tests, they also conducted *in vivo* tests with mice, and concluded that artocarpin can be a potential chemotherapeutic drug for the treatment of GBM cancer (Lee et al. 2018).

The work by Ruiz-Montañez et al. (2015) evaluated the antimutagenic and antiproliferative properties of Jackfruit pulp against two salmonella strains and found that Jackfruit contained phytochemicals with chemoprotective properties, thus preventing mutagenesis, as well as antiproliferative activity. However, they did not identify which compound or compounds were responsible for the observed biological effects, but confirmed that these effects were closely related to the antioxidant acidity of the extracts: R^2 = 0.724–0.747 for antiproliferative assay and R^2 = 0.706–0.758 for antimutagenic assay (Ruiz-Montañez et al. 2015).

Lectins are a class of glycoproteins (proteins that bind with carbohydrates), which can be found in jackfruit. The study by Zeng et al. (2018) evaluated the mitogenic activity of lectin as well as its apoptosis induction ability in *in vitro* essays and were able to establish the mechanisms through which this compound positively influences death of cancer cells, thus suggesting its therapeutic value for the treatment of leukemia (Zeng et al. 2018).

Because lung cancer is a very malignant type of cancer, for which conventional drugs are still not totally effective, new anticancer agents are still being investigated. Cycloartobiloxanthone is a flavonoid present in *Artocarpus* species, which has potential for anticancer therapy. Hence, the work by Losuwannarak et al. (2018) was dedicated to evaluate the cytotoxicity of cycloartobiloxanthone against four lung cancer cell lines. They observed that the studied compound exhibited potent cytotoxic effect, with a similar apoptosis-inducing capacity to that of other standard anticancer drugs. As a consequence, cycloartobiloxanthone proved to have anticancer activity against human lung cancer and for that reason might be promising for lung cancer therapy (Losuwannarak et al. 2018).

Liu et al. (2018) identified two new prenylated flavones which were tested, together with other flavones present in jackfruit, for their antiproliferative effects against two strains of human cancer cells in *in vitro* essays. One of the tested compounds, eleocharin A (10), showed particularly high cytotoxicity and inhibitory activity, being for that reason suggested as important for the development of novel anti-tumour agents.

The effect of a prenylated flavonoid extracted from *Artocarpus* species was studied by Etti et al. (2017) on a cellular model of breast cancer, evaluating the antiproliferative effect as well as the cell death and apoptosis induction mechanisms. Their results were very promising because they confirmed the potential of this compound to fight against the aggressive triple negative breast cancer.

In the work by Sun et al. (2017) artocarpin from jackfruit was identified as a promising agent with chemopreventive properties against colorectal cancer. Artocarpin exhibited selective cytotoxicity against human colon cancer cells, together with suppression of cancer cells growth and promotion of cell death through apoptosis and autophagia.

Gupta et al. (2011) found a high quantity of saponins (6.32 g/100 g) in jackfruit seeds, which have been known for their medicinal properties, such as antispasmodic activity and toxicity effect to cancer cells.

25.5.8 Others

Pectin obtained from jackfruit peels was used to produce nano scale bio composites with possible applications for bone healing. For that the materials were tested by cytocompatibility, alkaline phosphatase, fibroblast stem cells, anti-inflammatory and cell adhesion tests. The obtained materials showed good biocompatibility, thus suggesting that they could be used for bone grafting (Govindaraj et al. 2018).

Gupta et al. (2011) determined the alkaloid content in jackfruit seeds (1.16 g/100 g) and these molecules act in the human body as spasmolytic, anticholinergic and anesthetic agents.

Jacalin, which is the major protein isolated from the jackfruit seeds, has proved to possess immunological properties and constitutes a useful tool for the evaluation of immune status in patients infected with HIV[12] (Gupta et al. 2011; Kaushal & Sharma 2016).

25.6 Final Considerations

Jackfruit has long been identified as a healthful food, with many applications in traditional medicine. More recently, jackfruit and its diverse components have been investigated for their nutritional as well as health enhancing properties, thus confirming its usefulness as a nutritive food and providing bioactive components valuable for the treatment of many pathologies. Some of the areas in which jackfruit has proved valuable include cancer prevention and treatment, besides the anti-inflammatory, antidiabetic, antimicrobial or antioxidant activities.

Acknowledgments The authors would like to acknowledge CI&DETS and CERNAS Research centres from Polytechnic Institute of Viseu, Portugal.

References

Abreu RCR, Rodrigues PJFP (2005) Estrutura de populações de jaqueiras, subsídios para manejo e conservação da Mata Atlântica. In: Atas SBSEEI Categ 1 Trab Científicos Completo. Brasília; pp. 1–14

Adiani V, Gupta S, Padole R, Variyar PS, Sharma A (2014) SPME-GCMS integrated with chemometrics as a rapid non-destructive method for predicting microbial quality of minimally processed jackfruit (Artocarpus heterophyllus) bulbs. Postharvest Biol Technol 98:34–40

Ajiboye BO, Adeleke Ojo O, Adeyonu O, Imiere O, Emmanuel Oyinloye B, Ogunmodede O (2018) Ameliorative activity of ethanolic extract of Artocarpus heterophyllus stem bark on alloxan-induced diabetic rats. Adv Pharm Bull 8:141–147

[12] HIV: human immunodeficiency virus

Baliga MS, Baliga BRV, Kandathil SM, Bhat HP, Vayalil PK (2011) A review of the chemistry and pharmacology of the date fruits (Phoenix dactylifera L.). Food Res Int 44:1812–1822

Begum R, Aziz MG, Uddin MB, Yusof YA (2014) Characterization of jackfruit (Artocarpus Heterophyllus) waste pectin as influenced by various extraction conditions. Agric Agric Sci Procedia 2:244–251

Beidokhti MN, Jäger AK (2017) Review of antidiabetic fruits, vegetables, beverages, oils and spices commonly consumed in the diet. J Ethnopharmacol 201:26–41

de Abreu RCR, Rodrigues PJFP (2010) Exotic tree Artocarpus heterophyllus (Moraceae) invades the Brazilian Atlantic Rainforest. Rodriguésia 61:677–688

de Oliveira LF, de Godoy RLO, Borges SV (2011) Qualidade de jaca (Artocarpus heterophyllus, Lam.) desidratada sob diferentes condições de processo. Braz J Food Technol 14:241–248

Etti IC, Abdullah R, Kadir A, Hashim NM, Yeap SK, Imam MU, Ramli F, Malami I, Lam KL, Etti U et al (2017) The molecular mechanism of the anticancer effect of Artonin E in MDA-MB 231 triple negative breast cancer cells. PLoS One 12:e0182357

Falcão M d A, Clement CR, Gomes JBM, Flores WBC, Santiago FF, Freitas VP (2001) Fenologia e produtividade da fruta-pão (Artocarpus Altilis) e da Jaca (A. heterophyllus) na Amazônia Central. Acta Amaz 31:179–179

Faria AF, Rosso VV, Mercadante AZ (2009) Carotenoid composition of jackfruit (Artocarpus heterophyllus), determined by HPLC-PDA-MS/MS. Plant Foods Hum Nutr Dordr Neth 64:108–115

Feili R, Abdullah WNW, Yang TA (2013) Physical and sensory analysis of high fiber bread incorporated with jackfruit rind flour. Food Sci Technol 1:30–36

Fernandes F, Ferreres F, Gil-Izquierdo A, Oliveira AP, Valentão P, Andrade PB (2017) Accumulation of primary and secondary metabolites in edible jackfruit seed tissues and scavenging of reactive nitrogen species. Food Chem 233:85–95

Fonseca CMB (2016) Desidratação da jaca (Artocarpus heterophyllus Lam.) de São Tomé e Príncipe. Dissertação do Mestrado em Engenharia Agronómica – Agronomia Tropical. Instituto Superior de Agronomia, Lisboa

Foo KY, Hameed BH (2012) Potential of jackfruit peel as precursor for activated carbon prepared by microwave induced NaOH activation. Bioresour Technol 112:143–150

Govindaraj D, Rajan M, Hatamleh AA, Munusamy MA (2018) From waste to high-value product: jackfruit peel derived pectin/apatite bionanocomposites for bone healing applications. Int J Biol Macromol 106:293–301

Gupta D, Mann S, Sood A, Gupta RK (2011) Phytochemical, nutritional and antioxidant activity evaluation of seeds of jackfruit (Artocarpous heterolphyllus Lam.). Int J Pharma Bio Sci 2:336–345

Hu L, Wu G, Hao C, Yu H, Tan L (2016) Transcriptome and selected metabolite analyses reveal points of sugar metabolism in jackfruit (Artocarpus heterophyllus Lam.). Plant Sci 248:45–56

Indrowati M, Pratiwi R, null R, Astuti P (2017) Levels of blood glucose and insulin expression of beta-cells in streptozotocin-induced diabetic rats treated with ethanolic extract of Artocarpus altilis leaves and GABA. Pak J Biol Sci PJBS 20:28–35

Inoue M, Hitora Y, Kato H, Losung F, Mangindaan REP, Tsukamoto S (2018) New geranyl flavonoids from the leaves of Artocarpus communis. J Nat Med 72:632–640

Jagadeesh SL, Reddy BS, Basavaraj N, Swamy GSK, Gorbal K, Hegde L, Raghavan GSV, Kajjidoni ST (2007) Inter tree variability for fruit quality in jackfruit selections of Western Ghats of India. Sci Hortic 112:382–387

Jagtap UB, Waghmare SR, Lokhande VH, Suprasanna P, Bapat VA (2011) Preparation and evaluation of antioxidant capacity of jackfruit (Artocarpus heterophyllus Lam.) wine and its protective role against radiation induced DNA damage. Ind Crop Prod 34:1595–1601

Jayus DSG, Giyarto SD (2016) Physical and chemical characteristics of jackfruit (Artocarpus Heterophyllus Lamk.) seeds flour produced under fermentation process by Lactobacillus Plantarum. Agric Agric Sci Procedia 9:342–347

Karna S (2017) In-vitro wound healing effect of 15-hydroxyprostaglandin dehydrogenase inhibitor from plant. Pharmacogn Mag 13:S122–S126

Kaushal P, Sharma HK (2016) Osmo-convective dehydration kinetics of jackfruit (Artocarpus heterophyllus). J Saudi Soc Agric Sci 15:118–126

Kwankaew J, Phimnuan P, Wanauppathamkul S, Viyoch J (2017) Formulation of chitosan patch incorporating Artocarpus altilis heartwood extract for improving hyperpigmentation. J Cosmet Sci 68:257–269

Lee C-W, Hsu L-F, Lee M-H, Lee I-T, Liu J-F, Chiang Y-C, Tsai M-H (2018) Extracts of Artocarpus communis induce mitochondria-associated apoptosis via pro-oxidative activity in human glioblastoma cells. Front Pharmacol 9:411

Liu X, Kuang X-D, He X-R, Ren G, Wang Y, Xu L-Y, Feng L-H, Wang B, Zhou Z-W (2018) Prenylflavonoids from the Twigs of Artocarpus nigrifolius. Chem Pharm Bull (Tokyo) 66:434–438

Losuwannarak N, Sritularak B, Chanvorachote P (2018) Cycloartobiloxanthone induces human lung cancer cell apoptosis via mitochondria-dependent apoptotic pathway. Vivo Athens Greece 32:71–78

Madruga MS, de Albuquerque FSM, Silva IRA, do Amaral DS, Magnani M, Queiroga Neto V (2014) Chemical, morphological and functional properties of Brazilian jackfruit (Artocarpus heterophyllus L.) seeds starch. Food Chem 143:440–445

Mahomoodally MF, Ramalingum N (2015) An investigation into the consumption patterns, attitude, and perception of Mauritians towards common medicinal food plants. J Herb Med 5:99–112

Meera M, Ruckmani A, Saravanan R, Lakshmipathy Prabhu R (2017) Anti-inflammatory effect of ethanolic extract of spine, skin and rind of Jack fruit peel – a comparative study. Nat Prod Res:1–5

Nagala S, Yekula M, Tamanam RR (2013) Antioxidant and gas chromatographic analysis of five varieties of jackfruit (Artocarpus) seed oils. Drug Invent Today 5:315–320

Navarro-García V, David S, Rodríguez-Flores T, Abarca-Vargas R, Aldana-Llanos L (2012) Antimicrobial activity of artocarpesin from Artocarpus heterophyllus Lam. against methicillin-resistant Staphylococcus aureus (MRSA). J Med Plant Res 6:4879–4882

Pua CK, Sheikh Abd. Hamid N, Tan CP, Mirhosseini H, Abd. Rahman R, Rusul G (2008) Storage stability of jackfruit (Artocarpus heterophyllus) powder packaged in aluminium laminated polyethylene and metallized co-extruded biaxially oriented polypropylene during storage. J Food Eng 89:419–428

Pua CK, Hamid NSA, Tan CP, Mirhosseini H, Rahman RBA, Rusul G (2010) Optimization of drum drying processing parameters for production of jackfruit (Artocarpus heterophyllus) powder using response surface methodology. LWT – Food Sci Technol 43:343–349

Qu W, Liu X (2018) Identification of cytochrome P450 isoforms involved in the metabolism of artocarpin and assessment of its drug-drug interaction. Biomed Chromatogr BMC 32:1–8

Rahman AKMM, Huq E, Mian AJ, Chesson A (1995) Microscopic and chemical changes occurring during the ripening of two forms of jackfruit (Artocarpus heterophyllus L.). Food Chem 52:405–410

Resendiz-Vazquez JA, Ulloa JA, Urías-Silvas JE, Bautista-Rosales PU, Ramírez-Ramírez JC, Rosas-Ulloa P, González-Torres L (2017) Effect of high-intensity ultrasound on the techno-functional properties and structure of jackfruit (Artocarpus heterophyllus) seed protein isolate. Ultrason Sonochem 37:436–444

Ruas LP, Genaro LM, Justo-Junior AS, Coser LO, de Castro LF, Trabasso P, Mamoni RL, Roque-Barreira M-C, M-HSL B (2018) Effect of ArtinM on human blood cells during infection with Paracoccidioides brasiliensis. Front Microbiol 9:867

Ruiz-Montañez G, Burgos-Hernández A, Calderón-Santoyo M, López-Saiz CM, Velázquez-Contreras CA, Navarro-Ocaña A, Ragazzo-Sánchez JA (2015) Screening antimutagenic and antiproliferative properties of extracts isolated from Jackfruit pulp (Artocarpus heterophyllus Lam). Food Chem 175:409–416

Sarkar P, Lohith Kumar DH, Dhumal C, Panigrahi SS, Choudhary R (2015) Traditional and ayurvedic foods of Indian origin. J Ethn Foods 2:97–109

Saxena A, Maity T, Raju PS, Bawa AS (2015) Optimization of pretreatment and evaluation of quality of jackfruit (Artocarpus heterophyllus) bulb crisps developed using combination drying. Food Bioprod Process 95:106–117
Septama AW, Xiao J, Panichayupakaranant P (2017) A synergistic effect of artocarpanone from Artocarpus heterophyllus L. (Moraceae) on the antibacterial activity of selected antibiotics and cell membrane permeability. J Intercult Ethnopharmacol 6:186–191
Septama AW, Jantan I, Panichayupakaranant P (2018) Flavonoids of Artocarpus heterophyllus Lam. heartwood inhibit the innate immune responses of human phagocytes. J Pharm Pharmacol 70:1242–1252
Sivasankari B, Anandharaj M, Gunasekaran P (2014) An ethnobotanical study of indigenous knowledge on medicinal plants used by the village peoples of Thoppampatti, Dindigul district, Tamilnadu, India. J Ethnopharmacol 153:408–423
Souza LPF, Ramos ELP, Santana SS, Silva MV, Santiago FM, Mineo TWP, Mineo JR (2016) Lectins from Synadenium carinatum (ScLL) and Artocarpus heterophyllus (ArtinM) are able to induce beneficial immunomodulatory effects in a murine model for treatment of Toxoplasma gondii infection. Front Cell Infect Microbiol 6:164
Sreekeesoon DP, Mahomoodally MF (2014) Ethnopharmacological analysis of medicinal plants and animals used in the treatment and management of pain in Mauritius. J Ethnopharmacol 157:181–200
Sun G, Zheng Z, Lee M-H, Xu Y, Kang S, Dong Z, Wang M, Gu Z, Li H, Chen W (2017) Chemoprevention of colorectal cancer by artocarpin, a dietary phytochemical from Artocarpus heterophyllus. J Agric Food Chem 65:3474–3480
Suryadevara V, Lankapalli SR, Danda LH, Pendyala V, Katta V (2017) Studies on jackfruit seed starch as a novel natural superdisintegrant for the design and evaluation of irbesartan fast dissolving tablets. Integr Med Res 6:280–291
USDA (2018) National Nutrient Database for Standard Reference Legacy Release. [place unknown]: United States Department of Agriculture – Agricultural Research Service
Vargas-Torres A, Becerra-Loza AS, Sayago-Ayerdi SG, Palma-Rodríguez HM, de García-Magaña L, Montalvo-González E (2017) Combined effect of the application of 1-MCP and different edible coatings on the fruit quality of jackfruit bulbs (Artocarpus heterophyllus Lam) during cold storage. Sci Hortic 214:221–227
Xu S-Y, Liu J-P, Huang X, Du L-P, Shi F-L, Dong R, Huang X-T, Zheng K, Liu Y, Cheong K-L (2018) Ultrasonic-microwave assisted extraction, characterization and biological activity of pectin from jackfruit peel. LWT 90:577–582
Yeh C-J, Chen C-C, Leu Y-L, Lin M-W, Chiu M-M, Wang S-H (2017) The effects of artocarpin on wound healing: in vitro and in vivo studies. Sci Rep 7:15599
Zeng L, Li L, Zeng Q, Deng Y, Yin L, Liao L (2018) Mitogenic activity of Artocarpus lingnanensis lectin and its apoptosis induction in Jurkat T cells. J Nat Med 72:745–756
Zhang Y, Zhu K, He S, Tan L, Kong X (2016) Characterizations of high purity starches isolated from five different jackfruit cultivars. Food Hydrocoll 52:785–794
Zhu H, Zhang Y, Tian J, Chu Z (2018) Effect of a new shell material—jackfruit seed starch on novel flavor microcapsules containing vanilla oil. Ind Crop Prod 112:47–52

Chapter 26
Nopal Cactus (*Opuntia* ficus-*indica* (L.) Mill) as a Source of Bioactive Compounds

Haroon Elrasheid Tahir, Zou Xiaobo, Mahunu Gustav Komla, and Abdalbasit Adam Mariod

26.1 Introduction

Nowadays the world seems to be increasingly interested in health-promoting foods and has started to seek beyond the basic nutritional values of foodstuffs to disease control and prevention (Rahmatullah et al. 2009). Generally, the health benefits of the plant can be obtained from bioactive components found in the whole plant, part of the plant (e.g. seeds, fruit, flower, gum, etc.) or combinations of thereof, either in crude or processed formula (Rahmatullah et al. 2009). Bioactive compounds are compound with health benefits and are generally classified into polyphenol and non-polyphenol compounds and pigments (Martins et al. 2011). Polyphenol compounds include flavonoids, phenolic acids, and tannins with being the main compounds, accounting for over half of the eight thousand naturally occurring phenolic compounds (Baxter et al. 1998). Recently, the utilization of these bioactive compounds is of great interest in food, chemical, and pharmaceutical industries (Martins et al. 2011).

Cactus (*Opuntia* spp.) plants are an attractive source of natural bioactive compounds/ products (Sánchez et al. 2014; Barba et al. 2017). Their bioactive profile depends on species, cultivars, environment conditions (Paiva et al. 2016), thus a prospect for the detection of novel components from various opuntia cultivars. *Opuntia ficus indica* (OFI), also known as prickly pear is domesticated plant belongs

H. E. Tahir · Z. Xiaobo (✉)
School of Food and Biological Engineering, Jiangsu University, Zhenjiang, Jiangsu, China
e-mail: Zou_xiaobo@ujs.edu.cn

M. G. Komla
Department of Biotechnology, Faculty of Agriculture, University for Development Studies, Tamale, Ghana

A. A. Mariod
Indigenous Knowledge and Heritage Center, Ghibaish College of Science & Technology, Ghibaish, Sudan

A. A. Mariod (ed.), *Wild Fruits: Composition, Nutritional Value and Products*,
https://doi.org/10.1007/978-3-030-31885-7_26

to the family of Cactaceae. The cultivars belong to OFI have the highest economic value in the world (Griffith 2004). The consumption of OFI cultivars as a vegetable in the local market is approximately 6.4 kg per capita and there is an increasing demand for it in the US and Canada, where the product sales ranged between 31 and 2.86 million dollars, respectively (Isela et al. 2017). Apart of bioactive compounds, OFI has being used for various foodstuffs such as low-calorie marmalade, juices, beverages, jams and sweeteners (Saenz et al. 1996; Leopoldo et al. 2012; Barba et al. 2017; Aruwa et al. 2018). Also different parts of the plant used in cosmetics (Galati et al. 2003; Osuna-Martínez et al. 2014), food preservations of meat (Palmeri et al. 2018) and of high-fat cookies (Msaddak et al. 2015).

The current global trend in an increase of antimicrobial resistance cannot be over- emphasized. Therefore, several studies were conducted in order to discover a novel source of antimicrobial compounds with biological activity in the plant such as Opuntia spp. (Sánchez et al. 2014; Koubaa et al. 2017). With increases in chronic diseases and need of nutraceutical compounds, there is a search for new sources of food. In this respect, OFI is promising candidates.

Reviews on the effect of OFI on chronic diseases had shown that the various parts of OFI are effective for treatment of type 2 Diabetes Mellitus, colon cancer cells, hypertension, hypercholesterolemic, rheumatic pain, gastric mucosa diseases and asthma (González-Stuart 2013; Serra et al. 2013; Antunes-Ricardo et al. 2014; Osuna-Martínez et al. 2014; del Socorro Santos Díaz et al. 2017; Aragona et al. 2018) (Fig. 26.1).

Fig. 26.1 Various part of *Opuntia ficus indica* plant. Adapted from https://www.fotosearch.cn/CSP309/k24864250/ and https://www.picclickimg.com

26.2 Distribution

OFI is the most wide-spread and widely introduced as a commercial fruit and recently as part of agroforestry projects in developing countries. OFI is slow growing trunk-forming segmented cactus which can attain highest up 5–6 m. OFI cultivated in arid countries such as Saudi Arabia, Egypt, Pakistan, Middle East, Algeria, Iran, Libya, Mexico, South Africa, Spain (Aruwa et al. 2018). Semi-arid/steppe areas such as South Asia, India, New Zealand, Australia, North America, and North Africa (Feugang et al. 2006). The temperate zones where OFI may be cultivated including Tunisia (Yeddes et al. 2014a) Korea (Lee et al. 2002); whereas tropical include Brazil (de Souza et al. 2015) and Mediterranean country of Morocco (Bouzoubaâ et al. 2016) and in China which is tropical in south to subarctic in north (Zhong et al. 2010). The global distribution of OFI is presented in Fig. 26.2.

26.3 Bioactive Compounds

Nowadays, there has been great interest in the possible use of OFI as a source of natural antioxidants and it is potential in the prevention of chronic diseases have emerged worldwide. The various OFI parts contains considerable amount of bioactive compounds (phenols, flavonoids, anthocyanins betaxanthin and betacyanin), vitamin C, carotenoids, vitamin E and other antioxidant such lignins, sterols, esters, saponins, alkaloids which, have evidenced their protective effects against various disease (El-Mostafa et al. 2014; Costa et al. 2017; Xu et al. 2017; Jelena et al. 2018; Mendonça et al. 2019). In this respect several studies have proved health benefits of OFI such as antioxidant and antimicrobial activities (Sánchez et al. 2014;

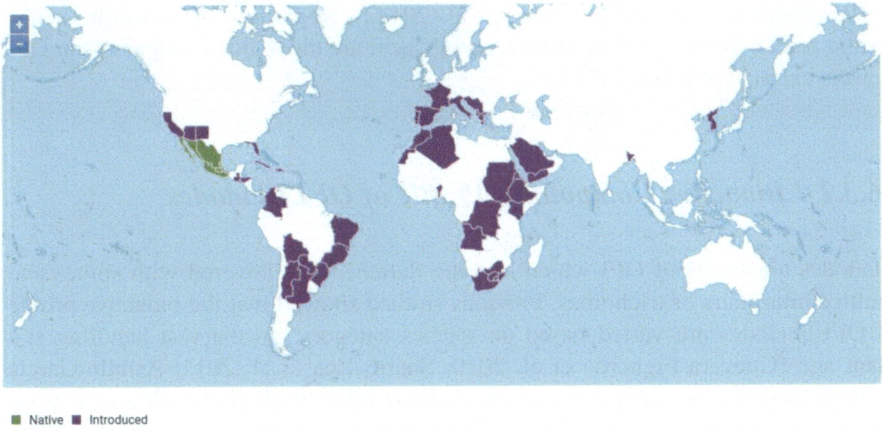

Fig. 26.2 Distribution of *Opuntia ficus-indica*. (Adapted from http://www.plantsoftheworldonline.org/taxon/urn:lsid:ipni.org:names:1151735-2#distribution-map)

Smida et al. 2017), antiviral (Rasoulpour et al. 2018), reduces oxidative stress (Petruk et al. 2017), anti-inflammatory activity (Filannino et al. 2016; Ammar et al. 2018), protective effect against chlorpyrifos CPF – induced immune damages (Smida et al. 2017), anti-proliferative and pro-apoptotic activity (Naselli et al. 2014), antiulcerogenic activity (Galati et al. 2003). Besides polyphenols, cactus seeds and peels can be used as source sterols and vitamin E (Ramadan and Mörsel 2003; El Mannoubi et al. 2009). Cladodes of OFI contain various bioactive compound including phenolic acids, flavonoids, lignins, anthocyanins, and vitamins and it is beneficial on human health, however up to date little is recognized as a potent source of bioactive compounds (Petruk et al. 2017). The major bioactive compounds in FOI flowers are flavonoids, particularly kaempferol and quercetin and their derivatives (Yeddes et al. 2014a; Mena et al. 2018). Additionally, the various parts of OFI fruit are rich in betalains pigments/compounds which consist of red-violet betacyanins and yellow-orange indicaxanthin compounds that are well-known for their benefits to human health (Coria Cayupán et al. 2011; Yahia and Mondragon-Jacobo 2011). The major phenolics/non-phenolic compounds detected in OFI are summarized in Tables 26.1, 26.2 and 26.3.

26.3.1 Bioactive Compounds Profile of OFI Flowers

The OFI flowers are a good source of antioxidants that might be utilized in food, pharmaceutical, and cosmetics, therefore helping to reduce the environmental effects of OFI by-products and it is valorization vaporization (Benayad et al. 2014). The polyphenol and biological profile are limited in the literature. Phenolic and flavonoids for the major bioactive compounds of OFI flowers (Table 26.1). Previous studies showed isorhamnetin glycosides being the predominant flavonoid compounds followed by quercetin and kaempferol glycosides (Fig. 26.3) (De Leo et al. 2010; Benayad et al. 2014; Yeddes et al. 2014a). Regarding the phenolic acids, ferulic acid, p-coumaric acid and rosmarinic acid are the major components in OFI flowers extract (Ines et al. 2017).

26.3.2 Bioactive Compounds Profile of OFI Cladodes

Cladodes are stems of OFI which become flattened and covered with spines and multicellular hairs or trichomes. Previous studied showed that the bioactive profile of OFI cladodes are varied based on species category, postharvest handling and plant age (Guevara-Figueroa et al. 2010; Santos-Zea et al. 2011; Astello-García et al. 2015; Figueroa-Pérez et al. 2018; da Cruz Filho et al. 2019). However, some studies have found that the phenolic profile of cladodes collected from various cultivars and geographical areas are varied which could beneficial for identification of OFI origin/products (Moussa-Ayoub et al. 2014). This is consistent with the results

Table 26.1 Polyphenol and non-polyphenol compounds reported in various parts of *Opuntia ficus-indica* (L.) Mill plant

Identified compounds	Cactus parts	References
Isorhamnetin; isorhamnetin glucosyl-dirhamnoside; isorhamnetin pentosyl-rutinoside; isorhamnetin pentosyl-glucoside; isorhamnetin pentosyl-rhamnoside; isorhamnetin rutinoside; 2-isorhamnetin-glucosyl-rhamnosyl-rhamnoside; isorhamnetin-glucosyl-rhamnosyl-pentoside; isorhamnetin-glucosyl-rhamnoside; isorhamnetin-3-O-rutinoside; isorhamnetin +1 hexose+1 pentose; isorhamnetin-3-glucoside; isorhamnetin glucoside; isorhamnetin-rutinoside; isorhamnetin 3-*O*-galactoside; isorhamnetin 3-*O*-robinobioside; isorhamnetin- rhamnose-rutinoside; isorhamnetin pentoside; kaempferol; kaempferol-glucosyl-rhamnoside; kaempferol-3-rutinoside; kaempferol 3-methyl ether; kaempferol-di-rhamnose-hexoside; Kaempferol-rutinoside; kaempferol 3-*O*-rutinoside; aromadendrin, (+)-Taxifolin; kaempferol 3-*O*-arabinoside; quercetin; quercetin 3-*O*-rutinoside; quercetin glucoside; quercetin-rhamnose-hexoside-rhamnose; quercetin-hexoside; quercetin-hexoside-pentoside; quercetin dicoumaryl glycoside; 3-O-methyl-quecetin; Rutin; rutin-pentoside; luteolin; naringin; naringenin-hexoside; benzyl-O-β-D-glucopyranoside; Picein; androsin; 1-O-feruloyl-β-D-glucopyranoside; myricetin-hexoside; syringaresinol; feruloyl piscidic acid; ferulic acid-hexoside; ferulic acid; eucomic acid; sinapic acid-hexoside; protocatechuic acid-hexoside; caffeic acid; ester of coumaric acid; dihydrosinapic acid hexoside; secoisolariciresinol-hexoside; Guaiacyl (8-O-4) syrinigyl (8–8) guaiacyl-hexoside; syrinigyl(t8-O-4) guaiacyl	Fruit/whole	Galati et al. (2003), Kuti (2004), Tesoriere et al. (2005), Castellanos-Santiago and Yahia (2008), Fernández-López et al. (2010), Santos-Zea et al. (2011), Dhaouadi et al. (2013), Jiménez-Aguilar et al. (2015b), Kim et al. (2016), Mata et al. (2016), Antunes-Ricardo et al. (2017), Suh et al. (2017), Mena et al. (2018)

(continued)

Table 26.1 (continued)

Identified compounds	Cactus parts	References
Isorhamnetin 3-O-rhamnoside-7-O-(rhamnosyl-hexoside; isorhamnetin 3-O-rutinoside; isorhamnetin 3-O-galactoside; isorhamnetin 3-O-glucoside; isorhamnetin-3-*O*-rutinoside; isorhamnetin diglycosides; isorhamnetin triglycosides; isorhamnetin glucosyl-rhamnosyl-rhamnoside; isorhamnetin; isorhamnetin-glucosyl-rhamnosyl-pentoside; isorahmnetin-3-*O*-glucosyl-pentoside; isorhamnetin-glucoside; isorhamnetin-3-O-glucosyl-rhamnoside; 3-O-glucosyl-rhamnosyl-rhamnoside; isorhamnetin 3-O-neohesperidoside; isorhamnetin+Xyl/Ara (xylose/arabinose); isorhamnetin+Xyl/Ara (xylose/arabinose); kaempferol 3-O-(rhamnosyl, galactoside)-7-O-rhamnoside; kaempferol 3-O-(rhamnosylglucoside)-7-O-rhamnoside; kaempferol 3-O-robinobioside-7-O-arabinofuranoside; kaempferol 3-O-rutinoside; kaempferol 3-O-glucoside; kaempferol 3-O-rutinoside; kaempferol 3-O-glucoside; kaempferol 7-O neohesperidoside; kaempferol-glucoside; kaempferol 3-Oarabinofuranoside; kaempferol; Quercetin 3-O-rhamnosyl- (1!2)-[rhamnosyl-(1!6)]-glucoside; quercetin 3-Oarabinofuranoside; quercetin 3-O-rutinoside; quercetin 3-O-dirhamnoside; quercetin-3-O-galactoside; quercetin 3-O-xylosylrhamnosyl-glucoside; quercetin 3-O-glucoside; quercetin; quercetin-glucoside; quercetin 3-O-rhamnosyl-(1!2)-[rhamnosyl-(1!6)]-glucoside; quercetin 3-O-xylosylrhamnosyl-glucoside; eucomic acid; benzoic acid; p-hydroxyphenylacetic acid; vanillic acid; caffeic acid; gallic acid; protocatechuic acid; tr- caffeic acid; p-coumaric acid; protocatechuic acid; caftaric acid; p-hydroxybenzoic acid; coutaric acid; syringic acid; sinapic acid; ferulic acid; cinammic acid; rosmarinic acid; ellagic acid; coumaric acid; piscidic acid; 4-hydroxy benzoic acid; salicylic acid; hydroxybenzoics; hydroxycinnamics acid; chlorogenic acid;3,3′,4′,5,7-Pentahydroxy-flavanone,2-hydroxy-4-(4′-hydroxyphenyl) butanoic acid; rutin; hesperidin; hyperoside; naringenin; apigenin; Rhamnetin; chrysin; catechin; epicatechin; epigallocatechin gallate; gallocatechin gallate; luteolin-glucoside; coumestrol apigenin-glucoside; isoflavonoids; procyanidin B1; procyanidin B2; cyanidin; cyanidin 3,5-O-diglucoside; cyanidin3-O-glucosyl-rutinoside; cyanidin 3-O-sambubiosyl 5-O-glucoside; cyanidin 3-O-xylosyl-rutinoside; delphinidin 3-O-galactoside; malvidin 3-O-(6″-caffeoyl-glucoside); pelargonidin; pelargonidin; peonidin; 3,5-O-diglucoside; pelargonidin 3-O-(6″-malonyl-glucoside); petunidin3-O-galactoside; cyanidin-glucoside; pelargonidin-glucoside; petunidin-glucoside; malvidin-glucoside; delphinidin-glucoside	Cladodes	Herrera-Hernández et al. (2011), Medina-Torres et al. (2011), Moussa-Ayoub et al. (2014), Ruiz-Gutiérrez et al. (2014), Antunes-Ricardo et al. (2015), Astello-García et al. (2015), Moussa-Ayoub et al. (2015), Msaddak et al. (2015), Allai et al. (2016), Figueroa-Pérez et al. (2018), Antunes-Ricardo et al. (2017), Bakari et al. (2017), Di Lorenzo et al. (2017), Msaddak et al. (2017), Ressaissi et al. (2017), Smida et al. (2017), De Santiago et al. (2018), Rocchetti et al. (2018)
Others		
Furofurans; dibenzylbutyrolactone; alkylphenols, hydroxybenzaldehydes; hydroxycoumarins, tyrosols; tannins; zataroside-A; vanillin		

Isorhamnetin; isorhamnetin-3-O-glucoside; isorhamnetin-3-O-galactoside; isorhamnetin-rutinoside; isoamericanol A; isorhamnetin pentoside; isorhamnetin- rhamnose-rutinoside; kaempferol; Kaempferol-di-rhamnose-hexoside; Kaempferol-rutinoside; Kaempferol-3-O-arabinoside; dihydrokaempferol; Quercetin; Quercetin-rhamnose-hexoside-rhamnose; Quercetin-hexoside-pentoside; Quercetin-3-O-rutinoside; Quercetin-3-O-glucoside; Quercetin-3-O-galactoside; Quercetin-hexoside; dihydroquercetin; myricetin-hexoside; catechin; amenthoflavone; rutin-pentoside; syrinigyl(t8-O-4) guaiacyl; dihydro sinapic acid hexoside; secoisolariciresinol-hexoside; syringaresinol; naringenin-hexoside; naringin; guaiacyl (8-O-4) syrinigyl (8–8) guaiacyl-hexoside; eucomic acid; rutin; narcissin; procyanidins; caffeic acid; cinnamic acid; ferulic acid; gallic acid; p-coumaric acid; rosmarinic acid; vanillic acid; syringic acid; ferulic, acid-hexoside; sinapic acid-hexoside	Flower	De Leo et al. (2010), Abdel-Hameed et al. (2014), Benayad et al. (2014), Yeddes et al. (2014a), Kang et al. (2016), Ben Saad et al. (2017), Ines et al. (2017), Mena et al. (2018)
Isorhamnetin; isorhamnetin pentosyl-rhamnoside; isorhamnetin glucosyl-di-rhamnoside; Isorhamnetin pentosyl-rutinoside; isorhamnetin pentosyl-glucoside; isorhamnetin pentosyl-rhamnoside; quercetin; quercetin-3-glucose-(1 → 6)-gallic acid; quercetin 3-O-β-d-glucoside; quercetin glucoside; quercetin 3-O-β-d-glucoside; kaempferol 3-O-β-d-glucoside; hydroxybenzoic acid; eucomic acid; ferulic acid glucoside; caffeic acid; rutin; 2-feruloyl piscidic acid; gallic acid; catechin; piscidic acid derivative II	Pulp	Abdel-Hameed et al. (2014), Mata et al. (2016)
Isorhamnetin-glucose- rhamnose- rhamnose; isorhamnetin- glucose- rhamnose- pentose; isorhamnetin-glucose- pentose; Isorhamnetin; isorhamnetin- glucose- rhamnose; Kaempferol; Kaempferol 3-O-β-d-glucoside; quercetin; quercetin 3-O-β-d-glucoside; quercetin-3-glucose-(1 → 6)-gallic acid; catechin; rutin; feruloyl glucose; eucomic acid; gallic acid; piscidic acid	Peel	Moussa-Ayoub et al. (2011), Abdel-Hameed et al. (2014), Chougui et al. (2015), Jiménez-Aguilar et al. (2015b), Koubaa et al. (2017)
Isorhapontin; quercetin; quercetin-3-galactoside; quercetin-3-glucoside; quercetin-mannoside; dihydroquercetin; syringic acid; cinnamic acid; ferulic acid; m-coumaric acid; vanillic acid; gallic acid; caffeic acid; chlorogenic acid, cinnamic acid; dihydroxy benzoic acid; dihydroxyphenolic acid; p-coumaric acid; syringic acid;3-phenylpropionic acid; sinapic acid; p-coumaroyl malic acid; hydroxycaffeic acid; psoralen; quercitrin; syringaldehyde,3-O-Methylgallic acid; sinapaldehyde; piceatannol; isopimpinellin; apigenin; butein; phloretin; catechin, enterodiol, 3′-O-methylcatechin, (+) -Gallocatechin; bisdemethoxycurcumin; astringin; catechol; juglone; pinoresinol; epigallocatechin 3-O-glucuronide;4′-O-methyl-(−) -epicatechin 3′-O-glucuronide; (−) - feruloyl-sucrose isomer 1; feruloyl-sucrose isomer 2; feruloyl-sucrose isomer 3; sinapoyl-diglucoside; amenthoflavone; epicatechin; rutin, sinapine; rutin; amenthoflavone; epicatechin; epigallocatechin; vanillin; 3,4′,5-Trihydroxysylbene-3-b-d-glucopyranoside; naringin; cyanidin; viscutin 1; cyanidin 3-O-(6″-malonyl-glucoside); procyanidin B2; cyanidin 3-O-rutinoside; cyanidin 3-O-(3″,6″-O-dimalonyl-glucoside); pelargonidin 3; cyanidin 3-O-diglucoside-5-O-glucoside	Seed	Tounsi et al. (2011), Chougui et al. (2013), Koubaa et al. (2017)

Table 26.2 Concentrations of sterols and saponins in the different parts of the *Opuntia ficus-indica* (L.) plants

Major compounds	Cactus parts	Concentrations	References
Sterols	Peel		
Ergosterol		0.68 g /kg	Ramadan and Moersel (2003), Ramadan and Mörsel (2003)
Campesterol		8.76 g /kg	
Stigmasterol		2.12 g /kg	
Lanosterol		1.66 g /kg	
β-Sitosterol		21.1 g /kg	
Δ5-Avenasterol		2.71 g /kg	
Ergosterol		0.68 g /kg	
Sterols	Seed		El Mannoubi et al. (2009)
Cholesterol		0.01 g /kg	
Campesterol		2.17 g /kg	
Stigmasterol		0.76 g /kg	
β-Sitosterol		11.50 g /kg	
Sitostanol		0.57 g /kg	
Δ5-Avenasterol		0.82 g /kg	
Δ7-Avenasterol		0.23 g /kg	
Sterols			
β-Sitosterol	Cladodes	Nq	Figueroa-Pérez et al. (2018)
β-Campesterol		Nq	
Campestanol		Nq	
Δ5-Avenastero		Nq	
Δ7-Stigmasterol		Nq	
Stigmastanol		Nq	
Saponins		Nq	Figueroa-Pérez et al. (2018), Msaddak et al. (2017)
Campesteryl 3β-D-glucopyranoside		Nq	
Stigmasteryl 3β-D-glucopyranoside		Nq	
Sitosteryl 3β-D glucopyranoside		Nq	

Nq not quantified

Table 26.3 Vitamins and betalains in the various parts of *Opuntia ficus-indica* (L.)

Compounds	Cactus parts	Concentrations per 100 g of fresh weight	References
Vitamin C	Fruit	26–72 mg	Fernández-López et al. (2010), Coria Cayupán et al. (2011), Albano et al. (2015), Cano et al. (2017)
β-carotene		2.6–478 μg	Tesoriere et al. (2005), Fernández-López et al. (2010), Cano et al. (2017)
Total vitamin E		0.11-.12 mg	Tesoriere et al. (2005)
Betacyanins		0.18–39.3 mg	Fernández-López et al. (2010), Coria Cayupán et al. (2011), Albano et al. (2015), Bouzoubaâ et al. (2016), Cano et al. (2017)
Betaxanthins		1.70–68 mg	Fernández-López et al. (2010), Coria Cayupán et al. (2011), Ruiz-Gutiérrez et al. (2014), Cano et al. (2017)
Indicaxanthin		Juice 5.65–19 mg	Fernández-López and Almela (2001), Butera et al. (2002), Stintzing et al. (2003), Khatabi et al. (2016)
Vitamin C	Cladodes	7–77 mg	Ramírez-Moreno et al. (2013), du Toit et al. (2018)
β-carotene		11.3-1787 μg	Leopoldo et al. (2012), Msaddak et al. (2015), du Toit et al. (2018)
Total folate		133.8 μg	Ortiz-Escobar et al. (2010)
Total Vitamin E		11.9 mg	Lanuzza et al. (2017)
Niacine		0.46 mg	du Toit et al. (2018)
Riboflavine		0.60 mg	
Thiamine		0.14 mg	
Betacyanins		0.32–1.72 mg	
Betaxanthins		0.332–1.7 mg	
Vitamin C	Pulp	2–30 mg	Butera et al. (2002), Kyriacou et al. (2016), Melgar et al. (2017)
Betacyanin		0.89–45 mg	Chavez-Santoscoy et al. (2009), Vergara et al. (2014), Robert et al. (2015), Palmeri et al. (2018)
Betaxanthin		0.52–19 mg/100 g	Chavez-Santoscoy et al. (2009), Vergara et al. (2014), Robert et al. (2015), Palmeri et al. (2018)
Betanin		1.04–5.12 mg	Butera et al. (2002)
Indicaxanthin		2.61–8.42 mg	
β-carotene	Peel	254 μg	Ramadan and Mörsel (2003)
Vitamin K1		109 μg	
Vitamin C		1.8–67 mg	Chaalal et al. (2013), Jiménez-Aguilar et al. (2015a), Méndez et al. (2015)
Betacyanins		58.8	Aparicio-Fernández et al. (2018)
Betaxantins		53.8	Aparicio-Fernández et al. (2018)
Betaxanthins		16.8 mg	Jiménez-Aguilar et al. (2015a)
Total vitamin E	Seed	44.74 mg/100 g total lipids	El Mannoubi et al. (2009)

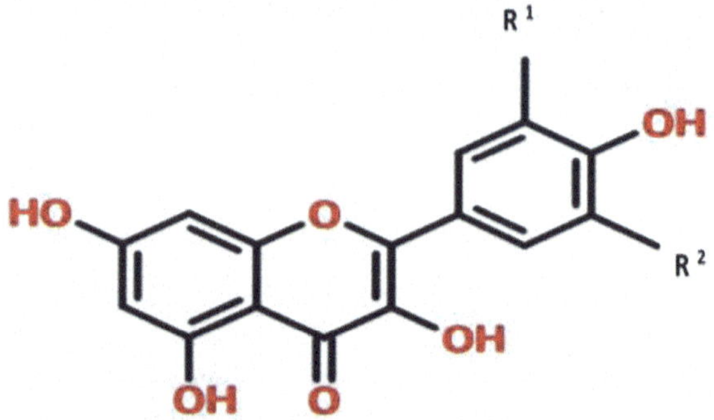

Quercetin (R^1 = OH; R^2= H; Kaempferol (R^1= R^2 = H); isorhamnetin (R^1= OCH_3; R^2=H

Fig. 26.3 Chemical structures of flavonoids (quercetin, kaempferol, and isorhamnetin) from various parts of *Opuntia ficus-indica* (L.) Mill plant

of Astello-García et al. (2015) who investigated different varieties of OFI cladodes (wild and domesticate). They found that the eucomic acid, kaempferol 3-O-robinobioside-7-O-arabinofuranoside, isorhamnetin 3-O-galactoside, and isorhamnetin 3-O-rhamnoside-7-O-(rhamnosyl-hexoside are the major phenolic compounds in both species and kaempferol 3-O-arabinofuranoside was the minor compound in wild while quercetin 3-O-rhamnosyl-(1 → 2)-[rhamnosyl-(1 → 6)]-glucoside) for domesticated. In Mexico OFI cladodes, Guevara-Figueroa et al. (2010) also were detected several flavonoid glycosides such isoquercitrin, kaempferol-3-*O*-rutinoside, isorhamnetin-3-*O*-glucoside, and isorhamnetin-3-*O*-rutinoside, while kaempferol was not identified. Studies showed that Several flavonols mainly isorhamnetin, quercetin, and kaempferol glycosylated derivatives were identified OFI cladodes (Astello-García et al. 2015; Ressaissi et al. 2017) (Table 26.1). The study has shown that OFI cladodes phenolic compounds profile change with the stages of maturity, young and medium cladodes indicated higher p-hydroxybenzoic acid, p-coumaric acid, ferulic acid, rutin, narcissism, and nicotiflorin (Figueroa-Pérez et al. 2018). Furthermore, chlorogenic acid, protocatechuic acid, sinapic acid, rosmarinic acid, ellagic acid, procyanidins B1 and B2, gallocatechin gallate, and epicatechin gallate were only identified in young and medium cladodes. This study is beneficial for foods and pharmaceutical industries to ascertain the ideal maturity stage of OFI cladodes in order to maximize their nutritional and pharmaceutical benefits. Interestingly, a study on phenolic profile of Italian OFI cladodes revealed that the cladode extracts are rich in anthocyanins with the conjugated forms of forms of cyanidin, pelargonidin, and petunidin as dominant compounds (Table 26.1) (Rocchetti et al. 2018).

UHPLC-ESI-MS analysis on cladodes, skin, and pulp of six cultivars of OFI showed that at different maturity stages young and old cladodes of showed that highest phenolic compounds were detected in young cladodes young cladodes > old cladodes > skin > pulp (Mena et al. 2018). Furthermore, about 26 phenolic compounds were determined in young cladodes with myricetin-hexoside being the major phenolic compounds. In this study, 41 phenolic compounds were detected with 23 phenolic compounds detected for the time in OFI plant. This study showed that the phenolic profile of young cladodes varied substantially among OFI varieties.

Isorhamnetin glycosides isolated from OFI cladodes have also presented in vitro anticancer activity against human colorectal adenocarcinoma (Caco2) and human colon adenocarcinoma (HT-29). The cladodes extracts and isolated isorhamnetin glycosides were more cytotoxic against HT-29 cells than Caco2 cells. OFI-30 showed the lowest IC50 activity against HT-29 (4.9 μg/mL) and against Caco2 (8.2 μg/mL). The authors found that isorhamnetin glucosyl-pentoside and glucosyl-rhamnoside were more cytotoxic than pure isorhamnetin aglycone or triglycosides when they were experienced in HT-29 cells (Antunes-Ricardo et al. 2014). This outcome could be valuable for the formulation of OFI-derived products with high contents of natural antioxidants.

The previous study showed that thermal treatments such as boiling could affect some important bioactive compounds (e.g. phenols and vitamin C) in cladodes (Ramírez-Moreno et al. 2013). In contrast, De Santiago et al. (2018) reported that the cooking methods mainly griddling and microwaving, improved the extraction of phenolic compounds (namely, quercetin, kaempferol, isorhamnetin, ferulic acid and 4-hydroxybenzoic acid) up to 3.2-fold higher than in raw samples, and accordingly their antioxidant activity. Another author reported that the heat treatment increased cryptoxanthin, carotene and lutein were detected in the cladodes, the last showing the highest value with a slight decrease in phenolic compounds (Jaramillo-Flores et al. 2003).

da Cruz Filho et al. (2019) compared the effect of lignins extracted from cladodes of OFI and *Opuntia cochenillifera* L. against mice splenocytes. Both lignins exhibited high cell viability (>96%) and cell proliferation. Activation signal was demonstrated for both lignins with an increase of reactive oxygen species and cytosolic calcium release and alterations in mitochondrial membrane potential. Coumestrol, a phytoestrogen is presented anticancer activity against prostate cancer, breast cancer and ovarian cancer (Dixon-Shanies and Shaikh 1999; Lim et al. 2017). Figueroa-Pérez et al. (2018) reported the presence of Coumestrol in OFI cladode during a different stage of maturity.

The different in OFI colors are attributed to variations in betalains. Betalains being accountable for the attractive color of OFI parts play an important role in the classification/authentication of cactus cultivars. Nowadays, numerous consumers desire to buy food with natural colorants due to toxicities of synthetic colorants, the betalain derivatives such as Red to purple color (betacyanins) and yellow to orange colors (betaxanthin) can be an excellent substitute to fulfill the demand of the food and pharmaceutical industries. Betalains are powerful with radical scavengers with an antioxidant activity 3–4 times higher than vitamin C rutin, and catechin (Cai et al. 2005; del Socorro Santos Díaz et al. 2017), higher than the determined in many fruits such as pear, apple, tomato, banana, and white grape, red grape, and orange (Strack et al. 2003). Study on the fresh and processed OFI cladodes showed the processed cladodes showed an excellent source of betacyanins and betaxanthins as presented in Table 26.3 (du Toit et al. 2018) (Fig. 26.4).

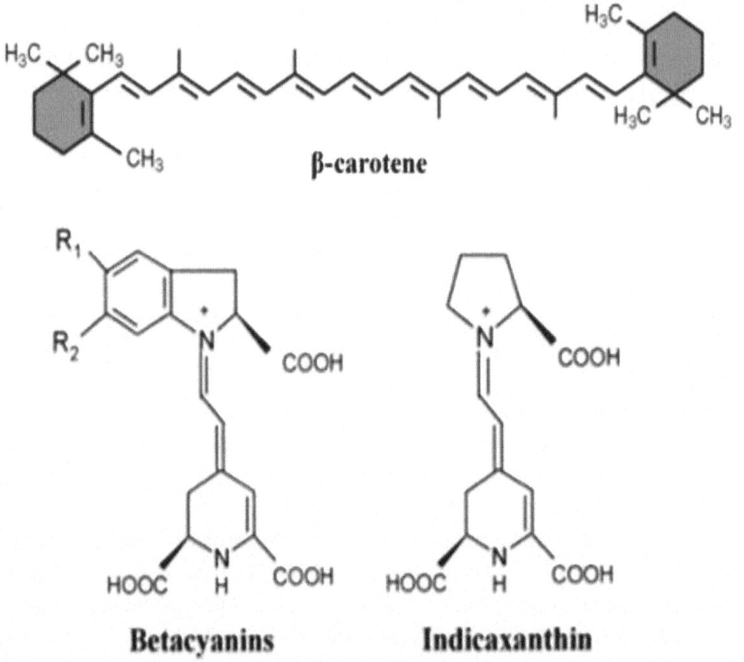

Betanidin: $R^1 = R^2 =$ OH. Betanin: (5-*O*-glucose benanindin); R^1 = glucose; R^2 = OH

Fig. 26.4 Structures of the main pigments in *Opuntia ficus-indica* (L.) Mill: Betacyanins (betanidin and betanin), the betaxanthin indicaxanthin and β-carotene

Several vitamins such as ascorbic acid carotenoids and total folate, Vitamin E content have been reported in the literature (Table 26.3) (Zafra-Rojas et al. 2013; Pignotti et al. 2016). Additionally, three carotenoids including β-carotene, α-cryptoxanthin, and lutein (Jaramillo-Flores et al. 2003; Leopoldo et al. 2012), have been detected in OFI cladodes. These compounds are essential for human metabolism because of their enhancement of gap junctional communication and increase immune response, inhibit lipid peroxidation and hemoglobin oxidation (Park et al. 2009; Chisté et al. 2014) besides to their role as modulators in processes related to mutagenesis, cell differentiation and proliferation (Waladkhani and Clemens 1998).

In another study, Figueroa-Pérez et al. (2018) indicated that cladodes of OFI are an excellent source of sterols and saponins. HPLC-DAD-MSD and LC-HRESIMS analyses of OFI cladodes extract allowed detection of several sterols including, β-sitosterol, β-campesterol, campestanol, Δ5-avenasterol, Δ7-stigmasterol and stigmastanol (Figueroa-Pérez et al. 2018; Msaddak et al. 2017). In addition to sterol compounds, several saponins including campesteryl 3β-D-glucopyranoside, stigmasteryl 3β-D-glucopyranoside, and sitosteryl 3β-D-glucopyranoside have been detected during various maturity stages. This study showed that the medium-age cladode indicated the highest saponins, sterols and antioxidant activity while young-cladodes indicated the highest content of tannins. These findings confirmed that maturity stages affect the nutraceutical properties of OFI cladodes.

26.3.3 Bioactive Compounds Profile of OFI Pulp/Whole Fruit

The OFI fruits have various colors: yellow, orange, red, purple and white for the combination of two betalain pigments (Aruwa et al. 2018). OFI fruits contain a considerable amount of bioactive compounds of betalains, ascorbic acid, carotenoids which have been reported to have health bandits such as hypoglycemic and hypolipidemic action, and antioxidant activities (El-Mostafa et al. 2014; Cruz-Cansino et al. 2015).

Several flavonols have been detected in OFI pulp including isorhamnetin, quercetin, and kaempferol glycosylated derivatives (Table 26.1). Moreover, hydroxybenzoic acid, eucomic acid, ferulic acid, gallic acid, and caffeic acid, piscidic acid derivative II have been identified in OFI pulp (Abdel-Hameed et al. 2014; Mata et al. 2016). Several studies reported that flavonols were only detected in the OFI fruit's peel and cladodes (Moussa-Ayoub et al. 2014; Moussa-Ayoub et al. 2015). In contrast, several isorhamnetin derivatives including, isorhamnetin glucosyl-di-rhamnoside, isorhamnetin pentosyl-rutinoside, isorhamnetin pentosyl-glucoside, isorhamnetin pentosyl-rhamnoside, isorhamnetin pentosyl-rhamnoside, isorhamnetin pentosyl-rhamnoside have been identified in pulp juice (Mata et al. 2016).

Analysis of OFI fruits has revealed considerable variation in total phenolics content. Approximately 482.26–492.71 mg GAE/100 g (Kuti 2004) and 89.2 mg GAE/100 g (Albano et al. 2015) have been detected. Higher contents of phenolic have also been reported in fruits (Lucía et al. 2018) and fruit pulp (Bouzoubaâ et al. 2016). Regarding total flavonoids content, several studies have shown OFI pulp/fruits are an excellent source of flavonoids (Saénz et al. 2009, Alimi et al. 2012; Cha et al. 2013; Jiménez-Aguilar et al. 2015b). Recently, acetophenones (picein, androsin) and 1-O-feruloyl-β-D-glucopyranoside were detected for the first time from OFI fruit (Kim et al. 2016). Furthermore, benzyl-O-β-D-glucopyranoside has been identified for the first time in the genus Opuntia (Fig. 26.5). These compounds are reported to have a significant health benefit to human health such as antioxidant activity and wound healing (Abreu et al. 2007; Walter and Merish 2015).

The betalain pigment contents in 10 cultivars of Mexican OFI fruits have been investigated by Castellanos-Santiago and Yahia (2008). In this study, 24 known/unknown betalains were detected, comprising 18 betaxanthins and 6 betacyanins. The purple fruits showed the highest betalain (8.1 mg/g dry fruit), whereas betacyanins content in OFI fruit's peel was about 1.4 higher than the pulp (Moussa-Ayoub et al. 2011). Furthermore, the analysis of pulp betalains by HPLC-DAD and HPLC-MS showed the main red pigment detected was the betacyanin. Three derivatives of betalains including isobetanin, 4′-malonil-betanin and 4′-malonil-isobetanin

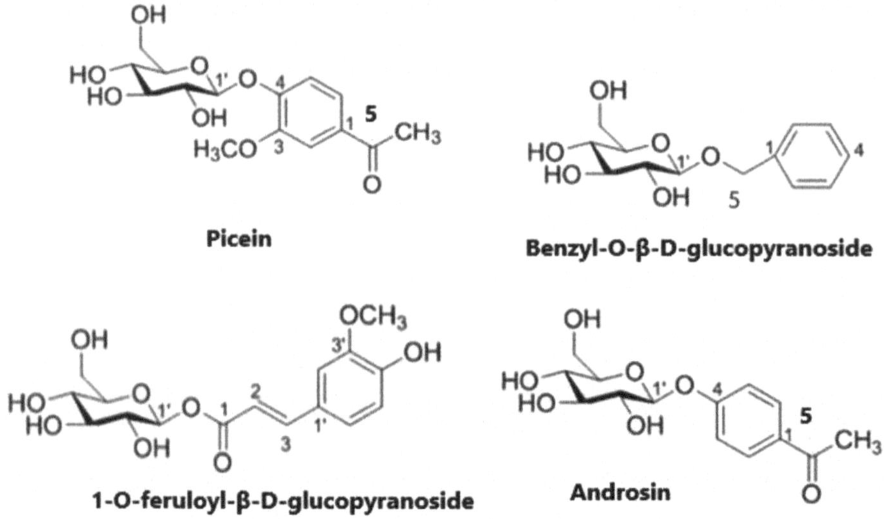

Fig. 26.5 The structures of compounds the new three components isolated from *Opuntia ficus-indica* (L.)

were also detected (de Souza et al. 2015). Indicaxanthin is greatly concentrated in the edible parts of the OFI fruits (Table 26.3). The impact of the extract from the fruit of OFI, and of its indicaxanthin, on the proliferation of human colon cancer Caco-2 cells have been studied (Naselli et al. 2014). The crude extract and indicaxanthin induced a dose-dependent apoptosis of proliferating cells at nutritionally related quantities, with IC_{50} 400 mg fresh pulp equivalents/mL, and 115 μM, respectively, without toxicity for post-confluent differentiated. Indicaxanthin represented ~80% of the impact of the whole extract.

Taurine is a free intracellular sulfur-containing β-amino acid, mostly found foods in fungi, algae, seafood, dairy and meat products (Aruwa et al. 2018). Therefore, it is considered an uncommon plant constituent. The positive effect of taurine on hypertension, cardiovascular diseases (Yamori et al. 2010), enhance the immune function during chemotherapy, reduce the toxicity of chemotherapy (Dong et al. 2017). improving growth, reduce inflammation and oxidative stress (Ramos et al. 2018; Thirupathi et al. 2018), have been proved in both experimental and epidemiological studies. Studies have shown that the taurine has potential to increase testicular function and quality of sperm as well as reducing the oxidative stress (Yang et al. 2013; Yang et al. 2015). Fernández-López et al. (2010) studies Spanish red-skinned various cactus fruits (*Opuntia ficus-indica, Opuntia undulata,* and *Opuntia stricta).* They observed that OFI fruit showed the highest antioxidant properties and taurine content. Similarly, the study of (Tesoriere et al. 2005) confirmed the presence of this compound (8.0–11.2 mg/100 g) in the tree cultivars of OFI fruits including yellow cultivar (Sulfarina), red cultivar (Sanguigna), white cultivar (Muscaredda) with the highest concentration in the yellow types. RP-HPLC analysis of South African and Mexican OFI fruits showed a significant amount of taurine (323.6–572 mg/L) (Stintzing et al. 1999). A lower concentration (6.80–11.96 mg/100 g) of taurine were detected in the edible pulp of Spain (Fernández-López et al. 2010) and Italian juice (Tesoriere et al. 2005).

Several authors have reported the antioxidant vitamins of OFI fruits from various cultivars (Table 26.3) (Butera et al. 2002; Cruz-Cansino et al. 2015; El Kharrassi et al. 2016; Cano et al. 2017). Various studies have reported the total carotenoid content in pulp/fruits from Spain, Morocco, and USA (Kuti 2004; Fernández-López et al. 2010; El Kharrassi et al. 2016). Some researches have been able to isolate some several individual carotenoids in purple, red, orange, yellow, and white OFI pulp (Mondragon-Jacobo et al. 2009). Fruit of all cultivars/ lines indicated the same carotenoid profile, with lutein being the major compound in 'Camuesa' whereas neoxanthin in Reyna. Recently, (all-E)-violaxanthin, (all-E)-neoxanthin, (9Z)-violaxanthin, (all-E)-anteraxanthin, (all-E)-lutein, (all-E)-zeaxanthin, lutein- 5,6- epoxide, (all-E)-α-carotene and (all-E)-β-carotene, (9Z)-β-carotene (Cano et al. 2017). (all-E-)-lutein is the predominant compound which constituted 71–72% of the total carotenoids while (all-E-)-β-carotene detected in lower concentration in pulp/fruit. Vitamin C is one of the main antioxidant compounds in OFI fruits. Various contents of vitamin C have been reported in the literature, which is reliant on the studies OFI variety. Cano et al. (2017) studied Spanish Sanguinos (red) and Verdal (orange) prickly pear fruits from Mexico and found that the higher vitamin C concentration is in red variety vs orange variety (71.94 mg/100 g fresh). In another study, El Kharrassi et al. (2016) compared the fruit and juice of two *cultivarsod Opuntia* species from different regions of Morocco and found that the highest vitamins concentrations in *Opuntia ficus indica* vs *Opuntia megacantha* (29.2 mg/L). A higher contents of vitamin C were also reported in purple cactus pear juice (464.23 mg /L) (Cruz-Cansino et al. 2015), cactus pear juice clarified by microfiltration (14.0–115.5 mg/L) (Cassano et al. 2010) and OFI beverage (336.0 mg /100 g) (Jiménez-Aguilar et al. 2015a). Kyriacou et al. (2016) studied the ripening behavior of two prominent OFI cultivars commonly grown in Turkish (Cyprus). They found that Vitamin C in the juice of 'Ntopia' and 'Hercules' increased during ripening and peaked near full maturity. Several studies have reported on the Vitamin C in OFI pulp/fruits (Piga et al. 2003; Coria Cayupán et al. 2011; Alimi et al. 2012). The variations among the study results might be attributed to environment and genotypic effects on carotenoids and vitamin C concentrations among *Opuntia* cultivars. In addition to vitamin C, small amounts of vitamin E were reported in three OFI cultivars Sulfarina (115 μg/100 g), red Sanguigna (111.5 μg/100 g) and Muscaredda (114 μg/100 g) (Tesoriere et al. 2005). The contents of vitamin E and vitamin K1 in pulp/fruits also were reported in the literature (El-Mostafa et al. 2014; Slimen et al. 2016).

26.3.4 Bioactive Compounds Profile of OFI Peel/Seeds

Regarding the use of OFI fruits for functional food formulation, one factor to take into consideration is that considerable phenolic concentrations are expected in the peel (Stintzing and Carle 2005; Yeddes et al. 2014b). Thus, from a functional perspective, processing both peel and pulp seems to be valuable. The weight proportional of OFI peel and seeds were respectively 37–67% and 2–10% a (Jiménez-Aguilar et al. 2015b). However, studies regarding the bioactive profile of peel and seeds are rare (Table 26.1). Although of these by-product (i.e. peel and seeds) accounting about 60% of the entire total weight of the fruit (Felker et al. 2005). Studies have proved the presence of antioxidant and antibacterial compounds in OFI peel and seeds (Iturriaga and Nazareno 2016; Ortega-Ortega et al. 2017). The previous study showed that the red peel of OFI contains higher total phenolic content (1152.97 mg/100 g) as compared to the peel of yellow cultivar (786.01 mg/100 g) (Abdel-Hameed et al. 2014). Jiménez-Aguilar et al. (2015b) studied the various parts of Mexican commercial varieties of cactus pear. The results showed that the Rojo San Martín peel indicated total phenolics over two times higher than the pulp, whereas antioxidant activity values of Rojo San Martín and Verde Villanueva peels were 4–9 higher than their corresponding pulp. Furthermore, Verde Villanueva peel presented the highest flavonoids (225.0 mg/kg), and Rojo Cenizo peels the lowest values (84.7 mg/kg). In addition to higher total phenolics (1512.58 mg GAE/100 g dry matter), sixteen specific phenolic compounds belonging to hydroxybenzoic acids, hydroxycinnamic acids and flavonoids were detected by HPLC–DAD-ESI-MS^n analysis in the peel extract (Chougui et al. 2015). The contents of quantified flavonoids ranged from 2.2 to 61.7 mg/100 g dry matters with isorhamnetin diglucoside pentoside being the most predominant component. In this study, the total amount of isorhamnetin glycosides in OFI peel was 216.4 mg/100 g dray matter, which equals to half of the content reported in cladode (Ginestra et al. 2009). Flavonols glucosyl including, isorhamnetin diglycoside, isorhamnetin diglycoside, isorhamnetin-3-O-rutinoside, and Isorhamnetin-3-O-glucoside were identified in the peel two varieties OFI (spiny and thornless forms) cultivated in Tunisia (Yeddes et al. 2014b). The peels of the thornless variety have more flavonoid than the spiny while quercetin 3-O-rutinoside was only identified in the spiny OFI. Study on the peel, pulp and whole fruit of two OFI Spanish cultivars also showed the OFI peel had the content carotenoids, vitamin C, betalains, and phenolic compounds (Cano et al. 2017).

OFI peels contain various water soluble and fat soluble vitamins (Ramadan and Moersel 2003; Ramadan and Mörsel 2003; Jiménez-Aguilar et al. 2015a). Considerable amounts of vitamin C (67.0 mg /100 g), vitamin E (21.8 g/ kg), vitamin K1 (1.09 g/ kg) and carotenoids (2.54 g/kg) were reported in OFI peels (Table 26.3). Apart from bioactive compounds, OFI peel powder has been used for improving the quality of the edible coating. The peel powder containing betacyanins (58.8 mg/100 g), betaxanthins (53.8 mg /100 g) and total phenolics (967.8 mg GAE/100 g). The authors found that a high concentration of aqueous extract and peel powder in carboxymethyl cellulose films increased the bioactive compounds content and antioxidant activities. The feasibility of using high hydrostatic pressure (HHP) treatments for extraction of phytochemical compounds OFI beverage was recently reported by Jiménez-Aguilar et al. (2015a). The results indicated that the HHP treatment preserves and increase the concentrations of betaxanthins (43.00 mg/100 g) and betacyanins (95.1 mg/100 g) and (2.20 mg/100 g) in OFI beverages when compared to untreated samples. Prakash Maran and Manikandan (2012) investigated the optimal condition of extraction of betacyanin betaxanthin. The authors found that the optimal extraction temperature, time, mass of sample were 40 °C, 115 min and mass of 1.44 g, respectively. According to this conditions, the total contents of betacyanin and betaxanthin were 13.4354 mg/100 g and 24.2922 mg/100 g respectively.

In addition to phenolic compounds, various sterols were reported in the peel and seeds of OFI (Aruwa et al. 2018). OFI peel lipid is described by the high quantity of un-saponifiables (128 g/ kg), in which about 29% were sterols (Ramadan and Mörsel 2003). Several sterol compounds have been identified in OFI peel, with β-sitosterol being the major compound (57.0% of the total sterol content) followed by campesterol (23.0% of the total sterol content). The concentrations of β-sitosterol in OFI peel were also reported (Ramadan and Moersel 2003). Stigmasterol and lanosterol are found in small amounts (Table 26.2). Generally, phytosterols are of great interest, because of their antioxidant properties and health benefits. Presently, sterols were incorporated into canola and flaxseed oils as an example of natural antioxidants compounds in frying oils (Singh 2013).

OFI seeds contain oil characterized by a higher content of health-promoting compounds such as saturated fatty acid, sterols, phenolic compounds, vitamin E, vitamin C and β-carotene (Koubaa et al. 2016). OFI seeds studies have shown the presence of total phenolic content −25.95- 40.90 mg GAE/100 g (Chaalal et al. 2013), 268.4 mg/100 g (Tlili et al. 2011) and 48–89 mg GAE/100 g (Chougui et al. 2013) as well as total flavonoids (4.44–27.20 mg QE/100) (Chaalal et al. 2013) and (1.5–2.6 mg QE/100 g) and tannins (4.1–6.6 mg CE/100 g) (Chougui et al. 2013). Several specific phenolic compounds have been reported in OFI seeds extracts (Tounsi et al. 2011; Chougui et al. 2013; Koubaa et al. 2017). A phenolic compound in OFI seeds showed significant antioxidant and antibacterial activities (e.g *Bacillus subtilis*, *Enterococcus faecalis*, *Bacillus thuringiensis*, *Klebsiella pneumoniae*, *Escherichia coli*, *Salmonella typhimurium*, and *Enterobacter sp.*) (Chougui et al. 2013; Berraaouan et al. 2015; Barba et al. 2017). Significant variations in antioxidant activity have been found in whole seeds as compared to ground OFI seeds which were ascribed to their high phenolic compounds and vitamins (Chaalal et al. 2013). OFI seeds contain significant vitamins, with vitamin E (447.380 mg/kg total lipids) being the major compound (Table 26.3). Recently research proved that vitamin E-enriched died prevents alloxan-induced diabetes in mice (Kamimura et al. 2013). This result was supported by Berraaouan et al. (2015) which proved that administration of vitamin E rich oil (2 mL/kg) significantly attenuated Allx — induced death and hyperglycemia in treated mice. Oil extracted from OFI seeds presented higher contents of β-sitosterol and campesterol but the levels of both compounds were lower than in peels (Table 26.3).

Furan-type lignans extracted from OFI seeds indicated substantial protective activities against ethanol-induced hepatotoxicity (Kim et al. 2017). Besides phenolic compounds, (Koubaa et al. 2017) reported that the seeds of OFI had flavonols and anthocyanins such as viscutin 1, naringin, cyanidin 3-O-(6″-malonyl-glucoside), cyanidin 3-O-rutinoside, cyanidin 3-O-(3″,6″-O-dimalonyl-glucoside) and cyanidin 3-O-diglucoside-5-O-glucoside.

26.4 Conclusion

In recent years, the *Opuntia ficus-indica (L.)* plant is an interesting source natural nutraceutical compounds which produced a large number of scientific papers analyzing the profile and/or bioactivity of crude extract or a specific purified cactus compounds such as lignins, indicaxanthin and isorhamnetin glycosides. Data from several scientific types of research indicated that the whole fruits and cladodes extracts are richest in bioactive compounds as well as widely evaluated for their biological activities. Furthermore, the available scientific reports proved that by-product of OFI plant (i.e. seeds, peels) also are a good source of bioactive compounds. The available information proved that the OFI plant particularly fruits, cladodes, and seeds could be an interminable source of functional compounds for food, and pharmaceutical industries. The information in this chapter could be valuable in understanding the phytochemical properties of OFI various parts and their biological uses in the development of functional foods or pharmaceutical medicines.

References

Abdel-Hameed E-SS, Nagaty MA, Salman MS, Bazaid SA (2014) Phytochemicals, nutritionals and antioxidant properties of two prickly pear cactus cultivars (*Opuntia ficus* indica Mill.) growing in Taif, KSA. Food Chem 160:31–38

Abreu PM, Braham H, Ben Jannet H, Mighri Z, Matthew S (2007) Antioxidant compounds from Ebenus pinnata. Fitoterapia 78(1):32–34

Albano C, Negro C, Tommasi N, Gerardi C, Mita G, Miceli A, De Bellis L, Blando F (2015) Betalains, phenols and antioxidant capacity in Cactus pear [Opuntia ficus-indica (L.) Mill.] fruits from Apulia (South Italy) genotypes. Antioxidants 4(2):269–280

Alimi H, Hfaeidh N, Bouoni Z, Sakly M, Ben Rhouma K (2012) Protective effect of Opuntia ficus indica f. inermis prickly pear juice upon ethanol-induced damages in rat erythrocytes. Alcohol 46(3):235–243

Allai L, Druart X, Öztürk M, BenMoula A, Nasser B, El Amiri B (2016) Protective effects of Opuntia ficus-indica extract on ram sperm quality, lipid peroxidation and DNA fragmentation during liquid storage. Anim Reprod Sci 175:1–9

Ammar I, Ben Salem M, Harrabi B, Mzid M, Bardaa S, Sahnoun Z, Attia H, Ennouri M (2018) Anti-inflammatory activity and phenolic composition of prickly pear (*Opuntia ficus-indica*) flowers. Ind Crop Prod 112:313–319

Antunes-Ricardo M, Moreno-García BE, Gutiérrez-Uribe JA, Aráiz-Hernández D, Alvarez MM, Serna-Saldivar SO (2014) Induction of apoptosis in colon cancer cells treated with isorhamnetin glycosides from Opuntia ficus-indica pads. Plant Foods Hum Nutr 69(4):331–336

Antunes-Ricardo M, Gutiérrez-Uribe JA, López-Pacheco F, Alvarez MM, Serna-Saldívar SO (2015) In vivo anti-inflammatory effects of isorhamnetin glycosides isolated from Opuntia ficus-indica (L.) Mill cladodes. Ind Crop Prod 76:803–808

Antunes-Ricardo M, Gutiérrez-Uribe JA, Guajardo-Flores D (2017) Extraction of isorhamnetin conjugates from *Opuntia ficus-indica* (L.) Mill using supercritical fluids. J Supercrit Fluids 119:58–63

Aparicio-Fernández X, Vega-Ahuatzin A, Ochoa-Velasco CE, Cid-Pérez S, Hernández-Carranza P, Ávila-Sosa R (2018) Physical and antioxidant characterization of edible films added with red prickly pear (Opuntia ficus-indica L.) cv. San Martín Peel and/or its aqueous extracts. Food Bioprocess Technol 11(2):368–379

Aragona M, Lauriano E, Pergolizzi S, Faggio C (2018) *Opuntia ficus-indica* (L.) miller as a source of bioactivity compounds for health and nutrition. Nat Prod Res 32(17):2037–2049

Aruwa CE, Amoo SO, Kudanga T (2018) Opuntia (Cactaceae) plant compounds, biological activities and prospects – a comprehensive review. Food Res Int 112:328–344

Astello-García MG, Cervantes I, Nair V, Santos-Díaz M d S, Reyes-Agüero A, Guéraud F, Negre-Salvayre A, Rossignol M, Cisneros-Zevallos L, Barba de la Rosa AP (2015) Chemical composition and phenolic compounds profile of cladodes from Opuntia spp. cultivars with different domestication gradient. J Food Compost Anal 43:119–130

Bakari S, Daoud A, Felhi S, Smaoui S, Gharsallah N, Kadri A (2017) Proximate analysis, mineral composition, phytochemical contents, antioxidant and antimicrobial activities and GC-MS investigation of various solvent extracts of cactus cladode. Food Sci Technol, Campinas 37(2):286–293

Barba FJ, Putnik P, Bursać Kovačević D, Poojary MM, Roohinejad S, Lorenzo JM, Koubaa M (2017) Impact of conventional and non-conventional processing on prickly pear (Opuntia spp.) and their derived products: from preservation of beverages to valorization of by-products. Trends Food Sci Technol 67:260–270

Baxter H, Harborne JB, Moss GP (1998) Phytochemical dictionary: a handbook of bioactive compounds from plants. CRC press, London

Ben Saad A, Dalel B, Rjeibi I, Smida A, Ncib S, Zouari N, Zourgui L (2017) Phytochemical, antioxidant and protective effect of cactus cladodes extract against lithium-induced liver injury in rats. Pharm Biol 55(1):516–525

Benayad Z, Martinez-Villaluenga C, Frias J, Gomez-Cordoves C, Es-Safi NE (2014) Phenolic composition, antioxidant and anti-inflammatory activities of extracts from Moroccan *Opuntia ficus-indica* flowers obtained by different extraction methods. Ind Crop Prod 62:412–420

Berraaouan A, Abderrahim Z, Hassane M, Abdelkhaleq L, Mohammed A, Mohamed B (2015) Evaluation of protective effect of cactus pear seed oil (Opuntia ficus-indica L. MILL.) against alloxan-induced diabetes in mice. Asian Pac J Trop Biomed 8(7):532–537

Bouzoubaâ Z, Essoukrati Y, Tahrouch S, Hatimi A, Gharby S, Harhar H (2016) Phytochemical study of prickly pear from southern Morocco. J Saudi Soc Agric Sci 15(2):155–161

Butera D, Tesoriere L, Di Gaudio F, Bongiorno A, Allegra M, Pintaudi AM, Kohen R, Livrea MA (2002) Antioxidant activities of Sicilian prickly pear (*Opuntia ficus indica*) fruit extracts and reducing properties of its betalains: Betanin and indicaxanthin. J Agric Food Chem 50(23):6895–6901

Cai Y-Z, Sun M, Corke H (2005) Characterization and application of betalain pigments from plants of the Amaranthaceae. Trends Food Sci Technol 16(9):370–376

Cano MP, Gómez-Maqueo A, García-Cayuela T, Welti-Chanes J (2017) Characterization of carotenoid profile of Spanish Sanguinos and Verdal prickly pear (*Opuntia ficus-indica,* spp.) tissues. Food Chem 237:612–622

Cassano A, Conidi C, Drioli E (2010) Physico-chemical parameters of cactus pear (Opuntia ficus-indica) juice clarified by microfiltration and ultrafiltration processes. Desalination 250(3):1101–1104

Castellanos-Santiago E, Yahia EM (2008) Identification and quantification of betalains from the fruits of 10 Mexican prickly pear cultivars by high-performance liquid chromatography and electrospray ionization mass spectrometry. J Agric Food Chem 56(14):5758–5764

Cha M-N, Jun H-I, Lee W-J, Kim M-J, Kim M-K, Kim Y-S (2013) Chemical composition and antioxidant activity of Korean cactus (Opuntia humifusa) fruit. Food Sci Biotechnol 22(2):523–529

Chaalal M, Louaileche H, Touati N, Bachir Bey M (2013) Phytochemicals, in vitro antioxidant capacity and antiradical potential of whole and ground seeds of three prickly pear varieties: a comparative study. Ind Crop Prod 49:386–391

Chavez-Santoscoy RA, Gutierrez-Uribe JA, Serna-Saldívar SO (2009) Phenolic composition, antioxidant capacity and in vitro cancer cell cytotoxicity of nine prickly pear (Opuntia spp.) juices. Plant Foods Hum Nutr 64(2):146–152

Chisté RC, Freitas M, Mercadante AZ, Fernandes E (2014) Carotenoids inhibit lipid peroxidation and hemoglobin oxidation, but not the depletion of glutathione induced by ROS in human erythrocytes. Life Sci 99(1):52–60

Chougui N, Tamendjari A, Hamidj W, Hallal S, Barras A, Richard T, Larbat R (2013) Oil composition and characterisation of phenolic compounds of *Opuntia ficus-indica* seeds. Food Chem 139(1):796–803

Chougui N, Djerroud N, Naraoui F, Hadjal S, Aliane K, Zeroual B, Larbat R (2015) Physicochemical properties and storage stability of margarine containing *Opuntia ficus-indica* peel extract as antioxidant. Food Chem 173:382–390

Coria Cayupán YS, Ochoa MJ, Nazareno MA (2011) Health-promoting substances and antioxidant properties of Opuntia sp. fruits. Changes in bioactive-compound contents during ripening process. Food Chem 126(2):514–519

Costa C, Tsatsakis A, Mamoulakis C, Teodoro M, Briguglio G, Caruso E, Tsoukalas D, Margina D, Dardiotis E, Kouretas D, Fenga C (2017) Current evidence on the effect of dietary polyphenols intake on chronic diseases. Food Chem Toxicol 110:286–299

Cruz-Cansino NDS, Ramírez-Moreno E, León-Rivera JE, Delgado-Olivares L, Alanís-García E, Ariza-Ortega JA, Manríquez-Torres JDJ, Jaramillo-Bustos DP (2015) Shelf life, physicochemical, microbiological and antioxidant properties of purple cactus pear (*Opuntia ficus indica*) juice after thermoultrasound treatment. Ultrason Sonochem 27:277–286

da Cruz Filho IJ, da Silva Barros BR, de Souza Aguiar LM, Navarro CDC, Ruas JS, de Lorena VMB, de Moraes Rocha GJ, Vercesi AE, de Melo CML, Maior AMS (2019) Lignins isolated from Prickly pear cladodes of the species *Opuntia fícus-indica* (Linnaeus) Miller and *Opuntia cochenillifera* (Linnaeus) Miller induces mice splenocytes activation, proliferation and cytokines production. Int J Biol Macromol 123:1331–1339

De Leo M, Abreu MBD, Pawlowska AM, Cioni PL, Braca A (2010) Profiling the chemical content of *Opuntia ficus-indica* flowers by HPLC–PDA-ESI-MS and GC/EIMS analyses. Phytochem Lett 3(1):48–52

De Santiago E, Domínguez-Fernández M, Cid C, De Peña M-P (2018) Impact of cooking process on nutritional composition and antioxidants of cactus cladodes (*Opuntia ficus-indica*). Food Chem 240:1055–1062

de Souza RLA, Santana MF, de Macedo EM, de Brito ES, Correia RT (2015) Physicochemical, bioactive and functional evaluation of the exotic fruits *Opuntia ficus-indica* and Pilosocereus pachycladus Ritter from the Brazilian caatinga. J Food Sci Tech 52(11):7329–7336

del Socorro Santos Díaz M, de la Rosa A-PB, Héliès-Toussaint C, Guéraud F, Nègre-Salvayre A (2017) Opuntia spp.: characterization and benefits in chronic diseases. Oxidative Med Cell Longev 2017:8634249

Dhaouadi K, Raboudi F, Funez-Gomez L, Pamies D, Estevan C, Hamdaoui M, Fattouch S (2013) Polyphenolic extract of Barbary-Fig (*Opuntia ficus-indica*) syrup: RP–HPLC–ESI–MS analysis and determination of antioxidant, antimicrobial and cancer-cells cytotoxic potentials. Food Anal Methods 6(1):45–53

Di Lorenzo F, Silipo A, Molinaro A, Parrilli M, Schiraldi C, D'Agostino A, Izzo E, Rizza L, Bonina A, Bonina F, Lanzetta R (2017) The polysaccharide and low molecular weight components of Opuntia ficus indica cladodes: structure and skin repairing properties. Carbohydr Polym 157:128–136

Dixon-Shanies D, Shaikh N (1999) Growth inhibition of human breast cancer cells by herbs and phytoestrogens. Oncol Rep 6(6):1383–1390

Dong J-F, Zheng X-Q, Rui H-B (2017) Effect of taurine on immune function in mice with T-cell lymphoma during chemotherapy. Asian Pac J Trop Biomed 10(11):1090–1094

du Toit A, de Wit M, Osthoff G, Hugo A (2018) Antioxidant properties of fresh and processed cactus pear cladodes from selected *Opuntia ficus-indica* and O. *robusta* cultivars. S Afr J Bot 118:44–51

El Kharrassi Y, Mazri MA, Benyahia H, Benaouda H, Nasser B, El Mzouri EH (2016) Fruit and juice characteristics of 30 accessions of two cactus pear species (*Opuntia ficus indica* and *Opuntia megacantha*) from different regions of Morocco. LWT Food Sci Technol 65:610–617

El Mannoubi I, Barrek S, Skanji T, Casabianca H, Zarrouk H (2009) Characterization of *Opuntia ficus indica* seed oil from Tunisia. Chem Nat Compd 45(5):616–620

El-Mostafa K, El Kharrassi Y, Badreddine A, Andreoletti P, Vamecq J, El Kebbaj MHS, Latruffe N, Lizard G, Nasser B, Cherkaoui-Malki M (2014) Nopal Cactus (*Opuntia ficus-indica*) as a source of bioactive compounds for nutrition, health and disease. Molecules 19(9):14879–14901

Felker P, Rodriguez S d C, Casoliba RM, Filippini R, Medina D, Zapata R (2005) Comparison of *Opuntia ficus indica* varieties of Mexican and Argentine origin for fruit yield and quality in Argentina. J Arid Environ 60(3):405–422

Fernández-López JA, Almela L (2001) Application of high-performance liquid chromatography to the characterization of the betalain pigments in prickly pear fruits. J Chromatogr A 913(1):415–420

Fernández-López JA, Almela L, Obón JM, Castellar R (2010) Determination of antioxidant constituents in cactus pear fruits. Plant Foods Hum Nutr 65(3):253–259

Feugang JM, Konarski P, Zou D, Stintzing FC, Zou C (2006) Nutritional and medicinal use of cactus pear (Opuntia spp.) cladodes and fruits. Front Biosci 11(1):2574–2589

Figueroa-Pérez MG et al (2018) Phytochemical composition and in vitro analysis of Nopal (O. Ficus-Indica) cladodes at different stages of maturity. Int J Food Prop 21(1):1728–1742

Filannino P, Cavoski I, Thlien N, Vincentini O, De Angelis M, Silano M, Gobbetti M, Di Cagno R (2016) Lactic acid fermentation of cactus cladodes (*Opuntia ficus-indica* L.) generates flavonoid derivatives with antioxidant and anti-inflammatory properties. PLoS One 11(3):e0152575

Galati EM, Mondello MR, Giuffrida D, Dugo G, Miceli N, Pergolizzi S, Taviano MF (2003) Chemical characterization and biological effects of sicilian *Opuntia ficus indica* (L.) Mill. fruit juice: antioxidant and antiulcerogenic activity. J Agric Food Chem 51(17):4903–4908

Ginestra G, Parker ML, Bennett RN, Robertson J, Mandalari G, Narbad A, Lo Curto RB, Bisignano G, Faulds CB, Waldron KW (2009) Anatomical, chemical, and biochemical characterization of cladodes from prickly pear [*Opuntia ficus-indica* (L.) Mill.]. J Agric Food Chem 57(21):10323–10330

González-Stuart AE (2013) Chapter 46 – prickly pear Cactus ('Nopal') for the treatment of type 2 diabetes mellitus. In: Watson RR, Preedy VR (eds) Bioactive food as dietary interventions for diabetes. Academic, San Diego, pp 601–607

Griffith MP (2004) The origins of an important cactus crop, *Opuntia ficus-indica* (Cactaceae): new molecular evidence. Am J Bot 91(11):1915–1921

Guevara-Figueroa T, Jiménez-Islas H, Reyes-Escogido ML, Mortensen AG, Laursen BB, Lin L-W, De León-Rodríguez A, Fomsgaard IS, Barba de la Rosa AP (2010) Proximate composition, phenolic acids, and flavonoids characterization of commercial and wild nopal (Opuntia spp.). J Food Compost Anal 23(6):525–532

Herrera-Hernández MG, Guevara-Lara F, Reynoso-Camacho R, Guzmán-Maldonado SH (2011) Effects of maturity stage and storage on cactus berry (Myrtillocactus geometrizans) phenolics, vitamin C, betalains and their antioxidant properties. Food Chem 129(4):1744–1750

Ines O, Hela H, Wissem AW, Iness BR, Majdi H, Brahim M, Moufida ST (2017) Phytochemical composition and antioxidant activity of Tunisian cactus pear (*Opuntia ficus indica* L.) flower. J Food Biochem 41(5):e12390

Isela V-AR, Elsa B-M, Silvia B-B, Fernando R-C (2017) Cactus stem (Opuntia ficus-indica Mill): anatomy, physiology and chemical composition with emphasis on its biofunctional properties. J Sci Food Agric 97(15):5065–5073

Iturriaga L, Nazareno M (2016) Functional components and medicinal properties of cactus products. In: Kristbergsson K, Ötles S (eds) Functional properties of traditional foods. Springer US, Boston, pp 251–269

Jaramillo-Flores ME, Gonzalez-Cruz L, Cornejo-Mazon M, Dorantes-Alvarez L, Gutierrez-Lopez GF, Hernandez-Sanchez H (2003) Effect of thermal treatment on the antioxidant activity and content of carotenoids and phenolic compounds of cactus pear cladodes (Opuntia ficus-indica). Food Sci Technol Int 9(4):271–278

Jelena CH, Giorgio R, Justyna G, Neda M-D, Natasa S, Artur B, Giuseppe G (2018) 3 - Beneficial effects of polyphenols on chronic diseases and ageing. Polyphenols: Properties, Recovery, and Applications. C. M. Galanakis. Woodhead Publishing, Austria, pp 69–102

Jiménez-Aguilar DM, Escobedo-Avellaneda Z, Martín-Belloso O, Gutiérrez-Uribe J, Valdez-Fragoso A, García-García R, Torres JA, Welti-Chanes J (2015a) Effect of high hydrostatic pressure on the content of phytochemical compounds and antioxidant activity of prickly pears (Opuntia ficus-indica) beverages. Food Eng Rev 7(2):198–208

Jiménez-Aguilar DM, López-Martínez JM, Hernández-Brenes C, Gutiérrez-Uribe JA, Welti-Chanes J (2015b) Dietary fiber, phytochemical composition and antioxidant activity of Mexican commercial varieties of cactus pear. J Food Compost Anal 41:66–73

Kamimura W, Doi W, Takemoto K, Ishihara K, Wang D-H, Sugiyama H, Oda S-i, Masuoka N (2013) Effect of vitamin E on alloxan-induced mouse diabetes. Clin Biochem 46(9):795–798

Kang J-W, Shin J-K, Koh E-J, Ryu H, Kim HJ, Lee S-M (2016) *Opuntia ficus-indica* seed attenuates hepatic steatosis and promotes M2 macrophage polarization in high-fat diet–fed mice. Nutr Res 36(4):369–379

Khatabi O, Hanine H, Elothmani D, Hasib A (2016) Extraction and determination of polyphenols and betalain pigments in the Moroccan prickly pear fruits (*Opuntia ficus indica*). Arab J Chem 9:S278–S281

Kim JW, Kim TB, Yang H, Sung SH (2016) Phenolic compounds isolated from Opuntia ficus-indica fruits. Nat Prod Sci 22(2):117–121

Kim JW, Yang H, Kim HW, Kim HP, Sung SH (2017) Lignans from *Opuntia ficus-indica* seeds protect rat primary hepatocytes and HepG2 cells against ethanol-induced oxidative stress. Biosci Biotechnol Biochem 81(1):181–183

Koubaa M, Barba FJ, Grimi N, Mhemdi H, Koubaa W, Boussetta N, Vorobiev E (2016) Recovery of colorants from red prickly pear peels and pulps enhanced by pulsed electric field and ultrasound. Innov Food Sci Emerg Technol 37:336–344

Koubaa M, Mhemdi H, Barba FJ, Angelotti A, Bouaziz F, Chaabouni SE, Vorobiev E (2017) Seed oil extraction from red prickly pear using hexane and supercritical CO2: assessment of phenolic compound composition, antioxidant and antibacterial activities. J Sci Food Agric 97(2):613–620

Kuti JO (2004) Antioxidant compounds from four Opuntia cactus pear fruit varieties. Food Chem 85(4):527–533

Kyriacou MC, Emmanouilidou MG, Soteriou GA (2016) Asynchronous ripening behavior of cactus pear (*Opuntia ficus-indica*) cultivars with respect to physicochemical and physiological attributes. Food Chem 211:598–607

Lanuzza F, Occhiuto F, Monforte MT, Tripodo MM, D'Angelo V, Galati EM (2017) Antioxidant phytochemicals of *Opuntia ficus-indica* (L.) Mill. Cladodes with potential anti-spasmodic activity. Pharmacogn Mag 13(Suppl 3):S424–S429

Lee J-C, Kim H-R, Kim J, Jang Y-S (2002) Antioxidant property of an ethanol extract of the stem of Opuntia ficus-indica var. saboten. J Agric Food chem 50(22):6490–6496

Leopoldo G-C, Santiago F-K, Arturo B-PL, Norma G-V, Aurea B-N (2012) Carotenoid content, antioxidant activity and sensory evaluation of low-calorie nopal (opuntia ficus-indica) marmalade. J Food Process Preserv 36(3):267–275

Lim W, Jeong M, Bazer FW, Song G (2017) Coumestrol inhibits proliferation and migration of prostate cancer cells by regulating AKT, ERK1/2, and JNK MAPK cell Signaling cascades. J Cell Physiol 232(4):862–871

Lucía A, Nallely N-J, Carbonell-Barrachina ÁA, Legua P, Hernández F (2018) Antioxidant properties and chemical characterization of Spanish *Opuntia ficus-indica* Mill cladodes and fruits. J Sci Food Agric 98(4):1566–1573

Martins S, Mussatto SI, Martínez-Avila G, Montañez-Saenz J, Aguilar CN, Teixeira JA (2011) Bioactive phenolic compounds: production and extraction by solid-state fermentation. A review. Biotechnol Adv 29(3):365–373

Mata A, Ferreira JP, Semedo C, Serra T, Duarte CMM, Bronze MR (2016) Contribution to the characterization of Opuntia spp. juices by LC–DAD–ESI-MS/MS. Food Chem 210:558–565

Medina-Torres L, Vernon-Carter EJ, Gallegos-Infante JA, Rocha-Guzman NE, Herrera-Valencia EE, Calderas F, Jiménez-Alvarado R (2011) Study of the antioxidant properties of extracts obtained from nopal cactus (*Opuntia ficus-indica*) cladodes after convective drying. J Sci Food Agric 91(6):1001–1005

Melgar B, Pereira E, Oliveira MBPP, Garcia-Castello EM, Rodriguez-Lopez AD, Sokovic M, Barros L, Ferreira ICFR (2017) Extensive profiling of three varieties of Opuntia spp. fruit for innovative food ingredients. Food Res Int 101:259–265

Mena P, Tassotti M, Andreu L, Nuncio-Jáuregui N, Legua P, Del Rio D, Hernández F (2018) Phytochemical characterization of different prickly pear (*Opuntia ficus-indica* (L.) Mill.) cultivars and botanical parts: UHPLC-ESI-MSn metabolomics profiles and their chemometric analysis. Food Res Int 108:301–308

Méndez LP, Flores FT, Martín JD, Rodríguez Rodríguez EM, Díaz Romero C (2015) Physicochemical characterization of cactus pads from Opuntia dillenii and Opuntia ficus indica. Food Chem 188:393–398

Mendonça RD, Carvalho NC, Martin-Moreno JM, Pimenta AM, Lopes ACS, Gea A, Martinez-Gonzalez MA, Bes-Rastrollo M (2019) Total polyphenol intake, polyphenol subtypes and incidence of cardiovascular disease: The SUN cohort study. Nutr Metab Cardiovasc Dis 29(1):69–78

Mondragon-Jacobo C, Yahia E, Castellanos E (2009) Identification and quantification of pigments in prickly pear fruit. VI international postharvest symposium, p 877

Moussa-Ayoub TE, El-Samahy SK, Kroh LW, Rohn S (2011) Identification and quantification of flavonol aglycons in cactus pear (*Opuntia ficus indica*) fruit using a commercial pectinase and cellulase preparation. Food Chem 124(3):1177–1184

Moussa-Ayoub TE, Abd El-Hady EA, Omran HT, El-Samahy SK, Kroh LW, Rohn S (2014) Influence of cultivar and origin on the flavonol profile of fruits and cladodes from cactus Opuntia ficus-indica. Food Res Int 64:864–872

Moussa-Ayoub TE, Youssef K, El-Samahy SK, Kroh LW, Rohn S (2015) Flavonol profile of cactus fruits (Opuntia ficus-indica) enriched cereal-based extrudates: authenticity and impact of extrusion. Food Res Int 78:442–447

Msaddak L, Siala R, Fakhfakh N, Ayadi MA, Nasri M, Zouari N (2015) Cladodes from prickly pear as a functional ingredient: effect on fat retention, oxidative stability, nutritional and sensory properties of cookies. Int J Food Sci Nutr 66(8):851–857

Msaddak L, Abdelhedi O, Kridene A, Rateb M, Belbahri L, Ammar E, Nasri M, Zouari N (2017) *Opuntia ficus-indica* cladodes as a functional ingredient: bioactive compounds profile and their effect on antioxidant quality of bread. Lipids Health Dis 16(1):32

Naselli F, Tesoriere L, Caradonna F, Bellavia D, Attanzio A, Gentile C, Livrea MA (2014) Anti-proliferative and pro-apoptotic activity of whole extract and isolated indicaxanthin from *Opuntia ficus-indica* associated with re-activation of the onco-suppressor p16INK4a gene in human colorectal carcinoma (Caco-2) cells. Biochem Biophys Res Commun 450(1):652–658

Ortega-Ortega M d l A, Cruz-Cansino N d S, Alanís-García E, Delgado-Olivares L, Ariza-Ortega JA, Ramírez-Moreno E, Manríquez-Torres J d J (2017) Optimization of ultrasound extraction of cactus pear (*Opuntia ficus indica*) seed oil based on antioxidant activity and evaluation of its antimicrobial activity. J Food Qual 2017(72):1–9

Ortiz-Escobar TB, Valverde-González ME, Paredes-López O (2010) Determination of the folate content in cladodes of Nopal (*Opuntia ficus indica*) by microbiological assay utilizing Lactobacillus casei (ATCC 7469) and enzyme-linked Immunosorbent assay. J Agric Food Chem 58(10):6472–6475

Osuna-Martínez U, Reyes-Esparza J, Rodríguez-Fragoso L (2014) Cactus (Opuntia ficus-indica): a review on its antioxidants properties and potential pharmacological use in chronic diseases. Nat Prod Chem Res 2(6):1–8

Paiva PMG, de Souza IFAC, Costa MCVV, Santos A d FS, Coelho LCBB (2016) Opuntia sp. Cactus: biological characteristics, cultivation and applications. Adv Res 7:1–14

Palmeri R, Parafati L, Restuccia C, Fallico B (2018) Application of prickly pear fruit extract to improve domestic shelf life, quality and microbial safety of sliced beef. Food Cosmet Toxicol 118:355–360

Park S-Y, Nomura AMY, Murphy SP, Wilkens LR, Henderson BE, Kolonel LN (2009) Carotenoid intake and colorectal cancer risk: the multiethnic cohort study. J Epidemiol 19(2):63–71

Petruk G, Di Lorenzo F, Imbimbo P, Silipo A, Bonina A, Rizza L, Piccoli R, Monti DM, Lanzetta R (2017) Protective effect of Opuntia ficus-indica L. cladodes against UVA-induced oxidative stress in normal human keratinocytes. Bioorg Med Chem Lett 27(24):5485–5489

Piga A, Del Caro A, Pinna I, Agabbio M (2003) Changes in ascorbic acid, polyphenol content and antioxidant activity in minimally processed cactus pear fruits. LWT Food Sci Technol 36(2):257–262

Pignotti GAP, Hook G, Ghan E, Vega-López S (2016) Effect of nopales (Opuntia spp.) on lipoprotein profile and oxidative stress among moderately hypercholesterolemic adults: a pilot study. J Funct Foods 27:115–121

Prakash Maran J, Manikandan S (2012) Response surface modeling and optimization of process parameters for aqueous extraction of pigments from prickly pear (*Opuntia ficus-indica*) fruit. Dyes Pigments 95(3):465–472

Rahmatullah M, Noman A, Hossan MS, Rashid M, Rahman T, Chowdhury MH, Jahan R (2009) A survey of medicinal plants in two areas of Dinajpur district, Bangladesh including plants which can be used as functional foods. Am Eur J Sustain Agric 3(4):862–876

Ramadan MF, Moersel J-T (2003) Oil cactus pear (*Opuntia ficus-indica L.*). Food Chem 82(3):339–345

Ramadan MF, Mörsel J-T (2003) Recovered lipids from prickly pear [*Opuntia ficus-indica* (L.) Mill] peel: a good source of polyunsaturated fatty acids, natural antioxidant vitamins and sterols. Food Chem 83(3):447–456

Ramírez-Moreno E, Córdoba-Díaz D, de Cortes Sánchez-Mata M, Díez-Marqués C, Goñi I (2013) Effect of boiling on nutritional, antioxidant and physicochemical characteristics in cladodes (*Opuntia ficus indica*). LWT Food Sci Technol 51(1):296–302

Ramos C d O, Campos KKD, Costa G d P, Cangussú SD, Talvani A, Bezerra FS (2018) Taurine treatment decreases inflammation and oxidative stress in lungs of adult mice exposed to cigarette smoke. Regul Toxicol Pharmacol 98:50–57

Rasoulpour R, Afsharifar A, Izadpanah K (2018) Antiviral activity of prickly pear (Opuntia ficus-indica (L.) Miller) extract: Opuntin B, a second antiviral protein. Crop Prot 112:1–9

Ressaissi A, Attia N, Falé PL, Pacheco R, Victor BL, Machuqueiro M, Serralheiro MLM (2017) Isorhamnetin derivatives and piscidic acid for hypercholesterolemia: cholesterol permeability, HMG-CoA reductase inhibition, and docking studies. Arch Pharm Res 40(11):1278–1286

Robert P, Torres V, García P, Vergara C, Sáenz C (2015) The encapsulation of purple cactus pear (Opuntia ficus-indica) pulp by using polysaccharide-proteins as encapsulating agents. LWT Food Sci Technol 60(2, Part 1):1039–1045

Rocchetti G, Pellizzoni M, Montesano D, Lucini L (2018) Italian *Opuntia ficus-indica* Cladodes as rich source of bioactive compounds with health-promoting properties. Foods 7(2):24

Ruiz-Gutiérrez MG, Amaya-Guerra CA, Quintero-Ramos A, de Jesús Ruiz-Anchondo T, Gutiérrez-Uribe JA, Baez-González JG, Lardizabal-Gutiérrez D, Campos-Venegas K (2014) Effect of soluble fiber on the physicochemical properties of cactus pear (*Opuntia ficus indica*) encapsulated using spray drying. Food Sci Biotechnol 23(3):755–763

Saenz C, Mecklenburg P, Estevez A, Sepulveda E (1996) Natural liquid sweetener from cactus pear: obtention and characteristics. III International congress on cactus pear and cochineal, p 438

Saénz C, Tapia S, Chávez J, Robert P (2009) Microencapsulation by spray drying of bioactive compounds from cactus pear (*Opuntia ficus-indica*). Food Chem 114(2):616–622

Sánchez E et al (2014) Antibacterial and antioxidant activities in extracts of fully grown cladodes of 8 cultivars of cactus pear. J Food Sci 79(4):M659–M664

Santos-Zea L, Gutiérrez-Uribe JA, Serna-Saldivar SO (2011) Comparative analyses of total phenols, antioxidant activity, and flavonol glycoside profile of cladode flours from different varieties of Opuntia spp. J Agric Food Chem 59(13):7054–7061

Serra AT, Poejo J, Matias AA, Bronze MR, Duarte CMM (2013) Evaluation of Opuntia spp. derived products as antiproliferative agents in human colon cancer cell line (HT29). Food Res Int 54(1):892–901

Singh A (2013) Sitosterol as an antioxidant in frying oils. Food Chem 137(1):62–67

Slimen IB, Najar T, Abderrabba M (2016) *Opuntia ficus-indica* as a source of bioactive and nutritional phytochemicals. J Food Sci Nutr 4(6):162–169

Smida A, Ncibi S, Taleb J, Ben Saad A, Ncib S, Zourgui L (2017) Immunoprotective activity and antioxidant properties of cactus (*Opuntia ficus indica*) extract against chlorpyrifos toxicity in rats. Biomed Pharmacother 88:844–851

Stintzing F, Carle R (2005) Cactus stems (Opuntia spp.): a review on their chemistry, technology, and uses. Mol Nutr Food Res 49(2):175–194

Stintzing FC, Schieber A, Carle R (1999) Amino acid composition and betaxanthin formation in fruits from *Opuntia ficus-indica*. Planta Med 65(7):632–635

Stintzing FC, Schieber A, Carle R (2003) Evaluation of colour properties and chemical quality parameters of cactus juices. Eur Food Res Technol 216(4):303–311

Strack D, Vogt T, Schliemann W (2003) Recent advances in betalain research. Phytochemistry 62(3):247–269

Suh S, Kim YE, Yang H-J, Ko S, Hong G-P (2017) Influence of autoclave treatment and enzymatic hydrolysis on the antioxidant activity of *Opuntia ficus-indica* fruit extract. Food Sci Biotechnol 26(3):581–590

Tesoriere L, Fazzari M, Allegra M, Livrea MA (2005) Biothiols, Taurine, and lipid-soluble antioxidants in the edible pulp of sicilian Cactus pear (*Opuntia ficus-indica*) fruits and changes of bioactive juice components upon industrial processing. J Agric Food Chem 53(20):7851–7855

Thirupathi A, Freitas S, Sorato HR, Pedroso GS, Effting PS, Damiani AP, Andrade VM, Nesi RT, Gupta RC, Muller AP, Pinho RA (2018) Modulatory effects of taurine on metabolic and oxidative stress parameters in a mice model of muscle overuse. Nutrition 54:158–164

Tlili N, El Guizani T, Nasri N, Khaldi A, Triki S (2011) Protein, lipid, aliphatic and triterpenic alcohol content of caper seeds “Capparis spinosa”. J Am Oil Chem Soc 88(2):265–270

Tounsi MS, Ouerghemmi I, Ksouri R, Wannes WA, Hammrouni I, Marzouk B (2011) HPLC-determination of phenolic composition and antioxidant capacity of cactus prickly pears seeds. Asian J Chem 23(3):1006

Vergara C, Saavedra J, Sáenz C, García P, Robert P (2014) Microencapsulation of pulp and ultra-filtered cactus pear (*Opuntia ficus-indica*) extracts and betanin stability during storage. Food Chem 157:246–251

Waladkhani A, Clemens MR (1998) Effect of dietary phytochemicals on cancer development. Int J Mol Cell Med 1(4):747–800

Walter TM, Merish S (2015) Pharmacognostical overview of “Gunma Kuligai”–a traditional siddha preparation for treating acid peptic disease (APD). Chin Med 1(2):91–97

Xu Z, Sun T, Li W, Sun X (2017) Inhibiting effects of dietary polyphenols on chronic eye diseases. J Funct Foods 39:186–197

Yahia EM, Mondragon-Jacobo C (2011) Nutritional components and anti-oxidant capacity of ten cultivars and lines of cactus pear fruit (Opuntia spp.). Food Res Int 44(7):2311–2318

Yamori Y, Taguchi T, Hamada A, Kunimasa K, Mori H, Mori M (2010) Taurine in health and diseases: consistent evidence from experimental and epidemiological studies. J Biomed Sci 17(1):S6

Yang J, Lin S, Feng Y, Wu G, Hu J (2013) Taurine enhances the sexual response and mating ability in aged male rats. Springer New York, New York

Yang J, Zong X, Wu G, Lin S, Feng Y, Hu J (2015) Taurine increases testicular function in aged rats by inhibiting oxidative stress and apoptosis. Amino Acids 47(8):1549–1558

Yeddes N, Chérif JK, Guyot S, Baron A, Trabelsi-Ayadi M (2014a) Phenolic profile of tunisian opuntia ficus indica thornless form flowers via chromatographic and spectral analysis by reversed phase-high performance liquid chromatography-UV-photodiode array and electrospray ionization-mass spectrometer. Int J Food Prop 17(4):741–751

Yeddes N, Chérif JK, Trabelsi Ayadi M (2014b) Comparative study of antioxidant power, polyphenols, flavonoids and betacyanins of peel and pulp of three Tunisian Opuntia forms. Pak J Biol Sci 17:650–658

Zafra-Rojas QY, Cruz-Cansino N, Ramírez-Moreno E, Delgado-Olivares L, Villanueva-Sánchez J, Alanís-García E (2013) Effects of ultrasound treatment in purple cactus pear (*Opuntia ficus-indica*) juice. Ultrason Sonochem 20(5):1283–1288

Zhong X-K, Jin X, Lai F-Y, Lin Q-S, Jiang J-G (2010) Chemical analysis and antioxidant activities in vitro of polysaccharide extracted from Opuntia ficus indica Mill. cultivated in China. Carbohydr Polym 82(3):722–727

Chapter 27
Antidesma montanum: Biochemistry and Bioactive Compounds

Norsharina Ismail, Nur Hanisah Azmi, Siti Nurulhuda Mastuki, Norazalina Saad, and Ahmad Faizal Abdull Razis

Abbreviations

DPPH	diphenyl-1-picrylhydrazyl
GCMS	gas chromatography mass spectrometry
ICP-AES	inductively coupled plasma emission spectroscopy
MTT	3-(4,5-dimethylthiazol-2-yl)-2,5-diphenyl tetrazolium bromide
MTD	maximum tolerable dose

27.1 Introduction

Antidesma is commonly found understorey of the tropical rain forest in the Old World topics, a dioecious shrubs and trees with homogeneous genus (Pax and Hoffmann 1922). The plant belongs to the Euphorbiaceae, under subfamily of Phyllanthoideae, the Antidesmeae tribe and Antidesminae subtribe (Webster 1994; Radcliffe-Smith 2001). There was different number of species identified by few researchers whereby 142 species were listed by Pax and Hoffmann (1922), 200 spe-

N. Ismail (✉) · N. H. Azmi · A. F. A. Razis
Laboratory of Molecular Biomedicine, Institute of Bioscience, Universiti Putra Malaysia, Serdang, Selangor, Malaysia
e-mail: norsharina@upm.edu.my; madfaizal@upm.edu.my

S. N. Mastuki
Laboratory of Natural Products, Institute of Bioscience, Universiti Putra Malaysia, Serdang, Selangor, Malaysia
e-mail: ctnurulhuda@upm.edu.my

N. Saad
Laboratory of UPM-MAKNA Cancer Research, Institute of Bioscience, Universiti Putra Malaysia, Serdang, Selangor, Malaysia
e-mail: norazalina@upm.edu.my

A. A. Mariod (ed.), *Wild Fruits: Composition, Nutritional Value and Products*,
https://doi.org/10.1007/978-3-030-31885-7_27

cies were estimated by Webster (1994), and 170 species were listed by Radcliffe-Smith (2001). As indicated by other investigations, the actual number of those species might be lower as high number of synonyms among them were noticed (Hoffmann 1999a, b).

Antidesma montanum is the most common and widespread of all *Antidesma* species. From the Greek, anti (against) and desma (Burman term for poison), and montanum (of the mountains). Many names are subsumed under *Antidesma montanum* that deserve taxonomic recognition at the level of variety (Hoffmann 1999a). The tree is found in the wild suited with various types of soils (i.e. peat, clay, sand, loam, limestone, sandstone, granite, volcanic and ultrabasic soil), and requires propagation with seeds. It is a dioecious species, meaning that if fruit and seed are required, both male and female forms need to be grown (http://www.nationaalherbarium.nl/euphorbs/). In Malay, it is called as "Madang lada", referring to its yellowish wood similar to "Madang" = Litsea (Lauraceae) and ("lada" = pepper) for the fruits, thus it is known as "Litsea with pepper fruits" (http://www.nationaalherbarium.nl/Euphorbs/specA/Antidesma.htm).

27.1.1 Plant Description

Antidesma montanum is a tree with a round crown of spreading branches, which grows up to 20 meters tall (Fig. 27.1). The stem can be clear of branches for up to 7 meters, though it often branches from near the base or below the middle; it can be up to 40 cm in diameter. The plant sometimes adopts a climbing habit. A very variable species found in a wide variety of habitats from deep forest shade to open

Fig. 27.1 Blume tree and fruits of *Antidesma montanum*. (Source: https://commons.wikimedia.org)

grassland and around human habitations; from the coast to montane regions; in dry soils to wet or seasonally flooded ground; at elevations to 2000 meters. The ellipsoid to globose or ovoid fruits can be 3– 8 mm long and 3–6 mm wide (http://www.nationaalherbarium.nl/euphorbs/). The plant possesses simple leaves, alternate, distichous; stipules in pairs, linear-lanceolate, acute, 1.2 cm long, pubescent, subpersistent; petiole 0.25–0.7 (−1.2) cm long, terete sometimes subcanaliculate, pubescent; lamina 8–22 × 2.5–7.7 cm, usually oblong to elliptic or oblanceolate, apex gradually acuminate with mucronate tip, base acute to rounded or cuneate, margin entire, chartaceous, glabrous except on midrib; secondary nerves prominent beneath, 5–9 pairs, ascending; tertiary nerves broadly reticulate. Inflorescence axillary or terminal racemes; flowers unisexual, dioecious; clusters of flowers in male more distant than in female (http://asianplant.net/Phyllanthaceae/Antidesma_montanum.htm).

27.1.2 *Plant Distribution*

The tree is found in East Asia throughout China, India, Indonesia, Malaysia, Philippines, Vietnam, Myanmar, Thailand, Cambodia, Laos, New Guinea and Northern Australia. It is distributed in the wild on peat, clay, sand, loam, limestone, sandstone, granite, volcanic and ultrabasic soil (http://www.nationaalherbarium.nl/euphorbs/).

27.2 *Biochemistry Composition* of *Antidesma montanum*

The major trace elements found in *Antidesma montanum* were Si, Fe, Mn, Sr, Cr, and Zn, which among 18 trace elements identified by microwave digestion/ inductively coupled plasma emission spectroscopy (ICP-AES) (Dan et al. 2012a). Thus, exploiting and utilizing of this plant may provide some pharmaceutical value out of its trace elements (Dan et al. 2012a).

27.3 Bioactive Compounds of *Antidesma montanum*

The methanolic extract of *Antidesma montanum* produced the highest yield when sequential extraction was carried out using hexane, ethyl acetate and methanol. The highest phenolic and flavonoid contents were noticed in methanol extract, followed by ethyl acetate and hexane extracts, respectively. Notably, phenols are bioactive secondary metabolites molecules that highly presence in more polar extraction solvent (Ratnadewi et al. 2018).

Another study by Dan et al. (2011) showed that the mixtures extracted from *Antidesma montanum* tree with purified water, 95% ethanol and petroleum ether contain effective constituents such as flavonoid, lactone, alkaloids, phenols, anthraquinone, and volatile oil as confirmed by chemical reaction identification method. Besides, alkaloid, terpene, saponin, tannin, glycoside, and quinone were present in 80% ethanol extracts from *Antidesma montanum* Blume (Elya et al. 2012). On other hand, sixteen compounds were found in the essential oil of *Antidesma montanum* extracted by hydrodistillation (HD). Their major components which has identified by gas chromatography mass spectrometry (GCMS) were 9-octadecenoic acid, n-hexadecanoic acid, and 9,12-octadecadienoic acid,(Z,Z) (Dan et al. 2012b). Earlier studies have reported some compounds from *Antidesma montanum* (formerly known as *Antidesma acuminatum, Antidesma pentandrum and Antidesma menasu*) to be canophyllal, canophyllol, friedelin, n-tritiacontane, and antidesmanol (3-keto-16a-hydroxy-friedelan) (Rizvi et al. 1980a). Also, Rizvi et al. (1980b) reported the presence of 3b-hydroxy-16-ketoisomultiflorene and 16a-hydroxy-3-ketoisomultiflorene in the same plant. The compounds of carpusin, geraniin and tannins antidesmin A were mentioned by Namba et al. (1991) and Yoshida et al. (1992), meanwhile two cyclopeptide alkaloids (Arbain and Taylor 1993) and lupeolactone (Kikuchi et al. 1983) were also found in *Antidesma montanum*. Interestingly, the antidesmone was stable over long periods as the compound was detectable in a sample of *Antidesma montanum* from the year 1935 (Buske et al. 2002).

27.4 Uses of Antidesma Montanum

The ripe fruit with a sour flavor are eaten locally as a pickle (*Das* et al. 2018). *Besides, the fruit is also used to adulterate black pepper* (http://www.nationaalherbarium.nl/euphorbs/). The wood can be brown, red, orange, yellow or white. The grain is fine; there is no discernible odour or taste. The wood is hard, somewhat brittle. It is used to build a house, to make as utensils for rice pounders and used for fuel (http://www.nationaalherbarium.nl/euphorbs/).

27.4.1 *Medicinal Uses of* Antidesma montanum

Antidesma montanum Blume has being used traditionally by the Indonesian indigenous people for healing of diabetic. Scientifically, the extract fraction of *Antidesma montanum* Blume leaves (i.e. hexane, ethyl acetate, and methanol) was reported to inhibit α-amylase and α-glucosidase activities (Ratnadewi et al. 2018). The highest inhibition activity of α-amylase was shown in ethyl acetate extract in comparison to the hexane and methanol extracts of *Antidesma montanum*. Meanwhile, the highest inhibition against α-glucosidase activity was noticed in hexane extract. Notably, the inhibition level of α-amylase and α-glucosidase activities by the extracts did not always correlate to the total phenol and flavonoid contents, thus indicated that non-phenolic and non-flavonoid bioactive constituents might be correlated to those inhibition activities (Ratnadewi et al. 2018). On the other study of α-glucosidase activity assay, 80% ethanol extract from *Antidesma montanum* had IC_{50} values of 2.83 μg/mL, which lower than 117.20 μg/mL acarbose (Elya et al. 2012). Furthermore, the *Antidesma montanum* leaves extract showed scavenging activities proportionally to the phenolic and flavonoid contents in three different antioxidant experiments (i.e. diphenyl-1-picrylhydrazyl (DPPH) assay, hydroxyl radical scavenging activity and superoxide anion radicals). In the DPPH assay, the ethyl acetate extract of *Antidesma montanum* had higher scavenging activity in comparison to its methanol and hexane extracts. In both hydroxyl radical and superoxide anion radicals scavenging activities, methanol extracts of *Antidesma montanum* indicated higher inhibition than its less polar extracts of hexane and ethyl acetate, respectively (Ratnadewi et al. 2018). Besides, 50% aqueous EtOH of *Antidesma montanum* leaves extract (0.1 mg/mL) inhibited 53.97% of nitric oxide production when activated with lipopolysaccharide and interferon-γ in a murine macrophage-like cell line, RAW264.7. At this concentration, no cytotoxity on the macrophage cells was noticed as determined by 3-(4,5-dimethylthiazol-2-yl)-2,5-diphenyl tetrazolium bromide (MTT) assay (Liao et al. 2005). Furthermore, the maximum tolerable dose (MTD) of *Antidesma montanum* ethanolic extracts in mice was found at 495.2 g/kg, thus indicated that this ethanolic extracts have no obvious toxicity (Li et al. 2012).

The Folin-Ciocalteu and aluminium chloride method were used to estimate the total polyphenol and flavonoid contents of *Antidesma montanum* fruits, respectively. As a result, more polyphenols were found in the fruits rather than flavonoids. In addition, tannins, polyphenols, flavonoids, saponins and steroid glycosides are the bioactive constituents that presence in the *Antidesma montanum* fruits. The fruits also exhibited higher antioxidant activity than trolox (1000 μM), ascorbic acid (50 μg/mL) and vitamin E (Myra E), as tested through DPPH assay (Barcelo 2015).

Meanwhile, the anti-inflammatory effect of *Antidesma montanum* was noticed from its ethyl acetate fraction, which could significantly inhibit xylene-induced ear edema and acetic acid-induced celiac capillary permeability increase (Li et al. 2012). *Antidesma montanum* Blume is a medicinal plants used by the people around Limu Mountains in China for the treatment of eye diseases (Zheng et al. 2013). The *Antidesma montanum* Blume root was prepared as tea and the leaves were made into paste before applying on chest for relieving chest pain (Kulip 2003). Roots of *Antidesma montanum* in water is used internally to treat measles, chickenpox and malaria, and the leaves were applied externally against headache and thrush in children (Eswani et al. 2010), for diuretic and removing kidney stone (Sunilchandra et al. 2008), anti-dermatitis and skin disease curing effect (Biswas et al. 2009). A tea from the *Antidesma montanum* leaves is used as a tonic for mothers after giving birth, and applied topically on lumbar pains and ulcers, while the roots are used in the treatment of stomach ache, and the fruits are used as a tonic for mothers after giving birth (http://www.nationaalherbarium.nl/euphorbs/).

References

Arbain D, Taylor WC (1993) Cyclopeptide alkaloids from *Antidesma montana*. Phytochemistry 33:1263–1267

Barcelo R (2015) Phytochemical screening and antioxidant activity of edible wild fruits in Benguet, Cordillera administrative region, Philippines. Electron J Biol 11(3):80–89

Biswas TK, Mitra A, Hazra A, Paul S, Debnath PK (2009) Medicinal plants used in Ayurveda. National Medicinal Plants Board. Govt. of India. pp. 3–63

Buske A, Schmidt J, Hoffmann P (2002) Chemotaxonomy of the tribe Antidesmeae (Euphorbiaceae): antidesmone and related compounds. Phytochemistry 60:489–496

Dan Z, Li H, Lianbo L, Zhaohui A (2011) Preliminary study on chemical components of *Antidesma montanum*. China Medical Herald, 13

Dan Z, Ke ZW, Yong LW, Li H (2012a) Determination of the trace elements in *Antidesma montanum*. J Hainan Med Coll 18(3):308–309

Dan Z, Zhao-Hui A, Juan L, Lian-Bo L (2012b) Analysis of the chemical constituents of essential oil extracted from *Antidesma montanum*. Lishizhen Medicine and Materia Medica Research, 1

Das KK, Dotaniya CK, Swamy GSK, Bunker RR (2018) Under utilized and under exploited fruits of Tripura – A review. Int J Pure App Biosci 6(1):1641–1644. https://doi.org/10.18782/2320-7051.5379

Elya B, Basah K, Mun'im A, Yuliastuti W, Bangun A, Septiana EK (2012) Screening of α-glucosidase inhibitory activity from some plants of Apocynaceae, Clusiaceae, Euphorbiaceae, and Rubiaceae. J Biomed Biotechnol 2012:281078, 6 pages. https://doi.org/10.1155/2012/281078

Eswani N, Abd Kudus K, Nazre M, Awang Noor AG (2010) Medicinal plant diversity and vegetation analysis of logged over hill forest of Tekai Tembeling Forest Reserve, Jerantut, Pahang. J Agric Sci 2(3):189

Hoffmann P (1999a) New taxa and new combinations in Asian *Antidesma* (Euphorbiaceae). Kew Bull 54:347–362

Hoffmann P (1999b) The genus *Antidesma* (Euphorbiaceae) in Madagascar and the Comoro Islands. Kew Bull 54:877–886

http://asianplant.net/Phyllanthaceae/Antidesma_montanum.htm. Retrieved on 20 May 2019

http://www.nationaalherbarium.nl/euphorbs/. Retrieved on 21 May 2019

http://www.nationaalherbarium.nl/Euphorbs/specA/Antidesma.htm. Retrieved on 21 May 2019

https://commons.wikimedia.org. Retrieved on 20 May 2019

Kikuchi H, Tensho A, Shimizu I, Shiokawa H, Kuno A, Yamada S, Fujiwara T, Tomita K-I (1983) Lupeolactone, a new beta-lactone from *Antidesma pentandrum* Merr. Chemistry Letters 12:603–606

Kulip J (2003) An ethnobotanical survey of medicinal and other useful plants of Muruts in Sabah, Malaysia. Telopea 10(1):2003

Li H, Yue-Li L, Xiao-Meng W, Dan Z (2012) Acute toxicity and anti-inflammatory effect of *Antidesma montanum*. Lishizhen Medicine and Materia Medica Research 6:30

Liao J-C, Lin K-H, Ho H-Y, Peng W-H, Yao X, Kitanaka S, Wu J-B (2005) Inhibitory effects of 87 species of traditional chinese herbs on nitric oxide production in RAW264.7 macrophages, activated with lipopolysaccharide and interferon-γ. Pharm Biol 43(2):158–163. https://doi.org/10.1080/13880200590919500

Namba O, Toshida T, Lu C, Yang L, Yen K-Y, Okuda T (1991) Antidesmin A, a new dimeric ellagitannin from *Antidesma pentadrum*. Planta Med 57(Suppl. 2):A125–A126

Pax F, Hoffmann K (1922) Euphorbiaceae-Phyllanthoideae-Phyllantheae. In: Engler A (ed) Das Pflanzenreich 81. Wilhelm Engelmann, Leipzig, pp 1–349

Radcliffe-Smith A (2001) Genera Euphorbiacearum. Royal Botanic Gardens, Kew

Ratnadewi AAI, Wahyudi LD, Rochman J, Susilowati A, Nugraha S, Siswoyo TA (2018) Revealing anti-diabetic potency of medicinal plants of Meru Betiri National Park, Jember – Indonesia. Arab J Chem. in press. https://doi.org/10.1016/j.arabjc.2018.01.017

Rizvi SH, Shoeb A, Kapil RS, Popli SP (1980a) Antidesmanol - a new pentacyclic triterpenoid from *Antidesma menasu*. Experientia 36:146–147

Rizvi SH, Shoeb A, Kapil RS, Popli SP (1980b) Two diuretic triterpenoids from *Antidesma menasu*. Phytochemistry 19:2409–2410

Sunilchandra U, Vijaykumar M, Sravanthi P (2008) Phytomedicines in veterinary practice. North East Veterinary 8(2):4–7

Webster GL (1994) Synopsis of the genera and suprageneric taxa of Euphorbiaceae. Ann Mo Bot Gard 81:33–144

Yoshida T, Namba O, Yang C-F, Lu L-L, Yen K-Y, Okuda T (1992) Tannins of Euphorbiaceous plants. X. Antidesmin A, a new dimeric hydrolyzable tannin from *Antidesma pentandrum* var. *barbatum*. Chem Pharm Bull 40:338–342

Zheng XL, Wei JH, Sun W, Li RT, Liu SB, Dai HF (2013) Ethnobotanical study on medicinal plants around Limu Mountains of Hainan Island, China. J Ethnopharmacol 148(3):964–974. https://doi.org/10.1016/j.jep.2013.05.051

Chapter 28
Calamus manillensis: Phytochemical Screening and Antioxidant Activity

Suzy Munir Salama

28.1 Introduction

In rural communities, indigenous people are considerably consuming the fruits of wild edible plants for their significant nutritional values and health benefits than any other cultivated plant species (Rodríguez 2014). Wild fruits can be considered as an interesting sources of nutrients and biologically-active ingredients that can be used in the industry of functional foods (Heinrich et al. 2006). At the pharmaceutical level, phytochemical screening plays important role in producing therapies against various ailments (Thakuria et al. 2018). Metabolic processes inside the different parts of plants result in the formation of bioactive secondary metabolites that increase the defensive abilities of plants against stress factors such as environmental conditions, pesticides and pathogens (Thakuria et al. 2018). Polyphenols including flavonoids, tannins, and phenolic acids are group of phytochemical compounds that possess significant biological activities such as antioxidant and antimicrobial (Denev et al. 2019), anti-inflammatory (Chen et al. 2019) and antiviral activities (Li et al. 2019). Natural and supplementary antioxidants taken from fruits and vegetables play pivotal role as primary line in optimizing health care and protecting against various diseases (Rani 2019).

S. M. Salama (✉)
Department of Molecular Medicine and Biomedical Science, Faculty of Medicine, University of Malaya, Kuala Lumpur, Malaysia
e-mail: suzymunir@um.edu.my

A. A. Mariod (ed.), *Wild Fruits: Composition, Nutritional Value and Products*,
https://doi.org/10.1007/978-3-030-31885-7_28

Fig. 28.1 *Calamus manillensis* edible fruits. (Source:http://www. commons.wikimedia.org)

28.1.1 Plant Distribution and Description

The genus *Calamus* is a group of rattan palms that belong to the family Arecaceae (Govaerts and Dransfield 2005) and native to the African, Australian and Asian tropical and sub-tropical forests (Govaerts and Dransfield 2005; Shengji et al. 2010). In Bangladesh, rattan fruits are one of the indigenous fruits that are rich in Vitamins A and C, and minerals such as Iron, Magnesium, Zinc, Calcium and Potassium. They are used in treating many diseases such as jaundice, dysentery and diabetes (Rahman and Rahman 2014). *Calamus manillensis*(Mart) (Fig. 28.1) that is commonly known as lituko is one of these species that is native to Philippines (Guangtian et al. n.d.). Filipinos used to cultivate this palm tree in their houses for its edible fruits which they traditionally use to treat coughing (Fernández-Ruiz et al. 2017). The fruit is eatable, utilized as topping, handled into wine, vinegar and sharp flavorings. The pulp or the sarcotesta is extricated and cooked with sugar to make confections. Whole fruits are additionally cured in vinegar, once in a while in saline solution. The fruit scales are gathered, leveled, dried and utilized for inventive impacts and accents for bamboo vases, pencil holders and ash plates.

28.2 Phytochemistry

The preliminary tests conducted on the presence of secondary metabolites (Aguinaldo et al. 2004) in the aqueous extract of *C. manillensis* fruits showed that it contains saponins, flavonoids, polyphenols, steroids and tannins (Fig. 28.2). The confirmatory tests oftotal flavonoids of the same extract recorded 5.9 mg Quercetin/100 g of fresh weight of fruitscompared to Quercetin as a standard. Additionally, Folin Ciocalteu test revealedthat the fruit extract is rich in polyphenol content. However, the total polyphenol constituent of the fruit extract was higher that its total flavonoid content (Barcelo 2015).

28.3 Antioxidant Activity

The antioxidant activity of the plant extract is directly related to its content of polyphenols which structures determine the chemical and biological properties of the phenolic compound constituents (Quideau et al. 2011). Based on Barcelo, 2015, the DPPH scavenging activity of the aqueous extract of *C. manillensis* fruits revealed significance compared to ascorbic acid and trolox along with insignificance compared to Vitamin E (Barcelo 2015). The correlation between the DPPH scavenging activity of the aqueous extract of *C. manillensis* and its polyphenol and flavonoid content reported negative correlation between the antioxidant power of the fruit extract and its total flavonoid content ($r = -0.06$) along with a positive correlation to its polyphenol content ($r = 0.50$) attributing the antioxidant activity of *C. manillensis* to the polyphenol contents of its fruit extract (Barcelo 2015).

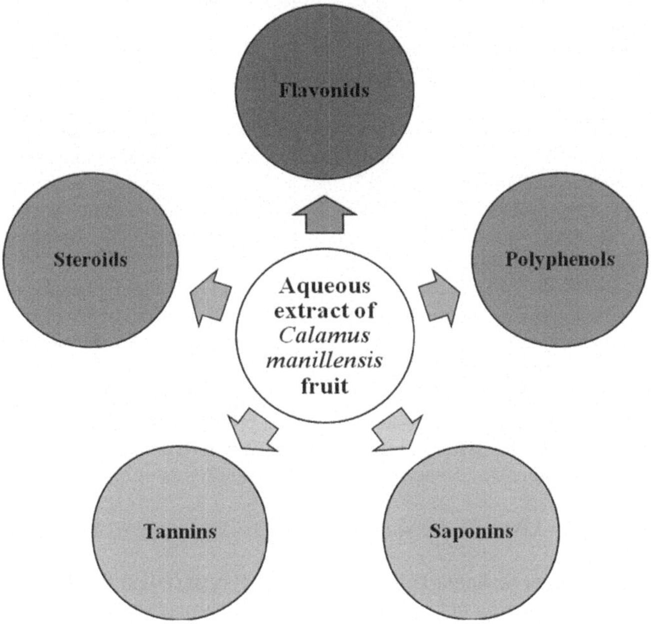

Fig. 28.2 Phytochemical screening of the aqueous extract of *Calamusmanillensis*fruits

28.4 Conclusion

The high polyphenol content of the aqueous extract of *Calamusmanillensis*fruits was the main phytochemical ingredient investigated by researchers till date. In addition, the highest antioxidant activity of the fruit extract was attributed to that significant polyphenol constituent. More studies are required to be conducted on isolating, characterizing and evaluating the different active bioactive compounds from the various plant extracts of *Calamusmanillensis*.

References

Aguinaldo A, Espeso E, Guevara B, Nonato M (2004) A guidebook to plant screening: phytochemical and biological: botany section. Research Center for the Natural Sciences, University of Santo Tomas, Philippines, pp 24–50

Barcelo RC (2015) Phytochemical screening and antioxidant activity of edible wild fruits in Benguet, Cordillera administrative region, Philippines. Electron J Biol 11(3):80–89

Chen Y, Zhang H, Liu R, Mats L, Zhu H, Pauls KP et al (2019) Antioxidant and anti-inflammatory polyphenols and peptides of common bean (*Phaseolus vulga* L.) milk and yogurt in Caco-2 and HT-29 cell models. J Funct Foods 53:125–135

Denev P, Číž M, Kratchanova M, Blazheva D (2019) Black chokeberry (*Aronia melanocarpa*) polyphenols reveal different antioxidant, antimicrobial and neutrophil-modulating activities. Food Chem 284:108–117

Fernández-Ruiz V, Morales P, Ruiz-Rodríguez BM, TorijaIsasa E (2017) Nutrients and bioactive compounds in wild fruits through different continents. Wild plants, mushrooms and nuts: functional food properties and applications, pp 263–314

Govaerts R, Dransfield J. World checklist of palms: Royal Botanic Gardens; 2005

Guangtian Y, Huangcan X, Weiliang Z, Jinggang F, Bingshan Z. Cultivation of rattan species. Notes.15(3.5):21.0

Heinrich M, Nebel S, Leonti M, Rivera D, Obón C (2006) 'Local Food-Nutraceuticals': bridging the gap between local knowledge and global needs. Local Mediterranean Food Plants and Nutraceuticals. Karger Publishers 59:1–17

Li H, Li M, Xu R, Wang S, Zhang Y, Zhang L et al (2019) Synthesis, structure activity relationship and *in vitro* anti-influenza virus activity of novel polyphenol-pentacyclic triterpene conjugates. Eur J Med Chem 163:560–568

Quideau S, Deffieux D, Douat-Casassus C, Pouysegu L (2011) Plant polyphenols: chemical properties, biological activities, and synthesis. Angew Chem Int Ed 50(3):586–621

Rahman M, Rahman J (2014) Medicinal value and nutrient status of indigenous fruits in Bangladesh. Nova J Med Biol Sci 2(6):1–19

Rani K (2019) Clinical role of antioxidants in the treatment of diseases. Nutraceutical and Functional Foods in Disease Prevention. IGI Global, pp 392–401

Rodríguez BMR (2014) Frutos silvestres de uso tradicional en la alimentación: evaluación de su valor nutricional, compuestos bioactivos y capacidad antioxidante. Universidad Complutense de, Madrid

Shengji P, Sanyang C, Lixiu G, Henderson A (2010) Arecaceae (Palmae). Flora of China 23:132–157

Thakuria P, Nath R, Sarma S, Kalita D, Dutta D, Borah P et al (2018) Phytochemical screening of medicinal plants occurring in local area of Assam. J Pharmacogn Phytochem 7(3):186–188

Chapter 29
Debregeasia longifolia: Biochemistry, Functions and Utilization

Nuha Mohamed Osman Mahmoud

29.1 Introduction

29.1.1 Plant Description

Debregeasia longifolia (Burm. F) Wedd., belongs to the family Urticaceae, is a shrub grown in a wide range in Asia, in India, Sirlanka, Japan, China, and Indonesia (Wilmot–Dear, 1989). Leaves are tight, serrulate, with green upper surface, glabrous when develop, have a white-tomentose on lower surface. Flowers orchestrated in axillary branched, conservative heads. Shrestha, (2014) described the tree as small tree about 5 m high, have whitish flower, it occurs in moist shady places (Fig. 29.1). Fruits are orange-yellow when ripe (Fig. 29.2). Usually in wet shaded valleys (Upreti et al. 2010). The plant is portrayed by woody stem, interchange leaf, catkin inflorescence with dense, white bloom. It is open fundamentally in the autumn time season (Padmakumar et al. 2015).

29.1.2 Plant Distribution

The plant sustains to grow in hard climate and has the ability to grow in acidic water, and become prevailing in East Java, Indonesia (Hapsari et al. 2014). The plant can also grow on the surface of rocks near the entrance of a dry limestone cave on the upper part of a karst hill in the county-level city of western Guangxi, China (Chen et al. 2008). Wang et al. (2018) studied the plant species abundance in the main forest on karst soils in China and revealed that *Debregeasia longifolia* is found in both forms of Deciduous tree or shrub. This tree is also grown beside the stream

N. M. O. Mahmoud (✉)
Sudanese Standards and Meteorology Organization, Khartoum, Sudan

A. A. Mariod (ed.), *Wild Fruits: Composition, Nutritional Value and Products*,
https://doi.org/10.1007/978-3-030-31885-7_29

Fig. 29.1 *Debregeasia longifolia* tree. (Source: https://commons.wikimedia.org)

Fig. 29.2 *Debregeasia longifolia* fruits (Source: https://commons.wikimedia.org)

sides in evergreen forests (Rajilesh et al. 2016). *Debregeasia longifolia* known locally as Awukhunain in Zunheboto, Nagaland, Northeast India, it spread from middle to higher height (600-2000 m) it is a cultivated crop (Sumi and Shohe 2018). In Nagaland, India the tree also is known as Orange Wild Rhea, it is abundant from June–November (Khruom and Deb 2018). It is a wild weed, it is also known as the pan and nicchia by the tribes of Tripura, northeast India (Majumdar and Datta 2013).

29.2 Chemical and Biochemical Constituents

Kumar et al. (2015) analyzed the polyphenls of *debregeasia longifolia* leaves using ultra high performance liquid chromatography (UHPLC) and found that the presence of chlorogenic acid (6.25 ug/g dw), gallic acid (0.02 ug/g dw), catechin (0.39 ug/g dw), coumaric acid (0.29 ug/g dw), epicatechin (1.99 ug/g dw), caffeic acid (3.09 ug/g dw), rutin (46.91 ug/g dw) umbelliferone (0.8 ug/g dw), kaempferol (16.07 ug/g dw), ellagic acid (0.93 ug/g dw) and total polyphenols (76.79 ug/g dw).

Debregeasia longifolia fruits grown in India were extracted using four different solvents (chloroform, benzene, acetone and methanol) and estimated the efficiency of each solvent on the antioxidants activity of the fruit. Flavonol, total phenolic and flavonoid contents of each extract were determined as well as 2,2′-azino-bis (3-eth ylbenzothiazoline-6-sulphonic acid) ABTS radical scavenging ability, reducing power capacity and 1,1-diphenyl-2-picryl hydrazyl (DPPH) radical scavenging activity. The results revealed the fruit as a potential source of antioxidant. Methanol is found to be more efficient in extraction, compared to other solvents (Seal and Chaudhuri, 2015).

Radhamani and Britto (2016), studied the preliminary phytochemicals and their free radical scavenging activities of both ethanolic extracts of *Debregeasia longifolia* leaf and stem and discovered that both extracts contains flavonoids, phenols, tannins, alkaloids and saponin, they also evaluated the role of both extracts in preventing conformation of hydroxyl radicals, 1, 1-diphenyl 2-picrl – hydrazyl free radical (DDPH), lipid peroxidation and superoxide anions and concluded that both ethanolic extract were prospective sources of natural antioxidants.

Devi et al. (2016a) studied the abundance of both ascorbic acid (vitamin C) and tocopherol (vitamin E) in *Debregeasia longifolia* and found that ascorbic acid content was found to be (26.49) while the tocopherol content was (3.36) and concluded that probably the cause of their using as healing traditionally is due to their anti-inflammatory property. Devi and Chongtham (2016) analyzed the methanolic extracts of *Deprgeasia longifolia* leaves and found that the DPPH radical scavenging activity was 65.41 inhibition, the reducing power ability was 1.66 as a ascorbic acid equivalents. The total phenol content was 72.11 mg/100 g as catechol equivalents. The flavonoid content was 45.15 mg/100 g as quercetin equivalent, the carotenoids content was 2.56 mg/100 g and the total alkaloids were 18.07 mg/100 g as caffeine equivalents. For that reason, it can be supportive in the prevention and healing of oxidative stress induced inflammatory diseases like diabetes, cardiovascular diseases, cancer, arthritis, gout, neurodegenerative diseases, respiratory tract infections and skin disorders. Besides the free radical scavenging activity, these plants also have high reducing power which provides the potentiality or ability of these medicinal plants to reduce and scavenge free radicals (Devi and Chongtham 2016).

29.3 Nutritional Value

The main compounds noticed in*Debregeasialongifolia* leaves USING GC-MS were cis-jasmone (19.03%) and benzene propanoic acid, 3, 5-bis (1,1-dimethyl ethyl) -4-hydroxy-, methyl ester (14.98%) (Kumar et al. 2016). The proximate analysis of *Debregeasia longifolia* leaves was the dry matter was found to be 91.42%, the ash content 18.65%, the ether extract (2.86%), the crude fiber (10.74%) and the crude ash (19.56%), P (1.98 $ug.g^{-1}$) Zn (30.18 $ug.g^{-1}$), Fe (753.4 $ug.g^{-1}$), Mg (4.031 $ug.g^{-1)}$, Ca (51.050 $ug.g^{-1)}$, Na (41.80 $ug.g^{-1}$), K (8.11 $ug.g^{-1}$), and S (2.131 $ug.g^{-1}$) (Lihong et al. 2007). The chemical contents of this plant are found to be: (19.15%) crude protein, (62.60%) neutral detergent fiber, (45.51%) acid detergent fiber and (8.48%) ADL while the total phenolic compounds was to be (2.52%) (Padmakumar et al. 2015).

The Proximate composition of *Debregeasia longifolia* fruit is analyzed by Seal and Chaudhuri (2014) and revealed that the fruit contains (16.19%) ash, (65.56%) moisture, (2.39%) crude fat, (1.26%) crude fiber, (11.99%) protein, (68.15%) carbohydrate and nutritive value (342.15) kcal/100 g, Na (0.38), K (20.46), Ca (19.49), Mn (0.12), Cu (0.016), Fe (3.000), Mg (0.91) and Zn (0.51) mg /g.

Nazarudeen (2010) found that the fruits of *Debregeasia longifolia* in Kerala, India contains 81.72% moisture, 3.01% protein, 2.31% fat, 7.15% reducing sugar, 2.7% non-reducing sugar, 9.9% total sugar, 2.0% fiber, 0.9% ash, 3.9 vitamin C, 7.3 mg/100 g iron, 8.5 mg/100 g sodium, 193.5 mg/100 g potassium and 72.51 Kcal energy content. Devi et al. (2016b) found that the calcium content of *Debregeasia longifolia* leaves was found to be 117.77 mg/100 g, the selenium content was 100.48 mg/100 g and the iron content was 112.37 mg/100 g and they concluded that this plant probably is a prospective resource of natural antioxidants.

29.4 Uses of *Debregeasia longifolia*

The Ethnobotanical details of *Debregeasia longifolia* were studied within 64 species in South India and revealed that its fine fiber taken from the bark and used as string to sew clothes (Sajeev and Sasidharan 1997). It is abundant as tree in India, its fruits and barks are used as shampoos and for digestion by the Sumi tribe in Zunheboto district in Nagaland (Jamir et al. 2015). The tree tender leaves are taken as a vegetable during dysentery, while the crushed leaves paste is used as poultice for arthritis (Majumdar and Datta 2013). The ripped fruits are also eaten raw (Khruom and Deb 2018). *Debregeasia longifolia* (Sindari) leaves are used as a veterinary medicine precisely for diarrhea and flatulence in Poonch Valley Azad Kash (Khan et al. 2012). The leaf of *debregeasia longifolia* was used as a treatment for Scabies (Shah et al. 2014). The tree which is known as Zangrushing found in Tree form in Tshothang Chiwog, Lauri Gewog, Bhutan forests, its fruits and leaves juice was used as a treatment for skin scabies (Jamba and Kumar 2018). This use for skin scabies was also reported by Shrestha, (2014). It is bark is used as shampoos by local tribes (Folk and Sidha system) and the fruits are palatable and used to enhance digestion (Sumi and Shohe 2018). In Indonesia the tree is known as daramanuk the bark is used as building material, and as firewood (Hidayat 2017). Purba et al. (2018) reported that, the ethnic group (Batak Karo) utilize the leaves of this shrub in a combination with 28 other plants for preparing their traditional food (Terites), which is considered as a therapy for digestive disorders. Kichu et al. (2015) reported that the leaf decoction of *Debregeasia longifolia* (Natsulawa) is taken orally by Chungtia villagers to treat diabetes, fever and high blood pressure and sometimes boiled leaves used as food while bark is used as a cure for bone fractures in Kumaun Himalaya, India (Joshi et al. 2010). The fruits are used as birds food (Srivastava 2009). Leaves are used as feed for pigs by farmers (Padmakumar et al. 2015). The tree of *Debregeasia longifolia* sold in local markets of Ukhrul in India for their remedial values and their leaves are cooked with pork (Salam et al. 2012). Chungtia villagers mix *Debregeasia longifolia* with *C. pareira* and *Gynura crepidioides* plants to treat diabetes disease as mentioned by Kichu et al. (2015). *Debregeasia longifolia* is used as a cure of skin illnesses, indigestion, and sunburn (Sameer et al. 2014). It is also used as bone fracture as well as enhancing lactation in animals in Himalaya in India (Pande et al. 2007). In Hariyali the fodder used in making ropes. It is also used as a treatment for scabies (Singh et al. 2017). *Debregeasialongifolia* leaves extract showed inhibition of the growth of *Staphylococcus aureus* using micro broth dilution but it has no effect when using the agar diffusion method as reported by Mariani et al. (2014).

References

Chen WH, Möller M, Shui YM, Zhang MD (2008) A new species of Paraboea (Gesneriaceae) from a karst cave in Guangxi, China, and observations on variationsin flower and inflorescence architecture. Bot J Linnean Soc 158:681–688

Devi OA, Chongtham C (2016) In-vitro antioxidant profile and reducing power of five medicinal plants of Manipur having Anti- inflammatory properties. Int J Current Res 8(05):31402–31408

Devi OA, Das M, Saikia A, Das P (2016a) Ascorbic acid and tocopherol content of ten medicinal plant extracts of Manipur having anti-inflammatory properties. Int J Home Sci 2(1):308–312

Devi O, Das A, Saikia M, Das A, Sharm PD (2016b) Evaluation of total mineral, calcium, selenium, iron content of ten medicinal plant extracts of Manipur having anti-inflammatory properties. J Med Plants Stud 4(3):189–194

Hapsari L, Basith A, Novitasiah HR (2014) Inventory of invasive plant species along the corridor of KawahIjen Nature Tourism Park, Banyuwangi, East Java. J Indo Tour Dev Stud 2(1):1–9

Hidayat S (2017) The use by local communities of plants from Sesaot Protected Forest, West Nusa Tenggara, Indonesia. Biodiversitas 18(1):238–247

Jamba N, Kumar BM (2018) Medicinal plants in the broad- leaf mixed coniferous forest of Tshothang Chiwog, Bhutan: Floristic Attributes, Vegetation Structure, Ethnobotany, and Socioeconomic Aspects. Front Env Sci 5(96):1–13

Jamir HK, Tsurho K, Zhimomi A (2015) Some indigenous medicinal plants and its uses in Zunheboto District. Nagaland Int J Dev Res 05(08):5195–5200

Joshi M, Kumar M, Bussmann RW (2010) Ethnomedecinal uses of plant resources of the Haigad watershed in Kumaun Himalaya, India. Medicinal and Aromatic Plant Science and Biotechnology 4(Special issue 1). Global Science books):43–46

Khan MA, Khan MA, Hussain M (2012) Ethno veterinary medicinal uses of plants of Poonch Valley Azad Kashmir. Pak J Weed Sci Res 18(4):495–507

Khruom N, Deb CR (2018) Indigenous wild edible fruits: Sustainable Resources for Food Medicine and Income Generation – A study from Nagaland, India. J Exper Bio Agri Sci 6(2):405–413

Kichu M, Malewska T, Aktera K, Imchen I, Harrington D, Kohen J, Subramanyam R, Vemulpad A, Joanne F, Jamie AN (2015) An ethnobotanical study of medicinal plants of Chungtiavillage, Nagaland, India. J Ethnopharmacol 166:5–17

Kumar V, Sharma A, Bhardwaj R, Thukral AK (2015) Polyphenol profiling in the leaves of plants from the Catchment Area of River Beas, India. Int J Pharm Bio Sci 6(4):(B) 1005–(B) 1012

Kumar V, Sharma A, Kumar A, Thukral B (2016) Rphytochemical profiling in the leaves of some medicinal plants. Int J Pharm 6(1):124–128

Lihong W, Liu L, Qian H, Jinguo Z, Li Z (2007) Analysis of nutrient components of food for Asian Elephants in the wild and captivity. Front Biol China 2(3):351–355

Majumdar K, Datta BK (2013) Practice pattern of traditional pharmaceutical formulations by the Tribes of Tripura, Northeast India. Glob J Pharm 7(4):442–447

Mariani R, Sukandar EY, Suganda AG (2014) Antimicrobial activity from Indonesian Urticaceae. Int J Pharm and Pharmaceut Sci 6(4):191–193

Nazarudeen A (2010) Nutritional composition of some lesser- known fruits used by the ethnic communities and local folks of Kerala. Indian J Tradit Knowl 9(2):398–402

Padmakumar V, Haque N, Sirie R, Khate K, Deka R, Rutsa V, Solomon K (2015) Availability and nutritional value of wild forages as feed for pigs and Mithun in Nagaland, India. ILRI Project Report. International Livestock Research Institute (ILRI), Nairobi

Pande PC, Tiwari L, Pande HC (2007) Ethnoveterinary plants of Uttaranchal, a review. Indian J Trad Know 6(3):444–458

Purba EC, Nisyawati, Silalahi M (2018) "terites" a traditional Batak Karo medicinal food: A ruminant's stomach content as a human food resource. J Ethnic Foods 5(2):114–120

Radhamani T, Britto JS (2016) Screening of preliminary phytochemicals and their free radical scavenging activities of ethanolic extracts of *Toddalia asiatica* (L.) Lam., *Debregeasia longifolia* (Burm. F.) Wedd. and *Polygala arillata* Buch.-Ham Ex Don. Int J Eng Res App 6(1, (Part – 4):):151–160

Rajilesh VK, Anoop KP, Madhusoodanan PV, Ansari R, Prakashkumar R (2016) A Floristic analysis of the aquatic, Marshy & Wetland plants of Idukki District, Kerala. India Int J Plant Animal Envir Sci 6(2):55–65

Sajeev KK, Sasidharan N (1997) Ethnobotanical observations on the tribes of Chinnar Wildlife Sanctuary. Anc Sci Life XVI(4):284–292

Salam S, Jamir NS, Singh PK (2012) Wild leafy vegetables sold in local markets of Ukhrul District of Manipur, India. Pleione 6(2):298–303

Sameer SM, Manigandan P, Sivaraj C, Sekar Babu H, Arumugam P, Sindhu S (2014) Antioxidant and antiproliferative activities of methanol extract. Int J Pharmacog Phytochem 6:567–572

Seal and Chaudhuri (2015) Antioxidant activities of five wild edible fruits of Meghalaya State in India and the effect of solvent extraction system. Int J Pharmac Sci 6(12):5134–5140

Seal T, Chaudhuri K (2014) Ethnobotanical importance and nutritional potential of wild edible fruits of Meghalaya State in India. J Chem Pharmac Res 6(10):680–684

Shah S, Ram J, Pala NA, Tripathi P, Kumar M (2014) Medicinal plant wealth of oak dominated forests in Nainital catchment area of Uttarakhand. Acad J Medicinal plants 2(1):006–013

Shrestha SS. (2014) Floristic study and vegetation analysis of Shivapuri national park central Nepal. Ph.D. Thesis submitted to the central department of botany institute of science and technology. Tribhuvan University Kathmandu, Nepal

Singh S, Youssouf M, Malik ZA, Bussmann RW (2017) Sacred groves: myths, beliefs, and biodiversity conservation—a case study from Western Himalaya, India. Int J Ecol 2017:1–12

Srivastava RC (2009) Adi community. Indian J Trad Know 8(2):146–153

Sumi A, Shohe K (2018) Ethnomedicinal Plants of Sumi Nagas in Zunheboto District, Nagaland, Northeast India. Acta Sci Pharmac Sci 2(8):15–21

Upreti K, Tewari LM, Pangtey YPS, Jalal JS (2010) Diversity and distribution of wild edible fruit plants of Uttarakhand. In: Biodiversity potentials of the Himalaya. Gyanodaya Prakashan, Nainital, pp 157–196

Wang J, Wen X, Zhang X (2018) Supplement of the strategies of water–carbon regulation of plants in a subtropical. Supp Biogeosci 15:4193–4203

Wilmot-Dear CM (1989) Debregeasia: a correction. Kew Bull 44:702

Chapter 30
Azanza garckeana L.: Distribution, Composition, Nutritive Value, and Utilization

Abdel Moneim El-Hadi Sulieman

30.1 Sudanese Wild Fruits

Sudan is viewed as one of the biggest countries in Africa with a region which includes diverse biological zones from the desert in the North to the tropical rainforest in the South (Ngambi et al. 2013). The savanna territories are immense, and occupy no less than 37% of the whole land in Sudan, and which are living spaces for various plant species (El-Amin 1990). The fruits of forest plant species assumed a vital part in the diet of people in the savannah belt of Sudan, particularly during food shortages and starvations (Kordufan and Darfur; 1983/84). During this time, the forest fruits were the primary ingredients given as a signal of hospitality and liberality (Abdel Muti 2002). The physical and financial availabilities of such food make it incredible significance in adapting to the unfriendly food conditions (FAO 1991). Likewise amid farming harvests off-seasons wild food plays a significant part in securing food and giving a substitution source of wage to the rustic family units (Msuya et al. 2010; Balemie and Kebebew 2006; Ali-Shtayeh et al. 2008; Ibrahim et al. 2012; Ogle et al. 2003). With respect to woody wild food trees, gathering for non-timber forest products, in general, were found of less natural constrained.

The local people of Sudan consume many kinds of edible fruits and appreciate few edibles over the others in their use. The most consumed types include; *Balanites aegyptiaca*, *Ziziphus spina-christi*, *Tamarindus indica*, *Adansonia digitata*, *T. indica*, *Grewia tenax*, *Sarcocephalus latifolius*, *Vangueria madagascariensis* and *Azanza garkeana L.* fruits, of most of the recorded species, are generally expended new as

A. M. El-Hadi Sulieman (✉)
Department oaf Biology, Faculty of Science, University of Hail,
Hail, Kingdom of Saudi Arabia

A. A. Mariod (ed.), *Wild Fruits: Composition, Nutritional Value and Products*,
https://doi.org/10.1007/978-3-030-31885-7_30

snacks without further processing or consumed as juices or prepared into porridges (Salih and Ali 2014; Sulieman et al. 2012)

With regard to financial significance, the market perceptions showed high attractiveness of some wild fruits, particularly the ones that have high eatable inclinations. Fruits from *A. digitata, B. aegyptiaca, Z. spina-christi* and *T. indica* were financially imperative with high attractiveness (Salih and Ali 2014). The other edibles were periodically sold at a nearby market. The majority of the nearby families are wild fruits collectors, to different degrees. Some of the population assembles for their own use just, while others gathers to be purchased too. The offer of the wild fruits was found to add to 50%, 60% and 75–100% of the aggregate yearly family income in Wad-Abid, Abu-Karshola and Rashad, respectively.

30.2 Jakjak Plant

Jakjak plant belongs to Malvaceae family and has got many other names. Snot apple, Morojwa, African chewing gum Jakjak or *Azanza garckeana* is an important consumable indigenous fruit tree species limited to east and southern Africa. This plant is found in Botswana, Kenya, Malawi, Mozambique, Namibia, South Africa, Tanzania, Zambia, Sudan and Zimbabwe (FAO 1983; Taylor and Kwerepe 1995). The species develops naturally in various sorts on Jabl Mara and streams under the Jabl Mara and backwoods of savanna. Increase to follow south, Southern Darfur state (Sudan) and South, Eastern Darfur state (Sudan), the level of tree relied upon annual rainfall. The plants of Jakjak trees are found on or close savanna region or close deserted areas (Personal contact 2010).

Jakjak is one of the important plant species that develops in Sudan. It is to a great extent spread in arid territory, such as sand and close mountain, exceptionally at Southern Darfur and Jabel Mara mountain region. It is the indigenous fruit trees although undomesticated assume various crucial roles in individuals living in provincial zones of Darfur states (Sudan). Indigenous fruit trees are critical conventional sources of nuts, fruits, spices, leafy vegetables, edible oil and beverages (Taylor 1995).

The Jakjak fruit might be eaten ready (biting) crisp or is kept for later use, since it comprises of the extraordinary extent of sugars, vitamins and minerals. Additionally, it is utilized as sauce before ripe, as utilized as porridge (madeda) and utilized as juice and for school boys chewing.

30.3 Description of *Azanza garckeana*

Azanza garckeana (Jakjak) tree develops naturally in savanna wooded grasslands, open forests. It is widespread in tropical eastern and southern Africa. Typically related tree species include *Berchemia discolor, Cassia abbreviata, Cassia singueana*, *Combretum molle*, *Dalbergia melanoxylon*, *Ehretia spp.*, *Grewia mollis*, and *Tamarindus indica*. The tree is evergreen in the hotter regions yet semi-deciduous in colder areas. *Azanza garckeana* is drought resistant yet flourishes with abundant water during the rainy season. It can withstand smooth ice (FAO 1983).

Jakjak tree can develop to a stature of 3–15 m depending upon nature in which it is creating. The tree is multi-stemmed with straight or odd stem. The bark is harsh and grayish-dull to dark colored, fibrous with longitudinal fissure. The twigs are bushy when youthful yet end up smooth with age and branches have wooly hairs. The leaves are particularly adjusted 8 by 12 cm on long stalks. They are constantly simple, alternate, and roundish. The leaves have 3–5 lobes, which are canvassed in star-molded hairs, and have longitudinal crevices. Blooms are large up to 6 cm long, solitary on long pedicels in axils of upper most leaves (Fig. 30.1).

The flowers have evident compartment like calyx with which the calyx is interlaced. The flowers have various stamens and five petals, which are yellow or purplish in color with diminish purple or dark red centre (Palgrave 2000). The flowers are androgynous with every flower part in five's (FAO 1983). They are demonstrate yellow, maturing to purplish in color with a maroon patch at the base of each petal. It flowers in wet season and fruits in dry season (Fig. 30.2).

The fruits are round and woody, 2.5–4 cm in diameter long with short hairs. The fruits are divided into 4–5 segment. They are yellowish to green and furry when mature and negligently dehiscent. In spite of their hard woody nature the fruits are palatable and bitten like a chewing gum delivering a sweet glutinous slime. Fruits start maturing from August. The seeds are hemispherical in shape, up to10 cm long and 7 cm thick with woolly floss (Mojeremane and Tshwenyane 2004).

Fig. 30.1 Azanza *Garckeana* plant

Fig. 30.2 *Azanza garkeana* flower. (Source: Dinesh Valke from Thane, India – Azanza, CC BY-SA 2.0, https://commons.wikimedia.org/w/index.php?curid=51587153)

30.4 Harvesting of Jakjak Fruits

Many of techniques are utilized to harvest fruits from wild developing trees in Sudan. The techniques include the picking of fruits from the beginning abscission, climbing trees to pick fruits, tossing items to oust fruits, hitting stems with heavy objects and shaking stems of branches to dislodge fruits. These crude techniques of harvesting fruits couldn't just harm the trees, yet in addition cause excessive wounding to the fruit that would diminish time span of usability and quality (Kadzere et al. 2002) (Fig. 30.3).

30.5 Marketing of Jakjak Fruits

Survey of local Market at Southern Darfur state Nyala (Sudan) reported that (2010) there is no characterized or formal mechanism for the setting of Jakjak fruits market prices. Producers relegate costs in view of data from individuals who have recently visited the market, from previous season costs and from figuring collecting costs. Retailers and wholesalers reported that they determine fruit prices costs as indicated by advertising costs and what their neighbours are charging.

Fig. 30.3 *Azanza garkeana* fruit. (Source: (**a**) https://commons.wikimedia.org/w/index.php?curid=51581453. (**b**) Personal photo)

30.6 Processing and Handling of Jakjak Fruit

The fruit is part open by hand to discharge the seed. Subsequent to cleaning, the seeds are dried in the sun. Ten kilograms of fruits deliver 1 kg of seeds or around 4000 seeds (http://www.worldagroforestry 2005). The fruit can be stored for up to 1 month without losing its sweetness.

30.7 Pests and Diseases

The tree gets plagued with leaf containers (Cicadellidae family) in both nursery and field. Control measures incorporate use of malathion and dichlorophos. The tree is a host to the cotton stainer and ought to thusly not be planted in cotton-creating territories.

30.8 Cytotoxicity of *Azanza garkeana*

Mshelia et al. (2016) and Omosa et al. (2016) investigated cytotoxicity activities of *A. garckeana*. Mshelia et al. (2016) found that the concentration killing 50% (LC_{50}) of the shrimps was 3.98 μg mL^{-1} for acetone extract, methanol extract exhibited LC_{50} of 47.66 μg mL^{-1}, ethyl acetate extracts (LC_{50} of 100 μg mL^{-1}), water extracts (LC_{50}

of 138.04 μg mL^{-1}) and petroleum ether extract exhibited LC_{50} value of greater than 1000 μg mL^{-1}. While Omosa et al. (2016) found that dichloromethane and methanol extract of *A. garckeana* stem bark displayed cytotoxicity towards leukemia CCRF-CEM cells with IC_{50} value of 85.0 μg mL^{-1}. They found that based on the results got from cytotoxic assessments show the likelihood that some plant parts of A. garckeana might be harmful or contain some cytotoxic compounds (Mshelia et al. 2016).

30.9 Nutritive Value of Jakjak Fruit

Chemical composition Jakjak (*Azanza garkeana*) has been documented by many investigators. Saka et al. (1994) reported that Jakjak has a high level of soluble carbohydrates (35.2%), fiber (45.3%), dry matter (52.8%), protein (12%), fat (1%), and ascorbic acid (21.5 mg/100 g). He also reported that the contents of the minerals P, Ca, Mg, Fe, K, and Na were 147.6 mg/100 g, 9.5 mg/100 g, 145.6 mg/100 g, 8.4 mg/100 g, 2619 mg/100 g, and 20.2 mg/100 g, respectively.

Sulieman et al. (2012) in his investigation the Nutritive Value of Jakjak (*Azanza garckeana*-L), he found that the contents of moisture, ash, protein, fat, total carbohydrate of Jakjak fruit study were 13.54%, 7.3%, 10.05%, 1.04% and 22.44%, respectively. The energy value of Jakjak fruits was 2313.081 Kk/100 g. Sulieman et al. (2012) also found the mineral contents Fe, K, Ca and Na were 6 mg/100 g, 1360 mg/100 g, 56 mg/100 g and 60 mg/100 g, respectively.

Djarova et al. (1998) investigated the nutritive value of some fruits including *Azanza garkeana,* he found that the fruit contained 9.6% protein and 3.61 mg/100 g ascorbic acid.

Mojeremane and Tshwenyane (2004) found that the ascorbic acid, fat, total carbohydrate, calcium, magnesium, iron potassium and sodium contents were 20.5 mg/100 g, 1.1%, 35.2%, 95 μg/g, 1453 μg/g, 84 μg/g, 26,190 μg/g and 202 μg/g in Botswana, respectively. Consumable parts of 16 indigenous wild fruits of Malawi were analyzed by Kalenga et al. (1991) for protein, fibre, carbohydrate and minerals (Ca, Mg, Fe, P, K and Na) and the most astounding mineral substance was found for *Adansonia digitata* and the least vitality content was found with *Azanza garkeana,* (810 kJ/100 g) (Table 30.1).

Nkafamiya et al. (2016) found that *Azanza garkeana* fruit is the most proteinous part (12.0%) and furthermore with the highest moisture content (6.50%), while the stem-bark is the least (4.91%, protein and 0.5% moisture content). The stem-bark contains the highest carbohydrate content (72.16%), nearly followed by roots (70.81%), while the minimum carbohydrate content is acquired from the fruits (28.40%). Crude fiber content was observed to be most elevated in fruits (45.00%) followed by leaves (25.00%), stem-bark (13.75%) and roots (11.89%). The highest total ash content was found in the leaves (11.00%), and the least in the fruits (6.70%) (w/w).

The vitamin content of the fruits and leaves of *A. garkeana* is exhibited in Table 30.2. The vitamin content except for vitamin C is higher contrasted with those

Table 30.1 Chemical composition of *Azanza garckena* fruits

Parameter	Value
Dry matter	52.8%
PH (at 2.5C)	5.96
Ascorbic acid	20.5 (1.8) (mg/100 g of fresh weight)
Crude protein	12.0%
Fat	1.1%
Fibre	45.3%
Total carbohydrate	35.2%
Energy value	810 kJ/100 g
Phosphorus	1476 μg/g
Calcium	95 μg/g
Magnesium	1453 μg/g
Iron	84 μg/g
Potassium	26,190 μg/g
Sodium	202 μg/g

Source: Saka et al. (1994)

Table 30.2 Vitamin content (mg/100 g) of *A. garckeana* fruits and leaves

Fruits/leaves	A	B1	B2	C	E
Fruits	75.00 ± 0.23	1.28 ± 0.97	1.18 ± 0.45	319.09 ± 0.45	3.08 ± 0.55
Leaves	28.75 ± 0.66	1.00 ± 0.67	0.95 ± 0.78	98.02 ± 0.65	2.09 ± 0/77

found in the fruits of *C. congoensis* and *N. latifolia*. Utilization of the fruits and leaves of *A. grackaena* can serve as vitamin supplements. Presence and measure of vitamin A and E in the fruits and leaves presents them as a potential solution for symptoms associable with the insufficiency of the vitamins e.g. visual deficiency and peroxidation. The value of vitamin C in both fruits and leaves of *A. garckeana* is low compared to the fruits of *C. congoensis* (410.50 ± 0.32 mg/100 g ripe) and *N. latifolia* (309.00 mg/100 g ripe). Despite the fact that the value of vitamin C of *A. garckeana* is low compared to those of the fruits of *Congoensis* and *N. latifolia*, the fruits and leaves may demonstrate vital in the counteractive action of scurvy and reduce side effects of regular cool (Nkafamiya et al. 2006).

The amino acid composition of *A. garckeana* is presented in Table 30.3. Essential amino acids were present in both the fruits and leaves (Nkafamiya et al. 2016). Leucine had the highest value for both the fruits and the leaves, while glycine had the lowest.

The difference in most of the chemical components because of contrasts in climate or presumably the soil environment in which it is developed. Jakjak fruit is a decent source of sugar that is required by an individual to keep the body healthy. Generally *Azanza garckeana* has a high mineral content additionally proximate chemical composition with the exception of the fat content which has a low value.

Table 30.3 Amino acid composition of the fruits and leaves of *A. garckeana* (g/100 g)

Amino acid Fruits Leaves	Fruits	Leaves
Alanine	3.30	4.00
Arginine	7.01	7.69
Cysteine	3.00	3.66
Glycine	1.00	1.23
Glutamic acid	10.79	11.09
Histidine	3.67	4.00
Isoleucine	4.98	4.98
Leucine	12.01	12.97
Lysine	11.78	12.85
Methionine	2.00	2.78
Phenylalanine	8.00	9.00
Proline	4.00	4.78
Serine	3.97	4.00
Threonine	4.78	4.97
Tyrosine	4.89	4.99
Valine	6.00	6.76

Table 30.4 Non-nutritional constituents of *Azanza garkeana*

Part of plant used	Alkaloids % (w/w)	Flavonoids % (w/w)	Saponins % (w/w)	Tannins % (w/w)	Phenols % (w/w)
Fruits	18.40 24.40	24.40	ND	15.05	ND
Leaves	13.60	26.50	30.00	ND	29.00
Roots	6.80	ND	24.50	ND	ND
Stem bark	12.80	ND	34.50	ND	ND

Source: Nkafamiya et al. (2016)

30.10 Non-nutritional Components of Jakjak

The non-nutritional component of the fruits, leaves, stem-bark and roots are presented in Table 30.4 as reported by Nkafamiya et al. (2016). These compounds can restrict the wide use of many tropical plants because of their pervasive occurrence as natural compounds capable of evoking pernicious impact on man and animals (Nkafamiya et al. 2010; Osagie 1998). Alkaloids, steroids, cardiovascular glycosides, terpenes, resins and saponins are available in the plant parts (fruits, leaves, roots and stem-bark). Volatile oils are available in the fruits, leaves and the roots. Flavonoids are available in the fruits and leaves. Phenols and tannins are available only in leaves and fruits. Non-nutritional compounds above some threshold intake can form an indigestible complex by binding to mineral elements. Oxalate for example tends to render calcium inaccessible by binding to the calcium ion to form complexes (calcium oxalate precious stones) which hinders the absorption and usage of the calcium, and may likewise hasten around the renal tubules accordingly causing renal stones (Nkafamiya et al. 2010).

30.11 Phytochemical Composition of *Azanza garckeana*

Numerous classes of mixes including alkaloids, amino acids, ascorbic acid, carotenoids, cyanogenic glucosides, flavonoids, lipids, phenols, saponins, and tannins have been secluded from *A. garckeana* fruits, leaves, roots, seeds and stem bark (Nkafamiya et al. 2015; Kanene 2016; Saka and Msonthi 1994). Zhang et al. (2015) contended that phytochemicals, for example, alkaloids, flavones, saponins, steroids, tannins and triterpenoids isolated from fruits, vegetables and grains apply a defensive impact against the development chronic illnesses, for example, cardiovascular diseases (CVD), diabetes and cancers. Nkafamiya et al. (2015) isolated amino acids from fruits and leaves of *A. garckeana* with aspartic acid, glutamic acid, leucine and lysine being the most bounteous amino acids establishing 9.67–12.97 g/100 g.

Letcher and Shirley (1992) isolated the following compounds from heartwood of *A. garckeana*, O-naphthoquinones 6, mansonones E 7, mansonones F 8, mansonones G 9, mansonones H 10, azanzone A 11 and azanzone B 12. Mutindi (2014) isolated the following phenolic compound disesquiterpene aldehydes from the crude root extract of A. garckeana, gossypol 1, 6, 6-Dimethoxygossypol 2, 6-Methoxygossypol 3, stigmasterol 4, E–docosyl 3-(3,4-dihydroxyphenyl) acrylate 5 and betulinic acid 13. Masila et al. (2015) isolated gossypol 1, 6, 6-Dimethoxygossypol 2, 6-Methoxygossypol 3 from the root extract of *A. garckeana* and the stem bark yielded stigmasterol 4, E-Docosyl-3-(3,4-dihydroxyphenyl) acrylate 5 and betulinic acid 13 (Table 30.5).

30.12 Product Uses of Jakjak

Ripe fruit carpels of Jakjak are palatable and have an energy content of 8.10 kJ/g. The fruit contains sweet mucilage which comes out upon chewing. All parts of Jakjak plant have their uses which include Maghembe et al. (1994):

1. The fruit might be eaten crude whenever gathered green and delicious and the skin is peeled off can likewise be bubbled, it is broadly utilized as a relish or to make porridge.
2. The leaves make a relish or can be singed to deliver salts.
3. Fodder: Browsed by amusement and in the dry season by cows
4. Fuel: Provides profitable kindling
5. Fiber: Good quality rope can be produced.
6. The profound dark colored mottled wood is utilized for making bows, device handles, and little household items,
7. Medicine: Adecoction is produced using the roots and taken orally for difficult feminine cycle and to treat hacks and chest torments.
8. An mixture produced using the roots and leaves is dropped into the ear to regard ear infection or taken orally as an antiemetic (Saka et al. 1994).

Table 30.5 Phytochemical chemical compounds isolated from *Azanza garckeana*

Compounds	Extract	Plant parts	References
Gossypol ethyl acetate in n-hexane; methanol in dichloromethane root	Ethyl acetate in n-hexane; methanol in dichloromethane	Root	Mutindi (2014), Masila et al. (2015)
6,6-Dimethoxygossypol	Ethyl acetate in n-hexane; methanol in dichloromethane	Root	Mutindi (2014), Masila et al. (2015)
Stigmasterol	Ethyl acetate in n-hexane; methanol in dichloromethane	Root	Mutindi (2014), Masila et al. (2015)
Acrylate	Ethyl acetate in n-hexane; methanol in dichloromethane	Root and stem bark	Mutindi (2014), Masila et al. (2015)
O-naphthoquinones	n-hexane	Heart wood	Letcher and Shirley (1992)
Mansonones F	n-hexane	Heart wood	Letcher and Shirley (1992)
Mansonones G	n-hexane	Heart wood	Letcher and Shirley (1992)
Mansonones H	n-hexane	Heart wood	Letcher and Shirley (1992)
Azanzone A	n-hexane	Heart wood	Letcher and Shirley (1992)
Azanzone B	n-hexane	Heart wood	Letcher and Shirley (1992)
Betulinic acid	Ethyl acetate in n-hexane; n-hexane; methanol in dichloromethane	Fruit pulp, root, stem bark	Dikko et al. (2016), Mutindi (2014), and Masila et al. (2015)

9. Shade or shelter: gathering of Jakjak trees can be planted in camps where shade is required for cows. It likewise makes an appealing greenhouse shade tree.
10. Decorative: Azanza garckeana makes a productive and fascinating pot plant anyway ought to be kept in full sun.

30.13 Uses of Jakjak in Pharmacological Activities

Jakjak (*A. garckeana)* has many pharmacological activities such as antibacterial (Dikko et al. 2016; Mutindi 2014; Masila et al. 2015), antifungal (Dikko et al. 2016; Mutindi 2014; Masila et al. 2015), antihyperglycemic (Amuri et al. 2017), antimalarial (Connelly et al. 1996), antioxidant (Mshelia et al. 2016) and iron absorption (Ahmed et al. 2016) activities.

30.13.1 Antibacterial

Mutindi (2014) evaluated antibacterial activities of *A. garckeana* roots and pure compounds isolated from the roots against many bacterial species. Compounds gossypol 1, 6, 6-Dimethoxygossypol 2 and 6-Methoxygossypol 3 exhibited antibacterial activities against *Enterococcus faecalis* and *Enterococcus faecium* with half maximal inhibitory concentration (IC_{50}), minimum inhibitory concentration (MIC) and minimum bactericidal concentration (MBC) values ranging from 0.89 to 20 μg mL^{-1} [43]. Gossypol 1 exhibited antibacterial activity against *Staphylococcus aureus* with IC_{50} value of 6.98 μg mL^{-1} [43]. In addition, Masila et al. (2015) evaluated antibacterial activities of compounds gossypol 1, 6, 6-Dimethoxygossypol 2 and 6-Methoxygossypol 3 isolated from *A. garckeana* against *Enterococcus faecium* and *Staphylococcus aureus* using ciprofloxin, methicillin and vancomycin as controls. He found that compound gossypol 1 exhibited strong activity against Enterococcus faecium with IC_{50}/MIC/MBC values of 1.71/4.82/19.31 μM. Compounds 6, 6-Ttramethoxygossy pol 2 and 6-Methoxygossypol 3 were less active with IC_{50}/MIC/MBC values of 2.73/4.70/9.40 μM and 6.14/18.32/18.32 μM against *Enterococcus faecium.* Compound gossypol 1 demonstrated modest activities against *Staphylococcus aureus* with IC_{50} value of 9.15 μM (Masila *et al.* 2015). In another study, Dikko et al. (2016) assessed antibacterial activities of fruit pulp ethyl acetate, n-hexane and methanol extracts of *A. garckeana* against *Enterococci, Escherichia coli, Helicobacter pylori, Proteus mirabilis, Pseudomonas aeruginosa* and *Staphylococcus aureus* using agar diffusion method. Ethyl acetate fraction of *A. garckeana* was the most active with MIC value of 0.625 mg mL^{-1} against *Escherichia coli,* while MIC and MBC values of fractions against the rest of bacteria species ranged between 1.25 and 2.5 mg mL^{-1} (Dikko et al. 2016).

30.13.2 Antifungal Activity

Azanza garckeana crude root extract exhibited strong antifungal activity of 100% inhibition against *Candida glabrata* at a concentration of 50 μg mL^{-1}. Compound 6-Methoxygossypol 3 of *Azanza garckeana* root extract exhibited strong antifungal activity against *Candida glabrata* with IC_{50} value of <0.8 μg mL^{-1} while gossypol 1 exhibited activity against *Candida glabrata* with IC_{50} value of 3.2 μg mL^{-1} (Mutindi 2014). Similarly, Masila et al. (2015) found that gossypol 1, 6, 6-Dimethoxygossypol 2 and 6-Methoxygossypol 3 isolated from *A. garckeana* had antifungal activities against *Candida glabrata* using amphotericin B as control. Compound gossypol 1 demonstrated modest activities against *Candida glabrata* with IC_{50} values of 0.73 μM (Masila et al. 2015). In another study, Dikko et al. (2016) revealed antifungal activities of fruit pulp ethyl acetate, n-hexane and methanol extracts of *A. garckeana* against *Candida albicans, Candida krusei* and *Candida tropicalis* using agar diffusion method.

30.13.3 Antihyperglycemic

The aqueous leaf extracts of *A. garckeana* exhibited hypoglycemic and antihyperglycemic activities by administering 500 mg kg^{-1} to guinea pigs (*Cavia porcellus*), both in glucose baseline conditions and in oral glucose tolerance test with follow-up over 210 min. (Amuri et al. 2017). In oral glucose tolerance test, *A. garckeana* was active with inhibition of glycemia increase of 36.9% compared with the hyperglycemic inhibition rate of glibenclamide (50%). This data support the traditional use of *A. garckeana* leaf decoction as herbal medicine for diabetes.

30.13.4 Antimalarial

The aqueous and organic fractions of *A. garckeana* against *Plasmodium falciparum. Azanza garckeana* demonstrated weak antimalarial activity with middle inhibitory concentration which was >3 µg mL^{-1} (Mutindi 2014). Antimalarial assessments conducted by Connelly et al. (1996) showed weak activities yet such discoveries may suggest that *A. garckeana* has bioactive constituents with potential in controlling mosquito vectors.

30.13.5 Antioxidant

Mshelia et al. (2016) assessed antioxidant potential of petroleum ether, ethyl acetate, acetone, methanol and water stem bark extracts of *A. garckeana*. He found that the methanol stem bark extracts exhibited antioxidant activity with IC_{50} value of less than 100 µg mL^{-1} while acetone extracts exhibited activity with IC_{50} value of 160 µg mL^{-1} against the standard ascorbic acid activity with IC_{50} value of 220 µg mL^{-121}. These antioxidants activities of stem bark are probably due to the presence of flavonoids and phenolics (Ndhlala et al. 2006). There is presently a worldwide pattern towards the utilization of common phenolics as antioxidants and useful fixings because of their apparent safety and predominance in wild edible fruits (Wu et al. 2013).

30.14 Jakjak Fruit Utilization in Sudan

The Jakjak fruits are collected from the trees upon ripening. Jakjak fruits may be eaten ripe (chewing) or left to dry for later consumption. A delicious Jakjak fruit is prepared by soaking or boiling and mixed the fruits, then hand pressed, sieved and sweetened. It can also be utilized for juice production as meal in Ramadan holy month. Or mixed with flour to make porridge. A light porridge called (madeda) is also prepared by addition of flour and custard to the Jakjak fruits extract (Sulieman et al. 2012).

In common, Jakjak is utilized during drought periods. In this regards, Jakjak was used by people of Western Darfur states as famine food especially when crops are not ready for harvesting (Sulieman et al. 2012). Similar indigenous fruits of Jakjak fruits were used at famine foods in Darfur state (Sudan) and other counties in Africa. Like *Grewia tenax* (Gudeim), *Bosciacoriacea* (mkhet), *Cordiasinensis* (andrab), *Tamarindus indica* (aradib), *Baobad* (tbldy) and others. These fruits assisted rural community. This all enhanced occupation in rural area, however today in urban territories likewise, as a result of nutritional value and commitment to the health status of rural communities. Despite the fact that there is limited trade of the fruit products at present, there is a broad potential scope for commercial and processing of these fruits Sulieman et al. (2012).

References

Abdel Muti OMS (2002) Nutritive value of wild plants of the Sudan. Arab J Food Nut 3(3):6–67

Ahmed RH, El Hassan MS, El Hadi HM (2016) Potential capability of *Azanza garckeana* fruits aqueous extract on enhancement of iron absorption in Wistar albino rats. Int J Adv Res Biol Sci 3:245–250

Ali-Shtayeh MS, Jamous RM, Al-Shafie JH, Elgharabah WA, Kherfan FA, Qarariah KH, Khdair IS, Soos IS, Musleh AA, Isa BA, Herzallah HM, Khlaif RB, Aiash SM, Swaiti GM, Abuzahra MA, Haj-Ali MM, Saifi NA, Azem HK, Nasrallah HA (2008) Traditional knowledge of wild edible plants used in Palestine. J Ethnobiol Ethnomed 4:13

Amuri B, Maseho M, Simbi L, Okusa P, Duez P, Byanga K (2017) Hypoglycemic and antihyperglycemic activities of nine medicinal herbs used as antidiabetic in the region of lubumbashi (DR Congo). Phytother Res 31:1029–1033

Balemie K, Kebebew F (2006) Ethnobotanical study of wild edible plants in Derashe and Kucha Districts, South Ethiopia. J Ethnobiol Ethnomed 2:53

Connelly MPE, Fabiano E, Patel IH, Kinyanjui SM, Mberu EK, Watkins WM (1996) Antimalarial activity in crude extracts of Malawian medicinal plants. Ann Trop Med Parasitol 90:597–602

Dikko YJ, Khan ME, Tor-Anyiin TA, Anyam JV, Linus, UA (2016) In vitro antimicrobial activity of fruit pulp extracts of Azanza garckeana (F. Hoffm.) Exell & Hillc. and isolation of one of its active principles. Betulinic Acid British Journal of Pharmaceutical Research 14(1):1–10

Djarova T, Kudanga T, Chirimba F (1998) Quantitative determination of protein, lipids and ascorbic acid in indigenous legumes and fruits of Zimbabwe. Trans Zimbabwe Sci Assoc 72:7–10

El-Amin MH (1990) Trees and shrubs of the Sudan. 1stst. Ithaca Press, England, p 106, 400, 451

FAO (1983) Food and Fruit bearing forest species. Examples from eastern Africa. Food and Agricultural Organisation, Forestry Paper 44/1. Rome

FAO (1991) Household food security and forestry an analysis of socio-economic issues. FAO, Rome

http://www.worldagroforestry (2005)

Ibrahim HA, Ali SGY, Halliru N, Usaini S, Abdullahi II (2012) Ethnobotanical survey of the wild edible food plants consumption among local communities in Kano State, North – West, Nigeria. Int J Sci Technol 2(10):713–717

Kadzere I, Hove L, Gatsi T, Masarirambi MT, Tapfumaneyi L, Maforimbo E, Magumise I (2002) Current status of post-harvest handling and traditional processing of indigenous fruits in Zimbabwe. Agroforestry Impacts on Livelihoods in Southern Africa: Putting Research into Practice. Is it journal or book?

Kalenga MK, de Gasparo M, De Hertogh R, Whitebread S, Vankrieken L, Thomas K (1991) Angiotensin II receptors in the human placenta are type AT1. Reprod Nutr Dev 31:257–267
Kanene KM (2016) Indigenous practices of environmental sustainability in the Tonga community of southern Zambia. Jamba: J Disaster Risk Stud 8:331. https://doi.org/10.4102/jamba.v8i1.331
Letcher RM, Shirley IM (1992) O-Naphthoquinones from the heartwood of Azanza garckeana. Phytochemistry 31:4171–4172
Maghembe JA, Kwesiga F, Ngulube M, Prins H, Malaya FM (1994) Domestication potential of indigenous trees of the miombo of southern Africa. In: Leakey RRB, Newton AC (eds) Tropical trees: the potential for domestication and rebuilding of forest resources. HMSO, London
Masila VM, Midiwo JO, Zhang J, Gisacho BM, Munayi R et al (2015) Anti-vancomycin-resistant Enterococcus faecium and E. faecalis activities of (-)-gossypol and derivatives from Thespesia garckeana. Nat Prod Commun 10:613–616. Check as above reference
Mojeremane W, Tshwenyane SO (2004) Azanza Garckeana: A valuable edible indigenous fruit tree of Botswana. Pak J Nutr 3(5):264–267. Check as above reference
Mshelia EH, Watirahyel EM, Maigari AU, Yohanna C, Ismail F (2016) Cytotoxicity and antioxidant activity of stem bark extracts of Azanza garckeana (kola of Tula). Eur J Pure Applied Chem 3:16–24. Check as above reference
Msuya TS, Kideghesho JR, Mosha TCE (2010) Availability, preference, and consumption of indigenous forest foods in the Eastern Arc Mountains, Tanzania. Ecol Food Nutr 49(3):208–227. Check as above reference
Mutindi MV (2014) Phytochemical investigation of Harrisonia abyssinica and Thespesia garckeana for antiplasmodial and antimicrobial compounds. M.Sc. Thesis. University of Nairobi, Nairobi. Dikko, Y.J., M.E. Khan, T.A. Tor-Anyiin, J.V. Anyam and U.A.
Ndhlala AR, Mupure CH, Chitindingu K, Benhura MAN, Muchuweti M (2006) Antioxidant potentials and degrees of polymerization of six wild fruits. Sci Res Essay 1:87–92. Check as above reference
Ngambi JW, Alabi O, Norris D (2013) Role of goats in food security, poverty alleviation and prosperity with special reference to Sub-Saharn Africa: a review. Indian J Anim Res 47(1):1–9
Nkafamiya II, Manji AJ, Modibbo UU, Umaru HA (2006) Biochemical evaluation of Cassipourea congoensis (Tunti) and *Nuclea latifolia* (Luzzi) fruits. Afr J Biotechnol 5(24):2461–2463
Nkafamiya II, Osemeahon SA, Modibbo UU, Amina A (2010) Nutritional status of non conventional leafy vegetables, *Ficus asperifolia* and *Ficus cycomorus*. Afr J Food Sci 4:104–108
Nkafamiya II, Ardo BP, Osemeahon SA, Akinterinwa A (2015) Evaluation of nutritional, non-nutritional, elemental content and amino acid profile of Azanza garckeana (Goron Tula). Br J Appl Sci Technol 12:1–10
Nkafamiya II, Ardo BP, Osemeahon SA, Akinterinwa A (2016) Evaluation of nutritional, non-nutritional, elemental content and amino acid profile of *Azanza garckeana* (Goron Tula). Br J Appl Sci Technol 12(6):1–10. 2016, Article no.BJAST.19811
Ogle BM, Tuyet HT, Duyet HN, Dung NNX (2003) Food, feed or medicine: the multiple functions of edible wild plants in Vietnam. Econ Bot 57:103–117
Omosa LK, Midiwo JO, Masila VM, Gisacho BM, Munayi R et al (2016) Cytotoxicity of 91 Kenyan indigenous medicinal plants towards human CCRF-CEM leukemia cells. J Ethnopharmacol 179:177–196
Osagie AU (1998) Anti-nutritional factors. In: Osagie AU, Offiong YE (eds) Nutritional quality of plant foods. University of Benin Press, Nigeria, pp 221–230
Palgrave KC (2000) Trees of Southern Africa. Struik Publishers, Cape Town
Saka JDK, Msonthi JD (1994) Nutritional value of edible fruits of indigenous wild trees in Malawi. For Ecol Manag 64:245–248
Saka J, Msothi JD, Maghembe JA (1994) The nutritional value of edible fruits of indigenous wild trees of Malawi. For Ecol Manag 64:245–248
Salih NKM, Ali AH (2014) Wild food trees in Eastern Nuba Mountains, Sudan: use, diversity, and threatening factors. J Agric Rural Dev Trop Subtrop 115(1):1–7

Sulieman AE, Difa IE, Saleh ZM (2012) The nutritive value of Jakjak (Azanza garkean L.) fruit and it utilization in juice production. Asian J Biol Sci 5(4):209–215

Taylor F, Kwerepe B (1995) Towards domestication of some indigenous fruit trees in Botswana. In: Maghembe JA, Ntupanyama Y, Chirwa PW (eds) Improvement of indigenous trees of miombo woodlands of southern Africa. ICRAF, Nairobi

Wu SB, Long C, Kennelly EJ (2013) Phytochemistry and health benefits of jaboticaba, an emerging fruit crop from Brazil. Food Res Int 54:148–159

Zhang YJ, Gan RY, Li S, Zhou Y, Li AN, Xu DP, Li HB (2015) Antioxidant phytochemicals for the prevention and treatment of chronic diseases. Molecules 20:21138–21156

Chapter 31
Grewia Species: Diversity, Distribution, Traditional Knowledge and Utilization

Rahul Dev, Venkatesan Kannan, M. Suresh Kumar, Devi Dayal, and Rohit Patel

31.1 Introduction

In India, about 120.40 mha area has been classified as degraded and wasteland category of which major part lies in Rajasthan and Gujarat (Kachchh and Saurashtra). The total degraded area in the Gujarat state is 3.12 m ha, (about 16% of the total geographical area (19.68 m ha) of state). As the Rann of Kachchh (Gujarat) lying on tropic of cancer, it experiences an intensely hot and arid climate (Priyanka et al. 2015). However, many indigenous and locally grown underutilized plant species are well adapted to such harsh climatic conditions including salt and drought stress. The genus "*Grewia*" (Family: Tiliaceae) are useful source of food, fodder, fiber, fuelwood, timber and a wide range of traditional medicines which cure the number of diseases. This genus was named in honour of Nehemiah Grew, an English physiologist, very famously known as the "Father of Plant Physiology" (1641–1712) by Carolus Linnaeus. Grewia consists of >400 species of shrubs/small trees and are distributed all over the tropical and subtropical regions of Africa, Asia and Australia (Bredenkamp 2000; Whitehouse et al. 2001). In India, 31 species of genus *Grewia* (includes *Grewia abutilifolia, G. asiatica, G. damine, G. flavescens, G. hirsute, G. optiva, G. rothii, G. serrulata, G. villosa, G. tiliifolia,*

R. Dev (✉) · M. S. Kumar · D. Dayal
ICAR-Central Arid Zone Research Institute, Regional Research Station, Kukma, Bhuj, Gujarat, India
e-mail: Devi.Dayal@icar.gov.in

V. Kannan
ICAR- Central Island Agricultural Research Institute, Port Blair, India

R. Patel
Government Science College-Dhanpur, Dahod, Gujarat, India

A. A. Mariod (ed.), *Wild Fruits: Composition, Nutritional Value and Products*,
https://doi.org/10.1007/978-3-030-31885-7_31

Table 31.1 Mineral composition of *Grewia* species

Parameters (mg/100 g)	*Grewia asiatica*	*Grewia tenax*	*Grewia villosa*	*Grewia flavescens*	*Grewia mollis*	*Grewia tiliaefolia*
K	372	817	966	877	–	1.7
Ca	136	790	536	269	–	1.09
Mn	–	5.1	0.1	0.1	1.89	83
Mg	–	–	–	–	287.06	0.90
Fe	1.08	20.8	29.6	26.9	47.94	270
Cu		1.5	1.2	1.1	0.27	10
Zn		1.9	2.5	1.1	0.37	–

Source: Yadav (1999), Murray et al. (2001), Elhassan and Yagi (2010), Singh et al. (2015), Sambo et al. (2015), Sati and Ahmed (2018)

G. tenax) are recorded and are distributed widely on sandy, clay, rocky and gravelly piedmont plains (BSI 2014). The presence of seven Grewia species (*Grewia damine* Sm., *Grewia flavescens* Juss, *Grewia subincqualis* DC, *Grewia tenax* (Forsk.) Fiori, *Grewia tiliaefolia* Vah and *Grewia villosa* Willd.) are reported in the Kachchh region of Gujarat (Patel et al. 2011). The *Grewia* fruits contain triterpenes, alkaloids, hydrocarbons, fatty acids, and leuco-anthocynidins. The fruit is locally used to make a drink. *Grewia* species are an excellent source of minerals (Table 31.1) particularly iron (Abdelmutti 1991). Relatively little scientific attention has been given to this species, which hampers its sustainable utilization. This paper is an attempt to summarize information with respect to its geographical distribution, morphological diversity, potential food values, fodder utilization, morphology, ethnobotanical and nutritional (Table 31.2) aspects for sustainable utilization of *Grewia* species.

31.2 Taxonomy of *Grewia*

Kingdom: Plantae
Division: Angiospermae
Sub-division: Dicotyledons
Class: Polypetalae
Series: Thalamiflorae
Order: Malvales
Family: Malvaceae (formerly: Tiliaceae)
Subfamily: Grewioideae
Genus: Grewia
Species: *tenax, villosa, flavescens, asiatica, damine, abutilifolia and tiliifolia*

Table 31.2 Nutrient composition of *Grewia* sp fruits (on DW basis)

Parameter	*Grewia asiatica*	*Grewia tenax*	*Grewia villosa*	*Grewia flavescens*	*Grewia mollis*
Protein (%)	1.57	–	–	–	–
Reducing sugar (%)	0.99	29.35	30.19	29.28	25.54
Sucrose content (%)	7.95	27.87	28.55	27.81	25.30
Ascorbic acid (%)	0.0043	0.035	0.048	0.019	0.058
Vitamin B3 (mg/100g)	0.82	–	–	–	–
Ascorbic acid (%)	4.38	0.035	0.048	0.019	0.058
Vitamin A (μg/100g)	419	–	–	–	–
Starch (%)		44.4	38.6	22.8	–

Source: Yadav (1999), Murray et al. (2001), Elhassan and Yagi (2010), Singh et al. (2015), Sati and Ahmed (2018)

31.3 Botanical Overview

31.3.1 Grewia asiatica *L.*

Grewia asiatica is a 4–5 m tall shrub with approximately 5–18 cm long and broad leaves. The flowers are arranged in cymes of several together, whereas, the individual flowers are yellowish in colour with five large (12 mm) sepals and five smaller (4–5 mm) petals and the flower has a diameter of about 2 cm (Sastri 1956). Leaves may be in ovate, suborbicular, acute, subacuminate or cuspidate, sharply and often coarsely, double serrate, subglabrous above, hoary-tomentose beneath and rounded or only slightly cordate at the base. Flower buds are broadly cylindric or clavate, peduncles are axillary, usually many, and long and slender. Bracts are present beneath the pedicles. The fruit is globose, 1.0–1.9 cm in diameter, 0.8–1.6 cm in vertical height, and 0.5–2.2 g in weight and is an edible portion of the plant. Plant flowers appear in January-February and fruits mature in May-June. During ripening, the fruit skin turns from light green to cherry red or purplish red, becoming dark purple or nearly black when fully ripened. The ripe fruit is soft and delicate and is covered with a very thin whitish blush (Tripathi et al. 2010; Kirtikar and Basu 2000; Gupta et al. 2006). The fruit is like a berry and has a sweet and sour acidic taste (Fig. 31.1).

Fig. 31.1 *Grewia asiatica* (**a**: plant; **b**: ripe fruits; **c**: mature fruits; **d**: juice). (Source: Dev et al. 2018b)

31.3.2 Grewia tenax *(Forsk.) Fiori*

G. tenax is a multi-stemmed sub-erect to erect shrub with 3 m tall. Its stem bark is smooth and ash-grey in colour. Leaves 3–5 costate, smooth, with a pointed or rounded apex and toothed margin, hairy on both sides. Flowers are generally white, rarely cream to yellow in colour, occur solitarily or rarely paired. Sepals are greenish and Petals are white in colour. Stamens yellow and numerous, filaments are white coloured slightly shorter than style or sub-equal. Ovary 4-lobed, glabrous or glabrescent; style 10 mm long, glabrous. Fruit is called as drupe usually 1–4 lobed, 8–12 mm wide. Fruit colour varied from orange, yellow to reddish tinge (Abdelmutti 1991; Gebauer et al. 2007; Whitehouse et al. 2001).

The *Grewia tenax* is a typical tropical plant species which can tolerate seasonal drought and withstand temperatures of more than 50 °C (Gebauer et al. 2007), this plant species is not only adapted to high temperatures, salinity and dry conditions, but also has deep roots to stabilize in sand dunes (Saied et al. 2007). The plant plays an effective role in the rehabilitation of wastelands (Teketay 1996). The *G.tenax*

Fig. 31.2 A view of *Grewia tenax* tree and fruits are taken from the experimental field at CAZRI – RRS, Bhuj, Gujarat, India

plant required only 200 mm of annual rainfall of the region and it is also well tolerant to soil salinity, the major problem in many parts of the world. There is a good opportunity for small farmers (women in particular) to take up cultivation of *G. tenax* under Horti-Silvi-pasture system, which can give them an edge by ensuring more earnings due to lack of competition vis a vis fodder security even in hottest months of the year. Plants of this species are having the aggressive root system to hold the soil firmly and thus protect it from water and wind erosion. Leaf litter falling from the shrub further improves the physical and chemical properties of the soil (Orwa et al. 2011). In addition to this, *G. tenax* does not require high inputs and can be grown in marginal and degraded lands and can contribute to increased agricultural production, fodder security, crop diversification and environment protection. But, its domestication was hampered due to limited documentation on the aspects such as production, nutritional value, consumption patterns and its use. Also, very few information is available on its various aspects such as economic benefits and market opportunities, improved planting materials, lack of improved production technologies, which are directly leading to low yields, lack of marketing channels as well as less attention on this species (Akundabweni et al. 2010) (Fig. 31.2).

31.3.3 Grewia villosa *Wild.*

The specific name (*villosa*) is latin, meaning-bearing long weak hairs. *Grewia villosa* locally known as Luska in Kachchh region of Gujarat, it is also known as sand Hairy-Leaf Cross Berry. It is usually a deciduous, much-branched shrub growing from 1 to 3 m height, though it sometimes becomes more tree-like and up to 4.5 m tall (Ruffo et al. 2002). It is a slow growing plant. Leaves are very distinctive covered with pale silky hairs. Leaves are almost round, opposite, cordate to subcordate at the base and more hairy below. Flowers are yellow-red-brown in colour. The inflorescence is cymes umbel, flowers come in small cluster (4–6) (Fig. 31.3), usually dioecious, though they are occasionally bisexual. Therefore, for getting fruits both male and female plants are need to be planted in the field. Each flowers has 5 sepals and petals and 25–30 stamens. Ovary is 4-lobed. Fruit is botanically drupes red-brown in colour and about 1 cm in diameter has 1–2 (4) seeds per in each fruit (Orwa et al. 2011). Fruits has brittle exocarp and concentrated sweet tasting mesocarp, before consumption the brittle exocarp has to be discarded (Ruffo et al. 2002).

Fig. 31.3 A view of *Grewia villosa* tree and fruits are taken from the experimental field at CAZRI – RRS, Bhuj, Gujarat, India

31.3.4 Grewia flavescens *Juss.*

Plant of *Grewia flavescens* is evergreen, frost-resistant with numerous pendulous branches. The specific name *flavescens* refers to the yellowish flower colour. The height of shrub varied from 3 to 8 m. Its branches clearly distinguish it from other Grewia species as they fluted and have a typical square-cross section (Mothogoane 2012). Leaves may be oblong, alternate, hairy, strongly acuminated towards the apex, with a round base, and toothed margins. Inflorescences type is cymes, which contains 2–3 flowers in the leaf axil. Flowers are beautiful, bright yellow, and sometimes fragrant with 5 slender petals often curling backwards (Fig. 31.4). Fruits are brown in colour on maturity, globose in shape, has 1–4 lobes with pubescent surface and edible, sweet, fibrous pulp. Each lobe contains 1–2 seeds and 8–14 mm in diameter. The flowering time is December-March (Mothogoane 2012; Kumar et al. 2017).

Fig. 31.4 A view of *Grewia flavescens* tree and flowers were taken from Tapkeshwari hills of Kachchh district, Gujarat, India

Fig. 31.5 A view of *Grewia damine* tree and flowers were taken from Tapkeshwari hills of Kachchh district, Gujarat, India

31.3.5 Grewia damine *Gaertn.*

G. damine is a shrub or small sized tree with 2–4 m height. Stem and branches are dark grey and rough. Leaves 4.5 × 2 cm, large ovate, upwardly dilated, apex acute to acuminate, base cordate, margins finely serrate. Petioles are densely hairy. The inflorescence is cyme, 2–3 flowered which born in axil. Flowers (1.5 × 1.5 cm) yellow in colour, pale hairy. Sepals are 8–12 mm long and 3 mm wide, oblong, pubescent; petals 2–4 mm, oblong, entire, with a gland at base; stamens indefinite. Ovary 1–2-lobed ovate, densely hairy, stigma two-lobed. Fruit is a drupe, slightly succulent, globose, and almost glabrous. Flowering and fruiting occur in July – September Sasidharan et al. (2018) (Fig. 31.5).

31.3.6 Grewia abutilifolia *Vent. ex Juss.*

Grewia abutilifolia are shrub species which grow up to 3 m high. Leaf lamina are broadly ovate to orbicular or broadly oblong with acute leaf apex. Inflorescence are axillary, pedunculate umbels usually 3-flowered. Flowers are white in colour; green

Fig. 31.6 A view of *Grewia abutilifolia* tree and new leaves were taken from Junagadh district, Gujarat, India

oblong sepals, petals oblong, emarginate or rounded at apex; filaments ca. 4 mm long, glabrous. The ovary ovoid, densely hirsute; styles longer than stamens. The stigma colour is green and has 5 lobes. Fruits usually 2–4 lobed, ca. 0.8 cm across, dark green, brown when dry, hirsute; pyrenes 2–4. Flowering and fruiting occur in throughout the year (Fig. 31.6).

31.3.7 Grewia tiliaefolia *Vahl*

Moderate sized to large deciduous tree up to 12 m height with a bole of 4–6 m and girth of 11.5 m. Barks are fissured, pale brown; young shoots densely pubescent. Leaves are simple, elliptic, ovate or rarely broadly ovate. Inflorescence are axillary, umbellate, 3 or more in a cluster; flowers yellow or creamy-white in colour with reddish or deep yellow anthers, fragrant. Drupes black at maturity, globose, usually bilobed, one to two-celled. It is a light demander, but tolerates shade when it remains in stunted form. It is known to coppice freely and produces root sucker (Luna 2005) (Fig. 31.7).

Fig. 31.7 A view of *Grewia tiliaefolia* tree and flowers were taken from Tapkeshwari hills of Kachchh district, Gujarat, India

31.4 Distribution and Diversity

31.4.1 Grewia asiatica *L.*

31.4.1.1 Distribution

The plants of *G. asiatica* grow widely distributed in tropical and subtropical climates, except at high altitude. It grows best in regions with distinct summer and winter seasons. Due to its very hardy nature and ability to tolerate drought, it is very suitable for arid regions. Its distribution is limited in only Southeast Asia (Bangladesh, Cambodia, India, Laos, Nepal, Pakistan, Philippines, Sri Lanka, Thailand and Vietnam). Its fruit found in all parts of India, especially Punjab, Bihar, Orissa, North of Bengal, Himalayan range, Maharashtra, Gujarat and South India. It can grow in salty regions also and up to a height of 1000 m from sea level.

31.4.1.2 Diversity

G. asiatica is diploid with a somatic chromosome number ($2n = 2\times$) of 32. It is indigenous Southeast Asia. In particular, phalsa is native to India, probably Vadodara, Gujarat (Singh 1974). Since it is mostly propagated through seed, therefore, a lot of variability of phalsa exist in central India, Rajasthan, Bihar, and drier parts of south India, where it is essential to conduct the exploration to collect the promising types.

31.4.2 Grewia tenax *(Forsk.) Fiori*

31.4.2.1 Distribution

Available previous reports (Collenette 1985; Konig 1986; Alrikain 2004; Sharma and Patni 2012), revealed that, *G. tenax* is most common in arid and semi-arid plains, lowlands and lower mountains up to a maximum altitude of about 1250 m from sea level and in regions with mean annual rainfall of 20–100 cm in sandy, rocky and lateritic soils. However, Gebauer et al. (2007) found that its presence up to an elevation of 1850 m above sea level at the upper edge of Wadi Muaydin. *G. tenax* is widely distributed in lowlands and mountains of Arabian countries such as Saudi Arabia and Egypt (Collenette 1985; Chaudhary 1999), United Arab Emirates (Jongbloed et al. 2003), Botswana, Chad, Djibouti, Ethiopia, Iran, Afghanistan, Kenya, Mali, Mauritania, Morocco, Namibia, Niger, Nigeria, Senegal, Somalia, Tanzania, Uganda, Zimbabwe (Al-Hubaishi and Muller-Hohenstein 1984), Oman, Sinai Peninsula, Yemen, Namibia, South Africa, Sudan, Mauritius Pakistan and Sri Lanka (BSI 2014 and GBIF Secretariat 2016). In India, *G. tenax* commonly found in rocky and gravelly piedmont plains of arid and semi-arid areas and abundantly distributed in the arid north-west region of India (Sharma and Patni 2012 and BSI 2014). Plant generally prefers eroded rocky, sandy, gravelly sandy and stony soils of piedmont plain, lower valley and lower hills, in the tropical forest habitats. The origin of *G. tenax* is less clear, however, Alrikain (2004) suggested that the plant may have originated in the southern part of Kordofan in Sudan. It has immense potential as important forage resources for animals and honeybees. Bees and other insects visit the flowers for pollen and nectar, thus, this plant could be used for apiculture. It is mostly found as natural wild in association with other species such as *Grewia villosa*, *Ziziphus nummularia*, *Salvadora oleoides*, *S. persica*, *Prosopis juliflora*, *Acacia senegal*, *Euphorbia caducifolia*, *Premna resinosa* etc, sometimes also seen on the boundaries of farmer's field. Being drought and salinity tolerant species, it is naturally grown in arid lands. It has the potential to meet the growing demand for fodder (Table 31.3) for animals in arid and semiarid regions.

31.4.2.2 Diversity`

Wide genetic diversity in *G. tenax* have been reported by various previous studies in the natural habitat (Alrikain 2004; Gebauer et al. 2007; Sohail 2009) but the systematic exploitation of diverse germplasms is very limited. The comprehensive botanical surveys conducted to collect diverse and valuable germplasm of *G. tenax* species for characterization and utilization from 40 sites of 5 blocks of Kachchh Gujarat

Table 31.3 Fodder value of *Grewia* species

Parameter	*Grewia asiatica*	*Grewia tenax*	*Grewia villosa*	*Grewia flavescens*	*Grewia mollis*	*Grewia hirsute*	*Grewia abutilifolia*	*Grewia tiliaefolia*
Moisture (%)	–	13	15	14	8.00	–	–	–
Fat (%)	–	0.46			1.63	–	–	–
Carbohydrate (%)	21.1	66	75	84	42.30	–	–	–
Crude Fiber (%)	12.50	20.5	42.8	25.5	33.33	–	–	–
Crude protein (%)	–	7.7	8.7	13.32	10.50	15.94	15.9	–
Crude protein(kg/ha)	–	585.01	530.77	–	533.29	–	–	–
Tannin content (%)	–	0.33	0.27	0.20	0.28	–	–	–
Phytic acid (%)	–	0.089	0.089	0.177	0.11	–	–	–
NDF (%)	63.42	2149.33 (kg/ha)	1250.70 (kg/ha)	61.25	1753.12 (kg/ha)	60.16	–	55.9
ADF (%)	48.68	1640.76 (kg/ha)	1390.88 (kg/ha)	38.48	1520.36 (kg/ha)	43.73	–	43.6
Cellulose	24.09	–	–	19.75	–	25.13	–	22.6
Lignin	20.86	–	–	18.42	–	16.68	–	
Ash (%)	8.95	5.2	3.4	9.44	4.30	8.89	–	8.0

Source: Murray et al. (2001), Khanal and Subba (2001), Elhassan and Yagi (2010), Singh et al. (2015), Sati and Ahmed (2018)

India (Dev et al. 2017b). It was found that *G. tenax* was distributed widely in the Kachchh region. Among surveyed locations, *G. tenax* was mostly distributed in Nakhatrana (37.5%) followed by Bhuj (32.5%), Mandvi (15.5%) and Bhachau (2.5%). Finally, it is concluded that there is a great diversity of *G. tenax* in Dhinodar hill region of Nakhatarana and Kotda Chakar piedomont region of Bhuj taluka. In addition to *G. tenax*, the fair diversity were also found on other *Grewia* species such as *G. villosa, G. tiliifolia, G. flavescens* in Kachchh region of Gujarat (Dev et al. 2017a). The genetic diversity studies among the 27 accessions of *Grewia tenax* revealed the presence of high amount of coefficient of variation for different morphological parameters. Range of variation for important traits of *G. tenax* viz., plant height (cm), number of fruits per 100 g, 1000-seed weight (g) and fruit thickness (mm) were observed with range (20.8–78.5, 690–1800, 60.3–150.7 and 4.0–6.7) and mean (46.8, 971.6, 113.8 and 5.1) value, respectively. Range of value of Pearson's correlation coefficient varied from – 0.85 between NF and TSW to 0.83 between PL and leaf width. Cluster analysis revealed the presence of three major clusters at coefficient value of Euclidian distance of 6.0. PCA analysis revealed the presence of 75.59% variability among *G. tenax* genotypes through first its five most informative components (Dev et al. 2017b).

31.4.3 Grewia villosa *Wild.*

31.4.3.1 Distribution

The *Grewia villosa* is commonly distributed in Tropical Africa, Mauritania, Egypt, Somalia, Namibia, Botswana, South Africa, Angola, Eritrea, Ethiopia, Kenya, Mozambique, Sudan, Swaziland, Tanzania, Uganda, Arabia, Pakistan, Cape Verde Isles, East Indies and India where it found mostly on uplands, lower valley, along roadside and field boundaries, up to 1200 m elevations (Orwa et al. 2009; von Maydell 1990; Kumar and Paul 2015). It can grow well in black, sedimentary, stony and rocky, ferruginous soils with a mean annual rainfall of 200–800 mm per year (Orwa et al. 2009; Ruffo et al. 2002; von Maydell 1990). In India, it is naturally reported in wild in Southern, Central and North-western region (Andhra Pradesh, Gujarat, Karnataka, Kerala, Madhya Pradesh, Maharashtra, Punjab, Rajasthan, Tamil Nadu) and also in Uttar Pradesh (BSI 2014). However, a more diverse population is found in Rajasthan (Thar desert) and Gujarat (Kachchh and Saurastra) (Dev et al. 2017a).

31.4.3.2 Diversity

A wide diversity of G. *villosa* is reported from Rajasthan and Gujarat state. The fair natural diversity of *Grewia villosa* were reported previously in Kachchh region of Gujarat *i.e.* Tapkeshwari hill region (Patel et al. 2010), Tharawada-Gandher Reserve Forest (Patel et al. 2013a) and Dhinodhar hill (Patel et al. 2013b). In Kachhch region, its occurence is reported in the scrub forest and foothill and valley of Dhinodhar region, Nani Aral, and on piedmont plain and field boundary of Moti Aral, Barapar, Devisar, Virani, whereas, its wild population are available in the region up-to Chari-Dhand Wetland Conservation Reserve which is located on the edge of arid Banni grasslands. Good diversity of *G. villosa* was also recorded in Mandvi and Naliya region (the stretch of Tapkeshwari hill scrub forest and on field boundary), lower hill scrub forest area (Kotada Chakar, Nana Reha and Mota Reha) (Dev et al. 2017a). The diversity study based on morphological and pomological traits of 20 *Grewia villosa* genotypes revealed the presence of a considerable amount of genetic variability between them. The cluster analysis classified genotypes into four different groups according to their pomological and morphological characteristics in which genotypes present in Cluster-I and II having better growth and fruit characters, whereas, Cluster-IV having genotypes with average growth and fruit characters. Principal component analysis (PCA) revealed that traits related to petiole length, fruit length, fruit thickness, number of fruits, thousand seed weight, stem girth, leaf length and width accounted for distinct variability among different genotypes. PCA analysis revealed the presence of 84.13% variability among *G. villosa* genotypes through the first six most informative components and it also confirmed the clustering pattern obtained through cluster analysis (Dev et al. 2018a).

31.4.4 Grewia flavescens *Juss.*

31.4.4.1 Distribution

The species is distributed throughout semiarid and sub-humid tropical areas of Africa, Arabia, Yemen and India (Hedberg and Edward 1989; von Maydell 1990). In Africa, its spread ranges from northern Kwa Zulu-Natal area to tropical Africa region (Sati and Ahmed 2018). In Sudan, it is found in Savanna Zone of Central Sudan, Red Sea Hills, Kassala and Equatoria (Abdelmutti 1991; Sati and Ahmed 2018). According to BSI (2014), it is commonly found in the following Indian states such as Gujarat, Rajasthan, Uttar Pradesh, Maharashtra, Delhi, Bihar, West Bengal, Orissa, Madhya Pradesh, Andhra Pradesh, Karnataka, Tamil Nadu and Kerala (BSI 2014). *Grewia flavescens* is the natural shrub species of scrub and deciduous forest, bushveld/open woodland type, rocky koppies/riverine fringes and the margins of forests (Arbonnier 2004) and it can grow better on various soil types such as rocky hilltops, fissured clay soils, sand and lateritic crusts (von Maydell 1990).

31.4.4.2 Diversity

Good diversity of *Grewia flavescens* is reported in the native range of the species like Tropical Africa, Sudan, India. Its wide genetic variation is reported in the Indian states such as Gujarat, Rajasthan, Uttar Pradesh, Maharashtra, Andhra Pradesh,

Karnataka, Tamil Nadu and Kerala (BSI 2014). *G. flavescens* has often been cited as a prime candidate plant species for domestication as a useful horticultural plant for the rural livelihood of the regions (Gebauer et al. 2007).

31.4.5 Grewia damine *Gaertn.*

31.4.5.1 Distribution and Diversity

Fair diversity of this species is reported in the native tropical ranges of the world. It is found in Pakistan (Sind, Punjab), India (Punjab, Madhya Pradesh, Peninsula), Sri Lanka, Nepal and tropical Africa. It is commonly growing in the dry arid plains and low hills up to 1200–1500 m. In India, it is frequently reported in the scrub forest of Andhra Pradesh, Bihar, Gujarat, Haryana, Karnataka, Kerala, Madhya Pradesh, Maharashtra, Punjab, Rajasthan, Tamil Nadu and West Bengal (BSI 2014).

31.4.6 Grewia abutilifolia *Vent. ex Juss.*

31.4.6.1 Distribution and Diversity

Generally, *G. abutilifolia* spreads across the regions as open dry mixed deciduous to semi-evergreen forests of India (except Manipur, Meghalaya and Nagaland), China and Indian sub-continent such as Cambodia, Vietnam, Laos and Malay peninsula.

31.4.7 Grewia tiliaefolia *Vahl*

31.4.7.1 Distribution and Diversity

The tree is well distributed throughout the major part of India extending from Ravi in the north sub Himalayan tract to Nepal and throughout the central and southern parts of India, ascending up to 1200 m elevation. It is found in across the Indian states such as Uttar Pradesh, Bihar, Orissa, Madhya Pradesh, Maharastra, South Gujarat, Andhra Pradesh, Western Ghats of Kerela, Karnataka and Tamil Nadu. According to Champion and Seth (1968), it is constituent of Tropical semi-evergreen, moist deciduous and dry deciduous forest of India and is also found in moist deciduous forest tracts of tropical countries like Sri Lanka, Myanmar and tropical Africa.

31.5 Traditional Knowledge

31.5.1 Grewia asiatica *L.*

The Goanese use roots of *G. asiatica* as the substitute of althea for treatment of rheumatism. The tribal communities from Singrauli, Madhya Pradesh, India use leaf paste for postural eruptions and their fruits as astringent and cooling (Chauhan 2017). It is used in ethnoveterinary treatments as a medicinal plant for easy delivery of cattle in Samahni valley located in Bhimber district of Pakistan. Its boiled stem with Ajwain (*Thymus serphyllum* L.) seeds are given for the prompt release of the placenta in cattle. Its root is used externally to hasten/cure suppuration and dressing for wounds and broken bones in cattle.

31.5.2 Grewia tenax *(Forsk.) Fiori*

In the encyclopedia of Ayurvedic medicinal plants, *G. tenax* is listed as one among 250 most important medicinal plants of India (Sharma and Patni 2012). The plant preparations made from this species are used for not only the treatment of bone fracture but also for bone strengthening. The use of root and fruits for the treatment of osteoporosis, tissue and wound healing are a well known household remedy. On the other hand, leaves and twigs are utilized as an integral part of folk medicine for the treatment of trachoma, tonsillitis, infections and are also used as a poultice to treat swelling (El-Ghazali et al. 1994). It is widely utilized for the treatment of various common diseases (stomachs upset, cough, fever, diarrhoea, dysentery, jaundice and rheumatism) of human beings (El-Ghazali et al. 1997). Fruits of *G. tenax* are significantly contributing to increasing Haemoglobin content of blood and considered as a simple safeguard against iron – deficiency and often provided as special diets for pregnant women and anaemic children to prevent from iron-deficiency (El-siddig et al. 2005; Gebauer et al. 2007; Ahmed et al. 2012). In Kenya, the bark is used as an anthelmintic and for the treatment of typhoid (Abdualrahman et al. 2011). The free radical scavenging activities of this plant provides therapeutic action against tissue damage. The plant gum found in this species could be a useful substitute binder in paracetamol and good suspending agent for Ibuprofen formulations (Martins et al. 2008; Ogaji and Okafor 2011; Ogaji and Hoag 2011).

31.5.3 Grewia villosa *Wild.*

Different parts of *Grewia villosa* are utilized by the local ethnic communities of the region in a different form to cure various ailments. The finely ground dry fruits diluted in water are orally taken for stomach problem like ache, dysentery etc. In

Tanzania, a mixture of the root extract of three *Grewia* species such as *G. villosa*, *G. tenax* and *G. flavescens* is used for the treatment of tuberculosis, syphilis and smallpox (Von Maydell 1986). The roots are used to treat body pains. The bark is used in the treatment of wounds, syphilis and smallpox (Von Maydell 1990 and Ruffo et al. 2002). Additionally, *Grewia villosa*, is also utilized for treatment of animal diseases viz. fresh fruits and leaves are given to animal for easy expel of placenta after the birth of a calf.

31.5.4 Grewia flavescens *Juss.*

Various parts of *G. flavescens* are utilized for fulfilling day to day requirements of the people in the arid region. Plant of *G. flavescens* possess various phytochemicals viz. tri-terpenoids, fatty acids, flavonoids, steroids, saponins and tannins which are useful in treating various ailments (Goyal 2012). The leaves of the plant are traditionally utilized to cure nose and eye problems, piles, rheumatism, joint pain (Dhawan et al. 1977; Joshi et al. 1980) and diabetus millitus (Bhakuni et al. 1971). The root bark extract is very useful in diarrhoea and act as an aphrodisiac agent (Von Maydell 1990). Fruits are important in making traditional medicines to cure stomach aches, respiratory disease, cardiac issues and blood disorders, treatment for diarrhoea, headache, eye complaints, sores and cholera (Bhakuni et al. 1971; Gebauer et al. 2013).

31.5.5 Grewia damine *Gaertn.*

Its wood is used to make strong agricultural implements. The fruits are edible. Various parts are used to make traditional medicines. The bark and roots are used for treating fractures, diarrhoea and skin diseases. The chemical investigation from the methanol extract of leaves of *Grewia damine* divulges the presence of sterols and glucosides such as lupeol, sitosterol β-D-glucoside, flavone C-glycosides vitexin, isovitexin glycosides (Jayasinghe et al. 2004).

31.5.6 Grewia abutilifolia *Vent. ex Juss.*

The leaves and bark of *G. abutilifolia* are used in medicated enema and also in traditional/folk medicine for the treatment of diabetic mellitus as its known α-amylase activity which lower postprandial blood glucose levels (Abdullah-Al-Mamun et al. 2017).

31.6 Utilization

31.6.1 Grewia asiatica *L.*

Fruits are a rich source of nutrients such as proteins, amino acids, vitamins, and minerals. It also contains various bioactive compounds, like anthocyanins, tannins, phenolics and flavonoids. Different plant parts possess different pharmacological properties and are: (1) Leaves are having antimicrobial, anticancer, antiplatelet and antiemetic activities; (2) fruits are possessing anticancer, antioxidant, radioprotective and antihyperglycemic properties and stem barks are possessing analgesic and anti-inflammatory activities.

31.6.1.1 Fruit

Fruits of *G. asiatica* are low in calories and fat, but high in vitamins, minerals, and fiber content (Yadav 1999). The detailed nutritional composition of fruit (100 g) are: Protein – 1.57 g; total lipid (fat) – <0.1 g; carbohydrate – 21.1 g; ash – 1.1 g; fibre – 5.53 g; calcium – 136 mg; iron – 1.08 mg; phosphorus – 24.2 mg; potassium – 372 mg; sodium – 17.3 mg; vitamin B1 – 0.02 mg; vitamin B3 – 0.825 mg; vitamin C – 4.385 mg; vitamin B2 – 0.264 mg and vitamin A – 16.11 g. Generally, the ripe fruits are eaten as a dessert and fetch high prices owing to their taste and flavour. A refreshing drink prepared from the juice (commonly known as phalsa syrup), is considered a delicacy during throughout the hot summer months in northern India. Sodium benzoate must be added as a preservative in this juice to prevent fermentation. The mucilaginous extract of the bark with water is used to clarify sugarcane juice during the preparation of 'gur', the traditional brown sugar made in India.

31.6.1.2 Fodder

The fresh leaves are valued as animal fodder owing to its nutritional content/composition. About 1.47 kg green biomass can be obtained from one plant for feeding purpose. The leaves also have good fodder value which contains 12.50% crude protein, 63.42% NDF, 48.68% ADF, 24.09% cellulose, 20.86% lignin and 8.95% ash content on a dry weight basis (Singh et al. 2015).

31.6.1.3 Medicine

Unripe fruits of Phalsa alleviates inflammation and is treated for respiratory, cardiac, and blood disorders, as well as in fever (Morton 1987). Whereas ripe fruits are consumed fresh, as desserts, or processed into refreshing fruit/soft drinks enjoyed in India during hot summer months because it overcomes thirst and burning sensation by its cooling tonic and aphrodisiac effects. The fruit is also good against throat trouble. The bark of the plant is used as a soap substitute in Burma. Further, an infusion of the bark is used as a demulcent, febrifuge, and also for treatment of diarrhoea. The leaves are applied on skin eruptions because of its antibiotic action.

31.6.2 Grewia tenax *(Forsk.) Fiori*

Generally flowering starts in the month of August – September and the main fruiting occurs in September – October. Generally, fruits are harvested manually based on fruit colour change from green to bright red or orange. Green fruits are more preferred by locals for consumption than ripened one by its taste and it can be stored for long duration (Dev et al. 2017a). Under adverse agro-climatic conditions, *G. tenax* can play a major role in food and fodder security for rural family and their livestock respectively. The plant has medicinal properties and other multiple uses with respect to health, income generation and environmental services. This *Grewia* species is used for various purposes in the daily life of rural and urban people. For example, the stem is used for traps, bows, stem and branches are used for fuel and wood product. It is also used to make handles of implements, spear and harpoon-shafts because it resists termites (EL-Amin 1995). The bark is used as cordage in hunt whereas park extractions are used for glueing tobacco leaves and the leaves are used in traditional medicine and fodder (Von Maydell 1990). The fruits are edible and consumed either fresh or dried for their high nutritive values especially during the dry periods of the year (Mabry – Hernandez 2009).

31.6.2.1 Fruit

The fruit of this species is nutritionally balanced because it is a rich source of carbohydrates, protein, vitamins and also minerals (Abdelmutti 1991; Aboagarib et al. 2015). According to Sharma and Patni (2012), the fruits also contain amino acids, minerals (K, Ca, Mn, Fe, Cu and Zn), tannin and pectic substances. Variations in chemical composition of fruit reported varying with collections from different locations (Abdalla 2001). The composition (in %) of *G. tenax* fruits are: protein (6.3), fat (0.4), crude fiber (8.1), ash (4.5), starch (15.1), sucrose (1.6), D-glucose (21.0), and D-fructose (24.3) (Abdelmutti 1991). However, Abdualrahman et al. (2011) reported higher protein and crude fibre than previous workers. A drink is prepared by soaking the fruit overnight followed by hand-pressing, sieving, and by adding sweetening agent, used as a refreshing drink during the hot summer season in Sudan as thirst quencher (Sharma and Patni 2012). A porridge called "Nesha" prepared from this drink by the addition of custard and flour is served during the fasting month of Ramazan and is also fed to lactating mothers to improve their health and lactation capacity after birth/delivery (Abdualrahman et al. 2011). Both fresh and dry *G.tenax* fruits are extensively consumed by the Oman, Wadi Muaydin and

Sudanese population (Ghazanfar 1992; Miller and Morris 1988). The fruit has commercial potential for consumption from beverages to ice cream, yogurt, porridge and confectionery.

31.6.2.2 Fodder

The palatability of leaves and twigs made this species as the good source of fodder for livestock especially during the dry spell/periods in the year. The slightly palatable young leaves with fairly good feed value are consumed by livestock at the end of dry seasons. *G. tenax* is important browse vegetation along with *Combretum acculatum, Dalbergia elanozylon, Geuera senegalensis, Balanites aegyptiaca* and *Maerua crassifolia* in Sudan. All species are browsed by at young stage. These species provide a good amount of bulk materials for stock grazing during the wet and dry season of the year (Abusuwar and Ahmed 2010). Anttila et al. (1994) determined to browse preferences through interviews in Eastern Kenya. They reported that, *Grewiatenax* was the second best preferable species among the plants found in that region. The reasons for preference were increased milk in production and fattening.

31.6.2.3 Industrial Utilization

G. tenax fruits can be used in making beverages/refreshing drinks, yoghurt, ice cream, and baby food. The mineral content of *G. tenax* fruits can be used to improve the nutritional properties of cereal and cereal products, especially flour for baked goods. The volatile compounds of fruits can be used as an additive to many foods or beverages to provide a better taste and flavour to them (Aboagarib et al. 2015). The water absorption capacity of pulp (3.3 ml/g) and foaming capacity of seed (8.6%) are high, which suggest that it could be useful for processing of food and food products in industries (Aboagarib et al. 2015).

Some effort is made to promote the industrial utilization of *G. tenax* plants. *G. tenax* fruit was utilized for flavouring ice cream. The organoleptic taste of *G. tenax* flavoured ice cream was as good as vanilla and strawberry – flavoured ice cream on basis of appearance, texture, colour and flavour (Abdualrahman et al. 2011).

31.6.2.4 Ecological Value

G. tenax plays a major role in maintaining the ecological balance of the system. Ecologically, it can withstand environmental stress like extreme temperatures, evapotranspiration etc., more easily than annual crops, which makes them produce yield in a sustainable manner without water and fertilizer. The aggressive root system of plants plays a major role in protecting the soil from severe erosion. Leaf litter falling from the shrub improves both the physical and chemical properties of the soil. Being a very useful source of pollen and nectar for visiting honey bees, it helps in sustainable biodiversity management in the arid region.

31.6.3 Grewia villosa *Wild.*

31.6.3.1 Fruit

The fruit of *G. villosa* are edible and may be found in some local markets of Africa (Sudan), as a substitute for *G. tenax* (Vogt 1995). Moreover, the fermented drink prepared form these fruits are available in Sudan and Southern Africa (FAO/WHO 1988). Its fruits are very useful for lactating mothers to improve their health and lactating abilities. Ripen fruits are preferred by animal herder as well as small kids during the fruiting season and for later use fruit is dried and stored. The fruit pulp extracted from sun-dried and the dry, papery hull loosened ripe brown fruits of *G. villosa* (by vigorous rubbing between the hands and winnowed by blowing) are eaten by the Hadza foragers in Tanzania region of Africa (Murray et al. 2001).

31.6.3.2 Fodder

The leaves and tender twigs of *G. villosa* are very palatable to livestock, making it a valuable fodder in the arid and semi-arid regions (Orwa et al. 2009).

31.6.4 Grewia flavescens *Juss.*

31.6.4.1 Fruit

The fruits of *Grewia flavescens* are edible and tasty. A refreshing drink are made from its mature ripened fruit by soaking the berries in water for 2 or 3 days (Gebauer et al. 2013). The branches are used for making bows, walking sticks and also for wickerwork. Leaves are used as fodder and the young twigs used for making baskets to carry vegetables and others in India (Wealth of India 1956). Its flower buds are utilized for making a vegetable (Sa'nchez-Monge 2001) in some part of the world. The fruits can be utilized either raw or they are mashed together with water and eaten as Porridge (Orwa et al. 2009; Gemishuizen and Meyer 2003). *G. flavescens* fruit mainly utilized during the dry season of the year (March to June) when other foods is limited in the field (Gebauer et al. 2013). *G. flavescens* fruits have good commercial potential in making beverages/refreshing drinks, ice cream, yoghurt and also as baby food (Gemishuizen and Meyer 2003).

31.6.4.2 Fodder

This species is highly valued for its green leaves fodder, because of high crude protein (6.7%), crude fibre (25.5%), and low in ash content (9.44%) and lignin (18.42%) content of green leaves and tender things, it becomes a quality fodder for livestock especially during dry season of the year (Gebauer et al. 2013; Singh et al. 2015). The high palatability of leaves is making it a good fodder in arid and semi-arid regions (Vogt 1995). Singh et al. (2015) reported that Grewia flavescens contains about 61.25% (NDF), 38.48% (ADF), 19.75% (Cellulose) and 18.42% (lignin) content in their leaves.

31.6.5 Grewia abutilifolia *Vent. ex Juss.*

31.6.5.1 Fruit and Fodder

Fruits are consumed by the local tribal communities living in the forest area as a raw (Deshmukh and Waghmode 2011). It is also a very good feed for wild animals and birds (Deshmukh and Waghmode 2011).

31.6.5.2 Medicine

Grewia abutilifolia showed high cytotoxicity, membrane stabilizing activities. Membrane stabilizing agents are the medications that reduce the hyper-excitability of nerves. In Thailand, the stems are used as a remedy for acne, and a root decoction as a remedy for fever (Sunny et al. 2017). Its fruits are utilized for weight gain, curing diarrhoea and gout (Deshmukh and Waghmode 2011). The bark yields good quality fibre that is made into a rope (Deshmukh and Waghmode 2011).

31.6.6 Grewia tiliaefolia *Vahl*

Grewia tiliaefolia is a multipurpose tree. The leaves are used as fodder, fruits are edible and it is a source of feed for wild bird and animal. The wood is recognized for its colour and strength, bole is used as timber and firewood.

31.6.6.1 Fruit

G. tiliaefolia fruits play a significant pivotal role in providing nutritional food supplement and also by generating additional income to the poor people. About 6600 fruits weigh 1 kg before depulping and about 15,000–16,000 seeds constituted 1 kg

Table 31.4 Ethnobotanical uses of *Grewia* species

Sl. No.	Parameter	Used for	Part used	Uses	Method of uses	Reference
1	***Grewia tenax***	**Human**	Fruit	Medicinal	Demulcent and cardiac tonic	Al-Qura'n (2006)
			Fruit pulp	Medicinal	Fruit pulp from ripe fruit applied externally on swelled parts thrice a day to treat swellings	Dev et al. (2017a), Patel et al. (2010)
			Root bark/ powder	Medicinal	Root bark/powder are boiled in water and taken orally twice in a day for 2–3 days to cure dysentery	Dev et al. (2017a), Patel et al. 2010
			Roots	Medicinal	Root decoction is given orally to treat female reproductive system problems	Dev et al. (2017a), Patel et al. 2010
			Leaves	Medicinal	Green leaves boiled in water and decoction is given in fever and Hepatitis two times a day.	Dev et al. (2017a)
			Root and leaves paste	Medicinal	The root and leaves paste applied externally to the fractured area of body.	Dev et al. (2017a)
			Green Fruit	Food	Unripe green fruits are eaten by cow herder and village kids.	Dev et al. (2017a)
		Livestock	Green leaves	Fodder	Green twigs and leaves are palatable fodder for goat, cattle and other animals.	Dev et al. (2017a), Patel et al. (2010)
			Seed and green leaves	Medicinal	Crushed green leaves given to animal for easy expel of placenta after delivery.	Dev et al. (2017a)
2	***Grewia villosa***	**Human**	Fruits	Food	Fruits acidic and eaten as raw also used in sarbat	Deshmukh and Shinde (2010)
			Whole plant	Medicinal	Antioxidant, antipyretic, analgesic	Amjad (2015)
			Fruits	Medicinal	Demulcent and cardiac tonic	Al-Qura'n (2006)
			Fruit	Food	Ripe fruit are eaten by villagers and kids.	Dev et al. (2017a)

(continued)

Table 31.4 (continued)

Sl. No.	Parameter	Used for	Part used	Uses	Method of uses	Reference
			Dry fruits	Medicinal	Dried fruit powder taken with water.	Dev et al. (2017a), Beche et al. (2016)
			Root	Medicinal	The 1–2 teaspoon roots powder are mixed in a glass of water and taken orally for diarrhea. Crushed root are boiled in water and juice taken orally to treat cough and body pains.	Dev et al. (2017a)
			Root and bark	Medicinal	Root and bark powder taken with water to treat urinary infections.	Dev et al. (2017a)
			Fruits	Medicinal	Used as part of mixtures for treatment of diabetes mellitus	Mondal et al. (2013)
			Fruits	Medicinal	Antibacterial activity	Tanira et al. (1994)
		Livestock	Fruit	Medicinal	Fresh fruit used after delivery for easy expel of placenta	Dev et al. (2017a)
			Leaves	Medicinal	Leaves are given to cattle especially during delivery for quick discharge of afterbirth. It is also given to young animals to induce puberty. Leaves are highly palatable for goat.	Amjad (2015)
			Leaves	Fodder	Directly browsing/ prepared for hay	Beche et al. (2016)
3	***Grewia flavescens***	Human	Flower buds	Vegetable	–	Sa'nchez-Monge (2001)
			Fruits	Food	Fruits are utilized by human for food purposes	Wealth of India (1956), Von Maydell (1990)
			Root bark	Medicinal	Root barks are used to treat diarrhea and act as aphrodisiac	Von Maydell (1990)

(continued)

Table 31.4 (continued)

Sl. No.	Parameter	Used for	Part used	Uses	Method of uses	Reference
			Young twigs	Economic use	Basket, bows, walking sticks making	Wealth of India (1956), Von Maydell (1990)
		Animal	Leaves	Fodder	Leaves are browsed by animals	Wealth of India (1956)
4	***Grewia abutilifolia***		Fruits	Food	A deciduous shrub. Fruits acidic and edible, also used in sarbat	Deshmukh and Shinde (2010)
				Medicinal	Cure abscesses and Rheumatism	Joshi et al. (1980)
				Medicinal	Respiratory, cardiac and blood disorders, treatment for diarrhoea, headache, eye complaints, sores and cholera	Bhakuni et al. (1971)
			Leaves	Medicinal	Nose and eye diseases, treating splenic enlargement, piles, rheumatism and relieving joint pain	Dhawan et al. (1977)
5	***Grewia Oppositifolia* (Beul)**	Human	Fruits	Food	Edible fruits	Yashwant et al. (2010), Jyoti et al. (2014)
			Timber/stem	Economic use	Fuel, fiber, making ropes	Yashwant et al. (2010), Jyoti et al. (2014)
		Animal	Leaves	Fodder	Leaves are browsed by animals	Yashwant et al. (2010), Jyoti et al. (2014)
6.	***Grewia optiva* J. R. Drumm. ex Burret. (Bhimal)**	Human	Fruits		Edible fruits	Pandey et al. (2017)
			Timber	Economic use	Fuel, fiber	Pandey et al. (2017)

(continued)

Table 31.4 (continued)

Sl. No.	Parameter	Used for	Part used	Uses	Method of uses	Reference
			Leaves	Medicinal	Preparation of concentrate decoction from fresh leaves by boiling in water at low temperatures. This paste is applied to cure joint pains.	Uniyal et al. (2006)
		Animal	Leaves	Fodder	Leaves are browsed by animals	Yashwant et al. (2010)
7	***Grewia asiatica***	Human	Fruits	Food	Edible fruits, and edible, also used or making sarbat in North India	Jyoti et al. (2014), Dev et al. (2018a)
			Fruit	Medicinal	To treat headache, diabetes and cardiovascular disorder, acidity, stomach ache, urinary infection.	Dev et al. (2018a)
			Timber/ stem	Economic use	Fuel, fiber, making ropes	Jyoti et al. (2014)
			Leaves	Fodder	Leaves are browsed by animals	Jyoti et al. (2014)
8	***Grewia damine* Gaertn.**	Human	Fruit	Human food	Fruits of *G. damine* are eaten as snacks	Neelo et al. (2015)
			Fruit	Food	Khadi, a local beer are prepared by mixture of these fruits and a solution of sugar.	Neelo et al. (2015)
			Wood	Economic use	Furniture	Neelo et al. (2015)
			Bark and roots	Medicinal	Fractures, diarrhoea and skin diseases.	Heywood et al. (2007)
		Animal	Leaves	Fodder	For fodder purposes	Neelo et al. (2015)

after removal of the pulp. Sengupta (1937) has reported 19,401 seeds/kg for freshly pulped seeds and 5291 seeds/kg for fruits with pulp. The average size of fresh fruits is of 0.8 × 0.5 cm. It contains Moisture (72.66%), Dry matter (17.33%), Carotenoid (2.95 g/100 g DW), Reducing sugar (2.18 g/100 g DW), Starch (11.3 g/100 g DW) and Total sugar (5.17 g/100 g DW) (Valvi et al. 2014). Fruits are rich in carbohydrates and in antioxidant enzymes.

31.6.6.2 Fodder

G. tiliaefolia leaves are used particularly for feeding cow, buffalo, goat and sheep. The chemical and fodder quality in *G. tiliaefolia* leaves analysed by Khanal and Subba (2001) were discussed briefly. The leaves contain 35.4% dry matter, 8.0% Ash and 15.9% crude protein. The leaves fodder quality were NDF (55.9% DM), ADF (43.6% DM) ADL (21.0% DM), Hemicellulose (12.3% DM) and Cellulose (22.6% DM). The macromineral content in leaves were calcium (1.09% DM), Phosphorous (0.41% DM), Magnesium (0.90% DM), Sodium (0.04% DM) and Potassium (1.7% DM). The micromineral content of leaves were Copper (10 mg/kg), Iron (270 mg/kg), Manganese (83 mg/kg) and Zn (17 mg/kg).

31.6.6.3 Medicine

The nutritional and phytochemical composition of fruits indicates that these fruits can be a valuable source of medicine. The bark of *G. tiliaefolia* is used for the treatment of healing wounds, curing kapha, vata, burning sensation, throat infections, dysentery biliousness and disease of the nose and blood (Goyal 2012). Its wood powder is an antidote to opium poisoning. The roots of *Grewia tiliaefolia* are used to treat skin diseases, hypertension, ulcers and diarrhoea. Lupenol isolated from this plant is known to cause apoptosis in several cancer cells. The bark contains Triterpenoids like lupeol, betulin and friedelin and root contains roots triterpenoids like lupeol and friedelin with high medicinal values (Goyal 2012).

31.6.6.4 Fibre

Among all the plant fibres, bark fibres are longer and possess the highest degree of compatibility between fibre and matrix. *G. tiliaefolia* bark provides excellent quality of fibre which contains Cellulose (62.80 wt. %), Hemicellulose (21.20 wt. %), Lignin (4.90 wt. %) and Moisture (2.3%) (Parthasarathy et al. 2018). The above chemical constituents of bark fibres are strongly dependent on the age, local environmental conditions of the plants.

31.7 Conclusion

Grewia species are generally well adapted to harsh/extreme climatic conditions including, salinity, drought and can grow on waste/marginal lands. There is a possibility of adopting these multipurpose plant species over a wide range of agroclimatic conditions owing to its hardy nature. They provide valuable food, fodder, nutrients and also more valuable medicines. All *Grewia* spp. can be considered as a rich source of iron (3.1–47.94 mg/100 g), potassium (375–966 mg/100 g) and phosphorus supplements. Therefore, their fruits are traditionally utilized for the treatment of various ailments (Table 31.4). *Grewia* spp. are also a good source of phytic acid (antioxidant) which can lower blood cholesterol. Its fruits are thirst quencher and are used as a refreshing drink during hot summer. Their plant parts such as leaves and twigs provide palatable fodder for livestock population especially during dry periods of the year. The plant preparation is used for the treatment of multiple problems like stomach upset, cough, fever, diarrhoea, dysentery, jaundice, hepatitis and rheumatism. Due to its paramount importance on above aspects, it deserves special emphasis on its intensive survey, genetic improvement and systematic utilization in arid and semiarid regions of India for the betterment of rural livelihoods of the region. Systematic work on its domestication, improvement, traditional knowledge and standardization of production technology are required for food and fodder supplement to people and their animals of the region.

References

Abdalla MA (2001) Propagation of *Grewia tenax* by seed, stem cuttings and tissue culture. Dissertation, University of Khartoum, Sudan

Abdelmutti OMS (1991) Biochemical of Nutritional Evaluation of Famine Food of Sudan. Dissertation, University of Khartoum, Sudan

Abdualrahman AMY, Osman AA, Abdalla MA, Suliman (2011) Nutritional evaluation of Guddaim fruits (*Grewia tenax*) and its utilization in ice cream production. J Sc Tech 12(03):38–43

Abdullah-Al-Mamun IMS, Sunny MI, Miah MAR, Hafsa U, Zabin SA, Roy PC (2017) Evaluation of alpha-amylase inhibitory activity of *Grewia abutilifolia* leaves. Eur J Pharm Med Res 4(11):25–29

Aboagarib EAA, Ruijin Y, Hua X (2015) Physicochemical, nutritional, and functional characteristics of seeds, peel and pulp of Grewia tenax (Forssk) Fiori fruits. Trop J Pharm Res 14(12):2247–2254

Abusuwar AO, Ahmed EO (2010) Seasonal variability in nutritive value of ruminant diets under open grazing system in the semi-arid rangeland of Sudan (South Darfur State). Agric Biol J N Am 1(3):243–249

Ahmad KS, Kayani WK, Hameed M, Ahmad F, Nawaz T (2012) Floristic diversity and ethnobotany of Senhsa, District Kotli, Jammu & Kashmir (Pakistan). Pak J Bot 44:195–201

Akundabweni LSM, Munene RW, Maina DM, Mangala JM (2010) Mineral micronutrient density characterization using energy dispersive X-ray fluorescence (xrf) analysis in four on-farm Kenyan wild African fruit tree germplasm. Afr J Food Agric Nutr Dev 8:2901–2938

Al-Hubaishi A, Muller-Hohenstein K (1984) An introduction to the vegetation of Yemen: ecological basis, floristic composition and human influence. Deutsche Gesellschaft fur Technische Zusammenarbeit (GTZ) GmbH. ASDruck, Schotten, Germany

Al-Qura'n S (2006) Verification of folk medicinal potentiality for some common plants in Jordan. Pak J Pharm Sci 219(4):303–308

Alrikain AEMS (2004) The yield and characteristics of gudeim (Grewia tenax) juice. Dissertation, University of Khartoum, Sudan

Amjad MS (2015) Ethnobotanical profling and floristic diversity of Bana Valley, Kotli. Pakistan Asian Pac J Trop Biomed 5(4):292–299

Anttila LS, Alakoski-Johansson GM, Johansson S et al (1994) Browse preference of Orma livestock and chemical composition of Prosopis juliflora and nine indigenous woody species in Bura, Eastern Kenya. East Afr Agric For J 58:83–90

Arbonnier M (2004) Trees, shrubs and lianas of West African dry zones. CIRAD, Margraf Publishers GmbH, Weikersheim

Beche D, Gebeyehu G, Feyisa K (2016) Indigenous utilization and management of useful plants in and around Awash National Park, Ethiopia. J Plant Biol Soil Health 3(1):1–12

Bhakuni DS, Dhar ML, Dhar MM, Dhawan BN, Gupta B, Srimal RC (1971) Screening of Indian plants for biological activity: part III. Ind J Exp Biol 9:91

Botanical Survey of India (2014) eFlora of India, List of Lower Taxa of Grewia. http://efloraindia.nic.in/efloraindia/lowerTaxaList.action?parentId=3482&parentName=Grewia#. Accessed 15 May 2016

Bredenkamp CL (2000) Tiliaceae. In: Leistner OA (ed) Seed plants of southern Africa: families and genera. Strelitzia 10. National Botanical Institute, Pretoria

Champion HG, Seth SK (1968) A revised survey of forest types of India. Govt. of India Press, New Delhi, p 404

Chaudhary SA (1999) Flora of the Kingdom of Saudi Arabia. Ministry of Agriculture and Water. National Herbarium, National Agriculture and Water Research Center, Riyadh

Chauhan RS (2017) Studies on medicinal plants used by tribal communities in district Singrauli of Madhya Pradesh. IJSRP 7(1):17–19

Collenette S (1985) An illustrated guide to the flowers of Saudi Arabia. Scorpion Publishing LTD, London

Deshmukh BS, Shinde V (2010) Fruits in the wilderness: a potential of local food resource. Int J Pharm Bio Sci 1(2):1–5

Deshmukh BS, Waghmode A (2011) Role of wild edible fruits as a food resource: traditional knowledge. Int J Pharm Life Sci 2(7):919–924

Dev R, Sharma GK, Singh T, Dayal D, Sureshkumar M (2017a) Distribution of Grewia species in Kachchh Gujarat, India: taxonomy, traditional knowledge and economic potentialities. Int J Pure App Biosci 5(3):567–574. https://doi.org/10.18782/2320-7051.5000

Dev R, Sureshkumar M, Dayal D, Venkatesan K (2017b) Genetic diversity among wild Grewia tenax accessions collected from Kachchh Region of Gujarat India. Indian J Plant Genet Resour 30(3):286–292. https://doi.org/10.5958/0976-1926.2017.00036.5

Dev R, Sureshkumar M, Venkatesan K, Singh T, Dayal D (2018a) Morphological and pomological diversity among hairy-leaf cross berry (Grewia villosa Willd) genotypes of Arid Kachchh, Gujarat, India. Int J Curr Microbiol App Sci 7(1):1163–1172. https://doi.org/10.20546/ijcmas.2018.701.141

Dev R, Vyas M, Machiwal D, Dayal D (2018b) Khanij padarthon se bharapur hai phalsa. Phalphool 39(3):25–27

Dhawan BN, Patnaik GK, Singh KK, Tandon JS (1977) Screening of Indian plants for biological activity: part VI. Ind J Exp Biol 15:208

EL-Amin KHM (1995) The industrial utilization of Guddiem "Grewia tenax" dissertation. University of Gezira, Sudan

El-Ghazali GEB, El Tohami MS, El Egami AAB (1994) Medicinal plants of the Sudan, medicinal plants of the White Nile provinces. Khartoum University Press, Khartoum

El-Ghazali GEB, El Tohami MS, El Egami AAB, Abdalla WS, Mohammed MG (1997) Medicinal plants of Sudan, part III medicinal plants of the eastern Nuba Mountains. Khartoum University Press, Khartoum

Elhassan GOM, Yagi SM (2010) Nutritional composition of Grewia species (Grewia tenax (Forsk.) Fiori, G. flavescens Juss and G. Villosa Willd) fruits. Adv J Food Sci Technol 2(3):159–162

El-Siddig K, Gebauer J, Inanaga S, Ludders P (2005) Auxin effects on rooting of stem cuttings of Grewia tenax: A potential alleviator of Iron-deficiency Anaemia in the Sudan, In: Deutscher Tropentag conference on The Global Food & Product Chain-Dynamics, Innovations, Conflicts, Strategies, 11-13th October, (Hohenheim)

FAO/WHO (1988) Traditional food plants. FAO, food nutrition paper 42. Rome. p 197

Gebauer J, Patzelt EA, Hammer ÆK, Buerkert ÆA (2007) First record of Grewia tenax (Forssk.) Fiori in northern Oman, a valuable fruit producing shrub. Genet Resour Crop Evol 54:1153–1158

Gebauer J, Bernholt H, Hammer K (2013) Grewia flavescens: a potential horticultural crop? Genet Resour Crop Evol 60:1915–1919

Gemishuizen G, Meyer NL (2003) Plants of southern Africa: An annotated checklist. Sterlitzia (PISAFr ed. 2)

Ghazanfar SA (1992) Quantitative and biogeographic analysis of the flora of the Sultanate of Oman. Glob Ecol Biogeogr Lett 2:189–195

Goyal PK (2012) Phytochemical and pharmacological properties of the genus Grewia: a review. Int J Pharm Pharm Sci 4:72–78

Gupta MK, Sharma PK, Ansari SH, Lagarkha R (2006) Pharmacognostical evaluation of *Grewia asiatica* fruits. Int J Plant Sci 1:249–251

Hedberg I, Edward S (1989) Flora of Ethiopia and Eritrea. J Botanical Taxonomy Geobotany 109:161–174

Heywood VH, Brummitt RK, Culham A, Seberg O (2007) Flowering plant families of the world. Firefly Books, Richmond Hill. ISBN 1-55407-206-9

Jayasinghe ULB, Balasooriya BAIS, Bandara AGD, Fujimoto Y (2004) Glycosides from Grewia damine and Filicium decipiens. Nat Prod Res 18(6):499–502. https://doi.org/10.1080/14786410310001620538

Jongbloed MVD, Feulner GR, Boer B, Western AR (2003) The comprehensive guide to the wild flowers of the United Arab Emirates. Emirates Printing Press, Dubai

Joshi MC, Patel MB, Mehta PJ (1980) Some folk medicines of Dangs, Gujarat State. Bull Med Ethno Bot Res 1:8–24

Jyoti SSS, Kishor K, Upreti BM, Bharati M, Bohra N, Sharma P, Tewari LM (2014) Diversity, distribution, indigenous use and conservation prioritization of the economically important floristic diversity of Nadaun block of Hamirpur district of Himachal Pradesh. Int J Biodivers Conserv 6(7):522–540

Khanal RC, Subba DB (2001) Nutrional evaluation of leaves from some major fodder trees cultivated in the hills of Nepal. Anim Feed Sci Technol 92:12–32

Kirtikar KR, Basu BD (2000) Indian medicinal plants. Lalit Mohan Publication, Allahabad

Konig P (1986) Zonation of vegetation in the mountainous region of South-Western Saudi Arabia ('Asir, Tihama). In: Ku¨rschner H (ed) Contributions to the vegetation of Southwest Asia. Beihefte zum Tu¨binger Atlas des Vorderen Orients. Dr. Ludwig Reichert Verlag, Wiesbaden

Kumar A, Paul TK (2015) Grewia villosa Willdenow (Malvaceae): a newrecord from Eastern India. Pleione 9(1):240–243

Kumar V, Uthappa AR, Srivastava M, Vijay D, Kumaranag KM, Manjunatha N, Rana M, Newaj R, Handa AK, Chaturvedi OP (2017) Floral biology of Grewia flavescens Juss.: an underutilized crop. Genet Resour Crop Evol 64:1789. https://doi.org/10.1007/s10722-017-0536-y

Luna RK (2005) *Grewia tiliaefolia,* plantation trees. International Book Distributers, Dehradun, pp 404–405

Mabry – Hernandez IR (2009) Screening for iron deficiency anemia – including iron supplementation for children and pregnant women. Am Fam Physician 79(10):897–898

Martins E, Christiana I, Olobayo K (2008) Effect of Grewia gum on the mechanical properties of paracetamol tablet formulations. Afr J Pharm Pharmacol 2:1–6

Miller AG, Morris M (1988) Plants of Dhofar, the Southern Region of Oman: traditional, economic and medicinal uses. Diwan of Royal Court, Muscat

Mondal P, Bhuyan N, Das S, Kumar M, Borah S, Mahato K (2013) Herbal medicines useful for the treatment of diabetes in north-east India: a review. J Ethnopharmacol 141:786

Morton JF (1987) Phalsa. In: Fruits of warm climates. Julia Morton, Miami, pp 276–277

Mothogoane MS (2012) Grewia flavescens Juss. National Herbarium, Pretoria. http://pza.sanbi.org/grewia-flavescens. Accessed on 16.10.2018

Murray SS, Margaret J, Schoeninger BTR, Pickering TR, Marlett JA (2001) Nutritional composition of some wild plant foods and honey used by Hadza Foragers of Tanzania. J Food Compos Anal 14:3–13. https://doi.org/10.1006/jfca.2000.0960

Neelo J, Kashe K, Teketay D, Masamba W (2015) Ethnobotanical survey of woody plants in Shorobe and Xobe villages, northwest region of Botswana. Ethnobot Res Appl 14:367–379

Ogaji IJ, Hoag SW (2011) Effect of Grewia gum as a suspending agent on Ibuprofen pediatric formulation. AAPS Pharm Sci Tech 12(2):507–513
Ogaji I, Okafor IS (2011) Potential of Grewia gum as film coating agent: some physicochemical properties of coated Praziquantel tablets. Int J Pharm Res 3(2):16–19
Orwa C, Mutua A, Kindt R, Jamnadass R, Anthony S (2009) Agroforestree database: a tree reference and selection guide version 4.0. http://www.worldagroforestry.org/sites/treedbs/treedatabases.asp
Orwa C, Mutua A, Kindt R, Jamnadass R, Simons A (2011) Agroforestree database: a tree reference and selection guide version 4.0. http://www.worldagroforestry.org/af/treedb/. Accessed 06.05.2016
Pandey NC, Joshi GC, Tewari LM, Pangtey YPS (2017) Plant diversity and utilization for sustainable development of agricultural field of Betalghat Region, Nainital District, Western Himalaya. J Curr Sci Human 4(4):25–48
Parthasarathy M, Senthamaraikannan P, Sanjay MR, Barile C (2018) Comparison of fibres properties of *Azadirachta indica* and *Acacia arabica* plant for lightweight composite applications. Struct Int Life 18(1):37–43
Patel YS, Joshi EP, Joshi PN (2010) Ethnobotanical study of Tapkeshwari Hill, Bhuj, Kachchh, India. Life Sci Leaflets 2:22–31
Patel YS, Patel RM, Joshi PN, Dabgar YB (2011) Study of angiospermic flora of Kachchh District, Gujarat, India. Life Sci Leaflets 19:739–768
Patel R, Roy Mahato AK, Vijay Kumar V, Asari RV (2013a) Status of the medicinal plants in Tharawada – Gandher Reserve Forest of Kachchh, Gujarat and the ethnomedicinal practices of local community. J Med Plants Stud 1(4):1–10
Patel Y, Patel RM, Roy Mahato AK, Joshi PN (2013b) Status and diversity of ethno-medicinal plants of Dhinodhar hill, Kachchh district, Gujarat. Int J Plant Animal Environ Sci 3(1):265–273
Priyanka SH, Meera, Lakhotiya MK (2015) Land degradation in Gujarat: an overview. J Indust Pollut Control 31(2):279–284
Ruffo CK, Birnie A, Tengnas B (2002) Edible wild plants of Tanzania. Regional Land Management Unit, Nairobi. ISBN 9966-896-60-0.
Sa'nchez-Monge E (2001) Diccionario de Plantas de Intere's Agrı'cola. Ministerio de Agricultura, Madrid
Saied A, Gebauer J, Sohail M, Buerkert AZ (2007) Ziziphus Spina-christi and Grewia tenax as promising fruit trees for afforestation in Northern Sudan. In: Tropentag conference on utilisation of diversity in land use systems: sustainable and organic approaches to meet human needs, 9-11th October, (Witzenhausen, Germany)
Sambo SH, Ahmed O, Shaltoe SM (2015) Phytochemical screening and mineral analysis of Grewia mollis stems bark. Int J Biochem Res Rev 6:75–81. https://doi.org/10.9734/ijbcrr/2015/14162
Sasidharan N, Renu G, Thilakar SJ, Narasimhan D (2018) Grewia damine Gaertn. [online] India biodiversity portal, species page: {Grewia damine}. Available at: https://indiabiodiversity.org/biodiv/species/show/229866. Accessed date 20 Oct 2018
Sastri BN (1956) The wealth of India raw material number 4 Grewia Linn. Council of Scientific and Industrial Research, New Delhi
Sati NME, Ahmed FAM (2018) Botanical overview and chemical composition of some Grewia spp. "Gudeim Plant" in Sudan. Open Sci J 3(1):1–12
Sengupta JN (1937) Seed weights, plant percents, etc for forest plants in India. Indian For Rec(NS) Silviculture 2(5):1–221
Sharma N, Patni V (2012) Grewia tenax (frosk.) Fiori.- a traditional medicinal plant with enormous economic prospectives. Asian J Pharm Clin Res 5(3):28–32
Singh JP (1974) Phalsa. Indian Hortic 19:15–16
Singh A, Dev R, Mahanta SK, Kumar RV (2015) Characterization of underutilized shrubs for forage potential in rainfed and dry areas. Range Mgmt Agroforestry 36(2):194–197
Sohail M (2009) Ecophysiology of seed dormancy and salt tolerance of Grewia tenax (Forssk.) Fiori and Ziziphus spina-christi (L.) Willd. Dissertation, Gomal University, Pakistan

Sunny MI, Abdullah-Al-Mamun BY, Islam MS, Hasan-Al-Jubayed M, Roy PC, Islam MS (2017) Evaluation of cytotoxic and membrane stabilizing activities of *Grewia abutilifolia* leaves. Eur J Biomedi Pharmaceu Sci 4(11):103–108

Tanira M, Bashir A, Dib R, Goodwin C, Wasfi I, Banna N (1994) Antimicrobial and phytochemical screening of medicinal plants of the United Arab Emirates. J Ethnopharmacol 41(3):201–205

Teketay D (1996) Germination ecology of twelve indigenous and eight exotic multipurpose leguminous species from Ethopia. For Ecol Manag 80:209–223

Tripathi S, Chaurey M, Balasubramaniam A, Balakrishnan N (2010) *Grewia asiatica* Linn. as a phytomedicine. Res J Pharm Tech 3:1–3

Uniyal SK, Singh KN, Jamwal P, Lal B (2006) Traditional use of medicinal plants among the tribal communities of Chhota Bhangal, Western Himalaya. J Ethnobiol Ethnomed 2(14):1–8. https://doi.org/10.1186/1746-4269-2-14

Valvi SR, Gadekar SS, Jadhav VD (2014) Phytochemical assessment of five wild edible fruits. Int J Life Sci 2(2):168–172

Vogt K (1995) Identification, propagation and uses of common trees and shrubs of dry land in Sudan-SOS. Sahel International UN Ford Foundation, London, pp 111–115

Von Maydell HJ (1986) Trees and shrubs of the Sahel. G.T.Z. Scientific Books, Weikersheim, p 525

Von Maydell HJ (1990) Trees and shrubs of the Sahel: their characteristics and uses. Deutsche Gesellschaft fu¨ r Technische Zusammenarbeit (GTZ) Germany. In: ISBN: 3-8236-1198-4

Wealth of India (1956) The wealth of India: a dictionary of Indian raw materials and industrial products, vol IV. F-G, Council of Scientific and Industrial Research, New Delhi

Whitehouse C, Cheek M, Andrews S, Verdcourt B (2001) Tiliaceae & Muntingiaceae. In: Beentje HJ (ed) Flora of tropical East Africa. A.A. Balkema, Rotterdam, Netherlands, 120 pp

Yadav AK (1999) Phalsa: a potential new small fruit for Georgia. In: Janick J (ed) Perspectives on new crops and new uses. ASHS Press, Alexandria, pp 348–352

Yashwant SR, Subash CR, Vishvakarma GB (2010) Diversity, distribution and utilization of fodder species in subtemperate, temperate and cold desert region of the Himachal Pradesh, northwestern, Himalaya. J Am Sci 6(6):72–81

Chapter 32
Hyphaene thebaica (Doum): Distribution, Composition and Utilization

Nawal Abdel-Gayoum Abdel-Raman

Abbreviations

NTFPs	Non-timber forest products
FAO	Food and Agriculture Organization
KCal	Kilo calorie
IU	International unit
H	Hyphaene

32.1 Introduction

Trees and shrubs play a significant role in maintaining the natural ecosystem and in preventing and combating desertification in the Sahel. They provide a multitude of useful products for the people, because the wild edible plants are good sources of nutrients for rural population.

Sudan has various natural forests; for eternity it has economically reliant on agriculture (Abdel Muti 2002). Thus, it is contains high substances in non-timber forest products (NTFPs) similar to food materials and pharmaceutical plants. Significant parts of the inhabitants in the belt of production of these plants as enterprises projects (Abdel-Rahman et al. 2018; FAO 2013).

Abdel-Rahman (2011) recorded that in Sudan there are five plant regions accord to temperature, rainfall and soil type. These regions are desert and semi-desert arid, low rainfall woodland Savanna, high rainfall woodland Savanna, flood regions and mountain regions. The requirement of the trees affects its distribution. Doum needs to rainfall from 100 to 600 mm, soil type tolerant to a wide range of soils and tolerant to very high temperatures. Doum wild fruit is distributed in high rainfall woodland savanna, flood and mountain regions.

N. A.-G. Abdel-Raman (✉)
National Food Research Centre, Shambat, Khartoum North, Sudan

A. A. Mariod (ed.), *Wild Fruits: Composition, Nutritional Value and Products*,
https://doi.org/10.1007/978-3-030-31885-7_32

32.2 Distribution

Hyphaene thebaica (Doum) is one of the 11 species of this genus, to be origin to the northern of Africa. Widespread in the Sahel of Africa, from Mauritania and Senegal in the far West, Kenya and Tanzania in the middle, to Egypt and Sudan in the East (Abdel-Rahman et al. 2018). As well as, found in some areas of the Arabian Peninsula. It is also found in part of the desert where water occurs, in oases and valleys. It is one of Sudanese edible forest fruits. It is spread throughout the rainy tropics in the world and to a limited extent in dry regions (FAO 2006). In Sudan, it is grown in the North, Central and West of the country.

32.3 Botanical Description

Doum (*Hyphaene thebaica*) fruit belongs to Arecaceae (*Palmae*) family. The tree is an evergreen, mostly a tropical plant; it can grow further away and is sometimes found on rocky slopes. It does not grow well in areas where stagnant water occurs. It has a Y-shaped trunk (Fig. 32.1), being very commonly dichotomously branched, most easily recognizable. The leaves are with the petiole often being more than a meter long. On each side of the petiole are many upwardly- curved hooks. The leaf itself is shaped like a large fan, with diameter reach to 120 and 180 cm in length and width, respectively. *Hyphaene thebaica* makes male and female flowers from different trees, the flower buds have red colour (Chang and Lo 2010). The palm is most often dioeciously, with the female palm producing shiny brown fruits (Fig. 32.2). The fruits are hard, woody, fibrous and sweety in taste with flavour of gingerbread, have size ranges from 5.00 to 8.29 cm in length, 4.62 to 6.43 cm in width and 5.12 to 6.15 cm in thickness (Abdel-Rahman et al. 2014). The colour is red–orange; this can stay on the tree for quite a long time. It contains one ivory-coloured seed. Doum tree may reach about 15 m in height (El-Ghazali et al. 2003). It flowers from February to April and fruits in November (El-Amin 1990).

Fig. 32.1 Doum (*Hyphaene thebaica* L. Mart) tree. (Source: Wikimedia 2018)

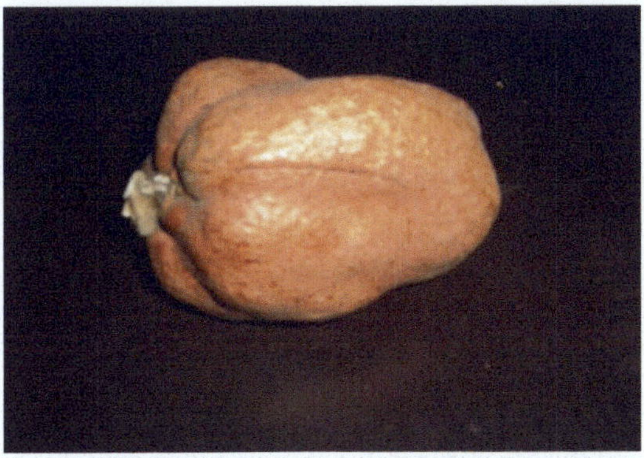

Fig. 32.2 Doum (*Hyphaene thebaica* L. Mart) fruit

32.4 Chemical Composition

The Doum fruits are a rich source of sugar, protein and fats. It contains some minerals such as calcium, phosphorus and high in iron content. The fruits are rich in thiamin, riboflavin and niacin (FAO 2006).

The edible part of the fruit of Doum contains 74.0% soluble sugars, 22.0% of its starchy substances and 37.0% sucrose, also Doum contain a high percentage of potassium (Abdel Muti 2002). Moreover, it contains quantities of proteins, fats, calcium, phosphorus and niacin, with traces of riboflavin and thiamin.

FAO (2006) recorded the chemical compositions (per 100 g) of African Doum of 4.00, 3.80, 0.80, 84.10 and 7.30% for moisture, protein, fats, carbohydrates and ash, respectively. In addition, energy value (390 KCal), thiamin (0.05 mg) and riboflavin (0.10 mg). As well as, the above organization reported the chemical composition of Indian of 9.26, 7.21, 75.81, 50.07 and 3.68%, for protein, fats, carbohydrates, fibre and ash, respectively, and energy value (4.6 KCal). While, Abdel-Rahman et al. (2014) recorded composition of Sudanese Doum fruit of 5.47, 3.80, 0.95, 18.36, 7.17, 69.72, 0.38, 50.00 and 4.67% for moisture, protein, fats, fibre, ash, carbohydrates, pectin, total and reducing sugars, respectively. Besides, 309.9 KCal (energy value), 31.74 mg (vitamin C, ascorbic acid) and 27.49 IU Vitamin A).

32.5 Minerals

Doum fruit has adequate amount of calcium (Ca), iron (Fe), phosphorus (P), sodium (Na), magnesium (Mg), manganese (Mn) and potassium (K) of minerals (Table 32.1). FAO (2006) mentioned levels of African Doum of Ca (34 mg) and P (110 mg), in addition levels of Ca (268 mg), P (224 mg) and Fe (38.24 mg) for Indian Doum. Okwu and Morah (2004) recorded similar result of Mg content for *Dennettia tripetala* fruit. The content of Mn within the range of 7–55 mg/kg was reported by Ertürk et al. (2006) for some fruit of chestnut cultivars. The content of Ca is lower

Table 32.1 Mineral contents of doum (*Hyphaene thebaica*) fruits

Minerals	Amount (mg/Kg)
Calcium	259.0
Iron	432.0
Phosphorus	24,270.0
Sodium	4,320.0
Magnesium	4,236.0
Manganese	51.0
Potassium	36,734.0

and the other minerals are higher than contents reported by Leterme et al. (2005) for peach palm (*Bactris gasipaes* Kunth). The level of P in Doum fruit is higher than many common fruits rich in this metal, such as strawberry (2900 mg/kg), guava (1354 mg/100 kg), fig (1110 mg/100 kg), tamarind (1080 mg/100 kg) and banana (1077 mg/100 kg). On the other hand, Doum fruit has Na content more than richest fruits in this metal, Hawthorn (*Crataegus azarolus*) 173.00 mg/100 kg and Sweet melon (*Cucumis melo*) of 1667 mg/100 kg (El-Nouri and El-Talbani 1981). Doum fruit is rich in magnesium, about four times compared to strawberry which is known to be 1000 mg/100 kg (Gil et al. 2006). But, contain about half compared with pineapple (Ensminger et al. 1983). The content of potassium is semi-similar to that reported by El-Nouri and El-Talbani (1981) for *Cucumis melo* which was 42,630 mg/kg.

Abdel Muti (2002) reported 1300 mg/kg (Ca), 17.00 mg/kg (Fe), 1400 mg/kg (P), 900 mg/kg (Na), 1800 mg/kg (Mg), 11.00 mg/kg (Mn) and 30,200 mg/kg (K) for Doum. These differences could be attributed to environmental conditions, or type of the soil.

32.6 Amino Acids

Abdel-Rahman et al. (2014) published that the levels of amino acids in Doum fruit were 88.66% from total protein. As well as, the fruit contains limited quantities of some nutritionally essential amino acids, such as leucine, valine, phenylalanine, threonine and arginine. On the other hand, this forest fruit contains some non-essential amino acids like glutamic acid, aspartic acid and alanine. The methionine and cysteine sulphur amino acids are limited in its levels (Table 32.2).

Glutamic acid recorded the highest percentage from total amino acids, (Osiecki et al. 2004) mentioned that glutamic acid has significant rule of metabolism of sugars and fats. Besides, assists the moving of potassium within the blood-brain barrier.

Table 32.2 Amino acids (mg/100 g) profile of doum (*Hyphaene thebaica*) fruits

Essential amino acids		Non-essential amino acids	
Histidine	32.40	Tyrosine	33.56
Isoleucine	73.85	Alanine	117.15
Leucine	125.01	Aspartic acid	186.49
Lysine	53.91	Glutamic acid	214.20
Methionine	27.59	Glycine	38.52
Phenylalanine	89.84	Serine	84.59
Threonine	81.70	Cystine	4.70
Tryptophan	ND	Proline	93.91
Valine	96.18	–	–
Arginine	75.30	–	–

ND not detected

32.7 Bioactive Compounds

The extracted oil of Doum fruit contains more than 55 substances with adequate percentages, use as medicine compounds. Several of those substances are fatty acids, incensole acetate and sapinene, as well as limonene, terpinen and octylacetete (El-Beltagi et al. 2018). In addition, phenolic substances such as: catechin, chlorogenic, metoxi-cinnamaci, hydroxicinnamic, and caffeic acids (Salih and Yahia 2015). Furthermore, Doum fruits have many flavonoid compounds, which included quercetin, naringin, glycosides and isoquercetrin (Aamer 2016; Farag and Pare 2013).

32.8 Utilization

The whole tree of Doum has many benefits, roots, wood, rinds, fibre, leaves, and any parts of the fruit nuts. Traditionally, people consumed it for food supply, building substances, medicine, for local and marketable proposes (Idohou et al. 2016).

32.8.1 Food Utilization

The powder of *H. thebaica* fruit is applied in some industrial stuff as a source of fiber and stabilizer. The edible part of the fruit is sharply aromatic (hence the name Gingerbread Palm) and can either be crushed and milled, or flaked. As well as mixed with flour and sugar to make *Madeda* drink. Which also use to produce concentrate or nectars (Abdel-Rahman et al. 2015). Sometimes, it is dried and added as a flavouring agent to food. In Turkana, Kenya, the powder is also used to make a mild alcoholic drink by adding water and leaving it to stand. In other countries, the sap is used to make palm wine (Vogt 1995). Mainly children are using doum fruits as a snack.

Mix of milled wheat with Doum fruit powder increasing the rheological properties of the dough. As well as, enhanced total phenolic, total flavonoid, and rising antioxidant action, and consequently improving the bread quality (Aboshora et al. 2015).

32.8.2 Medicinal Utilization

Doum wild fruit has adequate amount of many pharmacological compounds such as saponins, coumarins, hydroxyl cinnamates, essential oils and flavonoids (Aamer 2016). The aqueous fruit and leaf extracts of *H. thebaica* have potential application in pharmacology as anticancer, anti-inflammatory and antibacterial (Bello et al. 2017; Shalaby and Shatta 2013). Doum fruit also used to prepare a standardized extract antihypertensive drug (Khalil et al. 2018).

The extract of fruit is used for the treatment of gastric pains (El-Gazali et al. 1998). Local medicinal uses have also been reported, including uses of the roots in treatment for schisto-somiass, wounds and of charcoal from seed kernels which treat some eyes diseases in livestock (Vogt 1995). The roots are utilized against bilharziasis disease, whereas the gum used for diuretic and diaphoretic properties. The flavonoid extracts significantly increased adiponectin levels in diabetic rats, which arouse the hypoglycemic activity of insulin without changing the concentration of insulin in the blood (Salah et al. 2011). The blood glucose, cholesterol, triglycerides and total fats amounts were significantly decreased following a few days of ate of Doum fruits (Bayad 2016; Habib et al. 2014). The water extract of Doum fruits be able to decrease hyperlipidemia in nephrotic syndrome and guides to decline the danger of glomerulosclerosis and atherosclerosis and consequently the normal, safe and nontoxic *H. thebaica* fruit could be of huge value for usage as hypolipidemic treatments (Aboshora et al. 2014).

32.8.3 Other Uses

The fruit is used as natural source to produce dye-extract for dye-sensitized solar cells (Mohammed et al. 2015). As this palm has so many uses, it is often over exploited. The fruit is used to make a black dye often used in the tanning industry. The fruit's seed is very solid and ivory colour, which has been used commercially for producing buttons and small carvings (El-Beltagi et al. 2018). Carpets, baskets and writing paper made from leaves. Peels of Doum nuts are used to make molasses. The wood is occasionally used as firewood and for charcoal, but often in construction, providing support and rafters for horses, posts for fencing and for building rafts (Vogt 1995).

References

Aamer RA (2016) Characteristics of aqueous doum fruit extract and its utilization in some novel products. Ann Agri Sci 61:25–33

Abdel Muti OMS (2002) Nutritive value of wild plants of the Sudan. Arab J Fd Nutr 3(3):6–67

Abdel-Rahman NA (2011) A study of some Sudanese edible forest fruits and their nectars. Ph. D. Thesis in Food technology to Sudan Academy of Sciences. Khartoum, Sudan (Cited from: El-Khalefa KF (1996) Botanical of forest plants. 1st ed. University of Khartoum Press. Khartoum)

Abdel-Rahman NA, Ismail IA, Elshafe'a EBB (2014) Characterization of some Sudanese edible forest fruits. J Agri-Food Appl Sci 2(2):39–44

Abdel-Rahman NA, Ismail IA, Elshafe'a EBB (2015) Some quality attributes of four Sudanese forest fruits nectars. J Agri-Food Appl Sci 3(2):32–38

Abdel-Rahman NA, Ismail IA, Elshafe'a EBB (2018) Utilization of four Sudanese wild fruits to produce nectars. Fd Bio 7:01–07

Aboshora W, Lianfu Z, Dahir M, Qingran M, Qingrui S, Jing L (2014) Effect of extraction method and solvent power on polyphenol and flavonoid levels in *Hyphaene thebaica* L. Mart Arecaceae doum fruit, and its antioxidant and antibacterial activities. Tro J Pharm Res 13:2057–2063

Aboshora W, Lianfu z, Dahir M, Qingran M, Musa A, Mohammed Gasmalla MAA, Omar KA (2015) Influence of doum (*Hyphaene thebaica* L.) flour addition on dough mixing properties, bread quality and antioxidant potential. J Food Sci Technol 53(1):591–600

Bayad AE (2016) Influences of doum fruit *Hyphaene thebaica* extract on the reproductive parameters, blood picture, lipid profile and hepato-renal functions in rats. MRJMMS 4:384–391

Bello BA, Khan SA, Khan JA, Syed FQ, Mirza MB, Shah L, Khan SB (2017) Anticancer, antibacterial and pollutant degradation potential of silver nanoparticles from *Hyphaene thebaica*. Biochem Biophys Res Commun 490(3):889–894

Chang H, Lo YJ (2010) Pomergranate leaves and mulberry fruit as natural sensitizers for dye-sensitized solar cells. SEJ 84:1833–1837

El-Amin MH (1990) Trees and shrubs of the Sudan, 1st edn. Ithaca Press, England, p 106, 400, 451

El-Beltagi HS, Mohamed HI, Yousef HN, Fawzi EM (2018) Biological activities of the doum palm (*Hyphaene thebaica* L) extract and its bioactive components, Chapter 3. In: Shalaby E, Azzam GM, editors. Antioxidants in foods and its applications. Intechopen. pp. 49–66. Available in: https://doi.org/10.5772/intechopen.74772

El-Ghazali GEB, Abdalla WS, Subki H, Khalafalla MM, Hamad AA (2003) Medicinal plants of the Sudan: medicinal plants of Ingassana area. Medicinal of Aromatic Plants Research Institute. National Centre for Research, Khartoum

El-Ghazali GEB, El-Subki H, El-Tohami AAA, Abdalla WS, Yagi SMA (1998) Medicinal plants of the Sudan: medicinal of aromatic plants: uses medicinal plants in Khartoum State. New Life Press, Khartoum, Sudan, pp. 57, 157

El-Nouri FF, El-Talbani LG (1981) Human nutrition. House books province. El-Mawsil University, Iraq, p 431, 432, 433, 442, 443

Ensminger AH, Ensminger ME, Kondale JE, Robson JRK (1983) Foods and nutrition encyclopedia. Pegus Press, Clovis, California

Ertürk U, Mert C, Soylu A (2006) Chemical composition of fruits of some important chestnut cultivars. Braz Arch Biol Technol 49(2):183–188

FAO (2006) Composition and characteristics of selected palm products. FAO Corporate Document Repository. Forestry Department. Tropical Palm. FAO. Rome

FAO (2013) International conference on the role of forests in the food security and nutrition. Forestry Department. [FAO document/13 to 15 May 2013]. Rome

Farag MA, Pare PW (2013) Phytochemical analysis and anti-inflammatory potential of *Hyphaene thebaica* L. fruit. J Fd Sci 78:1503–1508

Gil MI, Aguayo E, Kader AA (2006) Quality changes and nutrient retention in fresh-cut versus whole fruits during storage. J Agri Fd Chem 54(12):4284–4296

Habib DF, Michael HN, Salib JY, Ahmed NM, Agaibyi MH (2014) Hypolipidemic efficacy of Hyphaene thebaica doum in experimental nephrotic syndrome. Int J Pharm 4:28–34

Idohou R, Peterson AT, Assogbadjo AE, Vihotogbe RL, Padonou E, KakaÏ RG (2016) Identification of potential areas for wild palm cultivation in the Republic of Benin through remote sensing and ecological niche modeling. Genet Resour Crop Ev 64(6):1383–1393

Khalil NA, Abdel Motaal A, Meselhy K, Abdel Khalek S (2018) Renin and angiotensin converting enzyme inhibition of standardized bioactive fraction of *Hyphaene thebaica* L. Mart growing in Egypt. J Pharmacogn 10(4):622–627

Leterme P, Garcia MF, Londoño AM, Rojas MG, Buldgen A, Souffrant WB (2005) Chemical composition and nutritive value of peach palm (*bactris gasipaes* kunth) in rats. J Sci Fd Agri 85(9):1501–1512

Mohammed IK, Isah KU, Yabagi JA, Taufiq S (2015) The effect on extracting solvents using natural dye extract from *Hyphaene thebaica* for dye-sensitized solar cells. Mater Sci Eng 5(1):1–3

Okwu DE, Morah FNI (2004) Mineral and nutritive value of *dennettia tripetala* fruits. J Fruits (Nigeria) 59:437–442

Osiecki H, Meeke F, Smith J (2004) The encyclopaedia of clinical nutrition: the nervous system, vol 1. Bio Conceps Publishing QLD. Available in: http://www.vitalhealthzone.com/nutrition/amino-acids/glutamic_acid.html

Salah SH, Abdou HS, Abd El Azeem AS, Abdel-Rahim EA (2011) The antioxidative effects of some medicinal plants as hypoglycemic agents on chromosomal aberration and abnormal nucleic acids metabolism produced by diabetes stress in male adult albino rats. J Diabetes Mellitus 1:6–14

Salih NK-EM, Yahia EM (2015) Nutritional value and antioxidant properties of four wild fruits commonly consumed in Sudan. Inte Fd Res J 22:2389–2395

Shalaby A, shatta A (2013) Potential antioxidant and anti-inflammatory effects of *Hyphaena thebaica* in experimentally induced inflammatory bowel disease. Afr J Pharm Pharmacol 7:1834–1840

Vogt K (1995) A field worker's guide to the identification, propagation and uses of: common trees and shrubs of dryland Sudan. SOS Sahel International (UK), London, p 100, 101, 125. Printed by: Antony Rowe Ltd. Chippenham

Wikimedia. 2018. Available in: https://www.google.com/search?q=wikimedia+images,+hyphaene+thebaica+tree

Chapter 33
Nauclea latifolia (Karmadoda): Distribution, Composition and Utilization

Nawal Abdel-Gayoum Abdel-Rahman

Abbreviations

AIDS	Acquired immune deficiency syndrome
CE	Catechin standard equivalents
CODEX	Codex Alimentarius Commission
FAO	Food and agriculture organization
GAE	Gallic acid standard equivalents
HIV	Human immunity virus
IU	International unit
Kcal	Kilo calorie
N	*Nauclea*
ND	Not detected
NWFPs	Non-wood forest products
QE	Quercetin standard equivalents
RDI	Recommended daily intake
WHO	World health organization
β	Beta

33.1 Introduction

It is usually recognized and agrees that plant resources form a necessary component of biodiversity. Plant resources are used as bioactive compounds against diseases and pests from generation to generation throughout the world (Enabulele et al. 2017). Furthermore, these plants are needed to guarantee the stability and safety of human food production. In addition, trees have protection among others by development of pharmaceuticals, cosmetics and pesticides. Non-wood forest products

N. A.-G. Abdel-Rahman (✉)
National Food Research Centre – Shambat, Khartoum North, Sudan
e-mail: ibreez2005@gmail.com

A. A. Mariod (ed.), *Wild Fruits: Composition, Nutritional Value and Products*,
https://doi.org/10.1007/978-3-030-31885-7_33

(NWFPs) are distributed and used in Sudan along many regions. Some processing of edible fruits and seeds has been practiced traditionally for a long time (Abdel-Rahman 2011).

Rubiaceae is a large family comprising more than 630 genera and about 13,000 species of plants. Nauclea is of one Rubiaceae genera. In Africa, the genus Nauclea includes seven species; *N. latifolia, N. pobeguinii, N. diderrichii, N. vanderguchtii, N. gilleti, N. nyasica and N. xanthoxylon* (Löfstrand et al. 2014). These species are well symbolized for many pharmaceutical utilization, especially *Nauclea latifolia* (Haudecoeur et al. 2018). Nworgu et al. (2010) reported that it is a small evergreen tree or straggling shrub with leaves rounded ovate (Fig. 33.1). In Sudan, *N. latifolia* called Karmadoda, whereas in another countries of African known in several names; koumkouma, marga and ubulu inu (Arise et al. 2012).

Locally in Africa Karmadoda (*N. latifolia*) fruit has been used as an anti-malarial, anti-hypertensive and anti-bacterial, as well as for treatment of some gastric diseases (Kabine et al. 2015; Fadipe et al. 2014a, b; Nworgu et al. 2008). The overall objective of this study is to investigate and give information about *N. latifolia* fruit as local resource in the food supply and management of some diseases.

33.2 Botanical Description

Karmadoda (*N. latifolia*) is a straggling shrub or tree up to 10 meters high, sometimes reaching 33 meters high (Gidado et al. 2008; El-Amin 1990). Bole crooked, bark grey or brown, deeply fissured. Branchlest stout, glabrous or minutely puberulous. Leaves broadly oblong-elleptic to obviate or nearly oblong-orbicular (Nworgu et al. 2010), 10–20 × 7–12 cm, glabrous; petioles red, 1.5–2 cm long; stipules broad,

Fig. 33.1 *Nauclea latifolia* tree. (Source: https://www.wikimedia.org)

ovate and persistent (Badiaga 2016). The flowers are white and the flower head about 5 cm across or more, fragrant, on stout peduncles in about 2 cm long.

Fruit globosely or ovoid, 5–7.5 cm across, red -brown pitted with pentagonal scars of the flowers, weighing 40–80 g. Karmadoda fruit is a berry fruit, bally, like pomegranate (*Punica grantum* L.) in size (James and Ugbede 2011, Abdel-Rahman 2011). The edible part is fibrous, high in water, crimson in colour (Figs. 33.2 and 33.3), acidic in taste. The seeds are numerous embedded in sweet edible pulps (Fadipe 2014), seeds colour is bright yellow and the weight is light. The peels are thin and its colour is crimson, easy to separate like blanched potato (Abdel-Rahman 2014a). The flowers tree March – August and fruits May – March. It is a spreading, evergreen, multi-stemmed shrub or small tree. It bears an interesting flower, large red ball fruit with long projecting stamens. It grows up to an altitude of 200 m. It grows rarely over 20 ft. high, bole crooked; or a larger tree over 100 ft. high and 8 ft. girth, in closed forest. The plant has rough bark and leaves that measure 7 by 4–5 inches (Vogt 1995). In most African society, Karmadoda take pleasure in huge popularity on account of its varied benefits for treating different illness. Many uses of these plants have been recorded in traditional treatment against diseases throughout East, middle and West African (Abbah et al. 2010).

33.3 Distribution

African plants constitute a rich and still underexplored source of natural products of potential medical interest. *N. latifolia* Smith is a straggling shrub or small tree (Ajoke et al. 2015) of about 4 m high abundantly spread in all inter-tropical Africa, especially in drylands.

Nauclea latifolia is spread in tropical regions of Africa and Asia (Gidado et al. 2005), in Savanna forest regions, such as Cameroon, Nigeria, Niger, Congo Tanzania, Malawi, Chad, Sudan, Ethiopia … etc. (Michel 2004). In Sudan, this Savanna covered all Southern regions and Southern parts of Kordufan, Darfur and

Fig. 33.2 *Nauclea latifolia* fruit. (Source: Author camera)

Fig. 33.3 External and cross-sectional structure of *N. latifolia* fruit. (Source: Author camera)

Blue Nile (El-Amin, 1990). The moderate temperature ranges between 18 and 38.8 °C and high relative humidity. The soil changes between clay to sandy and composite; near streams and rivers (Abdel-Rahman 2011).

33.4 Biochemistry and Composition

Nauclea latifolia is rich in soluble sugars, fibre and proteins with high lysine. As well as, it's richer than orange and baobob (*Adansonia digitata*) in vitamin C (ascorbic acid). The beta-carotene of this wild fruit is higher than pumpkin fruit. *N. latifolia* contains high amount of energy value. Moreover, the ash content is high in manganese, calcium and iron, while amino acids are rich in arginine, leucine and valine (Abdel-Rahman et al. 2014a). As well as, many pharmaceutical compounds mentioned subsequently.

33.4.1 Chemical Composition

Karmadoda fruit (*N. latifolia*) contains high percentages fiber, carbohydrates, ash, ascorbic acid and beta-carotene (Abdel-Rahman et al. 2014a; Agyare et al. 2006). The percentages of moisture, crude protein, fats, crude fibre, ash and Carbohydrates were obtained by many researchers. Abdel-Rahman (2014a, b) reported 60.52, 6.85, 1.05, 16.03, 2.48 and 73.59%, respectively. As well as, Eze and Ernest (2014) recorded 44.72, 15.42, 1.74, 35.88, 8.15 and 38.79%, respectively. While, Abdel Muti (2002) was reported 6.40% (crude protein), 3.20% (fat), 3.00% (crude fibre), 5.10 (ash) and 82.30 (carbohydrates). Moreover, Abdel-Rahman (2011) recorded further chemical composition of Karmadoda fruit illustrated elevate amounts of starch (39.96%), total sugars (79.86), ascorbic acid (389.80 mg/100 g), ß-carotene

(178.45 IU/100 g vitamin A), total polyphenols (2.76%), pectin (1.02%) and energy value (338.49 Kcal). Ascorbic acid content of Karmadoda is superior to contents of common fruits (orange, guava and tomato).

Besides, the level of ß-carotene of this wild fruit agrees with result obtained by Eze and Ernest (2014) to the same fruit (176.5 mg/100 g), and to that of green beans and pumpkins. According to CODEX/FAO/WHO (2009); the Karmadoda could provide about 30% and 34% of recommended daily intake (RDI) of vitamin A for children and adults, respectively.

33.4.2 Minerals

Karmadoda fruit contains variable quantities of calcium, iron, phosphorus, sodium, magnesium, manganese and potassium, respectively (Nkafamiya et al. 2006; Abdel Muti 1991). Abdel-Rahman (2014a, b) recorded of 6013, 193, 8180, 594, 7193, 50, 7700 mg/ Kg, respectively. Eze and Ernest (2014) reported of 851, 4300, 45 and 3690 mg/ Kg for Ca, P, Mg and K, respectively. Karmadoda fruit is richer in minerals than many common fruits and vegetables. Karmadoda wild fruit have calcium highest than varieties dates (300–1820 mg/kg) from Saudi Arabia (Alghamdi et al. 2018). As well as, it has superiority in iron than strawberry (100.00 mg/100 kg) and grapes (48.9 mg/100 kg). The level of phosphorus in Karmadoda fruit is higher than many common fruits rich in this metal, such as strawberry (2900 mg/kg), guava (1354 mg/100 kg) and fig (1110 mg/100 kg). This fruit contain high percentage of magnesium, about four times comparable to strawberry which is known to be 1000 mg/100 kg (Gil et al. 2006). It is highest than banana (440 mg/kg) in potassium (Kusuma et al. 2017).

33.4.3 Amino Acids

The protein of *N. latifolia* fruit is rich in nutritionally important amino acids (Table 33.1). The essential amino acids arginine, leucine and valine have highest values of 397.96, 318.59 and 214.93 mg/100 g respectively. Followed by lysine, isoleucine, phenylalanine and threonine (196.96, 167.28, 162.55 and 150.20 mg/100 g, respectively.

Histidine and methionine scored lower values of 100.73 and 12.05 mg/100 g, respectively (Abdel-Rahman et al. 2014b). According to FAO/WHO/UNU (2007) leucine and valine amino acids provide 11.65 and 11.81% from recommended daily amounts (for 70 kg body weight). Valine is essential in repair and growth of tissue and maintaining the nitrogen balance in the body and assists, with leucine, to regulate blood sugar and energy levels (Abdel-Rahman 2016). Lysine is one of the important essential amino acids, is required for growth and bone development in children and assists in calcium absorption and also in maintaining the correct nitrogen balance in the body (Osiecki et al. 2004).

Table 33.1 Amino acids profile of Karmadoda (*N. latifolia*) fruits

Essential amino acids (mg/100 g)		Non-essential amino acids (mg/100 g)	
Histidine	100.73	Tyrosine	42.54
Isoleucine	167.28	Alanine	237.46
Leucine	318.59	Aspartic acid	421.43
Lysine	196.96	Glutamic acid	782.76
Methionine	12.05	Glycine	183.70
Phenylalanine	162.55	Serine	156.23
Threonine	150.20	Cystine	20.39
Tryptophan	ND*	Proline	165.98
Valine	214.93	–	–
Arginine	397.96	–	–

* Not detected

Karmadoda fruit is very rich in nutritionally non-essential amino acids glutamic acid (782.76 mg/100 g), aspartic acid (421.43 mg/100 g) and alanine (237.46 mg/100 g). Medium amounts of glycine, proline and serine are represented by 183.42, 165.98 and 156.23 and 156.23 mg/100 g, respectively. It is poor in tyrosine and sulphur amino acids (Abdel Muti 2002).

33.5 Bioactive Compounds

The bioactive compounds isolated from *N. latifolia* fruit are phenols, alkaloids and flavonoids (Ukamaka et al. 2015). The fruits are containing abundant quantities of bioactive compounds, tannins (0.214%), flavonoids (0.433%), saponins (0.833%) and fatty acids esters (Eze and Ernest 2014; Karou et al. 2011). In addition, it has alkaloids (1.407%), phytates (0.377%) and cyanogenic glycosides (9.27 mg/kg) (Eze and Ernest 2014), moreover phosphate, anthraquinones, coumarins and monoterpenes (Antia and Okokon 2014; Bamidele et al. 2014). The water extract of fruits contain negligible values of polyphenol (1.75 mgGAE/g), flavonol (1.00mgQE/g) and flavanol (0.15mgCE/g), these compounds inhibiting capability of some free radicals (Ayeleso et al. 2014).

33.6 Food and Medicinal Utilization

The population in tropical regions relies on wild fruits. Which provide considerably to the diets consumption of rural societies during famine, because they supply many essential nutrients. *N. latifolia* fruits are precious sources of micronutrients and if people consumption is promoted them healthy and resolution some problems of malnutrition in Africa (Nhukarume et al. 2010; Fakim and Schmelzer 2007).

33.6.1 As Food

With a new understanding and focusing of the value of wild fruit trees in providing food security and meeting nutritional needs, the industrial uses of indigenous fruits are wide and increased through last millennium century, from traditional to industrial processing.

In Sudan, 5% from Southern Kordufan state people was used Karmadoda (*N. latifolia*) as food (Abdel Muti 2002). The fruit is eaten fresh, a juice with sugars or as soft drinks. In addition, after over-ripening the fruits are dried naturally as sheets under the sun and used as beverage later during long dry season (Abdel-Rahman et al. 2015; Abdel Muti 1991).

33.6.2 As Medicine

The natural vegetation in all regions is very rich with medicine substances, which give solutions to abundant number of health challenges (Balogun et al. 2016). About 80% of world inhabitants relies on herbal extract and smoke as source of healthcare (Chintamunnee and Mahomoodally 2012; WHO 2001, 2013; Akerele 1993). African plants include a rich and still undiscovered source of natural products of potential medical attention.

N. latifolia is very popular medicinal plant in Western countries of Africa, where it is used to treat jaundice, yellow fever, rheumatism, abdominal pains, hepatitis, diarrhea, dysentery, as well as diabetes (Haudecoeur et al. 2018; Donalisio et al. 2013; Gidado et al. 2008). In addition, hypertension, sleeping sickness, viral disease and prolong menstrual flow (Donalisio et al. 2013; Kerharo 1974). Moreover, *N. latifolia* fruit is recommended for piles, dysentery, colic, emetic, and have menstrual disorders and hypocholesterolemic effects (Akpanabiatu et al. 2016; Omale and Haruna 2011). One more, usage of the fruit for treating haemorrhoid and wound (Akiyana et al. 2011).

Karmadoa extract contain antioxidant potentials representing stronger activities. These antioxidants reduce some free radical (ferric ions); also, defenses lead against oxidants risk by reducing their destructive influence (Adejoh et al. 2016). Furthermore, fruit extract use against bacteria, virus and good effect against HIV/AIDS by increasing the immune deficiency (Eze and Ernest 2014; Hussaein et al. 1999).

In Western Sudan, the Karmadoda fruit is used as a traditional medicinal plant for the treatment of malaria and reduce the high blood-pressure. Furthermore, it used as anti-malaria (Fadipe et al. 2013), anti-diabetic, anti-hypertensive and to treatment the fever. The fruits of *N. latifolia* also utilized against hypocholesterolemic effects and dysentery (James and Ugbede 2011; Abbiw 1990).

The leaves extract of the *N. latifolia* has been used as antimicrobial, a therapy for diabetes, antihypertensive and laxative activities (Enabulele et al. 2017; Kubmarawa et al. 2007; Gidado et al. 2005; Akpanabiatu et al. 2016). The plant roots induced hypothermia, ant-malaria and had antipyretic effects in mice (Abbah et al. 2010; Taïwe et al. 2011). Also, In Nigeria, the roots used usually in mixture treatment of sexually transmitted diseases (Okoli and Iroegbu 2004). They have recommended using of the plant in such treatment.

Traditionally in Sudan (Southern Kordufan) the decoction of the leaves is recommended for stomach upset, especially in children, and the infusion of the root is also used as a remedy for stomach upset in adults. Nworgu et al. (2010) reported that, analysis of the root has shown the presence of sugar, saponins, and flavonoids. Traore-keita et al. (2000) and Emeje et al. (2005) studied a suitable tablet dosage form from leaves, roots and stem barks of *N. latifolia,* a potential antimalaria agent.

On the other hand, the Rubiaceae family is abundant in the tropical regions around the earth (Nino et al. 2006). Many plants belonging to Rubiaceae family are medicinal. The extract of plants was used to produce gold nanoparticles, it might be used to reduce cancer detection and treatment (Odzie-Nwachukwu et al. 2016).

33.7 Other Uses

The extracts from the leaves, seeds, heartwood, bark, roots and fruits of plants have been reported to inhibit metallic corrosion in acidic media (Okafor et al. 2010, Abiola and James 2010). The ethanol extracts from the root, leaves and bark of *N. latifolia* were found to inhibit the corrosion of mild steel in H_2SO_4 solutions (Uwah et al. 2013).

33.8 Conclusion

N. latifolia has been used traditionally in food consumption, but it mainly utilized in the treatment of numerous illness. A number of scientists studied pharmacologic promising of this fruit. However, additional research needed about utilize it in industrial field. As well as, clarify the active compounds and its mechanism when used pharmaceutical in medicine field.

References

Abbah J, Amos S, Chindo B, Ngazal I, Vongtau HO, Adzu B, Farida T, Odutola AA, Wambebe C, Gamaniel KS (2010) Pharmacological evidence favouring the use of *Nauclea latifolia* in malaria ethnopharmacy: effects against nociception, inflammation, and pyrexia in rats and mice. J Ethnopharmacol 127:85–90

Abbiw DK (1990) Useful plants of Ghana. The intermediate technology publications and the royal botanic gardens, kew, pp 154–157

Abdel Muti OMS (1991) Biochemical and nutritional evaluation of famine foods of the Sudan. Ph.D. Thesis, Sudan. University of Khartoum

Abdel Muti OMS (2002) Nutritive value of wild plants of the Sudan. Arab J Food and Nutr 3(3):6–67

Abdel-Rahman NA (2011) A study of some Sudanese edible forest fruits and their nectars. Ph. D. Thesis. Food technology to Sudan Academy of sciences, Khartoum. (Cited by: El-Khalefa, K. F. 1996. Botanical of forest plants. 1st ed. University of Khartoum Press. Khartoum. Sudan

Abdel-Rahman NA (2016) Introduction of amino acids, 1st edn. G. Town, Khartoum, p 49

Abdel-Rahman NA, Ismail IA, Elshafe'a EBB (2014a) Characterization of some Sudanese edible forest fruits. J Agri-Food Appl Sci 2(2):39–44

Abdel-Rahman NA, Ismail IA, Elshafe'a EBB (2014b) Nutritional value of some edible forest fruits. J Agri-Food Appl Sci 2(2):39–44

Abdel-Rahman NA, Ismail IA, Elshafe'a EBB (2015) Some quality attributes of four Sudanese forest fruits nectars. J Agri-Food Appl Sci 3(2):32–38

Abiola OK, James AO (2010) The effects of Aloe Vera extract on corrosion and kinetics of corrosion process of zinc in HCl solution. Corros Sci 52:661–665

Adejoh IP, Barnabas KO, Chiadikaobi OS (2016) Comparative anti-radical activity of five indigenous herbal plants and their polyherbal extract. Int J Biochem Res Rev 11(1):1–10

Agyare C, Mensah AY, Osei-Asante S (2006) Antimicrobial activity and phytochemical studies of some medicinal plants from Ghana. Boletin Latinoamericano y del Caribe de Plantas Medicinalesy Aromaticas 5(6):113–117

Akerele O (1993) Traditional medicine: Nature's medicinal bounty: don't throw it away. World Health Forum 14(4):390–395

Akiyana H, Fuju K, Yamaski O, Ono T, Iwatsuki K (2011) Antibacterial action of several tannins against *staphyloccus aureus*. J Antimicrob Chemother 48(4):487–491

Akpanabiatu MI, Akpan H, Bassey UE, Ufot UF (2016) Influence of *Nauclea latifolia* leaf on some haematological parameters of rats on coconut oil and non-coconut oil meals. Eur J Med Plants 17(4):1–6

Alghamdi AA, Awadelkareem AM, Hossain AMBS, Ibrahim NA, Fawzi M, Ashraf SAS (2018) Nutritional assessment of different date fruits (*Phoenix dactylifera* L) varieties cultivated in Hilal province, Saudi Arabia. Biosci Biotechnol Res Commun 11(2):263–269

Antia BS, Okokon JE (2014) Phytochemical composition and antidiabetic activity of ethanol root extract of Nauclea latifolia. J Phytopharmacol 3(1):52–56

Arise RO, Akintola AA, Olarinoye JB, Balogun EA (2012) Effects of *Nauclea latifolia* stem on lipid profile and some enzymes of rat liver and kidney. Int J Pharmacol 10(3):23–39

Ayeleso AO, Oguntibeju OO, Brooks NL (2014) *In vetro* study on the antioxidant of the leaves and fruits of *Nauclea latifolia*. Sci World J 2014:1–8

Badiaga M (2016) *Nauclea latifolia*: Ethnobotanique-phytochimie-activites biologiques. Éditions universitaires européennes, France

Balogun ME, Besong EE, Obu DC, Obu MSU, Djobissie SFA (2016) *Nauclea latifolia*: a medicinal, economic and pharmacological review. Int J Plant Res 6(2):34–52

Bamidele FA, Ogundipe FO, Shogeyinbo UA (2014) Determination of antibacterial activity and phytochemistry of three herbal plants on clinical isolates. Int J Sci Technol Res 3(12):355–360

Chintamunnee V, Mahomoodally MF (2012) Herbal medicine commonly used against infectious diseases in the tropical island of Mauritius. J Herb Med 2:113–125

CODEX/FAO/WHO (2009) Codex committee on nutrition of food dietary uses. Standard Program. CODEX Alimentarius Commission, Joint FAO/WHO Food Standards Program. Hall, RS. and Palast, MK. Thirty First Session. Düsseldorf

Donalisio M, Nana HM, Ngane RAN, Gasting D, Tchinda AT, Rovito R, Cagno V, Cagliero C, Boyom FF, Rubiolo P, Bicchi C, Lembo D (2013) In vitro anti-Herpes simplex virus activity of crude extract of the roots of *Nauclea latifolia* Smith (Rubiaceae). Complement Altern Med 13(266):1–8

El-Amin MH (1990) Trees and shrubs of the Sudan, 1st edn. pp. 106, 400, 451. Ithaca Press, England

Emeje MO, Isimi CY, Oqua DA, Kunle OO (2005) Some compaction characteristics of the hot water leaf extract of *Nauclea latifolia*: a potential antimalarial agent. J Pharm 5(4):23–30

Enabulele SA, Amusa O, Eseosa Uwadiae E (2017) Phytochemical, antimicrobial and nutritional properties of *Morinda lucida* benth and *Nauclea latifolia* leaf extracts. Int J Sci World 5(1):62–66

Eze OS, Ernest O (2014) Phytochemical and nutrient evaluation of the leaves and fruits of *Nauclea Latifolia* (Uvuru-ilu). Commun Appl Sci 2(1):8–24

Fadipe AL (2014) Some fatty acid esters of the ripe fruits of *Nauclea Latifolia* (Family Rubiaceae). Int J Res Pharm Chem 4(4):783–788

Fadipe AL, Haruna AK, Mohammed I, Ibikunle GF (2013) Phytochemical and *in-vitro* antibacterial evaluation of the extracts, portions and sub-portions of the ripe and unripe fruits of *Nauclea latifolia*. J Med Plants Res 7(11):629–636

Fadipe AL, Haruna AK, Mohammed I (2014a) Antibacterial activity-guided isolation of di-(2-ethylhexyl) phthalate from the acetone-soluble portion of the ripe fruits of *Nauclea latifolia*. J Phytochem Photon 115:245–252

Fadipe AL, Haruna AK, Mohammed I (2014b) Antibacterial activity of 1, 2- 12. Benzenedicarboxylic acid, dioctyl ester isolated from the ethyl acetate soluble sub-portion of the unripe fruits of *Nauclea latifolia*. Int J Pure Appl Biosci 2(1):223–230

Fadipe AL, Haruna K, Mohammed I, Pateh UU (2015) Some volatile constituents of the unripe fruits of *Nauclea latifolia* (Family: Rubiaceae). J Med Plants Stud 3(2):1–4

Fakim AG, Schmelzer GH (2007) Scope of medicinal plants in Africa: their contribution in ensuring access to food and medicine. International Workshop and Investors' Forum. September 24–26, 2007. PROTA, Nairobi, p 11

FAO/WHO/UNU (2007) Protein and amino acids in human nutrition. Joint FAO/WHO World Health Organization and Food and Agriculture Organization of the United Nations. WHO Press, Rome, p 150

Gidado A, Ameh DA, Atawodi SE (2005) Effect of *Nauclea latifolia* leaves aqueous extracts on bloods glucose levels of normal and alloxan-induced diabetic rats. Afr J Biotechol 4:91–93

Gidado A, Ameh DA, Sunday E, Atawodi SE, Ibrahim S (2008) Hypoglycaemic activity of *Nauclea latifolia* Sm. (Rubiaceae) in experimental animals. Afr J Tradit Complement Altern Med 5(2):201–208

Gil MI, Aguayo E, Kader AA (2006) Quality changes and nutrient retention in fresh-cut versus whole fruits during storage. J Agricult Food Chem 54(12):4284–4296

Haudecoeur R, Peuchmaur M, Peres B, Rome M, Taiwe GS, Boumendjel A, Boucherle B (2018) Traditional uses, phytochemistry and pharmacological properties of African *Nauclea latifolia* species: a review. J Ethnopharmacol 212:106–136

Hussein G, Miyashiro H, Nakamura N, Hattori M, Kawahata T, Otake T, Kakiuchi N, Shimotohno K (1999) Inhibitory effects of Sudanese plants on HIV-1 replication and HIV-1 protease. Phytother Res 13(1):31–36. https://www.wikimedia.org/

James O, Ugbede HH (2011) Hypocholesterolemic effects of *Nauclea latifolia* (Smith) fruits studied in albino rats. Am J Trop Med Public Health 1(1):11–21

Kabine O, Samba BM, fatoumata B, Namagan K, Luopou H, Mamadou BA (2015) Anti-oxidative activity of fruit extracts of some medicinal plants used against chronic diseases (diabetes, hypertension) in Kankan, Guinea. J Plant Sci 3(1–2):1–5

Karou SD, Chacondo TT, Ilboudo DP, Simpore J (2011) Sub-saharan Rubiaceae: a review of their traditional uses, phytochemistry and biological activities. Pak J Biol Sci 14(3):149–169

Kerharo J (1974) Historic and ethnopharmacognosic review on the belief and traditional practices in the treatment of sleeping sickness in West Africa. Bull Soc Med Afr Noire Lang Fr 19:400–420

Kubmarawa D, Ajoku GA, Enwerem NM, Okorie DA (2007) Preliminary phytochemical and antimicrobial screening of 50 medicine plants from Nigeria. Afr J Biotechnol 6(14):1690–1696

Kusuma SAF, Mita SR, Firdayani I, Mustarichie R (2017) Study on the antibacterial activity of fruit extracts of klutuk banana (*musa balbisiana* colla) against *shigella dysenteriae* ATCC 13313. Asian J Pharm Clin Res 10(7):220–223

Löfstrand SD, Krüger A, Razafimandimbison SG, Bremer B (2014) Phylogeny and generic delimitations in the sister tribes Hymenodictyeae and Naucleeae (Rubiaceae). Syst Bot 39:307–315

Michel A (2004) Trees, shrubs and lianas of West African dry zones. Margraf Publishers GMBH, Paris, pp 135–515

Nhukarume L, Chikwambi Z, Muchuweti M, Chipurura B (2010) Phenolic content and antioxidant capacities of *Parinari curatelifolia*, *Strychnos spinosa* and *Adansonia digitata*. J Food Biochem. Special Issue: Antioxidants in Foods: Content, Characterization and Physiological Roles 34(1):207–221

Nino J, Narvaez DM, Mosquera OM, Correa YM (2006) Antibacterial, antifungal and cytotoxic activities of Eight Asteraceae and Rubiaceae plants from Colombian biodiversity. Braz J Microbiol 37:566–570

Nkafamiya II, Manji AJ, Modibbo UU, Umaru HA (2006) Biochemical evaluation of Cassipourea congoensis (Tunti) and *Nuclea latifolia* (Luzzi) fruits. Afr J Biotechnol 5(24):2461–2463

Nworgu ZAM, Onwukaeme DN, Afolayan AJ, Amaechina FC, Ayinde BA (2008) Preliminary studies of blood pressure lowering effect of *N latifolia* in rats. Afr J Pharm Pharmacol 2(2):037–041

Nworgu ZAM, Owolabi OJ, Atomah JE (2010) Effect of the ethanolic extract of *Nauclea latifolia* (Family: Rubiaceae) on the isolated uterus of non-pregnant rats. Int J Pharm 4(1):48–53

Odzie-Nwachukwu SO, Etuk-Udo G, Obayemi JD, Anuku N, Ododusanya OS, Malatesta K, Chi C, Soboyeji WO (2016) Biosynthesis of gold nanoparticles from *Nauclea latifolia* leaves. Adv Mater Res 1132:36–50

Okafor PC, Ebenso EE, Ekpe UJ (2010) *Azadirachta indica* extracts as corrosion inhibitors for mild steel in acid medium. Int J Electrochem Sci 5:978–982

Okoli AS, Iroegbu GU (2004) Evaluation of extracts of *Anthocleista djalonensis*, *Nauclea latifolia* and *Uvaria afzalii* for activity against isolates from cases of non-gonococcal urethritis. J Ethnopharmacol 92:135–144

Omale J, Haruna HU (2011) Hypocholesterolemic effects of Nauclea latifolia(Smith) fruit studied in albino rats. Am J Trop Med 1(1):11–21

Osiecki H, Meeke F, Smith J (2004) The encyclopaedia of clinical nutrition: the nervous system, vol 1. Bio Conceps Publishing, Eagle Farm

Taïwe GS, Bum EN, Talla E, Dimo T, Weiss N, Sidiki N, Dawe A, Moto FC, Dzeufiet PD, De Waard M (2011) Antipyretic and antinociceptive effects of *Nauclea latifolia* root decoction and possible mechanisms of action. Pharm Biol 49(1):15–25

Traore-Keita F, Gasquet M, Di Giorgio C, Ollivier E, Delmas F (2000) Antimalarial activity of four plants used in traditional medicine in Mali. Phytother Res 14:45–47

Ukamaka O, Paul N, Ikechi E (2015) Antiviral effects of *Nauclea latifolia* on Newcastle disease virus (NDV). Sky J Microbiol Res 3(1):1–5

Uwah IE, Okafor PC, Ebiekpe VE (2013) Inhibitive action of ethanol extracts from *Nauclea latifolia* on the corrosion of mild steel in H_2SO_4 solutions and their adsorption characteristics. Arab J Chem 6(3):285–293

Vogt K (1995) A field worker's guide to the identification, propagation and uses of: Common trees and shrubs of dryland Sudan. pp. 100, 101, 125. SOS Sahel International (UK), London. Printed by: Antony Rowe Ltd. Chippenham

WHO (2001) In: Chaudhury RR, Rafei UM (eds) Traditional medicine in Asia. WHO, World Health Organization. Reginoal Office for South Asia. SEARO Regional Publication No. 39, New Delhi

WHO (2013) Traditional medicine strategy 2014–2023. World Health Organization, Geneva

Wikimedia (2016) Available in: https://www.google.com/search?tbm=isch&q=wikimedia,+Nauclea+latifolia+tree+image&chips=q:wikimedia+nauclea+latifolia+tree+image

Chapter 34
Physalis peruviana Linnaeus, an Update on its Functional Properties and Beneficial Effects in Human Health

Luis Puente, Diego Nocetti, and Alejandra Espinosa

34.1 Introduction

34.1.1 Plant

Physalis peruviana, whose botanical name is *Physalis peruviana Linnaeus*, corresponds to a herbaceous species, semi shrubby and native from South America, belonging to the family Solanaceae (Othman et al. 2014; García-arias et al. 2016). The cultivation of this plant has spread throughout the world receiving various denominations, Colombia and South Africa, Peru, Venezuela, Ecuador and in English speaking countries it is called Golden berry, Goose berry or Cape goose berry (Puente et al. 2011). The plant without protective stakes can reach a height of 1.0–1.5 m, while with stakes reaches and exceeds 2.0 m in height (Fischer et al. 2014).

L. Puente (✉)
Departamento de Ciencias de los Alimentos y Tecnología Química, Facultad de Ciencias Químicas y Farmacéuticas, Universidad de Chile, Santiago, Chile
e-mail: lpuente@uchile.cl

D. Nocetti
Departamento de Tecnología Médica, Facultad de Ciencias de la Salud, Universidad de Tarapacá, Arica, Chile

Programa de Doctorado en Ciencias Médicas, Universidad de la Frontera, Temuco, Chile

A. Espinosa
Departamento de Tecnología Médica, Facultad de Medicina, Universidad de Chile, Santiago, Chile

A. A. Mariod (ed.), *Wild Fruits: Composition, Nutritional Value and Products*,
https://doi.org/10.1007/978-3-030-31885-7_34

34.1.2 Fruit

The plant has the potential to produce around 300 fruits, inside which are between 100 and 200 small flattened seeds. The fruits are ovoid, 1.25 and 2.50 cm in diameter, weight 4 to 10 g (Puente et al. 2011; Tapia and Fries 2007). Due to changes in the composition of the fruits during their ripening process, they modify their color from green to a yellow-orange due to the reduction of chlorophyll and an increase in the content of β-carotenes (Gutierrez et al. 2008).

34.1.3 Calyx

Calyx corresponds to a structure that completely covers and surround the fruit and since it hangs down it has the appearance of a Chinese lantern (García-arias et al. 2016). Its functions are essentially to protect the fruit during its growth and maturation, avoiding the attack of pathogens, insects and birds, as well as the action of adverse weather conditions. In addition, it constitutes a source of carbohydrates during the first 20 days of fruit ripening, contributing to its conservation once harvested (Franco et al. 2014). This structure is of high importance for the fruit, since it extends up to six times its shelf life time, reaching 30 days when it is present (Cerdeño and Montenegro 2004). Usually the calyces are considered waste products, however, in traditional medicine has been used to infuse them to reduce the symptoms of arthritis (Toro et al. 2014).

34.2 Chemical Composition Issues

34.2.1 Proximal Composition

Table 34.1 shows the proximal characterization of the *Physalis peruviana* fruit in terms of the typical chemical approach. Among them, highlights the low fat content of the fruit, which on average did not exceed 1.0% of the total weight of this, a situation that contrasts with the high moisture of the fruit (77.3% to 85.5%) which reflects a high water content, which added to its high concentration of carbohydrates, gives the fruit greater protection in structural terms (Cortés Díaz et al. 2015). It is necessary to emphasize that, on average the pH of the fruit is 4.4 which is propitious to ensure the activity of vitamin C (Cortés Díaz et al. 2015).

It is possible that the differences observed in some reports are due to variations between one crop region and another, since the climatic conditions, the characteristics of the soil and other multiple factors intervene directly in the qualities of the fruit.

Table 34.1 Proximal analysis of the fruit of *Physalis peruviana* L.

Component	Repo de Carrasco and Encina (2008)	Rodrigues et al. (2009)	Ramadan (2011)	Briones-Labarca et al. (2013)	Yıldız et al. (2015)	Cortés Díaz et al. (2015)
Moisture (%)	78.9	81.0	78.9	77.3	–	85.5
Ash (%)	1.0	0.8	1.0	1.9	3.0	0.8
Protein (%)	1.9	1.9	0.1–0.3	2.0	1.7	1.5
Fat (%)	<0.1	3.2	0.2	0.9	0.2	0.5
Carbohydrates (%)	17.3	13.2	19.6	12.2	13.9	11.9
pH	3.4	–	–	3,8	6,1	–
Total energy (kcal/100 g)	76.8	88.7	–	–	–	58.0

Table 34.2 Fatty acid composition in fruit pulp of *Physalis peruviana L.*

Fatty acid	Amount of fatty acids in mg per 100 g of pulp	
	Ramadan and Moersel (2007)	Rodrigues et al. (2009)
Palmitic (C16:0)	19.30	9.38
Palmitoleic (C16:1 n-7)	7.52	0.71
Stearic (C18:0)	1.87	2.67
Oleic (C18:1 n9)	22.20	10.03
Linoleic (C18:2 n6)	22.70	72.42
α-Linolenic (C18:3 n3)	0.63	0.32
Arachic (C20:0)	0.21	1.36
Behenic (C22:0)	–	0.26
Lignoceric (C24:0)	0.65	0.24

34.2.2 Fatty Acids

The general structure of a fatty acid corresponds to a hydrocarbon chain that has a carboxyl group at one end and a methyl group at the other (Calder 2005). In the case of *Physalis peruviana*, the fat comes mainly from the seeds and the fat are mostly in the form of saturated and polyunsaturated fatty acids, which have one or more double bonds between their carbons, of them, the highest concentrations reported correspond to palmitic acid and linoleic acid as can be seen in Table 34.2.

Palmitic or hexadecanoic acid corresponds to a saturated long chain fatty acid with a skeleton of 16 carbons. Because the endogenous biosynthesis, by way *of novo* lipogenesis, of this compound may be insufficient, its level in the body must be compensated through dietary intake. Palmitic acid fulfills a structural role in the cellular phospholipid bilayer, acts as a surfactant at pulmonary level (Carta et al. 2017) and participates in the palmitoylation of proteins, a process involved in neuronal regeneration and synaptic plasticity (González Polo 2011). On the other hand, linoleic acid corresponds to a polyunsaturated long chain fatty acid, formed by 18 carbons and that has double bonds in carbons 9 and 12. This compound cannot be biosynthesized and

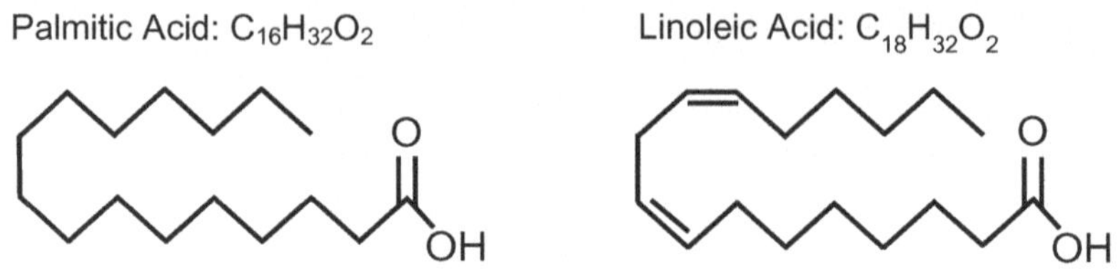

Fig. 34.1 Structural formula of the main fatty acids present in *Physalis peruviana*

therefore, it must be consumed since it is an essential component of ceramides and is involved in the maintenance of the transdermal water barrier of the epidermis (see Fig. 34.1) (Whelan and Fritsche 2013).

34.2.3 Vitamins

Vitamins are organic compounds whose consumption is necessary for the normal physiological function of the human organism, especially at the metabolic level, so that insufficient intake could cause some diseases (Thomas 2006). Because vitamins are not synthesized endogenously by the body, they must be obtained from food (Kennedy 2016). The fruit of *Physalis peruviana* is considered an important source of vitamin A (retinol) and pro-vitamin A (β-carotenes), also has high amounts of vitamin B_3, B_6 and C (ascorbic acid) as can be seen in Table 34.3, and Fig. 34.2) .

Vitamin A: is a lipo-soluble compound with antioxidant activity and plays an important role in vision, reproduction, embryogenesis, integrity of membranous structures and in the growth and differentiation of epithelia (Hamishehkar et al. 2016). The main precursor of vitamin A is β-carotene, a compound found mainly in vegetables and fruits, such as *Physalis peruviana*, giving it an orange color and antioxidant properties (Thomas 2006). Vitamins B or vitamin B complex, corresponds to a group of vitamins that, even though they do not have chemical structural similarity, were grouped by their water-soluble character and by playing coenzyme functions at the cellular level (Kennedy 2016). Of these, the fruit of *Physalis peruviana* presents detectable amounts of vitamin B_1, B_2, B_3 and B_6, with the last two being present in a greater proportion (Table 34.3). Vitamin B_3 is involved in the production of energy at the cellular level, acts as a modulator of immune cells, in addition to presenting antioxidant properties. Vitamin B_6 is a speed-limiting cofactor in the synthesis of neurotransmitters such as dopamine, serotonin, gamma-aminobutyric acid (GABA), noradrenaline and the hormone melatonin; it also has a direct effect on immune function and transcription / gene expression (Kennedy 2016). Vitamin C: is a water-soluble compound widely found in citrus fruits and some vegetables, whose consumption is essential due to its role in the synthesis of collagen and neurotransmitters, as well as its immunomodulatory and antioxidant properties (Thomas 2006; Hamishehkar et al. 2016). This fact reduces the damage that reactive oxygen and nitrogen species induce in macromolecules such as lipids, proteins and DNA (Puente et al. 2011).

Table 34.3 Vitamin content of the fruit *Physalis peruviana* L.

	Milligrams of vitamin per 100 g of pulp				
Vitamins	Ramadan and Mörsel, (2004)	Humberto (2012)	Cortés Díaz et al. (2015)	Vega-Gálvez et al. (2016)	Llano et al. (2018)
Carotenes (mg/100 g)	1.60	–	–	–	0.68
Thiamin B_1 (mg/100 g)	0.10	0.18	0.01	ND	–
Riboflavin B_2 (mg/100 g)	0.03	0.03	0.17	–	–
Niacin B_3 (mg/100 g)	1.70	–	0.80	26.59	–
Pyridoxine B_6 (mg/100 g)	–	–	–	24.80	–
Retinol A (mg/100 g)	–	–	0.52	–	–
Ascorbic acid C (mg/100 g)	43.00	–	20.00	–	16.51

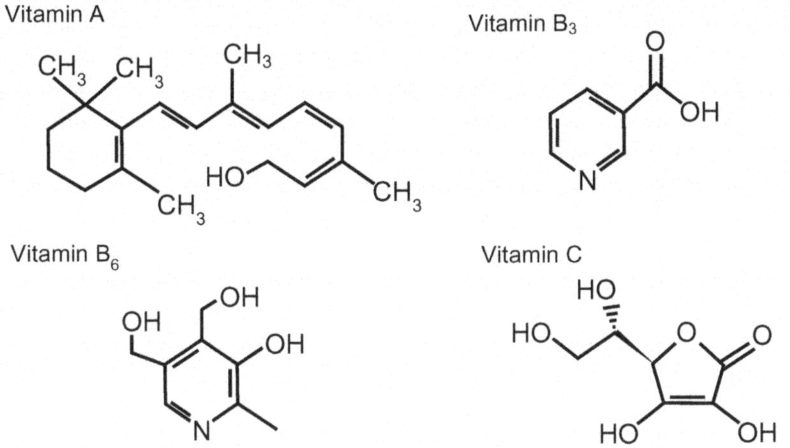

Fig. 34.2 Structural formula of the main vitamins present in Physalis peruviana

Additionally, the different parts of the plant and the fruit of the Physalis peruviana, have a variable content of vitamin E (tocopherol), substance with great antioxidant capacity and protective role against various forms of cancer and pathologies such as rheumatoid arthritis and cataracts (Restrepo et al. 2009). A study by Vega-Gálvez et al., (Vega-Gálvez et al. 2016) reported an α-tocopherol content of 10.70 ± 0.28 g / kg in the lipid fraction of the *Physalis peruviana* fruit (Vega-Gálvez et al. 2016). The main components in whole berries and *Physalis peruviana* seed oils are β-tocopherol and γ-tocopherol, while in the pulp and oil of the fruit skin, is mainly δ-tocopherol and α-tocopherol (Olivares-Tenorio 2017). The content of tocopherol and carotenes plays an important antioxidant role at the lipid level, mainly thanks to its ability to break chain reactions by interacting with the peroxide radicals present in the fatty acids (Ramadan and Mörsel 2004).

34.2.4 Minerals

The minerals correspond to inorganic substances that are naturally found in a wide variety of foods. Given that minerals are essential for a wide variety of metabolic and physiological processes in the human body, it is that about twenty different minerals are required for proper metabolism (Williams 2005). In general, the physiological functions of minerals are given by their participation in muscle contraction, the acid-base balance of the blood, regulating the heart rate, the conduction of the nervous impulse, oxidative phosphorylation, the activation of enzymes and the transport of oxygen (Speich et al. 2001). Minerals that have been reported in greater quantity in the pulp of the fruit of the *Physalis peruviana* are calcium, potassium, phosphorus and magnesium, in average concentrations of 31.7; 292.6; 51.1 and 43.0 mg per 100 g of pulp (see Table 34.4). Calcium is a mineral necessary for the development and growth of bone structures, and it also has possible benefits in the regulation of blood pressure and in the control of body weight (Ryan-Harshman and Aldoori 2005; Eken et al. 2016). Potassium plays an important role in the protection against hypertension and, possibly, in the improvement of bone health (Weaver 2013). Phosphorus has an important role as an enzymatic component; it also participates in phosphorylation processes allowing the activation of certain compounds such as hormones and in the acid-base regulation of the organism (Takeda et al. 2012). Finally, magnesium has an important role as a stabilizer of the nervous system and participates in muscle contraction as an activator of alkaline phosphatase, and some studies indicate a possible benefit in the control of cardiovascular pathologies, osteoporosis and diabetes (Ryan-Harshman and Aldoori 2005; Eken et al. 2016).

Table 34.4 Mineral content determined in the fruit of Physalis peruviana L.

	Content in mg by 100 g of pulp				
Mineral	Leterme et al. (2006)	Musinguzi et al. (2007)	Rodrigues et al. (2009)	Erkaya et al. (2012)	Eken et al. (2016)
Iron	0.09	–	1.47	3.70	36.00
Magnesium	19.00	7.00	34.70	9.24	145.00
Zinc	0.28	–	0.49	9.75	11.40
Calcium	23.00	28.00	9.00	79.53	19.10
Potassium	467.00	210.00	347.00	146.50	–
Phosphorus	27.00	34.00	–	92.30	–
Sodium	6.00	–	1.10	48.90	–
Copper	0.64	–	0.28	–	–
Manganese	0.20	–	0.26	0.27	–

34.3 Specific Bioactive Substances of the Fruit

34.3.1 Withanolides

Its name derives from "withan" referred to Withania and "-olide" associated with the presence of steroidal lactones in these compounds (Zhang et al. 2012). The withanolides were isolated mainly in members of the Solnaceae family, belonging to the genera: Acnistus, Datura, Dunalia, Jaborosa, Lycium, Nicandra, Tubocapsicum, Physalis and Withania (Fang et al. 2012; Zhang and Timmermann 2016). However, of the 24 known structural types, more than 13 of them (I, III, V, VIII, XI, XVII, XIX, XX, XXI, XXII) have been reported in *Physalis peruviana* (Zhang et al. 2014).

Witanolides have gained great interest in the scientific community due to their biological and pharmacological activity, which gives them immunomodulatory (Zhang and Timmermann 2016), antitumor (Zhang et al. 2014), anti-inflammatory (Sang-ngern et al. 2016) and antibacterial properties (Göztok and Zengin 2013). One of the most studied witanolides in *Physalis peruviana* corresponds to 4β-hydroxywitanolide E (see Fig. 34.3), which has significant antiproliferative activity, with low IC50 values, in liver cancer cell lines (Hep G2 and Hep 3B), breast cancer MDA-MB-231 and MCF-7) and lung cancer (A549) (Lan et al. 2009). The same behavior was reported for Witanolide E, possibly due to its scarce structural differences, which do not cause significant changes in biological activity (Lan et al. 2009).

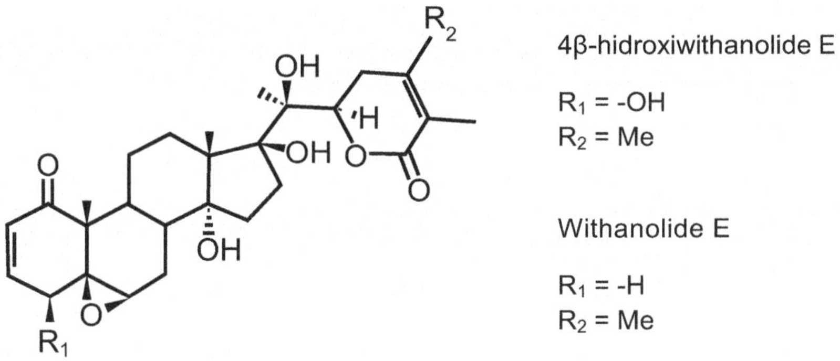

Fig. 34.3 Structural formula of the main withanolides present in *Physalis peruviana* (Ramadan 2011)

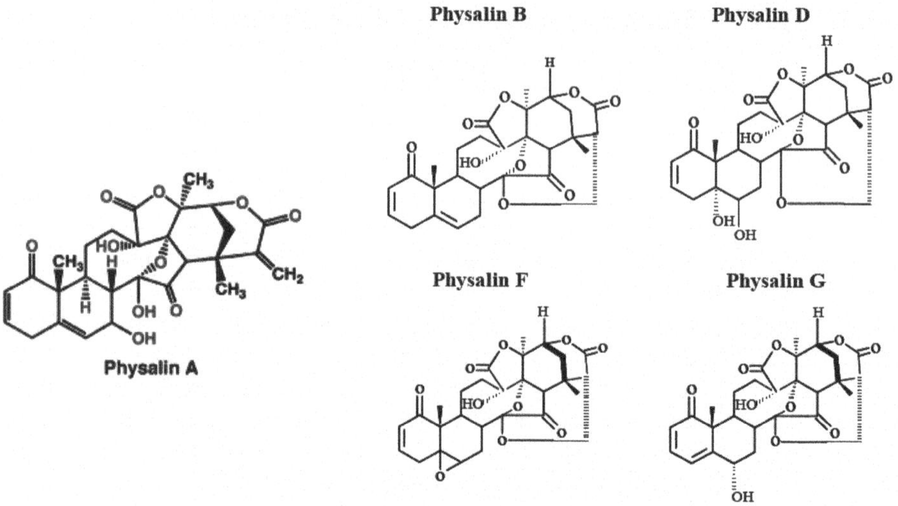

Fig. 34.4 Structural formula of the Physalins in *Physalis peruviana*

34.3.2 *Physalins*

Correspond to a group of pseudo-steroids with a high level of oxidation, present in species of the genus Physalis, 60 of them have been studied up to this moment (Sun et al. 2017). Several studies have evaluated the properties of these compounds in isolation, determining their anti-tumor, antibacterial and immunosuppressive power (Puente et al. 2011; Olivares-Tenorio et al. 2016). Likewise, isoforms A, B, D and F extracted from *Physalis Angulata* (Wu et al. 2004a) have shown antitumor activity in hepatoma, cervical cancer and nasopharyngeal cancer cell lines (Chiang et al. 2018). Figure 34.4 shows structural formula of Physalins.

34.3.3 *Peruvioses*

They are aliphatic sucrose esters, present in several varieties of the Physalis family. This class of compounds exhibits biological properties such as insecticide, antibacterial, anti-inflammatory, modulator of multidrug resistance and inhibitor of plant growth (Cicchetti et al. 2018). The first two variants of this group of compounds were reported for the first time by Franco et al. (2014) in 2014, those who detected isoforms A and B in extracts of the calyces of *Physalis peruviana* (Franco et al. 2014). Subsequently, a study by Bernal et al. 2018 (Bernal et al. 2018), described peruvioses C to E in fruits of *Physalis peruviana*, determining the hypoglycemic activity of these compounds. Finally, the work of Cicchetti et al. (Cicchetti et al. 2018) in 2018, allowed knowing the varieties F to M present in the calyces of

Physalis peruviana, with potential as an anti-aging cosmetic compound. Figure 34.5 shows structural formula of Peruvioses.

34.3.4 Saponins

They correspond to glycosides present in various species of the plant kingdom, characterized by containing in its structure a triterpene and one or more chains of sugars (Güçlü-Üstündağ and Mazza 2007). The effects of saponins on biological systems that can range from membrane-permeabilising, immunostimulant, hypocholesterolaemic and anticarcinogenic properties and are being subjected to research in the last 20 years (Francis et al. 2002).

Regarding to this substances there are at least 14 spices of saponins present in *Physalis peruviana* fruits. A detailed description of its chemical and physical properties can be found in (Ahmad and Basha 2007).

34.4 Effects on Health

The fruit of *Physalis peruviana* has traditionally been used as antispasmodic, diuretic, antiseptic, sedative and analgesic (Puente et al. 2011). These diverse uses are based on the ancestral knowledge of several Andean cultures and their folklore.

34.4.1 Antiproliferative Effect

Among the multiple in vitro studies conducted to evaluate the properties of *Physalis peruviana* and its components, highlights the Yen and Cols. in 2018, whose objective was to evaluate the effects of administering 4β-hydroxywithanolide E obtained from the crude extract of *Physalis peruviana*, in a lung cancer cell line (H1299), demonstrating significantly the antiproliferative action of the compound through cell damage induced by apoptosis and arrest of the cell cycle in G2 / M (Yen et al. 2010). The same effect was evaluated by Zavala et al. (2013) in 2013, administering an ethanolic extract of the plant to colon cancer cells (colo-205) and chronic myeloid leukemia cells (K562), proving the cytotoxic effect on the cell lines studied (Zavala et al. 2013), this effect had previously been reported by other study groups (Chang et al. 2016). Wu et al. administered an ethanolic extract of the plant to hepatoma cells (Hep G2) demonstrating an antihepatoma effect and anti-inflammatory properties (Wu et al. 2004b), these results are mainly explained by the mechanism of apoptosis through the mitochondrial pathway (Wu et al. 2004a). The antiproliferative effect has also been evaluated in other cancer cell lines, such as prostate (Xu et al.

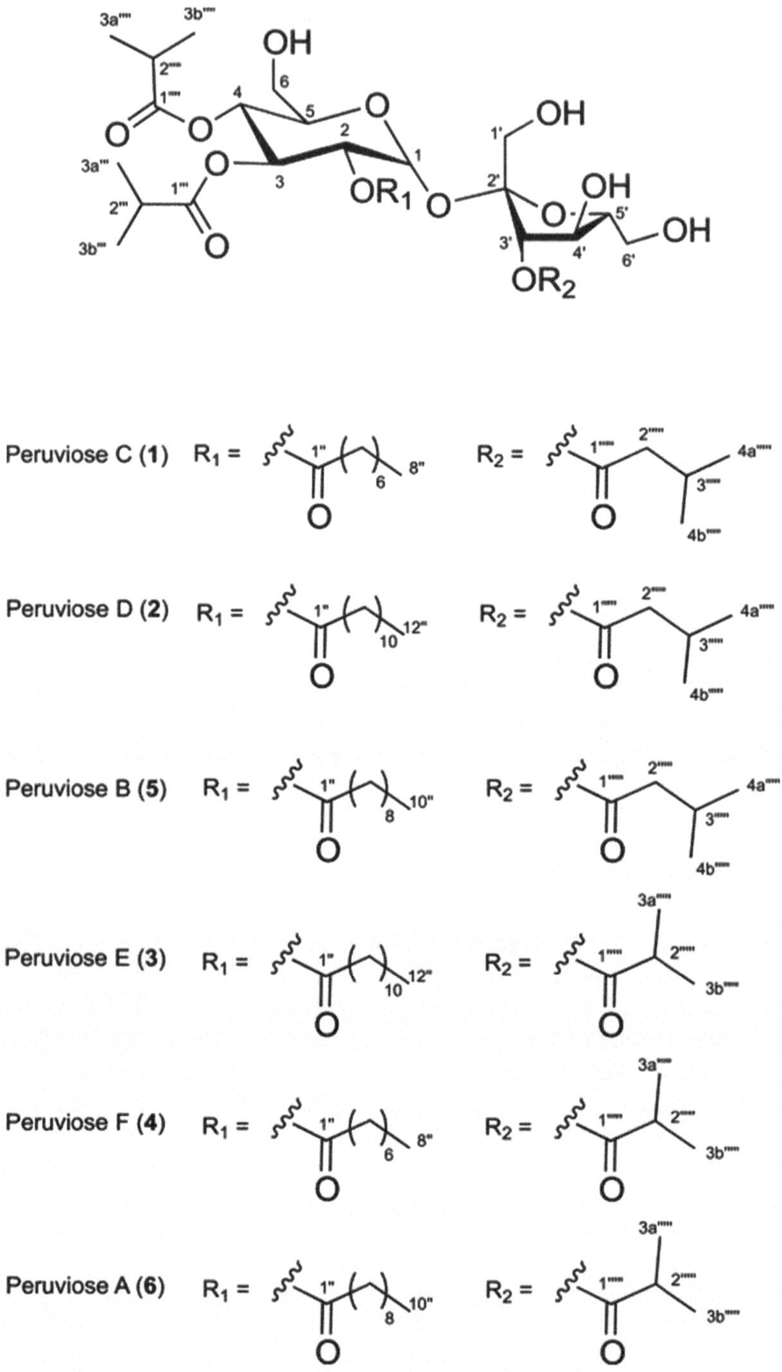

Fig. 34.5 Structural formula of the Peruvioses in *Physalis peruviana*

2017), renal (Xu et al. 2017), colorectal (Quispe-Mauricio et al. 2018) and mammary (Peng et al. 2016).

34.4.2 Antibacterial Effect

Bernarda et al. reported the antibacterial action of the ethanolic extract of the *Physalis peruviana* plant and its leaves, after its application on *Listeria* spp. cultures (Bernarda et al. 2017).

34.4.3 Hypoglycemic Effect

Physalis peruviana has the potential to act as hypoglycemic and antidiabetic, this property has been widely studied, determining that there is an action of the ethanolic extract of *Physalis peruviana* inhibiting carbohydrases at intestinal level (Rey et al. 2015), the same effect was achieved by administering a lyophilized formulation of ethanolic extract of *Physalis peruviana* (Whelan and Fritsche 2013) and recently it was determined that the inhibitory action on α-amylase was mainly produced by the action of the peruvioses (A - F) that the fruit and plant contains (Bernal et al. 2018).

34.4.4 Antioxidant Effect

On the other hand, the studies carried out in animal models have been mainly based on the use of rats as subjects of study. In that sense, Wu et al. tested the antioxidant action of *Physalis peruviana* in an ethanolic extract of the plant, which was administered to male Wistar rats, determining that the flavonoid content of the plant was the agent that demonstrated an antioxidant role (Wu et al. 2005). Hassan et al. (2017) evaluated the antioxidant activity of *Physalis peruviana* in a model of hepatocellular carcinoma with Rattus rattus males, showing the compound's power as a free radical antagonist and an antioxidant cellular response enhancer, which increased the hepatic biomarkers (Franco et al. 2007).

34.4.5 Anti-Inflammatory Effect

A study, conducted in female Wistar rats, showed, through the administration of peruvioses A and B isolated from the calyces of the *Physalis peruviana*, the anti-inflammatory properties of these compounds whose mechanism of action was determined through the inhibition of rust nitrate and prostaglandin E2 (Franco et al. 2014). Franco et al. (2007) in 2007 they demonstrated the potent anti-inflammatory effect of *Physalis peruviana* calyx extracts through a murine model of atrial edema induced by 12-tetradecanoylphorbol 13-acetate, where a decrease in inflammation was achieved in around 70%.

34.4.6 Hypolipidemic Effect

Campos et al. (Campos Florián et al. 2011) induced hyperlipidemia in *Mus musculus* var. swis males and demonstrated the power of *Phsyalis peruviana* extract to reduce blood lipid levels.

34.4.7 Hypoglycemic Effect

Oral administration of *Physalis peruviana* ethanolic extracts showed an increase in insulin sensitivity and a reduction of hyperglycemia in rats with type 2 diabetes induced (Sathyadevi et al. 2014).

34.4.8 Human Studies

Reyes-Beltrán et al. (2015) studied the effect of *Physalis peruviana* consumption on the lipid profile of patients with hypercholesterolemia during 8 weeks, detecting a decrease in serum levels of total cholesterol (9.93%) and LDL-C (14.79%).

34.5 Final Remarks and Future Trends

Physalis peruviana has been known and widely used for centuries, extending its cultivation from the region of the Ecuadorian Andes in South America, to other latitudes of the world. Its high content of vitamins, minerals, essential fatty acids and antioxidants, make the interest in studying this plant and its fruit, has been increasing in recent decades, allowing both a greater direct consumption of the fruit and juice made with it, as the development of ointments and oils for medicinal purposes. Maximize the bioavailability of the components of the *Physalis peruviana* fruit and its performance in terms of economic cost and health benefits, is of great relevance to further massify its use and consumption, however, it is necessary to develop new forms of cultivation, processing, storage and handling of *Physalis peruviana* and the byproducts derived from it. The high nutritional content of *Physalis peruviana*, make it a potential component for the preparation of functional foods and beverages, also, allow it to be a good candidate as a compound used in phytomedicine as an adjuvant or for the treatment of pathologies such as hypertension and diabetes, fields in which more evidence is still required. Along with the above, it is essential that the fruit be evaluated with greater emphasis on humans, since there are few epidemiological studies conducted in this field.

References

Ahmad V, Basha A (2007) Spectroscopic data of steroid glycosides: Cholestanes, Ergostanes, Withanolides, Stigmastane. Springer New York, New York

Bernal C-A, Castellanos L, Aragón DM, Martínez-Matamoros D, Jiménez C, Baena Y, et al (2018) Peruvioses A to F, sucrose esters from the exudate of Physalis peruviana fruit as α-amylase inhibitors. Carbohydr Res [Internet] [cited 2018 Sep 3] 461:4–10. Available from: http://www.ncbi.nlm.nih.gov/pubmed/29549750

Bernarda M, Cueva R, Angelica R, León T, Magdalena López M, Yanchaliquín A et al (2017) Antibacterial Effects of Uvilla (Physalis peruviana L.) extracts against Listeria spp. Isolated from Meat in Ecuador. Int J Curr Microbiol App Sci [Internet.] [cited 2018 Sep 3 6(4):1146–1153. Available from. https://doi.org/10.20546/ijcmas.2017.604.141

Briones-Labarca V, Giovagnoli-Vicuña C, Figueroa-Alvarez P, Quispe-Fuentes I, Pérez-Won M (2013) Extraction of β-carotene, vitamin C and antioxidant compounds from Physalis peruviana (Cape Gooseberry) assisted by high hydrostatic pressure. Food Nutr Sci [Internet] [cited 2018 Oct 2] 04(08):109–18. Available from: http://www.scirp.org/journal/doi.aspx?DOI=10.4236/fns.2013.48A014

Calder PC (2005) Polyunsaturated fatty acids and inflammation. Biochem Soc Trans [Internet] [cited 2018 Oct 6] 33(2):423–427. Available from: http://www.ncbi.nlm.nih.gov/pubmed/15787620

Campos Florián J, Bobadilla Villa D, Huamán Bermeo M, Bazán Vásquez M (2011) Effect of Physalis peruviana "tomatillo" fruit extract in Mus musculus var. swis with induced hyperlipidemi. Sci Agropecu [Internet] [cited 2018 Sep 4] 2(2):83–9. Available from: http://www.revistas.unitru.edu.pe/index.php/scientiaagrop/article/view/47

Carta G, Murru E, Banni S, Manca C (2017) Palmitic Acid: physiological role, metabolism and nutritional implications. Front Physiol [Internet] [cited 2018 Oct 6] 8:902. Available from: http://www.ncbi.nlm.nih.gov/pubmed/29167646

Cerdeño MM, Montenegro DM (2004) Plan exportador, logístico y de comercialización de uchuva al mercado de Estados Unidos paraa Frutexpo S.C.I. Ltda. [Internet]. Pontificia Universidad Javeriana [cited 2018 Oct 4]. Available from: http://www.javeriana.edu.co/biblos/tesis/ingenieria/tesis135.pdf

Chang LC, Sang-Ngern M, Pezzuto JM, Ma C (2016) The Daniel K. Inouye College of Pharmacy Scripts: Poha Berry (Physalis peruviana) with Potential Anti-inflammatory and Cancer Prevention Activities. Hawaii J Med Public Health [Internet] [cited 2018 Sep 10] 75(11):353–9. Available from: http://www.ncbi.nlm.nih.gov/pubmed/27920947

Chiang HC, Jaw SM, Chen PM (2018) Inhibitory effects of physalin B and physalin F on various human leukemia cells in vitro. Anticancer Res [Internet] [cited 2018 Sep 10] 12(4):1155–62. Available from: http://www.ncbi.nlm.nih.gov/pubmed/1503404

Cicchetti E, Duroure L, Le Borgne E, Laville R (2018) Upregulation of skin-aging biomarkers in aged NHDF cells by a sucrose ester extract from the agroindustrial waste of Physalis peruviana calyces. J Nat Prod [Internet] [cited 2018 Sep 3] acs.jnatprod.7b01069. Available from: http://pubs.acs.org/doi/10.1021/acs.jnatprod.7b01069

Cortés Díaz GM, Prieto Suárez GA, Rozo Nuñez WE (2015) Bromatological and physicochemical characterization of Physalis peruviana L., and its potential as a nutraceutic food. Cienc en Desarro [Internet] 20(1):87–97. Available from: http://www.scielo.org.co/scielo.php?script=sci_arttext&pid=S0121-74882015000100011&lng=en&nrm=iso&tlng=es

Eken A, Ünlü-Endirlik B, Baldemir A, İlgün S, Soykut B, Erdem O, et al (2016) Antioxidant capacity and metal content of Physalis Peruviana L. fruit sold in markets. J Clin Anal Med [Internet] 7(3):291–4. Available from: http://www.jcam.com.tr/files/KATD-2709.pdf

Erkaya T, Dağdemir E, Şengül M (2012) Influence of Cape gooseberry (Physalis peruviana L.) addition on the chemical and sensory characteristics and mineral concentrations of ice cream. Food Res Int [Internet] [cited 2018 Oct 2] 45(1):331–335. Available from: https://www-sciencedirect-com.ezproxy.biblio.uta.cl/science/article/pii/S0963996911005436

Fang S-T, Liu J-K, Li B (2012) Ten new withanolides from Physalis peruviana. Steroids [Internet] [cited 2018 Sep 2] 77(1–2):36–44. Available from: http://www.ncbi.nlm.nih.gov/pubmed/22037277

Fischer G, José Almanza-Merchán P, Miranda D (2014) Importancla y cultlvo de la uchuva (Physalis peruviana l.) 1. Rev Bras Frutic, Jaboticabal-SP [Internet] [cited 2018 Sep 10] 1(1):1–015. Available from: https://doi.org/10.1590/0100-2945-441/13

Francis G, Kerem Z, Makkar H, Becker K (2002) The biological action of saponins in animal systems: a review. Br J Nutr 88(06):587–605

Franco LA, Matiz GE, Calle J, Pinzón R, Ospina LF, Ospina LF (2007) Actividad antinflamatoria de extractos y fracciones obtenidas de cálices de Physalis peruviana L. Biomedica [Internet] [cited 2018 Sep 9] 27(1):110. Available from: http://www.revistabiomedica.org/index.php/biomedica/article/view/237, 115

Franco L, Ocampo Y, Gómez H, De la Puerta R, Espartero J, Ospina L (2014) Sucrose esters from Physalis peruviana calyces with anti-inflammatory activity. Planta Med [Internet] [cited 2018 Sep 3] 80(17):1605–1614. Available from: http://www.ncbi.nlm.nih.gov/pubmed/25338213

García-arias FL, Osorio-guarín JA, Zarantes VMN, Calderón JM, Ruiz N, Castellanos L, et al (2016) Antioxidative and hepatoprotective effects of Physalis peruviana extract against acetaminophen-induced liver injury in rats. Food Chem [Internet] [cited 2018 Jul 22] 2(4):724–731. Available from: https://doi.org/10.1016/j.foodres.2010.12.042

González Polo V (2011) Estudio del papel de las palmitoil-proteínas en la regeneración del sistema nervioso, la importancia de la palmitoilación y la ubicación en dominios lipídicos de la membrana en su función [Internet]. http://digital.bl.fcen.uba.ar. Universidad de Buenos Aires [cited 2018 Oct 6]. Available from: http://digital.bl.fcen.uba.ar/gsdl-282/cgi-bin/library.cgi?a=d&c=tesis&d=Tesis_4919_GonzalezPolo

Göztok F, Zengin F (2013) The antimicrobial activity of Physalis peruviana L. [cited 2018 Sep 2] Available from: http://journals.beu.edu.tr/Medya/science_296_2013_7_8_3b93a5d7-657c-4f91-b58e-35bd236e50fd.pdf

Güçlü-Üstündağ Ö, Mazza G (2007) Saponins: properties, applications and processing. Crit Rev Food Sci Nutr [Internet] [cited 2018 Oct 1] 47(3):231–258. Available from: http://www.ncbi.nlm.nih.gov/pubmed/17453922

Gutierrez MS, Trinchero GD, Cerri AM, Vilella F, Sozzi GO (2008) Different responses of goldenberry fruit treated at four maturity stages with the ethylene antagonist 1-methylcyclopropene. Postharvest Biol Technol [Internet] [cited 2018 Sep 9] 48(2):199–205. Available from: https://www.sciencedirect.com/science/article/pii/S0925521407003456?via%3Dihub

Hamishehkar H, Ranjdoost F, Asgharian P, Mahmoodpoor A, Sanaie S (2016) Vitamins are they safe? Adv Pharm Bull [Internet] [cited 2018 Oct 4] 6(4):467. Available from: https://www.ncbi.nlm.nih.gov/pmc/articles/PMC5241405/

Hassan HA, Ghareb NE, Azhari GF (2017) Antioxidant activity and free radical-scavenging of cape gooseberry (Physalis peruviana L.) in hepatocellular carcinoma rats model. Hepatoma Res [Internet.] [cited 2018 Sep 9 3(2):27. Available from: http://hrjournal.net/article/view/1753

Humberto J, de Rodríguez SA, Millán P (2012) Caracterización físico química de la Uchuva (Physalis peruviana) en la región de Silvia Cauca. Biotecnol en el Sect Agropecu y Agroindustrial [Internet] [cited 2018 Oct 2] 10(2):188–96. Available from: http://www.scielo.org.co/scielo.php?script=sci_arttext&pid=S1692-35612012000200022

Kennedy DO (2016) B Vitamins and the brain: mechanisms, dose and efficacy–a review. Nutrients [Internet] [cited 2018 Oct 6] 8(2):68. Available from: http://www.ncbi.nlm.nih.gov/pubmed/26828517

Lan Y-H, Chang F-R, Pan M-J, Wu C-C, Wu S-J, Chen S-L, et al (2009) New cytotoxic withanolides from Physalis peruviana. Food Chem [Internet] [cited 2018 Oct 8] 116(2):462–469. Available from: https://www-sciencedirect-com.ezproxy.biblio.uta.cl/science/article/pii/S0308814609002507

Leterme P, Buldgen A, Estrada F, Londoño AM (2006) Mineral content of tropical fruits and unconventional foods of the Andes and the rain forest of Colombia. Food Chem [Internet] [cited 2018 Oct 3] 95(4):644–652. Available from: https://www.sciencedirect.com/science/article/pii/S030881460500138X

Llano SM, Muñoz-Jiménez AM, Jiménez-Cartagena C, Londoño-Londoño J, Medina S (2018) Untargeted metabolomics reveals specific withanolides and fatty acyl glycoside as tentative metabolites to differentiate organic and conventional Physalis peruviana fruits. Food Chem [Internet] [cited 2018 Sep 9] 244:120–127. Available from: https://www-sciencedirect-com.ezproxy.biblio.uta.cl/science/article/pii/S0308814617316564

Musinguzi E, Kikafunda JK, Kiremire BT (2007) Promoting indigenous wild edible fruits to complement roots and tuber crops in alleviating vitamin A deficiencies in Uganda [Internet] [cited 2018 Oct 3]. Available from: http://www.istrc.org/images/Documents/Symposiums/Thirteenth/MusinguziAbs.pdf

Olivares-Tenorio M (2017) Exploring the potential of an Andean fruit: An interdisciplinary study on the cape gooseberry (Physalis Peruviana L.) value chain [Internet]. Wageningen University [cited 2018 Sep 10]. Available from: https://www.wur.nl/en/activity/Exploring-the-potential-of-an-Andean-fruit-An-interdisciplinary-study-on-the-cape-gooseberry-Physalis-Peruviana-L.-value-chain-1.htm

Olivares-Tenorio M-L, Dekker M, Verkerk R, van Boekel MAJS (2016) Health-promoting compounds in cape gooseberry (Physalis peruviana L.): Review from a supply chain perspective. Trends Food Sci Technol [Internet] [cited 2018 Sep 10] 57:83–92. Available from: https://www-sciencedirect-com.ezproxy.biblio.uta.cl/science/article/pii/S0924224416303053

Othman MS, Nada A, Zaki HS, Abdel Moneim AE (2014) Effect of Physalis peruviana L. on cadmium-induced testicular toxicity in rats. Biol Trace Elem Res [Internet] [cited 2018 Jul 22] 159(1–3):278–287. Available from: http://www.ncbi.nlm.nih.gov/pubmed/24728876

Peng CY, You BJ, Lee CL, Wu YC, Lin WH, Lu TL, et al (2016) The roles of 4β-hydroxywithanolide E from Physalis peruviana on the Nrf2-anti-oxidant system and the cell cycle in breast cancer cells. Am J Chin Med [Internet] [cited 2018 Sep 10] 44(03):617–636. Available from: http://www.ncbi.nlm.nih.gov/pubmed/27109152

Puente LA, Pinto-Muñoz CA, Castro ES, Cortés M (2011) Physalis peruviana Linnaeus, the multiple properties of a highly functional fruit: a review. Food Res Int [Internet] [cited 2018 Jul 22] 44(7):1733–1740. Available from: https://www.sciencedirect.com/science/article/pii/S0963996910003571

Quispe-Mauricio A, Callacondo D, Rojas J, Zavala D, Posso M, Vaisberg A. (2018) Cytotoxic effect of physalis peruviana in cell culture of colorectal and prostate cancer and chronic myeloid leukemia. Rev Gastroenterol Peru [Internet] [cited 2018 Sep 10] 29(3):239–46. Available from: http://www.ncbi.nlm.nih.gov/pubmed/19898596

Ramadan MF (2011) Bioactive phytochemicals, nutritional value, and functional properties of cape gooseberry (Physalis peruviana): an overview. Food Res Int [Internet]. [cited 2018 Oct 2] 44(7):1830–1836. Available from: https://www.sciencedirect.com/science/article/pii/S0963996911000147

Ramadan MF, Moersel JT (2007) Impact of enzymatic treatment on chemical composition, physicochemical properties and radical scavenging activity of goldenberry (Physalis peruviana L.) juice. J Sci Food Agric [Internet] [cited 2018 Oct 2] 87(3):452–460. Available from: http://doi.wiley.com/10.1002/jsfa.2728

Ramadan M, Mörsel J-T (2004) Goldenberry, a novel fruit source of fat-soluble bioactives. Inform [Internet] [cited 2018 Oct 2] 15(2):129–30. Available from: https://www.researchgate.net/publication/236147470_Goldenberry_a_novel_fruit_source_of_fat-soluble_bioactives

Repo de Carrasco R, Encina C (2008) Determinación de la capacidad antioxidante y compuestos bioactivos de frutas nativas peruanas [Internet], vol. 74. Revista de la Sociedad Química del Perú. Sociedad Química del Perú; [cited 2018 Oct 2]. pp 108–124. Available from: http://www.scielo.org.pe/scielo.php?script=sci_arttext&pid=S1810-634X2008000200004

Restrepo AM, Cortés Rodriguez M, Márquez CJ (2009) Uchuvas (Physalis peruviana L.) mínimamente procesadas fortificadas con vitamina E. VITAE, Rev la Fac Química Farm [Internet] 16(1):19–30. Available from: http://www.scielo.org.co/pdf/vitae/v16n1/v16n1a03

Rey DP, Fernando Ospina L, Aragón DM (2015) Inhibitory effects of an extract of fruits of Physalis peruviana on some intestinal carbohydrases [Internet]. vol. 44, Rev Colomb Cienc Quím Farm [cited 2018 Sep 9]. Available from: www.farmacia.unal.edu.coArtículodeinvestigacióncientífica

Reyes-Beltrán M, Guanilo-Reyes C, Ibáñez-Cárdenas M, García-Collao C, Idrogo-Alfaro J, Huamán-Saavedra J (2015) Efecto del consumo de Physalis peruviana L. (aguaymanto) sobre el perfil lipídico de pacientes con hipercolesterolemia. Acta Med Per [Internet] [cited 2018 Sep 4] 32(4):195–201. Available from: http://www.scielo.org.pe/scielo.php?script=sci_arttext&pid=S1728-59172015000400002

Rodrigues E, Rockenbach II, Cataneo C, Gonzaga LV, Chaves ES, Fett R (2009) Minerals and essential fatty acids of the exotic fruit Physalis peruviana L. Ciência e Tecnol Aliment [Internet]. [cited 2018 Oct 2] 29(3):642–5. Available from: http://www.scielo.br/scielo.php?script=sci_arttext&pid=S0101-20612009000300029&lng=en&nrm=iso&tlng=en

Ryan-Harshman M, Aldoori W (2005) Health benefits of selected minerals. Can Fam Physician [Internet] [cited 2018 Oct 7] 51(5):673–5. Available from: http://www.ncbi.nlm.nih.gov/pubmed/15934270

Sang-ngern M, Youn UJ, Park E-J, Kondratyuk TP, Simmons CJ, Wall MM, et al (2016) Withanolides derived from Physalis peruviana (Poha) with potential anti-inflammatory activity. Bioorg Med Chem Lett [Internet] [cited 2018 Sep 2] 26(12):2755–9. Available from: http://www.ncbi.nlm.nih.gov/pubmed/27210437

Sathyadevi M, Suchithra ER, Subramanian S (2014) Physalis peruviana Linn. fruit extract improves insulin sensitivity and ameliorates hyperglycemia in high-fat diet low dose STZ-induced type 2 diabetic rats. J Pharm Res [Internet] [cited 2018 Sep 9] 8(4):625–32. Available from: http://jprsolutions.info

Speich M, Pineau A, Ballereau F (2001) Minerals, trace elements and related biological variables in athletes and during physical activity. Clin Chim Acta [Internet] [cited 2018 Oct 7] 312(1–2):1–11. Available from: http://www.ncbi.nlm.nih.gov/pubmed/11580904

Sun C-P, Qiu C-Y, Zhao F, Kang N, Chen L-X, Qiu F (2017) Physalins V-IX, 16,24-cyclo-13,14-seco withanolides from Physalis angulata and their antiproliferative and anti-inflammatory activities. Sci Rep [Internet] [cited 2018 Sep 9] 7(1):4057. Available from: http://www.ncbi.nlm.nih.gov/pubmed/28642618

Takeda E, Yamamoto H, Yamanaka-Okumura H, Taketani Y (2012) Dietary phosphorus in bone health and quality of life. Nutr Rev [Internet] [cited 2018 Oct 7] 70(6):311–321. Available from: http://www.ncbi.nlm.nih.gov/pubmed/22646125

Tapia ME, Fries AM (2007) Guía de campo de los cultivos andinos [Internet]. Available from: http://www.fao.org/docrep/010/ai185s/ai185s.pdf

Thomas DR (2006) Vitamins in aging, health, and longevity. Clin Interv Aging [Internet] [cited 2018 Oct 4] 1(1):81. Available from: https://www.ncbi.nlm.nih.gov/pmc/articles/PMC2682456/

Toro RM, Aragón DM, Ospina LF, Ramos FA, Castellanos L (2014) Phytochemical analysis, antioxidant and anti-inflammatory activity of calyces from Physalis peruviana. Nat Prod Commun [Internet]. [cited 2018 Sep 9] 9(11):1573–5. Available from: http://www.ncbi.nlm.nih.gov/pubmed/25532284

Vega-Gálvez A, Díaz R, López J, Galotto MJ, Reyes JE, Perez-Won M, et al (2016) Assessment of quality parameters and microbial characteristics of Cape gooseberry pulp (Physalis peruviana

L.) subjected to high hydrostatic pressure treatment. Food Bioprod Process [Internet] [cited 2018 Sep 10] 97:30–40. Available from: https://www-sciencedirect-com.ezproxy.biblio.uta.cl/science/article/pii/S0960308515001170

Weaver CM (2013) Potassium and health. Adv Nutr [Internet] [cited 2018 Oct 7] 4(3):368S–377S. Available from: http://www.ncbi.nlm.nih.gov/pubmed/23674806

Whelan J, Fritsche K (2013) Linoleic acid. Adv Nutr [Internet] [cited 2018 Oct 6] 4(3):311–2. Available from: http://www.ncbi.nlm.nih.gov/pubmed/23674797

Williams MH (2005) Dietary supplements and sports performance: minerals. J Int Soc Sports Nutr [Internet] [cited 2018 Oct 7] 2(1):43–9. Available from: http://www.ncbi.nlm.nih.gov/pubmed/18500950

Wu S-J, Ng L-T, Lin D-L, Huang S-N, Wang S-S, Lin C-C (2004a) Physalis peruviana extract induces apoptosis in human Hep G2 cells through CD95/CD95L system and the mitochondrial signaling transduction pathway. Cancer Lett [Internet] [cited 2018 Sep 10] 215(2):199–208. Available from: http://www.ncbi.nlm.nih.gov/pubmed/15488639

Wu S-J, Ng L-T, Chen C-H, Lin D-L, Wang S-S, Lin C-C (2004b) Antihepatoma activity of Physalis angulata and P. peruviana extracts and their effects on apoptosis in human Hep G2 cells. Life Sci [Internet.] [cited 2018 Sep 2 74(16):2061–2073. Available from: http://www.ncbi.nlm.nih.gov/pubmed/14967200

Wu S-J, Ng L-T, Huang Y-M, Lin D-L, Wang S-S, Huang S-N, et al (2005) Antioxidant activities of Physalis peruviana. Biol Pharm Bull [Internet] [cited 2018 Sep 3] 28(6):963–966. Available from: http://www.ncbi.nlm.nih.gov/pubmed/15930727

Xu Y-M, Wijeratne EMK, Babyak AL, Marks HR, Brooks AD, Tewary P, et al (2017) Withanolides from Aeroponically Grown Physalis peruviana and their selective cytotoxicity to prostate cancer and renal carcinoma cells. J Nat Prod [Internet] [cited 2018 Sep 10] 80(7):1981–91. Available from: http://www.ncbi.nlm.nih.gov/pubmed/28617598

Yen C-Y, Chiu C-C, Chang F-R, Chen JY-F, Hwang C-C, Hseu Y-C, et al (2010) 4β-Hydroxywithanolide E from Physalis peruviana (golden berry) inhibits growth of human lung cancer cells through DNA damage, apoptosis and G2/M arrest. BMC Cancer [Internet] [cited 2018 Sep 2] 10(1):46. Available from: http://www.ncbi.nlm.nih.gov/pubmed/20167063

Yıldız G, İzli N, Ünal H, Uylaşer V (2015) Physical and chemical characteristics of goldenberry fruit (Physalis peruviana L.). J Food Sci Technol [Internet] [cited 2018 Oct 1] 52(4):2320–2327. Available from: http://www.ncbi.nlm.nih.gov/pubmed/25829615

Zavala D, Quispe A, Posso M, Rojas J, Vaisberg A, Vaisberg A (2013) Efecto citotóxico de Physalis peruviana (capulí) en cáncer de colon y leucemia mieloide crónica. An la Fac Med [Internet] [cited 2018 Sep 4] 67(4):283. Available from: http://revistasinvestigacion.unmsm.edu.pe/index.php/anales/article/view/1309

Zhang H, Timmermann BN (2016) Withanolide structural revisions by 13 C NMR spectroscopic analysis inclusive of the γ-gauche effect. J Nat Prod [Internet] [cited 2018 Sep 2] 79(4):732–742. Available from: http://www.ncbi.nlm.nih.gov/pubmed/26894655

Zhang H, Samadi AK, Cohen MS, Timmermann BN (2012) Antiproliferative withanolides from the Solanaceae : a structure – activity study. Pure Appl Chem 84(6):1353–1367

Zhang H, Cao C-M, Gallagher RJ, Timmermann BN (2014) Antiproliferative withanolides from several solanaceous species. Nat Prod Res [Internet] [cited 2018 Sep 2] 28(22):1941–1951. Available from: http://www.ncbi.nlm.nih.gov/pubmed/24871278

Chapter 35
Borassus flabellifer L. Palmyra Palm (Daleeb): Biochemistry, Functions and Utilization

Isam M. Abu Zeid and Abdul Rahman A. FarajAlla

35.1 Introduction

Since antiquity, man depends on plants for his existence. Plants played a very important role in supplementing humans' diets and were also used as fuel, wood, and medicine. Plants are excellent sources of many bioactive, medicinal, nutraceutical and pharmaceutical compounds that are important in the prevention of various diseases (Iwu 1993; Prance and Nesbitt 2005; Aljaghthmi et al. 2017; Shikov et al. 2017). Worldwide, humanity relies on a handful of plant species widely cultivated for food security (FAO 1999; Bharucha and Pretty 2010). Throughout history, about 12,000 plant species have been used as food, but only 2000 have been domesticated, and only about 150 have been commercially grown. It is known that 30 kinds of plants cover about 90% of the world's nutrient needs (WWF 1993), and only 12 plants are known that contribute between 85–90% of the world calorie intake (Bukenya-Ziraba 1996; Pieroni et al. 2007). On the other hand, wild plants offer greater dietary, nutritional and medical diversity (Bharucha and Pretty 2010; Tairo et al. 2011).

The high population growth, high levels of poverty, malnutrition, scarcity of fertile land for cultivation, high prices for staple foods, limited access to nutrient-rich foods, limited access to different diets and social insecurity are the main constraints in many developing countries (Kumari et al. 2004; Aboagye et al. 2007; FAO 2008; Essex 2010; Anywar et al. 2014; Ngome et al. 2017). The number of malnourished people in the world remains unacceptable. It is estimated that 925 million people are

---

I. M. Abu Zeid (✉)
Department of Biological Sciences, Faculty of Science, King Abdulaziz University, Jeddah, Saudi Arabia

A. R. A. FarajAlla
Batterjee Medical College for Science & Technology, Jeddah, Kingdom of Saudi Arabia

A. A. Mariod (ed.), *Wild Fruits: Composition, Nutritional Value and Products*,
https://doi.org/10.1007/978-3-030-31885-7_35

malnourished in 2010 (FAO 2010; Watkins 2010). The majority of malnourished people live in rural areas of developing countries (FAO 2011).

The number of wild edible plant species was reported to be much higher than those cultivated. This has created a growing fear of excessive human dependence on a narrow food base (Treweek et al. 2006). This situation led to documentation, collection, conservation and consumption of wild edible plants or wild crop relatives (Ong et al. 2011; Agea et al. 2011). Wild plants may be defined as those plants that grow spontaneously in self-maintaining populations in natural or semi-natural ecosystems and can exist independently of direct human actions (FAO 1999). The folkloric knowledge on the use of plants as medicine is well documented (Maheswari 2000). However, the information on the use of wild plants as food is very limited (Sudhakar and Vedavathy 2000; Uniyal et al. 2002). Recently, a lot of interest has been focused on the evaluation of various wild foods because they are important in diets, ensure food security and nutrition enhancement among rural communities in many developing countries (Balemie and Kebebew 2006; Bohra et al. 2017; Seal et al. 2017). These plants are considered rich genetic trove and also contributes some beneficial products like medicine, fiber, fodder, dyes, etc. (Seal et al. 2017). Wild plants are potential sources of species for domestication and provide valuable genetic, phytochemical and pharmacological characteristics for the development of new crops through breeding and selection (Ford-Lloyd et al. 2011).

Wild edible plants are underutilized in most poor communities either due to ignorance, shame or misconception about these foods (Tabuti 2007; Watkins 2010; FAO 2015; Loek and Maxwell 2011; Termote et al. 2012; Cocks et al. 2008). Despite society's primary reliance on cultivated plants, the tradition of eating wild plants has not yet completely reached the threshold of being reduced or disappeared (Pardo-de-Santayana et al. 2007). In many societies, a large number of wild edible plants are regarded as emergency or hunger gap or famine foods (Watkins 2010; Pieroni et al. 2007; Agea et al. 2011). However, the nutritional value awareness and culinary uses of most species of wild edible plants are still limited to specific cultures and regions (Ngome et al. 2017). Rural communities accept them through mere customs and habits as appropriate and desirable diets (Pardo-de-Santayana et al. 2007).

Wild fruits have been an important food source in many developing countries since early times (Morales et al. 2013). Wild fruits are nutrient-rich foods that are sources rich with carbohydrates, nutrients, proteins, fats, vitamins, and minerals that may be lacking or deficient in a great number of assorted common human diets. Wild fruits contain many bioactive compounds than many cultivated species (Saka and Msonthi 1994; Le Torheim et al. 2004; Drewnowski 2005; Barros et al. 2010a, b; Guimarães et al. 2010; Magaia et al. 2013). Wild edible fruits can be eaten raw or cooked in jams and can be included in alcoholic beverages (Tardío et al. 2006; Guarrera et al. 2005; Scherrer et al. 2005; González et al. 2010). People in many rural communities are unaware of the nutritional value of wild edible fruits. For example, they often only eat pulp of *Sclerocarya birrea* and *Adansonia digitata* while discarding the seeds, which contain a kernel with a higher protein and fat contents than peanuts (Ogbobe 1992).

35.2 *Borassus flabellifer* Taxonomic Classification

The plant belongs to the family *Arecaceae*, subfamily *Borassoideae*, genus *Borassus* and species: *flabellifer* Linn., (Gummadi et al. 2016; Saranya and Poongodi Vijayakumar 2016). The synonyms of the plant included Asian Palmyra, Jiggery palm, longer palm, Lontar Sugar palm, Toddy palm, Wine palm and Cambodian palm (Kumar et al. 2012a; Sahni et al. 2014; Gummadi et al. 2016; Saranya and Poongodi Vijayakumar 2016). *Borassus flabellifer* Linn. has a common English name "Sugar palm" (Dej-adisai et al. 2017).

35.3 Distribution and Ecology

Palmyra Palm, *Borassus flabellifer* Linn. is widely distributed and cultivated in Africa, South Asian countries (e.g., Sri Lanka and India), Southeast Asia (e.g., Myanmar, Cambodia, Malaysia, Indonesia, Vietnam, and Thailand) (Jansz et al. 1994; Ariyasena et al. 2000; Ariyasena et al. 2001; Artnarong et al. 2016). Globally, India ran ahead of all countries in tapping the riches of Palmyrah palms surpassing all other producing countries with a population nearly 122 million palms (Vengaiah et al. 2012). The Palmyra is valuable and important Indian tree no wonder it is the official tree of Tamil Nadu. The plant is highly respected in Tamil culture, Hindus and Buddhists both venerate this tree because sacred writings were inscribed on its leaves in olden times. The plant is also a natural symbol tree of Cambodia (Krishna Mohan et al. 2016).

35.4 Botanical Features

Borassus is derived from a Greek word describing the leathery covering of the fruit and flabellifer means "fan-bearer" (Kumar et al. 2012a, b; Krishna Mohan et al. 2016). It is a tall, erect tree attaining a height of about 20–30 m, with a black stem and crown of leaves at the top (Fig. 35.1). The growth of the tree is very slow and it takes from 15 to 30 years (Krishna Mohan et al. 2016), however no distinguishing features to identify the sex until reaching the flowering stage (Vengaiah et al. 2017). Trees can live up to more than 100 years. The rate of growth has been estimated at about 3 cm per year. Hot, sunny and well-drained soil conditions are suitable for plant to grow. The tree is drought tolerant but cold sensitive. Propagation is usually done using seeds (Krishna Mohan et al. 2016). The tree trunk is black, and look like cylinders. The trunk is substantiated with major hard pulp with a distinct circumference reaching about 1.5 meters at its lower base. It is also corrugated by the semi-circular scars of fallen leaves (Krishna Mohan et al. 2016; Vengaiah et al. 2017).

Fig. 35.1 Trees of *Borassus flabellifer*. (Source: https://commons.wikimedia.org)

There may be about 25–40 fresh leaves which are leathery, gray-green and folded along their midribs, and they spring at the top (Vengaiah et al. 2017). The leaves are about 0.9–1.5 m in diameter, palmately fan-shaped. Moreover, The trees can easily be recognized by their large, fan-shaped leaves which are quite unlike the pinnate leaves of other palms (Kumar et al. 2012a, b; Krishna Mohan et al. 2016). Moreover, the leaves are usually very tough and have thick stalks (Vengaiah et al. 2017). There are two kinds of the palmyra palm trees, the male, and the female. Two different trees hold the male and female flowers, never in one.

Both male and female trees produce spikes of flowers, but only the female trees bear fruits. The flowers are small and appear in densely clustered spikes, developing into large, brown, roundish fruits. The male flowers are smaller than the female flowers (Vengaiah et al. 2017). The onset of flowering occurs only after 12–15 years of maturity (Saranya and Poongodi Vijayakumar 2016). Flowers are unisexual; fruits are large. In order to support large fruits, male Inflorescences can be described as interfoliar, stout, tighty and intact, terete branches, whereas the female inflorescence are, to a lesser degree, more sparingly branched (Kumar et al. 2012a, b; Sahni et al. 2014; Saranya and Poongodi Vijayakumar 2016).

35.5 Fruits

Although a lot of research has been done on wild fruits, but still many wild fruits are considered to be edible and have not yet been documented, and many communities are not aware of the nutritional value of these fruits. However, most of these studies focused on medical species, and little emphasis has been paid to the wild edible plants. The fruit of palmyra palm (*Boraassus flabellifer* Linn.) (Fig. 35.2) is considered among the less well documented tropical fruits (Saranya and Poongodi Vijayakumar 2016). The major constraints of not reaching the full expected benefits of Palmyra fruits is due, primarily to their short shelf-life under the abundance of favorable moisture content (Artnarong et al. 2016), separation of the pulp from the fiber and its bitter taste (Saranya and Poongodi Vijayakumar 2016). Needless to mention that bitter taste and difficulty to separate pulp from fiber, both deter against easy use hence, contemporary practices embraces a variety of debittering taste techniques that have been enacted encompassing heating fruits over hot coals or using catalytic heat stable enzymes to reduce the bitterness (Jeyaratnam 1986; Ariyasena 2002). The fruits of *Borassus flabellifer* are large and fibrous, usually containing three nuts like portions each of which encloses a seed (Gummadi et al. 2016). The fruits are three-sided when young becoming semi-globose to globose or more or less oval, hence they will be capped at the base with the overlapping sepals. The color of the fruit is deep brown to black when ripe (Morton 1988; Ramachandran et al. 2004).

The cover of the fruit is thin, smooth with a leathery texture and brown, turning nearly black after harvest. Moreover, the inside of the fruit kept is a juicy mass made of long, tough, coarse, white fibers immersed and covered with yellow or orange pulp. Finally, the rich tasty pulp of ripe fruits is made of delicious, rich, tasty treat to be sucked directly from the wiry fibers of roasted, peeled fruits (Ramachandran

Fig. 35.2 A bunch of young *Borassus flabellifer* fruit. (Source: https://commons.wikimedia.org)

et al. 2004). The mesocarp is sweet, rich in carotenoid, it turned yellow-orange as it becomes ripe and can be used for foods such as cakes, jellies, ice creams, jams, and coffee (Ariyasena et al. 2000; Chakraborty et al. 2011). About 40% of the fruit is undiluted pulp, which has a dark yellow color with its characteristic flavor and bitterness. The pulp is sweet, creamy and is delicious to eat. The fresh pulp contains vitamins A and C. The pulp of palmyrah fruit could be used commercially to produce food and animal feed (Vengaiah et al. 2012). The kernel is soft, jelly-like and translucent like ice, and it has a watery sweetish liquid (Morton 1988). The mature fruit is usually tossed over low burning fire or embers to cook them mildly, and then the skin is peeled off to expose the juicy fruit (Vengaiah et al. 2015).

35.6 Biochemistry and Nutritional Value

The nutritional analysis of the fresh pulp of *Borassus flabellifer* showed that the pulp contains calcium, ascorbic acid, maltose, starch, reducing and non-reducing sugars and carbohydrate. The water absorption capacity of Palmyra pulp powder is 18% (2.5 ml/g), The fat was found to be 2.8%, bulk density is 0.78 cm^{-3}, the powder has swelling power value of 4 and the percent foam capacity is about 2.5%, moisture content is 74–77%, ash content is 1.2 g, pH range between 5.5 and 6, total soluble Solids (TSS) of the fresh pulp is 16.50 Brix, (Vijayakumari et al. 2014). In a similar study, Vengaiah et al. (2015) reported that the fresh pulp powder of *Borassus flabellifer* contains protein, carbohydrate, caloric value, ash and fat contents. These nutritional values can be utilized directly or combined with other pulps for the preparation of food (Vengaiah et al. 2015). On the other hand, the roots of *Borassus flabellifer* contain appreciable amount of protein content, carbohydrates, crude fiber, Vitamin E and negligible fat, and some amount of metals (Sahni et al. 2014). The sap of unfermented palmyrah contain sugars (mainly in the form of sucrose), while the fermented palmyrah palm wine contains ethanol (Theivendirarajah and Chrystopher 1986). It has also been reported to possess immunosuppressant properties (Revesz et al. 1999).

35.7 Uses of *Borassus flabellifer*

35.7.1 Therapeutic Applications

The different parts of *Borassus flabellifer* have several commercial and medicinal values (Ghosh et al. 2012). The ash obtained by burning the inflorescence is used as anti-acid, antiperiodic, and also useful in heart burn, splenomegaly and bilious fever (Kapoor 2000; Pramod et al. 2013; Krishna Mohan et al. 2016). The young plants have essential health benefits that are said to relieve biliousness, dysentery, and gonorrhea (Goyal et al. 2015). The roots are diuretic and antihelmintic, the decoction of the root is used for gastritis, hiccup and it is taken by labor class people for the treatment of diabetes (Nadkarni and Nadkarni 2000). The decoction of the root is also given in certain respiratory diseases (Ladeji et al. 2003). The plant roots and juice are useful in inflammatory reactions (Vaidyaratnam et al. 1994; Kapoor 2000). The cabbage, petioles, and dried male flower spikes have diuretic activity (Morton 1988). The flour obtained from young shoots of the *Borassus flabellifer* was tested for mutagenicity (Andersen and Poulsen 1985), mitogenic activity (Kangwanpong et al. 1981), neurotoxic effect (Sumudunie et al. 2004). The fruits are stomachic, sedative, laxative, aphrodisiac, hyperdipsia, dyspepsia, flatulence and used for the treatment of skin disease, hemorrhages, fever, and general debility (Pramod et al. 2013; Krishna Mohan et al. 2016). The rich, dense delicious sap of the plant has been used as a sweetener for diabetic patients (Masayuki et al. 2007). The pulp of the mature fruit relieves dermatitis (Morton 1988). The flowers have analgesic and antipyretic effects (Paschapur et al. 2009a), anti-inflammatory and hematological activity (Paschapur et al. 2009b, c), and immunosuppressant property (Révész et al. 1999). The pellets of *Borassus flabellifer* reduced and delayed hypersensitivity (DTH) (Devi et al. 1985). The endosperm of *Borassus flabellifer* contains a high proportion of mucilage. The mucilage did not produce any dermatological reactions and were well tolerated by the guinea pig (Kumar et al. 2012b). The authors concluded that the mucilage of *Borassus flabellifer* can be used as a pharmaceutical excipient in gel formulations; Therefore, it is able to replace some synthetic gelling polymers upon further modifications and found that they are ideal and comparable with a commercial preparation (Voltaren gel®). The mucilage has a potential gelling characteristics and can be used to develop gel formulations because of its good release profile, water-soluble nature, physical stability and good spreadability (Kumar et al. 2012b).

35.7.2 Pharmacological and Medicinal Uses

35.7.2.1 Anti-Inflammatory Activity

The leaves extract of *Borassus flabellifer* has anti-inflammatory activity (Krishna Mohan et al. 2016). The ethanolic extract of male flowers (inflorescences) of the plant showed a significant anti-inflammatory effect in the rat when compared to control (Turner and Harborn 1971; Paschapur et al. 2009a). In another study, Kumar et al. (2012b) documented the anti-inflammatory activity of the gel formulation extracted from *Borassus flabellifer* endosperm in rat hind paw edema model.

35.7.2.2 Anti-Bacterial Activity

The methanolic extract of *Borassus flabellifer* L. seed coat showed significant inhibitory activity on gram-positive bacteria, i.e., *Staphylococcus aureus, Bacillus subtilis* and gram-negative bacteria, i.e., *Klebsiella pneumonia* and *Serratia marces-*

cens (Duddukuri et al. 2011). In another study, Jamkhande et al. (2016) investigated the antimicrobial potential of *Borassus flabellifer* leaves on eight pathogenic strains of bacteria and fungi using agar well diffusion method. All the samples showed an inhibitory effect on fungal strains with inhibition zone (10–17). Also, the methanolic extract and acetone extract had potent antibacterial activities. The authors concluded that the plant exhibits a broad spectrum antimicrobial and antioxidant activity and could be used in treating various infectious diseases.

35.7.2.3 Anti-Oxidant Activity

The root/rhizome extracts of *Borassus flabellifer* showed good scavenging activity on 2, 2-diphenyl-1-picrylhydrazyl (DPPH) free radical and less on hydroxyl free radical (Talluri et al. 2017). In a similar study Kommu et al. (2011) reported that the leaves and roots of *Borassus flabellifer* possess antioxidant activity. The authors concluded that the antioxidant activity of the methanolic extracts of leaves and roots might be due to the presence of phytochemical constituents like flavonoids, saponins, tannins, and phenolic compounds. Thus, the leaves and roots of *Borassus flabellifer* might have potential as "nutraceuticals" for the preparation of functional foods. The methanolic extract of *Borassus flabellifer* leaves showed strong antioxidant effect, which might be due to presence of phenolic compounds. Furthermore, the authors concluded that this antioxidant activity might be used in the development of natural antioxidants for agro-food and pharmaceutical industries (Jamkhande et al. 2016). The bioactive compounds like flavonoids and phenolic acids have been recognized largely as beneficial antioxidants that can scavenge harmful active oxygen species, including O_2-, H_2O_2, -OH, and O_2 (Sakihama et al. 2002). The role of oxygen radicals has been implicated in several dangerous diseases, including cancer, diabetes, cardiovascular disease and aging (Halliwell and Gutteridge 1999).

35.7.2.4 Antidiabetic

Different concentrations (150, 300, and 600 mg/kg) of leaves extract of a *Borassus flabellifer* showed a significant decrease in blood glucose level. Continuous treatment with the ethanol leaves extract (150, 300, and 600 mg/kg) for 28 days resulted in a significant decrease in blood glucose level in the diabetic rat. Maximum reduction of blood glucose level occurred at the dose of 600 mg/kg (Goyal et al. 2015). *Borassus flabellifer* was reported to have anti-diabetic activity in animal model and in inhibition of glucose transportation in *in vitro* study (Uluwaduge et al. 2005; Uluwaduge et al. 2006). Many phytochemical substances from natural products were found to inhibit α-glucosidase which is one of the mechanisms for anti-diabetes treatment (Yin et al. 2014). *Borassus flabellifer* extracts can reduce the blood sugar level in an animal model (Uluwaduge et al. 2006) and the clinical pilot study (Uluwaduge et al. 2008).

35.7.2.5 Antihyperlipidemic

A regular administration of 600 mg/kg body weight of *Borassus flabellifer* leaf extract enhanced the cardioprotective lipid HDL level (Goyal et al. 2015).

35.7.3 Other Uses of Borassus flabellifer

Traditionally the different parts of *Borassus flabellifer* are used for various human disorders, food, and other multiple purposes. In Tamil India, the plant is called "karpaha" or celestial tree because all its parts without exception have different uses and valuable benefits (Krishna Mohan et al. 2016). The hard outer woody layer of the plant is used in posts, rafters and many other useful domestic purposes in housing and architectural establishments (Krishna Mohan et al. 2016). The leaves of the plant have many diverse uses such as making roofs, fences, brooms, umbrellas, thatching, mats, baskets, fans, hats and valuable artifacts. The fibres from the stalks of the leaves and mid-ribs are used in brushes. The web-like substance at base of young leaf stalks is used for making torches (Kapoor 2000; Kumar et al. 2012a, b; Krishna Mohan et al. 2016; Gummadi et al. 2016; Saranya and Poongodi Vijayakumar 2016). The fruit pulp of *Borassus flabellifer* has been used in recipes traditional dishes (Masayuki et al. 2007). The pulp of the fruit and the soft kernels of young fruit are pleasant to eat, while the germinated nuts, with their fleshy embryos are cooked and eaten as vegetables (Kommu et al. 2011). The pulp of ripe fruits is sucked directly from the fibers of roasted, peeled fruits. It is also extracted to prepare a product called punatoo that can be eaten by itself or together with the starch from the Palmyra seedlings. The mucilage of *Borassus flabellifer* is used as a pharmaceutical adjuvant, where as the extracted juice is used in the preparation of health drinks and jellies (Kumar et al. 2012a, b). Therefore most or practically all the different plant parts serve as food (fruit, sap, young shoots) and as a building material (the stem, the leaves). As mentioned earlier the pulp can be used for foods such as cakes, jellies, ice creams, jams, cordials, beverages and coffee (Ariyasena et al. 2000; Chakraborty et al. 2011). Currently, new food techniques mix Palmyrah pulp with other fruits to make different foods such as cordial, jam and cream (Vengaiah et al. 2015). The tree is also used in the pharmacopeia (roots, male inflorescence) for the treatment of various diseases such as secondary syphilis, antiperiodic, heart burns, liver and spleen enlargement and inflammatory reactions (Kumar et al. 2012a, b). Some yeast strains were isolated from Palmyra palm fruit and used in cake production (Tuntiwongwanich and Leenanon 2009). Theivendirarajah and Chrystopher (1986) found that fermented palmyra palm fruit with *Saccharomyces cerevisiae* was used to make alcohol. Merugu et al. 2014 reported that toddy palm was capable of producing silver nanoparticles ($Ag+$) from an aqueous solution of $Ag+$. The authors concluded that the application of such ecofriendly nanoparticles synthesis can be exploited for the production of nanoparticles.

35.8 Phytochemistry

The phytochemicals extracted from *Borassus flabellifer* include polyphenols, organo-sulfur compounds, carotenoids, alkaloids and saponins (Lampe and Chang 2007). The bioactive compounds are important for the prevention of many diseases such as cancer, atherosclerosis as well as other several health benefits (Sahni et al. 2014). Sandhya et al. (2010) reported that *Borassus flabellifer* contains fats, abuminoids, fats, steroidal glycosides, steroidal saponins, carbohydrate like sucrose and gums. It also contains spirostane type steroids like brassosides and dioscin (Sandhya et al. 2010). Also, several studies have documented that *Borassus flabellifer* possesses several steroidal saponins (Jansz et al. 1994: Ariyasena et al. 2000; Ariyasena et al. 2001), a polysaccharide (Awal et al. 1995), and a triterpene (Révész et al. 1999). The roots of *Borassus flabellifer* possess high amount of antioxidant properties and phenolic content, these compounds are a good source of dietary essential micronutrients and polyphenolics in the human diet (Sahni et al. 2014). The authors concluded that the root of the plant could be considered as a new source of valuable ingredient for diet and nutraceutical applications in the promotion of health. Also, the roots contain catechin, flavonoids, alkaloids, saponins, phenolic compounds, fatty acids, alkanes, alkenes, ketones, aldehydes, diterpenes, phytols, and sterols. (Sahni et al. 2014). Subashini et al. 2015 identified 28 chemical constituents from the ethanol root extract of *Borassus flabellifer* by Gas Chromatogram-Mass spectrometry (GC-MS) analysis namely 2-Furanmethanol, Propane, 1-(1-methylethoxy), 2-Cyclopentene-1-one, 2-hydroxy-, 2,4-Dihydrox y-2,5-dimethyl-3(2H)-furan-3-one, Glycerin, 1,3-Propanediamine, 1,2-Propanediol 2-acetate, Butane, 1-(ethenyloxy)-3-methyl-, Propane, 1,1-diethoxy-,1H-Imidazole-4-carboxamide, 5-amino-, 4H-Pyran-4-one, 2,3-dihydro-3,5-dihydroxy-6-methyl-, Resorcinol, Phenol, 2,6-dimethoxy-,6H-Purin-6-one, 2-amino-1,7-dihydro-,6H-Purin-6-one, 2-amino-1,7-dihydro-, 1,4-Benzenediol, 2-methoxy-, Phenol, 3,4-dimethoxy-, Benzene, 1-(1,5-dimethyl-4-hexenyl)-4-methyl-, Phenol, 4-[2-(dimethylamino) ethyl]-, 1-Butanol, 2-amino-, 3-Hydroxy-4-methoxy benzoic acid, Phenol, 3,4,5-trimethoxy-, Phenol, 5-(1,5-dimethyl-4-hexenyl)-2-methyl-, (R)-, 7H-Furo[3,2-g] benzopyran-7-one, n-Hexadecanoic acid, Pentanoic acid, 10-undecenyl ester and octadecanoic acid (Subashini et al. 2015).

Contemporary novel technologies show that the raw palmyrah palm fruit pulp (RPFP) and thermally processed palmyrah palm fruit pulp (PPFP) consist of all phytochemicals except protein. Appreciable amounts of inherent plant chemicals such as glycosides, carbohydrate, steroids, sterols and terpenes are present in the pulp (Saranya and Poongodi Vijayakumar 2016). The authors noticed that there was no major alteration between raw and thermal processed palmyrah palm fruit pulp in the qualitative phytochemical compounds, and these phytochemical compounds can be extracted for the drug development. In a similar study, four compounds were isolated from the pulp of *Borassus flabellifer* namely glucosyl-(6-1)-glycerol, a mixture of sitosterol and stigmasterol, 5-hydroxymethyl-furfural and tyrosol and two isolated compounds namely, tyrosol and glucosyl-(6-1)-glycerol. These compounds showed moderate and mild α-glucosidase inhibitory activities (Dej-adisai et al. 2017). The male inflorescence of *Borassus flabellifer* constitutes spirostane-type steroid saponins, like brass sides and dioscin (Morton 1988). The male inflorescence also contains some steroidal glycosides (Yoshikawa et al. 2007) and carbohydrates like- sucrose (Kapoor 2005). The extracted mucilage from *Borassus flabellifer* endosperm contains two major polysaccharides galactomannan and mannan (Kumar et al. 2012a).

References

Aboagye ML, Obirih-Opareh N, Amissah L, Adu-Dapaah H (2007) Analysis of existing policies that enable or inhibit the wider use of underutilized plant species in Ghana. Council for Scientific and Industrial Research Ghana

Agea JG, Okia AC, Atalla R, Abohassan A, Kimondo JM, Obua J, Hall J, Teklehaimanot Z (2011) Wild and semi-wild food plants of Bunyoro-Kitara Kingdom of Uganda: growth forms, collection niches, parts consumed, consumption patterns, main gatherers and consumers. Environ Res J 5(2):74–86

Aljaghthmi OH, Heba HM, Abu Zeid IM (2017) Antihyperglycemic properties of mangrove plants (*Rhizophora mucronata* and *Avicennia marina*): an overview. Adv Biol Res 11(4):161–170

Andersen PH, Poulsen E (1985) Mutagenicity of flour from the palmyrah palm (*Borassus flabellifer*) in Salmonella typhimurium and Escherichia coli. Cancer Lett 26(1):113–119

Anywar G, Oryem-Origa H, Mugisha MK (2014) Wild plants used as nutraceuticals from Nebbi District, Uganda. Eur J Med Plants 4(6):641–660

Ariyasena DD (2002) The diversity, bioactivity and structural studies on the flabelliferins of palmyrah (*Borassus flabellifer* L.) fruit pulp. M.Phil. Thesis, University of Sri Jayewardenepura, Sri Lanka, p 12–26.

Ariyasena DD, Jansz ER, Jayesekera S, Abeysekara AM (2000) Inhibitory effect of bitter principle of palmyra (*Borassus flabellifer* L.) fruit pulp on the growth of mice: evidence using bitter and non-bitter fruit pulp. J Sci Food Agr 80:1763–1766

Ariyasena DD, Jansz ER, Abeysekera AM (2001) Some studies directed at increasing the potential use of palmyra (*Borassus flabellifer* L.) fruit pulp. J Sci Food Agr 81:1347–1352

Artnarong S, Masniyom P, Maneesri J (2016) Isolation of yeast and acetic acid bacteria from palmyra palm fruit pulp (*Borassus flabellifer* Linn.). IFRJ 23(3):1308–1314

Awal A, Haq QN, Quader MA, Ahmed M (1995) Structural study of a polysaccharide from the seeds of *Borassus flabellifer* Linn. Carbohydr Res 277:189–195

Balemie K, Kebebew F (2006) Ethnobotanical study of wild edible plants in Derashe and Kucha Districts, South Ethiopia. J Ethnobiol Ethnomed 2:53. https://doi.org/10.1186/1746-4269-2-53

Barros L, Carvalho AM, Ferreira ICRF (2010a) Exotic fruits as a source of important phytochemicals: improving the traditional use of *Rosa canina* fruits in Portugal. Food Res Int 44:2233–2236

Barros L, Carvalho AM, Morais JS, Ferreira ICRF (2010b) Strawberry-tree, blackthorn and rose fruits: detailed characterisation in nutrients and phytochemicals with antioxidant properties. Food Chem 120:247–254

Bharucha Z, Pretty J (2010) The roles and values of wild foods in agricultural systems. Philos Trans R Soc Lond Ser B Biol Sci 365:2913–2926

Bohra N, Tewari LM, Tewari A (2017) Ethnobotany of wild edible plants traditionally used by the local people in the Ramnagar region from Nainital District, Uttarakhand, India. Biolife 5(1):12–19

Bukenya-Ziraba R (1996) The non-cultivated edible plants of Uganda, Monograph series No 9. NAPRECA, Addis Ababa

Chakraborty I, Chaurasiya AK, Saha J (2011) Quality of diversified value addition from some minor fruits. J Food Sci 48(6):750–754

Cocks ML, Bangay L, Shackleton CM, Wiersum KF (2008) Rich man poor man' inter-household and community factors influencing the use of wild plant resources amongst rural households in South Africa. Int J Sustain Dev World Ecol 15(3):198–210

Dej-adisai S, Pitakbut T, Wattanapiromsakul C (2017) Alpha-glucosidase inhibitory activity and phytochemical investigation of *Borassus flabellifer* Linn. Afr J Pharm Pharmacol 11(3):45–52

Devi S, Arseculeratne SN, Pathmanathan R, McKenzie IF, Pang T (1985) Suppression of cell-mediated immunity following oral feeding of mice with palmyrah (*Borassus flabellifer* L) flour. Aust J Exp Biol Med Sci 63(4):371–379

Drewnowski A (2005) Concept of a nutritious food: toward a nutrient density score. Am J Clin Nutr 82(4):721–732

Duddukuri GR, Nagendra Sastry YN, Kaladhar DSVGK, Kamalakara Rao K, Krishna Chaitanya K (2011) Antibacterial activity of methanolic seed coat extract of *Borassus flabellifer* L. Int J Pharm Sci Res 2(9):2435–2438

Essex J (2010) Sustainability, food security, and development aid after the food crisis: assessing aid strategies across donor contexts. Sustainability 2:3354–3382

FAO (1999) Use and potential of wild plants in farm households. FAO Farm Systems Management Series version 15

FAO (2008) The state of food insecurity in the world. In: High food prices and food security – threats and opportunities. FAO, Rome

FAO (2010) The state of food insecurity in the world. In: Addressing food insecurity in protracted crises. FAO, Rome

FAO (2011) The state of food insecurity in the world. In: How does international price volatility affect domestic economies and food security? FAO, Rome

FAO, IFAD, WFP (2015) The state of food insecurity in the world 2015. Meeting the 2015 international hunger targets: taking stock of uneven progress. FAO, Rome

Ford-Lloyd BV, Schmidt M, Armstrong SJ, Barazani O, Engels J, Hadas R, Hammer K, Kell SP, Kang D, Khoshbakht K, Li Y, Long C, Lu BR, Ma K, Nguyen VT, Qiu L, Ge S, Wei W, Zhang Z, Maxted N (2011) Crop wild relatives – undervalued, underutilized and under threat? BioScience 61:559–565

Ghosh S, Chakraborty R, Raychaudhuri U (2012) Optimizing process conditions for palm (*Borassus flabelliffer*) wine fermentation using response surface methodology. Int Food Res J 19:1633–1639

González JA, García-Barriuso M, Amich F (2010) Ethnobotanical study of medicinal plants traditionally used in the Arribes del Duero, western Spain. J Ethnopharmacol 131:343–355

Goyal P, Agarwal AK, Lakashminarasimhaiah, Singh GK (2015) Antidiabetic and antihyperlipidemic effect of *Borassus flabellifer* strepozotocin (STZ) induced diabetic rats. World J Pharm Pharm Sci 4(1):1172–1184

Guarrera PM, Forti G, Marignoli S (2005) Ethnobotanical and ethnomedicinal uses of plants in the district of Acquapendente (Latium, Central Italy). J Ethnopharmacol 96:429–444

Guimarães R, Barros L, Barreira JCM, Sousa MJ, Carvalho AM, Ferreira ICFR (2010) Targeting excessive free radicals with peels and juices of citrus fruits: grapefruit, lemon, lime and orange. Food Chem Toxicol 48:99–106

Gummadi VP, Battu GR, Keerthana DMS, Manda K (2016) A review on Palmyrah palm (*Borassus flabellifer*). Int J Curr Pharm Res 8(2):17–20

Halliwell B, Gutteridge JMC (1999) Free radicals in biology and medicine. Oxford University Press, Oxford

Iwu MM (1993) Handbook of African medicinal plants. CRC Press, Boca Raton

Jamkhande PG, Suryawanshi VA, Kaylankar TK, Patwekar SL (2016) Biological activities of leaves of ethnomedicinal plant, *Borassus flabellifer* Linn. (Palmyra palm): an antibacterial, antifungal and antioxidant evaluation. Bull Fac Pharm Cairo Univ 54:59–66

Jansz ER, Nikawela JK, Gooneratne J (1994) Studies on the bitter principle and debittering of Palmyrah fruit pulp. J Sci Food Agr 65:185–189

Jeyaratnam M (1986) Studies on the chemistry and biochemistry of palmyrah products. M.Phil. Thesis, University of Jaffna, Sri Lanka, p 1–200.

Kangwanpong D, Arseculeratne SN, Sirisinha S (1981) Clastogenic effect of aqueous extracts of palmyrah (*Borassus flabellifer*) flour on human blood lymphocytes. Mutat Re 89(1):63–68
Kapoor LD (2000) Handbook of Ayurvedic medicinal plants. Herbal reference library, vol 82. CRC Press, Boca Raton
Kapoor LD (2005) Handbook of ayurvedic medicinal plants. CRC Press, New York, p 82
Kommu S, Chilukaa VL, Gowri Shankara NL, Matsyagiria L, Shankar MA, Sandhyab S (2011) Anti oxidant activity of methanolic extracts of female *Borassus flabellifer* leaves and roots. Der Pharmacia Sinica 2(3):193–199
Krishna Mohan C, Soundarya V, Vasanth Kumar R, Kiran Kumar L, Vamshi Sharathnath K, Narender B (2016) *In vitro* anti-inflammatory activity of dried leaves of *Borassus flabellifer*. Indo Am J P Sci 3(8):809–813
Kumar R, Rajarajeshwari N, Narayana Swamy VB (2012a) Isolation and evaluation of *Borassus flabellifer* mucilage as a natural suspending agent. Int J PharmTech Res 4(4):1614–1630
Kumar R, Rajarajeshwari N, Narayana Swamy VB (2012b) Exploitation of *Borassus flabellifer* fruit mucilage as novel natural gelling agent. Pharm Lett 4(4):1202–1213
Kumari M, Gupta S, Lakshmi A, Prakash J (2004) Iron bioavailability in green leafy vegetables cooked in different utensils. Food Chem 86:217–222
Ladeji O, Omekarah I, Solomon M (2003) Hypoglycemic properties of aqueous bark extract of *Ceiba pentandra* in streptozotocin-induced diabetic rats. J Ethnopharmacol 3(84):139–142
Lampe JW, Chang JL (2007) Interindividual differences in phytochemical metabolism and disposition. Semin Cancer Bioln.d.; 17(5): 347–353.
Le Torheim LE, Ouattara F, Diarra MM, Thiam FD, Barikmo I, Hatloy A, Oshaug A (2004) Nutrient adequacy and dietary diversity in rural Mali: association and determinants. Eur J Clin Nutr 58:594–604
Loek EAP, Maxwell DG (2011) Characteristics and strategies favouring sustained food access during Guinea's food-price crisis. Dev Pract 21(4–5):613–628
Magaia T, Uamusse A, Sjöholm I, Skog K (2013) Proximate analysis of five wild fruits of mozambique. Sci World J 2013:1–6. https://doi.org/10.1155/2013/601435
Maheswari JK (ed) (2000) Ethnobotany and medicinal plants of Indian subcontinent. Scientific Publishers, Jodhpur, p 672
Masayuki Y, Fengming X, Toshio M, Yutana P, Seikou N, Yasunobu A, Akira K, Hisashi M (2007) Medicinal flowers. XII.(1) New spirostane-type steroid saponins with antidiabetogenic activity from *Borassus flabellifer*. Chem Pharm Bull 55:308–316
Merugu R, Garimella S, Kudle KR, Prashanthi Y, Balla D, Preethi KS (2014) Green synthesis and characterization of silver nanoparticles using Palm Toddy (*Borassus flabellifer*). Nanosci Nanotechnol Int J 4(2):25–27
Morales P, Ferreira ICFR, Carvalho AM, Fernández-Ruiz V, Sánchez-Mata MC, Cámara M, Morales R, Tardío J (2013) Wild edible fruits as a potential source of phytochemicals with capacity to inhibit lipid peroxidation. Eur J Lipid Sci Technol 115(2):176–185
Morton JF (1988) Notes on distribution, propagation and products of Borassus palms (Arecaceae). Econ Bot 42(3):420–441
Nadkarni KM, Nadkarni AK (2000) Indian materia medica, 2nd edn. Popular Prakashan, Bombay, pp 209–210
Ngome PIT, Shackleton C, Degrande A, Tieguhong JC (2017) Addressing constraints in promoting wild edible plants' utilization in household nutrition: case of the Congo Basin forest area. Agric Food Secur 6:20. https://doi.org/10.1186/s40066-017-0097-5
Ogbobe O (1992) Physico-chemical composition and characterization of the seed and seed oil of *Sclerocarya birrea*. Plant Foods Hum Nutr 42(3):201–206
Ong HC, Chua S, Milow P (2011) Traditional knowledge of edible plants among the Temuan Villagers in Kampung Jeram Kedah, Negeri Sembilan, Malaysia. Sci Res Essays 6(4):694–697
Pardo-de-Santayana M, Tardío JM, Blanco E, Carvalho AM, Lastra JJ, Miguel E, Morales R (2007) Traditional knowledge of wild edible plants used in the Northwest of the Iberian

Peninsula (Spain and Portugal): a comparative study. J Ethnobiol Ethnomed 3:27. https://doi.org/10.1186/1746-4269-3-27
Paschapur MS, Patil MB, Kumar R, Patil SR (2009a) Evaluation of anti-inflammatory activity of ethanolic extract of *Borassus flabellifer* L. male flowers (inflorescences) in experimental animals. J Med Plants Res 3(2):49–54
Paschapur MS, Patil MB, Kumar R, Patil SR (2009b) Influence of ethanolic extract of *Borassus flabellifer* L. male flowers (inflorescences) on chemically induced acute inflammation and poly arthritis in rats. Int J PharmTech Res 1(3):551–556
Paschapur MS, Patil S, Patil SR, Kumar R, Patil MB (2009c) Evaluation of the analgesic and antipyretic activities of ethanolic extract of male flower (inflorescences) of *Borassus flabellifer* L. (Arecaceae). Int J Pharm Pharm Sci 1(2):98–106
Pieroni A, Houlihan L, Ansari N, Hussain B, Aslam S (2007) Medicinal perceptions of vegetables traditionally consumed by south-Asian migrants living in Bradford, Northern England. J Ethnopharmacol 113:100–110
Pramod HJ, Yadav AV, Raje VN, Mohite M, Wadkar G (2013) Antioxidant activity of *Borassus flabellifer* (Linn.) fruits. Asian J Pharm Tech 3(1):16–19
Prance G, Nesbitt M (eds) (2005) The cultural history of plants. Routledge, London
Ramachandran VS, Swarupanandan K, Renuka C (2004) A traditional irrigation system using palmyra palm (*Borassus flabellifer*) in Kerala, India. Palms 48(4):175–181
Revesz L, Hiestand P, La Vecchia L, Naef R, Naegeli HU, Oberer L (1999) Isolation andsynthesis of a novel immunosuppressive 17 alpha- substituted dammarane from the flour of the Palmyrah palm (*Borassus flabellifer*). Bioorg Med Chem Lett 9:1521–1526
Sahni C, Shakil NA, Jha V, Gupta RK (2014) Screening of nutritional, phytochemical, antioxidant and antibacterial activity of the roots of *Borassus flabellifer* (Asian Palmyra Palm). J Pharmacogn Phytochem 3(4):58–68
Saka JDK, Msonthi JD (1994) Nutritional value of edible fruits of indigenous wild trees in Malawi. For Ecol Manag 64(2–3):245–248
Sakihama Y, Cohen MF, Gace SC, Yamasaki H (2002) Plant phenolic antioxidant and prooxidant activities: phenolics induced oxidative damage mediated by metals in plants. Elsevier 177:67–80
Sandhya S, Sudhakar K, Banji D, Rao KNV (2010) Pharmacognostical standardization of *Borassus flabellifer* root. Ann Biol Res 1(4):85–94
Saranya P, Poongodi Vijayakumar T (2016) Preliminary phytochemical screening of raw and thermally processed palmyra palm (*Borassus flabellifer* Linn.) fruit pulp. JIPBS 3(1):186–193
Scherrer AM, Motti R, Weckerle CS (2005) Traditional plant use in the areas of Monte Vesole and Ascea, Cilento National Park (Campania, Southern Italy). J Ethnopharmacol 97:129–143
Seal T, Pillai B, Chaudhuri K (2017) Evaluation of nutritional potential of five unexplored wild edible plants consumed by the tribal people of Arunachal Pradesh State in India. J Food Nutr Res 5(1):1–5. https://doi.org/10.12691/jfnr-5-1-1.
Shikov AN, Tsitsilin AN, Pozharitskaya ON, Makarov VG, Heinrich M (2017) Traditional and current food use of wild plants listed in the Russian Pharmacopoeia. Front Pharmacol 8:841. https://doi.org/10.3389/fphar.2017.00841
Subashini S, Ramesh Kannan V, Mani P (2015) Phytochemical and GC-MS analysis of bioactive compounds from *Borassus flabellifer* Linn root. Eur J Mol Biol Biochem 2:148–152
Sudhakar A, Vedavathy S (2000) Wild edible plants used by the tribals of Chittoor district (Andhra Pradesh), India. In: Ethnobotany and medicinal plants of Indian subcontinent. Scientific Publishers, Jodhpur, p 321
Sumudunie KA, Jansz ER, Jayasekera S, Wickramasinghe SM (2004) The neurotoxic effect of palmyrah (*Borassus flabellifer*) flour re-visited. Int J Food Sci Nutr 55:607–614
Tabuti JRS (2007) Status of non-cultivated food plants in Bulamogi County, Uganda. Afr J Ecol 45(Suppl.1):96–101

Tairo VE, Njoka JT, Lukhoba WC, Lyaruu HVM (2011) Nutritive and anti-nutritive qualities of mostly preferred edible woody plants in selected dry lands of Iringa district, Tanzania. Pak J Nutr 10(8):786–791

Talluri MR, Gummadi VP, Battu GR (2017) Antioxidant activity of *Borassus flabellifer*. Int Res J Pharm 8(3):18–22

Tardío J, Pardo-de-Santayana M, Morales R (2006) Ethnobotanical review of wild edible plants in Spain. Bot J Linnean Soc 52:27–72

Termote C, Bwama MM, Dhed'a DB, Huybregts L, Lachat CA (2012) Biodiverse rich environment does not contribute to a better diet: a case study from DR Congo. PLoS One 7(1):1–10

Theivendirarajah K, Chrystopher RK (1986) Alcohol from palmyrah palm *(Borassus flabellifer* L.) fruit pulp. J Sci 1:44–46

Treweek JR, Brown C, Bubb P (2006) Assessing biodiversity impacts of trade. Impact Assess Proj Appraisal 2:299–309

Tuntiwongwanich S, Leenanon B (2009) Morphology and identification of yeasts isolated from toddy palm in Thailand. J Micro Soci of Thailand 23(1):34–37

Turner RA, Harborn P (1971) Screening methods in pharmacology, vol 1. Academic Press Inc, New York, pp 302–304

Uluwaduge I, Punya AA, Senadheera SN, Jansz ER (2005) Studies on the natural hydrophobic blinder of flabelliferins and their effect on some bioactivities. J Natl Sci Found Sri Lanka 33:187–191

Uluwaduge I, Thabrew MI, Janz ER (2006) The effect of flabelliferins of palmyrah fruit pulp on intestinal glucose uptake in mice. J Natl Sci Found Sri Lanka 34:37–41

Uluwaduge I, Parera A, Janszn E, Thabrew I (2008) A pilot study on palmyrah pinattu (dried fruit pulp) as an anti– diabetic food component. Int J Biol Chem Sci 1:250–254

Uniyal SK, Awasthi A, Rawat GS (2002) Current status and distribution of commercially exploited medicinal and aromatic plants in upper Gori valley, Kumaon Himalaya, Uttaranchal. Curr Sci 82(10):1246–1252

Vaidyaratnam PS, Varier's Arya Vaidya Sala (1994) Indian medicinal plants: a compendium of 500 species, vol 4. Orient Longman, Chennai, pp 293–296

Vengaiah PC, Murthy GN, Prasad KR, Kumari KU (2012) Post-harvest technology of Palmyrah (*Borassus flabellifer* L) present practises and scope. International Conference on Food Processing by Omics Group, India.

Vengaiah PC, Vijaya Kumara B, Murthy GN, Prasad KR (2015) Physio-chemical properties of Palmyrah palm pulp (*Borassus flabellifer* L). J Nutr Food Sci 5:5. https://doi.org/10.4172/2155-9600.1000391

Vengaiah PC, Murthy GN, Sattiraju M, Maheswarappa HP (2017) Value added food products from Palmyrah palm (*Borassus flabellifer* L). JNHS 4(1):105. https://doi.org/10.15744/2393-9060.4.105

Vijayakumari, Vengaiah PC, Kiranmayi P (2014) Physicochemical and functional characteristics of powder prepared from palmyra fruit pulp (*Borassus flabellifer* L.). Int J Curr Microbiol App Sci 3(9):352–356

Watkins TY (2010) The prevalence of wild food knowledge among Nomadic Turkana of Northern Kenya. J Ethnobiol 30(1):137–152

World Wide Fund for Nature (WWF) (1993) The vital wealth of plants: WWF and the conservation of plants, WWF CH-1196. WWF, Gland

Yin Z, Zhang W, Feng F, Zhang Y, Kang W (2014) Alpha – glucosidase inhibitors isolated from medicinal plants. Food Sci Hum Wellness 3:136–174

Yoshikawa M, Xu F, Morikawa T, Pongpiriyadacha Y, Nakamura S, Asao Y et al (2007) Medicinal flowers. XII. New spirostane-type steroid saponins with antidiabetogenic activity from *Borassus flabellifer*. Chem Pharm Bull 55:308–316

Chapter 36
Ceratonia siliqua (Carob-Locust Bean) Outgoing and Potential Trends of Phytochemical, Economic and Medicinal Merits

Dalia M. Rasheed, Dina M. El-Kersh, and Mohamed A. Farag

Abbreviation

LBG Locust bean gum

36.1 Introduction

Ceratonia siliqua, (Carob, locust bean, St John's bread) is an evergreen tree, hardy, drought tolerant and prominent member of the pea family in the Mediterranean region. Arabs have contributed in tree domestication along North Africa, Levant and South Europe and it is notarized that the gemstones measuring unit "*carat*" holds its name from carob seeds (Turnbull et al. 2006). The tree attains a develop stature and spread of about 6–15 m high, with branches spreading out to the ground (Fig. 36.1a). Its evergreen leaves are compound pinnate with 6–10 opposite oval, dark-green, leathery leaflets with rounded apex (Morton 1987) (Fig. 36.1b). The carob flowers are tiny, red, numerous and spirally arranged along a male, female, hermaphrodite or a polygamous inflorescence which sprout in Autumn (Custódio et al. 2004) (Fig. 36.1c). The inflorescence has a distinct offensive odor, pollination is

D. M. Rasheed
Pharmacognosy Department, Faculty of Pharmacy, October 6 University, Sixth of October, Egypt

D. M. El-Kersh
Pharmacognosy Department, Faculty of Pharmacy, The British University in Egypt (BUE), Cairo Governorate, Egypt

M. A. Farag (✉)
Pharmacognosy Department, Faculty of Pharmacy, Cairo University, Cairo, Egypt

Chemistry Department, School of Sciences & Engineering, The American University in Cairo, New Cairo, Egypt
e-mail: Mohamed.farag@pharma.cu.edu.eg; Mohamed.alifarag@aucegypt.edu

A. A. Mariod (ed.), *Wild Fruits: Composition, Nutritional Value and Products*,
https://doi.org/10.1007/978-3-030-31885-7_36

Fig. 36.1 Photographs of *Ceratonia siliqua*. (**a**) Mature tree, (**b**) Compound leaf, (**c**) Inflorescence and (**d**) Ripe pods. (Source: https://www.wikimedia.org/) (∗Sources of all photographs: Wikimedia website)

accomplished by both wind and insects and it's recommended to plant 3–5 trees to ensure a mix of sexes and successful fruiting (Custódio et al. 2004). Carob fruits are indehiscent, elongated, straight or curved compressed pods, glossy dark brown in color with a thickened margin on full maturation, and their size may reach 20 cm in length (Morton 1987) (Fig. 36.1d). The pods are filled with shiny seeds (10–12 per pod), characterized by a remarkable uniformity in size and weight of about 0.2 g each justifying their use in weighing systems (Harper et al. 1970). The carob main parts are the pulp (90%) noted as kibble "deseeded broken carob legumes" and seeds "locust kernel gum" (8–10%) where the seeds consists of the husk, the endosperm and the germ (Loullis and Pinakoulaki 2017; Dakia et al. 2007; Ayaz et al. 2009). The chemical composition of carob fruits depends mainly on cultivar, origin, climate, stage of maturation and time of collection (Battle and Tous 1997; Dakia et al. 2007). Carob is enriched with several primary metabolite classes, including sugars *viz.*, *D*-pinitol "sugar alcohol" and a galactomannan gum, aside from proteins, fatty acids, minerals and dietary fibers. Carob secondary metabolites also grab attention for its use as a functional food, mostly for its abundance in phenolic acids, tannins and flavonoids. In the next sections, both primary, secondary metabolites of carob and its health attributed effects shall be discussed in details.

36.2 *Ceratonia siliqua* Metabolism

36.2.1 Nutritive Primary Metabolites

36.2.1.1 Carbohydrates

Sugars

Carob pulp is characterized by its high sugar content ranging from 40–60% dominated by disaccharides *i.e*., sucrose 32–38% followed by monosaccharides e.g. fructose 5–7% and glucose 5–6% (Sigge et al. 2011; Rtibi et al. 2017a; Ayaz et al. 2009). Sugar levels vary among carob pulp of different origins, varieties, ripening stage and or processing methods, though with sucrose amounting for the major form. Turkish carob samples characterized by sucrose as the predominant sugar while presence of glucose and fructose was equivalent (Tetik et al. 2011) or even low content of fructose (Ayaz et al. 2007). Cultivated varieties of Turkish carob encompassed higher sugar levels exemplified in sucrose, albeit with almost similar levels of glucose and fructose (Biner et al. 2007; Goulas et al. 2016). Likewise, the "*House hold flour*" prepared domestically by sun drying and grinding to fine powder of Turkish carob kibbles varies in its sugar content from commercially *"carob powder"* processed industrially at high temperature drying ovens and mechanical milling instruments. The commercial carob powder encompasses almost double of its original sugar content (Ayaz et al. 2009). Two varieties of Jordanian *C. siliqua* pods (*siliqua* and *macrocarpa*) were assessed for its sugar content, with *macrocarpa* variety comprising higher mono and disaccharides content than *siliqua* (Shawakfeh and Ereifej 2005). Italian carob pods, seeds and germ "obtained after removal of brown coat of the seed and the endosperm either by treatment with acids or boiling water" (Dakia et al. 2007) from Sicily were analyzed for their carbohydrates content, with pods made up of up to 45% carbohydrates and low fats (0.6%). In contrast, germ and seeds were opposite to the pods in composition with more proteins, fats and less carbohydrates (Dakia et al. 2007; Avallone et al. 1997). In terms of individual sugars composition, pods recorded the highest sucrose level (35%) followed by fructose (6%) and glucose (4%). (Avallone et al. 1997). In Morocco, the influence of different geographical origin was observed in sugars variation as well as in the syrup yield of the different carob pulps. The total sugar contents ranged from 31–50 g/100 g pulps (El Batal et al. 2016). Absolute determination of sugar levels in carob has yet to be reported with most results expressed in literature as percentile values.

Polyols (Sugar Alcohols)

Polyols "cyclitols" as *myo*-inositol, *chiro*-inositol and *D*-pinitol are low molecular weight carbohydrates usually abundant in legumes. Polyols differ from cyclic sugars in being of low-calorie intake due to their low bioavailability which favors their use for diabetic patients or in slimming products. *D*-pinitol is specifically present in family Fabaceae dominantly in soy beans and carob pods (Baumgartner et al. 1986). A previous study suggested that 10 mg/kg body weight of pure *D*-pinitol efficiently present in 10 g of carob syrup quiet sufficient to lower elevated blood glucose levels from ½-2 hours post administration (Narayanan et al. 1987) asides from other health benefits to include anti-inflammatory, anti-asthmatic, anti-hyperlipidemic and antioxidant effects (Loullis and Pinakoulaki 2017). It's worth to mention that the highest content of *D*-pinitol is present in Turkish samples of carob pods ranging between 65–95 g/kg dry weight in contrast to soy beans recording 5–20 g/kg dry weight. *D*-pinitol as a unique and rich component of carob pods can be employed as a biomarker for detection of carob adulteration or admixture (Baumgartner et al. 1986; Tetik et al. 2011). *D*-pinitol content was found higher in wild Anatolian carob than cultivated one (Goulas et al. 2016; Turhan 2014), however that has yet to be reported from other origin. Considering *D*-pinitol effective health benefits, proper extraction methods were assessed for its maximal extraction yield in its rich sources *viz.* carob and Soya beans. Not only ultrasound assisted extraction was more efficient in *D*-pinitol extraction than conventional hot water but also in lowering extraction temperature, time as well as energy consumption (Tetik and Yüksel 2014).

Locust Bean Gum (LBG)

Compared to fruit richness in sugars, carob seeds are recognized for their richness in the locust bean gum (LBG) also named *tragasol*. It constitutes about the third of seed weight besides the husk and the germ. Chemically this polymer is a galactomannan used as food additive known as E 410 in ice creams and bakery products for its gelling and thickening effects. In pharmaceutical industry, it is also used in cosmetics and emulsions. LBG is prepared from seeds endosperm after removal of seed coat "dehusking" made either by thermo-mechanical treatment *ca.* acid peeling or thermal peeling. Acid peeling revealed more whitish LBG than the one obtained by thermal peeling due to the roasting factor. (Loullis and Pinakoulaki 2017; Dakia et al. 2007; Battle and Tous 1997). LBG purification process by precipitation of impurities and endogenous enzymes using isopropanol recorded higher mannose/galactose ratios than crude ones which enhances its rheological properties (Bouzouita et al. 2007).

36.2.1.2 Proteins & Amino Acids

Proteins of carob augmented in the seeds embryo amounts for up to 50% of seeds constituents "*germ seed flour*" valued for the preparation of gluten free-food products for celiac patients owing to its caroubin. Caroubin is a protein mixture of similar rheological properties to that of gluten, albeit with no hypersensitivity reactions (Youssef et al. 2013; Loullis and Pinakoulaki 2017). Protein content of carob germ flour obtained from the seeds (48.4%) have been recorded to be higher than those of pea (18.8%) and soy bean (34.4%) (Bengoechea et al. 2008). Except for tryptophan an essential amino acid found at trace levels in carob, it contains all the seven essential amino acids with acceptable levels that meet world health organization (WHO) standards (Dakia et al. 2007). 18 Amino acids were detected in carob germ *viz.*, aspartic, glutamic acids, arginine, asparagine, alanine, leucine and valine are the predominant amino acids comprising about 57% of total amino acids (Goulas et al. 2016; Ayaz et al. 2007; Bengoechea et al. 2008; Dakia et al. 2007). Method of preparation of carob pods either home-made or commercially could vary in amino acids composition. Aspartic acid followed by alanine were the major amino acids in home-made carob pods, whereas glutamic acid followed by aspartic acid were predominant in commercially prepared carob suggestive that preparation method affects macronutrients composition in carob (Ayaz et al. 2009).

36.2.1.3 Fatty Acids

Carob germ seed oil comprised low fat content dominated by saturated palmitic and stearic acids and unsaturated fatty acids *viz.* linoleic, linolenic and oleic acids. The low content of fats may be beneficial in production of low fat food products but unlikely low nutritional quality oil (Youssef et al. 2013; Loullis and Pinakoulaki 2017; Dakia et al. 2007).

36.2.1.4 Minerals and Vitamins

Carob fruits mainly the seeds contain high content of potassium and calcium as a portion of carob fruit could be equivalent to a cup of cow milk (Goulas et al. 2016; Loullis and Pinakoulaki 2017). Macro-minerals found in fruits are phosphorous and magnesium while the micro-minerals present are copper, zinc, selenium as antioxidant enzymes cofactors as well as iron, manganese, nickel, barium, cobalt (Goulas et al. 2016; Youssef et al. 2013; Loullis and Pinakoulaki 2017; Rtibi et al. 2017a; Ayaz et al. 2007; Ayaz et al. 2009). Blend of vitamins exist in carob pods such as vitamins E, D, C, B_6, Niacin and folic acid as major vitamins while A, B_2 and B_{12} are present at lower levels (Youssef et al. 2013).

36.2.1.5 Dietary Fibers

Dietary fibers comprised about 30–40% of the carob pulps. Structurally they are heterogeneous substances of insoluble and soluble fractions. The insoluble fraction "70%" chemically consists of cellulose, hemicellulose and lignin. Insoluble polyphenols, proteins and minerals included within the fibers uniquely recognize it from other dietary fibers. The other lower percentile soluble fraction consists mainly of simple carbohydrates as pentosans, pectin and mucilage (Goulas et al. 2016; Loullis and Pinakoulaki 2017; Rtibi et al. 2017a). Carob fibers proved its antioxidant activity, lowered serum triglycerides and cholesterol whereas the polyphenols increased energy expenditure and fat oxidation. Both carob fibers and polyphenols could be efficient in lipids metabolism (Gruendel et al. 2006).

36.2.2 Secondary Polyphenolic Metabolites

A myriad of polyphenols are present in the different carob parts mostly in the fruit pulp followed by the leaves and wood. Secondary metabolites classes present in carob include phenolic acids, flavonoids and tannins as in Fig. 36.2. Expectedly, polyphenols level depends massively on genetic, environmental factors, extraction process and cultivar type (Loullis and Pinakoulaki 2017; Custódio et al. 2011b). Carob trees of different gender and cultivars revealed that total phenolic content (TPC) were higher in hermaphrodite trees than female ones recording 41.3 mg GAE/g and 18 mg GAE/g, respectively (Custódio et al. 2011b). The content of catechins in such samples ranged from 6–10 mg/g while gallic acid ranged from 1–2 mg/g of different Portuguese cultivars (Custódio et al. 2011b) in contrary to other previous studies recorded gallic acid as the major phenolic acid (Papagiannopoulos et al. 2004; El Ansari Zineb et al. 2017; Corsi et al. 2002; Owen et al. 2003; Ayaz et al. 2007) and suggestive for growing habitat impact on polyphenols composition in carob. Variability among carob phenolic contents of Sicilian origin (Avallone et al. 1997) versus Anatolian one was reported (Ayaz et al. 2007). The antibacterial activity of carob pod and leaf against phytopathogenic bacteria *Pectobacterium atrosepticum* was attributed mostly to phenolics *viz.* gallic acid, quercetin and kampferol glycosides as well as tannins (Meziani et al. 2015). A study of sapwood and heart wood polyphenols content from Turkey using GC/MS revealed that catechin and gallic acid were major component of heart wood versus sapwood abundance in chalcones and flavones (Balaban 2004).

36.2.2.1 Phenolic Acids

A variety of phenolic acid structures are present in carob pulps either of benzoic or cinnamic acid structural motif. Gallic acid Fig. 36.2(1) and its derivatives as methyl gallate comprised the major fraction of phenolic acids likely attributed for gallic acid release from tannins during the extraction procedures (Torun et al. 2013) ranging from 24–165 mg/100 g dry weight, other less abundant phenolic acids include gentisic, ellagic, vanillic, syringic acids and pyrogallol Fig. 36.2(2–6) (Loullis and Pinakoulaki 2017; Eldahshan 2011; Owen et al. 2003). The phenolic acids of cinnamates origin *viz.* cinnamic, *p*-coumaric, ferulic and chlorogenic acids Fig. 36.2(7–10) are present in general at lower levels compared to benzoic acid analogues (Roseiro et al. 2013a; Custódio et al. 2011b; Ayaz et al. 2007; Rtibi et al. 2017a).

36.2.2.2 Flavonoids

Carob extract possessed a comparable antioxidant activity to that of tea, with an advantage of absence of caffeine and theobromine stimulant action (Karim and Azlan 2012). Carob is a rich source of dietary flavonols subclass *viz.* myricetin,

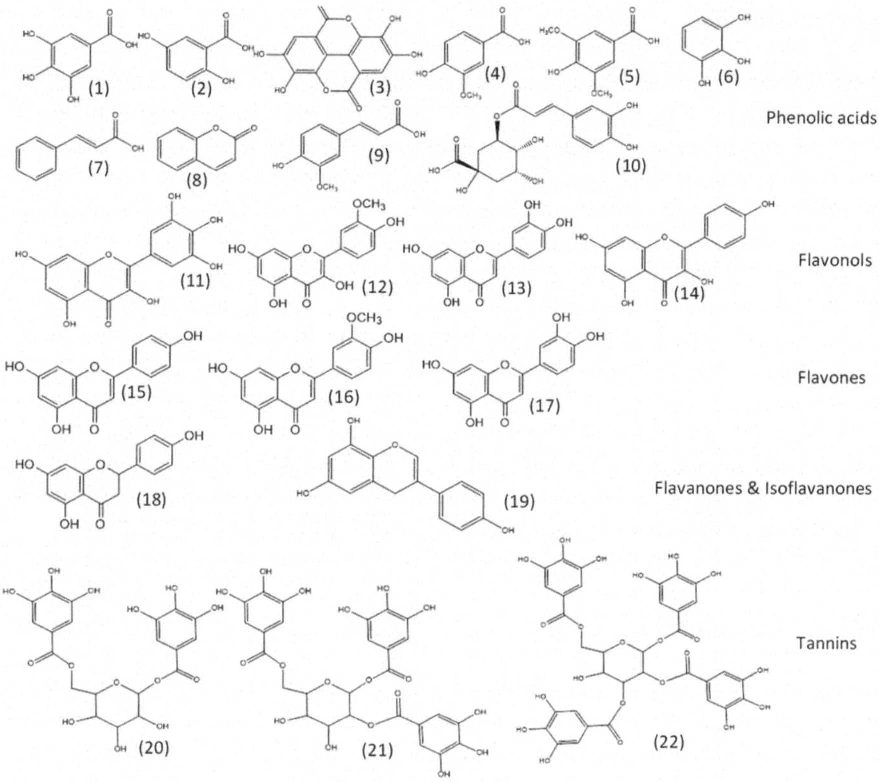

Fig. 36.2 Identified phenolics from Carob; Phenolic acids: **1** gallic acid, **2** gentisic acid, **3** ellagic acid, **4** vanillic acid, **5** synirgic acid, **6** pyrogallol, **7** cinnamic acid, **8** *p*-coumaric acid, **9** Ferulic acid, **10** chlorogenic acid, flavonoids **11** myricetin, **12** isorhmnetin, **13** quercetin, **14** kaempferol, **15** apigenin, **16** chrysoeriol, **17** luteolin, **18** naringenin, **19** genistein, tannins **20** 1,6 digalloyl glucose, **21** 1,2,6 trigalloyl glucose and **22** 1,2,3,6 tetragalloyl glucose

isorhmnetin, quercetin and kaempferol Fig. 36.2(11–14) and their glycosides (Owen et al. 2003; Owis and El-Naggar 2016; Eldahshan 2011) whereas flavones, flavanones and isoflavones are present at much lower levels represented by apigenin, chrysoeriol, luteolin, naringenin and genistein Fig. 36.2(15–19) (Goulas et al. 2016; Papagiannopoulos et al. 2004; Hsouna et al. 2011; Vaya and Mahmood 2006). Roasting of carob is a prerequisite step used to intensify its aroma during beverage preparation but can also affect polyphenols composition either *via* degradation by heating or release from another polymers (Loullis and Pinakoulaki 2017). Different *C*-flavonoid glycosides of different substitution patterns were detected in Spanish carob seeds (Picariello et al. 2017), as determined *via* HPLC coupled to high resolution MS/MS containing apigenin flavone, luteolin and chrysin aglycones.

36.2.2.3 Tannins

Tannins are the major secondary metabolite class found in carob and accounting for its astringent taste. The ripening stage affects tannins level being most enriched in unripe fruit and dropping with ripening (Ayaz et al. 2007). Both classes of tannins are present in carob pods with much higher levels of condensed tannins versus hydrolysable ones. The hydrolysable class being esters of gallic or ellagic acids with sugars *ca.* glucose noted as ellagitannins. The most predominant ellagitannins widely present in carob pods of different geographical origins *viz.* di, tri and tetra galloyl glucose Fig. 36.2(20–22) (Owen et al. 2003; Hsouna et al. 2015). Condensed tannins, proanthocyanidins chemically are flavan-3-ol nucleus with galloyl esters, catechin, epicatechin gallate or epigallocatechin gallate. Levels of tannins vary according to its location. The pods comprised 3.7 mg/g total tannins while the germ of much higher levels 19.2 mg/g (Goulas et al. 2016; Papagiannopoulos et al. 2004). HPLC analysis of carob pods and leaves from eastern Sicily (Italy) revealed presence of gallic acid, epicatechin gallate, epigallocatechin gallate where the concentration in leaves were higher than pods recording 6.3 and 1.4 mg/g, respectively (Corsi et al. 2002).

36.2.2.4 Essential Oil

Despite extensive reports on *Ceratonia* non-volatile metabolites, much less is known regarding its volatile flavor compound (Farag and El-Kersh 2017) identified 31 volatiles in unroasted and roasted carob pod belonging to *viz.* short chain fatty acids, esters and aldehydes. The volatiles were identified using head space solid phase micro extraction technique coupled to Gas chromatography Mass spectrometry (HS-SPME/GC-MS) a better alternative than steam distillation for volatile profiling in case of low amounts as in carob aside in that it does not involve heat treatment which results in artifacts formation (Farag et al. 2013; Dakia et al. 2007). Unroasted carob pods were generally more rich in short chain fatty acids *viz.* caproic and pentanoic acids whereas roasted ones were abundant in Millard products such as pyranone. Monitoring of these volatiles seems empirical as they could influence the organoleptic characters of *C. siliqua* beverages or food products.

36.3 Economical Uses

Carob is considered an important economic Mediterranean vegetation. Global production of carob trees is about 160,000 tons per year collected from 80,000 hectare (Goulas et al. 2016). The major world exporter of carob pods is Spain, followed by other Mediterranean countries including Italy, Portugal and Cyprus (Tous and Ferguson 1996). Carob is commonly used as a beverage or food additives in bakery and confectionery, feed for livestock, production of bioethanol, mannitol, citric and lactic acids. The deseeded carob powder is used mainly as Cocoa substitute, *carob cocoa*", after processing and is sometimes more favored than cocoa due to the absence of theobromine and caffeine alkaloids (Tetik et al. 2011; Ayaz et al. 2007). Carob pulp is enriched in nutritive primary constituents due to its high sugar content (40–50%) (Turhan et al. 2006) widely utilized in different food products as in *pekmez* production, a concentrated carob syrup well known in Turkey (Tetik et al. 2011). The sugar abundance mainly in the carob pulps (Dakia et al. 2007; Rtibi et al. 2017a) and its relatively low extraction costs poses carob among the top horticultural crops used for alcohol and molasses production in several Mediterranean countries (Orwa et al. 2009). The nutritive value of carob pulp is attributed to the presence of amino acids, fatty acids, minerals and dietary fibers. Interestingly, the processing of pulps depends on whether it will be consumed as a staple feed or in human food products. In animal feed, carob pods are grounded to different sizes based on type of animals. With regards to carob processing for human consumption is either processed to syrup or powder (Loullis and Pinakoulaki 2017). Carob syrup is a traditional product native to the Mediterranean region, containing a high concentration of sugar, phenolic compounds and minerals. *D*-Pinitol cyclitol "3-*O*-methyl-*D*-chiro-inositol" is a low molecular weight carbohydrate commonly present in legumes mainly carob, (Tetik et al. 2011). Traditionally carob pulp was the only recognized nutritive part but now the benefit extends to cover seeds owing to its content of locust bean gum (LBG) which is widely used as gelling agent in food products where patents registered for novel uses in baby and jelly foods. The carob fibers being predominantly of insoluble fraction have been utilized in dough rheology when used in bakery products (Goulas et al. 2016).

36.4 Folk Medicinal Uses

Extracts of *C. siliqua* have been employed by Mediterranean and Middle East populations for the treatment of various gastrointestinal ailments. In traditional medicine, brewer of carob powder is recommended as an anti-diarrheal remedy especially for infants (Loeb et al. 1989). Carob soft drink is particularly effective for the treatment of GIT disorders asides from a mild diuretic action based on its richness in minerals and electrolytes (Tetik and Yüksel 2014). Traditional uses of pods include their application as an anti-tussive agents and against heartburn and warts (Amico and Sorce 1997). Further traditional uses of carob include remedies for herpes lip sores, menorrhagia, intestinal parasites as well as viral and bacterial infections in some ethno-veterinary practices (Azab 2017; Farag and El-Kersh 2017).

Carob fiber preparation obtained by water extraction of carob pulps, is rich in insoluble dietary fibers, mainly cellulose and hemicellulose, polyphenols and proteins and free from soluble carbohydrates (Farag and Wessjohann 2012). Regular consumption of carob fiber has been associated with lowering cholesterol and triglyceride levels, increased fat oxidation as well as anti-proliferative effects on cancer cells in humans contingent to the preparation enrichment with polyphenolic compounds (Kumazawa et al. 2002; Farag et al. 2013; Turhan 2014).

36.5 Biological Activities

Pharmacological actions of *C. siliqua* proven either *in vitro* or *in vivo* have been attributed mainly to their secondary metabolite's composition in several studies. Secondary metabolites profile of carob extracts encompasses a polyphenolic rich matrix that mediate for several biological effects and health attributes *viz.* anti-oxidant, anti-cancer, anti-diabetic, anti-microbial as well as anti-hyperlipidemic activities. An outline of *C. siliqua* organs main chemical constituent, economic and medicinal uses is presented in Fig. 36.3. The following sections will provide a highlight on the main pharmacological studies conducted for verifying anticipated biological actions of carob extracts and/or products.

36.5.1 Anti-oxidant and ROS-Scavenging Effects

Several researchers reported on the prominent anti-oxidant and radical scavenging actions of carob extracts established by various *in-vitro* and *in-vivo* models. Aqueous and alcoholic extracts of leaves, ripe and unripe pods were assessed for their capacity to scavenge different forms of reactive oxygen species and overall anti-oxidant potential. *In-vitro* quenching assays of free radicals *viz.* 2,2-diphenyl-1-picrylhydrazyl (DPPH) and 2,2′-azino-bis [3-ethylbenzthiazoline-6-sulphonic acid] (ABTS) have reported potent effects for carob extracts (Amessis-Ouchemoukh et al. 2017; Kumazawa et al. 2002; Romano et al. 2007; Roseiro et al. 2013c; Sebai et al. 2013).

36.5.2 Cytotoxic and Anti-proliferative Effects

The search for drugs from natural sources that can exhibit a chemopreventive or cytotoxic effect is on the rise (Amessis-Ouchemoukh et al. 2017; Custódio et al. 2009, 2011a, b; Klenow et al. 2009; Roseiro et al. 2013b). Furthermore, carob preparations rich in pods fibers hold an European patent registered as anti-inflammatory products and chemopreventive agents (Haber 2003).

36.5.3 Anti-diabetic Effects

Considering the dominance of sugars in carob primary metabolites profile (Biner et al. 2007), the assumption that carob could help modulate blood glucose levels and assist in dietary management protocols for patients with type II diabetes is very unlikely. However, studies demonstrated that carob gum (Locust bean gum), the

Fig. 36.3 Diagramtic sketch *of carob* organs main chemical composition, economic and medicinal uses

natural thickening agent much employed in food industry, appeared to suppress blood glucose level rise in rats (Forestieri et al. 1989). *In-vivo* determination of the glycemic index of carob powder alone and in a herbal dietary supplement was ranked as low (GI = 40 ± 0.05) (Son et al. 2010; Milek Dos Santos et al. 2015).

Two very characteristic constituents in carob products have posed a fundamental difference on its anticipated glycemic pattern, *viz.* the dietary fiber and cyclitols content. Dietary fibers exert a variety of pharmacological actions throughout the GIT including regulating intestinal capacity, reduction of blood pressure and reducing blood glucose and cholesterol levels (Anderson et al. 2009). Cyclitols, on the other hand are cyclic polyols synthesized in considerable amounts by several plants in the Fabaceae family, including carob (Baumgartner et al. 1986; Phillips et al. 1982). *D*-pinitol (1*D*-3-*O*-methyl-chiro-inositol), the principle cyclitol identified in carob pods, is a compound of considerable merit for its established role in regulating blood glucose levels by increasing insulin sensitivity in type II diabetic patients (Bates et al. 2000; Kim et al. 2007; Sivakumar et al. 2010). Optimization of extraction method that can capture both bioactive agents should now follows for consideration of marketing carob products for treatment of diabetes.

Asides, aqueous decoctions of leaves, germ flour, pods, gum and stem bark of carob inhibited the actions of α-amylase and α-glucosidase *in-vitro*, offering additional post-prandial delay in glucose digestion and absorption (Custódio et al. 2015). Recently, the lyophilized aqueous extract of immature carob pods appeared to reduce intestinal glucose absorption by inhibiting the electrogenic sodium-dependent glucose transporter, SGLUT-1 (Rtibi et al. 2017b). Unlike ripe carob pods, the immature pods afford a richer composition in polyphenolic compounds such as chlorogenic acid, epicatechin and gallic acid conjugates as well as flavonoid conjugates (Papagiannopoulos et al. 2004), with established anti-diabetic activities (Pandey and Rizvi 2009; Ranilla et al. 2010).

36.5.4 Anti-microbial and Anti-diarrheal Effects

Secondary metabolite profiles of carob extracts comprise several types of gallic acid conjugates, condensed tannins and other phenolic compounds imposing an astringent and anti-microbial effects, a prerequisite for the control of diarrhea (Avallone et al. 1997; Tetik and Yüksel 2014; Hsouna et al. 2011; Papagiannopoulos et al. 2004). A dietary preparation of carob pods particulate, containing at least 20%, of water-insoluble tannins by weight is patented in the US for the treatment of diarrhea (Wursch 1991). Two clinical trials verified the efficacy and safety of carob juice and a tannin-rich carob pod preparation in management of acute-onset diarrhea in infants (Akşit et al. 1998; Loeb et al. 1989). On the other hand, carob leaves extract and fractions thereof demonstrated strong anti-bacterial effects against different strains of *Escherichia coli* and *Staphylococcus* species (Kivçak et al. 2002). Inhibition of these pathogens decisively encourages the adoption of carob products in treatment of toxic diarrhea incidences.

Alcoholic and acetone extracts of carob leaves possessed effective inhibitory actions against the food-borne and phytopathogens, *Listeria monocytogenes* and *Pectobacterium atrosepticum*, respectively (Aissani et al. 2012; Meziani et al. 2015). Analysis of these active fractions by HPLC/MS led to the characterization of (−)-epigallocatechin-3-gallate, catechin, chlorogenic acid, malic acid, myricitrin and isoquercitin as the major active constituents (Aissani et al. 2012). Methanol extract of carob leaves was assessed for controlling citrus sour rot disease caused by the fungal pathogen *Geotrichum candidum* (Talibi et al. 2012). Study revealed that this anti-fungal activity was attributed to alkaloids and polyketides, metabolites derived from the carob endophyte, *Penicillium citrinum* (El-Neketi et al. 2013). As for the anti-viral effects, very limited reports have addressed such effect except for partial inhibition of Newcastle disease virus by treatment with ethanol extract of carob leaves (Al-Hadid 2016).

36.5.5 Anti-hyperlipidemic Effects

Anti-hyperlipidemic effects imparted by statins and other similar medications come with serious ramifications including myopathy and polyneuropathy (Andrade et al. 1995; Moosmann and Behl 2004). In this context, the search for natural remedies to prevent and/or treat hyperlipidemia and hypercholesterolemia has become a growing quest recently.

Biological studies exploring the anti-hyperlipidemic effects of carob are primarily focused on pods, seeds and products thereof (Ershoff and Wells 1962; Yamamoto et al. 2000). Carob powder, the ground powder of deseeded roasted carob pods, improved the lipid profile makers in rats fed on a hyperlipidemic diet in a concentration-dependent manner (Hassanein et al. 2015). Carob pulp, with its enriched composition of insoluble dietary fibers, reduced serum levels of cholesterol and triglycerides and also minimized several structural and functional markers for the development of atherosclerosis in rabbits (Valero-Muñoz et al. 2014). A patented dietary food supplement recommended for promoting cardiovascular functions is a combined formulation of carob insoluble fibers and n-3 fatty acids (Haber et al. 2006).

Clinical trials also confirmed the anti-hyperlipidemic effects of carob. Insoluble dietary fiber preparations of carob pulp rich in polyphenols, reduced total serum cholesterol, LDL and triglycerides in hypercholesterolemic subjects (Ruiz-Roso et al. 2010; Zunft et al. 2003). Adults and children subjects with familial hypercholesterolemic were fed Locust beans gum (LBG) enriched products for 3 months. This approach resulted in significant improvement in the lipid profiles of the treated subjects, with about 10–19% diminution in serum cholesterol and LDL levels, and no significant adverse effects were observed over the course of the study (Zavoral et al. 1983).

36.6 Conclusion

C. siliqua (carob) tree is a drought tolerant, evergreen ornamental tree of the Mediterranean region that is well suited for cultivation in arid and temperate geographical localities. Carob pods and seeds hold a considerable nutritional and economic importance, whereas the leaves and unripe pods exhibit evident pharmacological and medicinal effects. The edible sweet pulps are a rich source of insoluble dietary fibers, sugars and polyphenols and comprises a cheap substrate for the production of valuable economic products as bioethanol and citric acid. Carob powder is a cocoa substitute free of the caffeine and theobromine and less fats, in fact carob fibers impart an evident anti-hyperlipidemic effect. Carob bean gum (Locust bean gum) obtained from the endosperm of dehusked carob seeds, is a galactomannan enriched preparation of growing application in food industry. Whilst carob is well recognized in many industrial purposes (Fig. 36.3), its mergence as a functional food is yet to be fully exploited, especially considering its many health benefits. Although *C. siliqua* is indeed a very competitive target for phytochemical studies, global variation in its secondary metabolome in context of its origin, growth stage, processing method and cultivar type has yet to be determined.

References

Aissani N, Coroneo V, Fattouch S, Caboni P (2012) Inhibitory effect of carob (*Ceratonia siliqua*) leaves methanolic extract on Listeria monocytogenes. J Agric Food Chem 60:9954–9958

Akşit S, Çag˘layan S, Cukan R, Yaprak I (1998) Carob bean juice: a powerful adjunct to oral rehydration solution treatment in diarrhoea. Paediatr Perinat Epidemiol 12:176–181

Al-Hadid KJ (2016) Evaluation of antiviral activity of different medicinal plants against Newcastle disease virus. Am J Agric Biol Sci 11(4):157–163

Amessis-Ouchemoukh N, Ouchemoukh S, Meziant N, Idiri Y, Hernanz D, Stinco CM, Rodríguez-Pulido FJ, Heredia FJ, Madani K, Luis J (2017) Bioactive metabolites involved in the antioxidant, anticancer and anticalpain activities of Ficus carica L., Ceratonia siliqua L. and Quercus ilex L. extracts. Ind Crop Prod 95:6–17

Amico F, Sorce E (1997) Medicinal plants and phytotherapy in Mussomeli area (Caltanissetta, Sicily, Italy). Fitoterapia 68:143–159

Anderson JW, Baird P, Davis RH, Ferreri S, Knudtson M, Koraym A, Waters V, Williams CL (2009) Health benefits of dietary fiber. Nutr Rev 67:188–205

Andrade SE, Walker AM, Gottlieb LK, Hollenberg NK, Testa MA, Saperia GM, Platt R (1995) Discontinuation of antihyperlipidemic drugs—do rates reported in clinical trials reflect rates in primary care settings? N Engl J Med 332:1125–1131

Avallone R, Plessi M, Baraldi M, Monzani A (1997) Determination of chemical composition of carob (Ceratonia siliqua): protein, fat, carbohydrates, and tannins. J Food Compos Anal 10:166–172

Ayaz FA, Torun H, Ayaz S, Correia PJ, Alaiz M, Sanz C, Gruz J, Strnad M (2007) Determination of chemical composition of anatolian carob pod (Ceratonia siliqua L.): sugars, amino and organic acids, minerals and phenolic compounds. J Food Qual 30:1040–1055

Ayaz FA, Torun H, Glew RH, Bak ZD, Chuang LT, Presley JM, Andrews R (2009) Nutrient content of carob pod (Ceratonia siliqua L.) flour prepared commercially and domestically. Plant Food Hum Nutr 64:286–292

Azab A (2017) Carob (Ceratonia siliqua): health, medicine and chemistry. Eur Chem Bull 10:456–469

Balaban M (2004) Identification of the main phenolic compounds in wood of Ceratonia siliqua by GC-MS. Phytochem Anal 15:385–388

Bates SH, Jones RB, Bailey CJ (2000) Insulin-like effect of pinitol. Br J Pharmacol 130:1944–1948

Battle I, Tous J (1997) Carob tree: *Ceratonia siliqua* L. – promoting the conservation and use of underutilized and neglected crops. 17. Bioversity International, Rome

Baumgartner S, Genner-Ritzmann R, Haas J, Amado R, Neukom H (1986) Isolation and identification of cyclitols in carob pods (Ceratonia siliqua L.). J Agric Food Chem 34:827–829

Bengoechea C, Romero A, Villanueva A, Moreno G, Alaiz M, Millán F, Guerrero A, Puppo M (2008) Composition and structure of carob (Ceratonia siliqua L.) germ proteins. Food Chem 107:675–683

Biner B, Gubbuk H, Karhan M, Aksu M, Pekmezci M (2007) Sugar profiles of the pods of cultivated and wild types of carob bean (Ceratonia siliqua L.) in Turkey. Food Chem 100:1453–1455

Bouzouita N, Khaldi A, Zgoulli S, Chebil L, Chekki R, Chaabouni M, Thonart P (2007) The analysis of crude and purified locust bean gum: a comparison of samples from different carob tree populations in Tunisia. Food Chem 101:1508–1515

Corsi L, Avallone R, Cosenza F, Farina F, Baraldi C, Baraldi M (2002) Antiproliferative effects of Ceratonia siliqua L. on mouse hepatocellular carcinoma cell line. Fitoterapia 73:674–684

Custódio L, Nogueira JMF, Romano A (2004) Sex and developmental stage of carob flowers affects composition of volatiles. J Hortic Sci Biotechnol 79:689–692

Custódio L, Fernandes E, Escapa AL, López-Avilés S, Fajardo A, Aligué R, Alberício F, Romano A (2009) Antioxidant activity and in vitro inhibition of tumor cell growth by leaf extracts from the carob tree (Ceratonia siliqua). Pharm Biol 47:721–728

Custódio L, Escapa AL, Fernandes E, Fajardo A, Aligue R, Albericio F, Neng N, Nogueira JM, Romano A (2011a) Phytochemical profile, antioxidant and cytotoxic activities of the carob tree (Ceratonia siliqua L.) germ flour extracts. Plant Foods Hum Nutr 66:78–84

Custódio L, Fernandes E, Escapa AL, Fajardo A, Aligué R, Alberício F, Neng NR, Nogueira JMF, Romano A (2011b) Antioxidant and cytotoxic activities of carob tree fruit pulps are strongly influenced by gender and cultivar. J Agric Food Chem 59:7005–7012

Custódio L, Patarra J, Alberício F, Neng NR, Nogueira JMF, Romano A (2015) In vitro antioxidant and inhibitory activity of water decoctions of carob tree (Ceratonia siliqua L.) on cholinesterases, α-amylase and α-glucosidase. Nat Prod Res 29:2155–2159

Dakia PA, Wathelet B, Paquot M (2007) Isolation and chemical evaluation of carob (Ceratonia siliqua L.) seed germ. Food Chem 102:1368–1374

El Ansari Zineb BM, Alain B, Ahmed L (2017) Total polyphenols and gallic acid contents in domesticated carob (Ceratonia siliqua L.) pods and leaves. Int J Pure App Biosci 5:22–30

El Batal H, Hasib A, Ouatmane A, Dehbi F, Jaouad A, Boulli A (2016) Sugar composition and yield of syrup production from the pulp of Moroccan carob pods (Ceratonia siliqua L.). Arab J Chem 9:S955–S959

Eldahshan O (2011) Isolation and structure elucidation of phenolic compounds of carob leaves grown in Egypt. Curr Res J Biol Sci 3:52–55

El-Neketi M, Ebrahim W, Lin W, Gedara S, Badria F, Saad H-EA, Lai D, Proksch P (2013) Alkaloids and polyketides from Penicillium citrinum, an endophyte isolated from the Moroccan plant Ceratonia siliqua. J Nat Prod 76:1099–1104

Ershoff BH, Wells AF (1962) Effects of gum guar, locust bean gum and carrageenan on liver cholesterol of cholesterol-fed rats. Proc Soc Exp Biol Med 110:580–582

Farag MA, El-Kersh DM (2017) Volatiles profiling in Ceratonia siliqua (Carob bean) from Egypt and in response to roasting as analyzed via solid-phase microextraction coupled to chemometrics. J Adv Res 8:379–385

Farag MA, Wessjohann LA (2012) Metabolome classification of commercial Hypericum perforatum (St. John Wort) preparations via UHPLC-qTOF-MS and chemmometrics. Planta Med 78(5):488–496

Farag MA, El-Ahmady SH, Elian FS, Wessjohann LA (2013) Metabolomics driven analysis of artichoke leaf and its commercial products via UHPLC–q-TOF-MS and chemometrics. Phytochemistry 95:177–187

Forestieri A, Galati E, Trovato A, Tumino G (1989) Effects of guar and carob gums on glucose, insulin and cholesterol plasma levels in the rat. Phytother Res 3:1–4

Goulas V, Stylos E, Chatziathanasiadou MV, Mavromoustakos T, Tzakos AG (2016) Functional components of carob fruit: linking the chemical and biological space. Int J Mol Sci 17:1875

Gruendel S, Garcia AL, Otto B, Mueller C, Steiniger J, Weickert MO, Speth M, Katz N, Koebnick C (2006) Carob pulp preparation rich in insoluble dietary fiber and polyphenols enhances lipid oxidation and lowers postprandial acylated ghrelin in humans. J Nutr 136:1533–1538

Haber B (2003) Carob product based antiinflammatory or chemopreventative agent. Eur Patent 366:A3

Haber, B., Kiy, T., Hausmanns, S., Ruesing, M. (2006). Dietary foodstuff for positively influencing cardiovascular health. Google Patents

Harper JL, Lovell P, Moore K (1970) The shapes and sizes of seeds. Annu Rev Ecol Evol Syst 1:327–356

Hassanein KM, Youssef MKE, Ali HM, El-Manfaloty MM (2015) The influence of carob powder on lipid profile and histopathology of some organs in rats. Comp Clin Pathol 24:1509–1513

Hsouna AB, Saoudi M, Trigui M, Jamoussi K, Boudawara T, Jaoua S, El Feki A (2011) Characterization of bioactive compounds and ameliorative effects of Ceratonia siliqua leaf extract against CCl4 induced hepatic oxidative damage and renal failure in rats. Food Chem Toxicol 49:3183–3191

Hsouna AB, Trigui M, Jarraya RM, Damak M, Jaoua S (2015) Identification of phenolic compounds by high performance liquid chromatography/mass spectrometry (HPLC/MS) and

in vitro evaluation of the antioxidant and antimicrobial activities of Ceratonia siliqua leaves extracts. J Med Plant Res 9:479–485
Karim AA, Azlan A (2012) Fruit pod extracts as a source of nutraceuticals and pharmaceuticals. Molecules 17:11931–11946
Kim MJ, Yoo KH, Kim JH, Seo YT, Ha BW, Kho JH, Shin YG, Chung CH (2007) Effect of pinitol on glucose metabolism and adipocytokines in uncontrolled type 2 diabetes. Diabetes Res Clin Pract 77:S247–S251
Kivçak B, Mert T, Öztürk HT (2002) Antimicrobial and cytotoxic activities of Ceratonia siliqua L. extracts. Turkish. J Biol 26:197–200
Klenow S, Jahns F, Pool-Zobel BL, Glei M (2009) Does an extract of carob (Ceratonia siliqua L.) have chemopreventive potential related to oxidative stress and drug metabolism in human colon cells? J Agric Food Chem 57:2999–3004
Kumazawa S, Taniguchi M, Suzuki Y, Shimura M, Kwon M-S, Nakayama T (2002) Antioxidant activity of polyphenols in carob pods. J Agric Food Chem 50:373–377
Loeb H, Vandenplas Y, Würsch P, Guesry P (1989) Tannin-rich carob pod for the treatment of acute-onset diarrhea. J Pediatr Gastroenterol Nutr 8:480–485
Loullis A, Pinakoulaki E (2017) Carob as cocoa substitute: a review on composition, health benefits and food applications. Eur Food Res Technol:244(6)1–19
Meziani S, Oomah BD, Zaidi F, Simon-Levert A, Bertrand C, Zaidi-Yahiaoui R (2015) Antibacterial activity of carob (Ceratonia siliqua L.) extracts against phytopathogenic bacteria Pectobacterium atrosepticum. Microb Pathog 78:95–102
Milek Dos Santos L, Tomzack Tulio L, Fuganti Campos L, Ramos Dorneles M, Carneiro Hecke Krüger C (2015) Glycemic response to carob (Ceratonia siliqua L) in healthy subjects and with the in vitro hydrolysis index. Nutr Hosp 31:482–487
Moosmann B, Behl C (2004) Selenoprotein synthesis and side-effects of statins. Lancet 363:892–894
Morton JF (1987) Carob. In: Fruits of warm climates. JF Morton, Miami
Narayanan C, Joshi D, Mujumdar A, Dhekne V (1987) Pinitol—a new anti-diabetic compound from the leaves of Bougainvillea spectabilis. Curr Sci 56:139–141
Orwa C, Mutua A, Kindt R, Jamnadass R, Simons A (2009) Agroforestree database: a tree species reference and selection guide version 4.0. World Agroforestry Centre ICRAF, Nairobi
Owen R, Haubner R, Hull W, Erben G, Spiegelhalder B, Bartsch H, Haber B (2003) Isolation and structure elucidation of the major individual polyphenols in carob fibre. Food Chem Toxicol 41:1727–1738
Owis AI, El-Naggar E-MB (2016) Identification and quantification of the major constituents in Egyptian carob extract by liquid chromatography–electrospray ionization-tandem mass spectrometry. Pharmacogn Mag 12:S1
Pandey KB, Rizvi SI (2009) Plant polyphenols as dietary antioxidants in human health and disease. Oxidative Med Cell Longev 2:270–278
Papagiannopoulos M, Wollseifen HR, Mellenthin A, Haber B, Galensa R (2004) Identification and quantification of polyphenols in Carob Fruits (Ceratonia siliqua L.) and derived products by HPLC-UV-ESI/MS n. J Agric Food Chem 52:3784–3791
Phillips DV, Dougherty DE, Smith AE (1982) Cyclitols in soybean. J Agric Food Chem 30:456–458
Picariello G, Sciammaro L, Siano F, Volpe MG, Puppo MC, Mamone G (2017) Comparative analysis of C-glycosidic flavonoids from Prosopis spp. and Ceratonia siliqua seed germ flour. Food Res Int 99:730–738
Ranilla LG, Kwon Y-I, Apostolidis E, Shetty K (2010) Phenolic compounds, antioxidant activity and in vitro inhibitory potential against key enzymes relevant for hyperglycemia and hypertension of commonly used medicinal plants, herbs and spices in Latin America. Bioresour Technol 101:4676–4689
Romano A, Custódio L, Fernandes E (2007) Study of the antioxidant activity of extracts from carob tree (Ceratonia siliqua L.). II International symposium on human health effects of fruits and vegetables: FAVHEALTH 2007(841):507–510

Roseiro LB, Duarte LC, Oliveira DL, Roque R, Bernardo-GIL MG, Martins AI, Sepúlveda C, Almeida J, Meireles M, Gírio FM (2013a) Supercritical, ultrasound and conventional extracts from carob (Ceratonia siliqua L.) biomass: effect on the phenolic profile and antiproliferative activity. Ind Crop Prod 47:132–138

Roseiro LB, Duarte LC, Oliveira DL, Roque R, Bernardo-GIL MG, Martins AI, Sepúlveda C, Almeida J, Meireles M, Gírio FM, Rauter AP (2013b) Supercritical, ultrasound and conventional extracts from carob (Ceratonia siliqua L.) biomass: effect on the phenolic profile and antiproliferative activity. Ind Crop Prod 47:132–138

Roseiro LB, Tavares CS, Roseiro JC, Rauter AP (2013c) Antioxidants from aqueous decoction of carob pods biomass (Ceretonia siliqua L.): optimisation using response surface methodology and phenolic profile by capillary electrophoresis. Ind Crop Prod 44:119–126

Rtibi K, Selmi S, Grami D, Amri M, Eto B, El-Benna J, Sebai H, Marzouki L (2017a) Chemical constituents and pharmacological actions of carob pods and leaves (Ceratonia siliqua L.) on the gastrointestinal tract: a review. Biomed Pharmacother 93:522–528

Rtibi K, Selmi S, Grami D, Saidani K, Sebai H, Amri M, Eto B, Marzouki L (2017b) Ceratonia siliqua L.(immature carob bean) inhibits intestinal glucose absorption, improves glucose tolerance and protects against alloxan-induced diabetes in rat. J Sci Food Agric 97:2664–2670

Ruiz-Roso B, Quintela JC, De la Fuente E, Haya J, Pérez-Olleros L (2010) Insoluble carob fiber rich in polyphenols lowers total and LDL cholesterol in hypercholesterolemic sujects. Plant Foods Hum Nutr 65:50–56

Sebai H, Souli A, Chehimi L, Rtibi K, Amri M, El-Benna J, Sakly M (2013) In vitro and in vivo antioxidant properties of Tunisian carob (Ceratonia siliqua L.). J Med Plants Res 7:85–90

Shawakfeh K, Ereifej K (2005) Pod characteristics of two Ceratonia siliqua L. varieties from Jordan. Ital J Food Sci 17:187–194

Sigge G, Lipumbu L, Britz T (2011) Proximate composition of carob cultivars growing in South Africa. S Afr J Plant Soil 28:17–22

Sivakumar S, Palsamy P, Subramanian SP (2010) Impact of D-pinitol on the attenuation of proinflammatory cytokines, hyperglycemia-mediated oxidative stress and protection of kidney tissue ultrastructure in streptozotocin-induced diabetic rats. Chem Biol Interact 188:237–245

Son D-W, Lee J-W, Lee P-J, Bae K-H (2010) Glycemic index of Insu 100® herbal preparation containing Korean red ginseng, carob, mulberry, and banaba. J Ginseng Res 34:89–92

Talibi I, Askarne L, Boubaker H, Boudyach E, Msanda F, Saadi B, Ait Ben Aoumar A (2012) Antifungal activity of Moroccan medicinal plants against citrus sour rot agent Geotrichum candidum. Lett Appl Microbiol 55:155–161

Tetik N, Yüksel E (2014) Ultrasound-assisted extraction of d-pinitol from carob pods using response surface methodology. Ultrason Sonochem 21:860–865

Tetik N, Turhan I, Oziyci HR, Karhan M (2011) Determination of D-pinitol in carob syrup. Int J Food Sci Nutr 62:572–576

Torun H, Ayaz FA, Colak N, Grúz J, Strnad M (2013) Phenolic acid content and free radical-scavenging activity of two differently processed carob tree (Ceratonia siliqua L.) pod. Food Nutr Sci 4:547

Tous J, Ferguson L (1996) Mediterranean fruits. Prog New Crops:416–430

Turhan I (2014) Relationship between sugar profile and D-pinitol content of pods of wild and cultivated types of carob bean (Ceratonia siliqua L.). Int J Food Prop 17:363–370

Turhan I, Tetik N, Aksu M, Karhan M, Certel M (2006) Liquid–solid extraction of soluble solids and total phenolic compounds of carob bean (Ceratonia siliqua L.). J Food Process Eng 29:498–507

Turnbull LA, Santamaria L, Martorell T, Rallo J, Hector A (2006) Seed size variability: from carob to carats. Biol Lett 2:397–400

Valero-Muñoz M, Martín-Fernández B, Ballesteros S, Lahera V, de las Heras N (2014) Carob pod insoluble fiber exerts anti-atherosclerotic effects in rabbits through sirtuin-1 and peroxisome proliferator-activated receptor-γ coactivator-1α–3. J Nutr 144:1378–1384

Vaya J, Mahmood S (2006) Flavonoid content in leaf extracts of the fig (Ficus carica L.), carob (Ceratonia siliqua L.) and pistachio (Pistacia lentiscus L.). Biofactors 28:169–175

Wursch P (1991) Treatment of diarrhoea with compositions derived from carob pod. Google Patents

Yamamoto Y, Sogawa I, Nishina A, Saeki S, Ichikawa N, Iibata S (2000) Improved hypolipidemic effects of xanthan gum-galactomannan mixtures in rats. Biosci Biotechnol Biochem 64:2165–2171

Youssef MKE, El-Manfaloty MM, Ali HM (2013) Assessment of proximate chemical composition, nutritional status, fatty acid composition and phenolic compounds of carob (Ceratonia Siliqua L.). Food Public Health 3:304–308

Zavoral JH, Hannan P, Fields DJ, Hanson MN, Frantz ID, Kuba K, Elmer P, Jacobs DR Jr (1983) The hypolipidemic effect of locust bean gum food products in familial hypercholesterolemic adults and children. Am J Clin Nutr 38:285–294

Zunft H, Lüder W, Harde A, Haber B, Graubaum H, Koebnick C, Grünwald J (2003) Carob pulp preparation rich in insoluble fibre lowers total and LDL cholesterol in hypercholesterolemic patients. Eur J Nutr 42:235–242

Chapter 37
Garcinia binucao: Biochemistry, Functions and Utilization

Abdalbasit Adam Mariod

Abbreviations

GBE *Garcinia* binucao extract
HCA Hydroxycitric acid
ZOI Zone of inhibition

37.1 Introduction

Garcinia genus member of family Clusiaceae has been engaged with ayurvedic medications to cure different pathophysiological diseases. The bioactive compounds like hydroxycitric acid (HCA), flavonoids, terpenes, polysaccharides, procyanidines and polyisoprenylated benzophenone derivatives like garcinol, xanthochymol and guttiferone isoforms have been obtained from the genus Garcinia. The genus has gotten the consideration of pharmaceutical enterprises because of their enormous healing characteristics (Hemshekhar et al. 2011). The fruits are a food source for several animals, (monkeys) of tropical eastern Asia which relish the sap of overripe mangosteens.

The family includes a wide range of tree species with approximately 50 genera and 1200 species (Paull and Duarte 2011). Some of them are very popular, have potential or economic significance, and are extensively used for their medicinal value. Batuan is one of the indigenous, lesser-known, edible fruit-bearingtreethatis widelydistributedthroughoutthePhilippinesand Vietnam Also the fruits are commercially important as a rich source of the much valued antiobesity phytochemical hydroxycitric acid (HCA) and several industrial units for extracting the value added product from the fruits (Hemshekhar et al. 2011).

A. A. Mariod (✉)
Indigenous Knowledge and Heritage Center, Ghibaish College of Science & Technology, Ghibaish, Sudan

A. A. Mariod (ed.), *Wild Fruits: Composition, Nutritional Value and Products*, https://doi.org/10.1007/978-3-030-31885-7_37

37.2 Botanical Overview

37.2.1 Garcinia *Genus*

Garcinia is a large genus of the family *Clusiaceae (Syn: Guttiferae)*. The genus name was proposed in honor of a French botanist Dr. Laurent Garcin a French botanist Dr. Laurent Garcin, who served in the Dutch Indies company in India (Glen 2004). Trees belonging to this genus are commonly known as sap trees, mangosteens, garcinias or popularly as monkey fruit (Yapwattanaphum et al. 2002). *Garcinia* consists of over 200 species distributed in the tropics of the world, chiefly in Asia, Africa, and Polynesia. They are evergreen polygamous trees, shrubs and herbs. About 35 species are reported to exist in India, 17 of which are endemic and economically important (Rameshkumar 2016).

37.2.2 Garcinia binucao *Description*

Garcinia binucao (Batuan) is an evergreen tree, bole around 40 cm in width. Leaves obovate oval, 5–12 cm × 4–7 cm. It is a wild tree around 15–20 m tall that develops in the backwoods of the Islands of Panay and Negros. Batwan bears fruits liberally amid the wet season and develop fruits hold tight the primary trunk and parts of the tree until harvested. Fruit is gathered for both individual use and available to be purchased. A beneficial tree can give 50–100 kg of fruits in a half year, and there are assessed to be 5000 profitable trees on the two islands. There are two sorts of batwan, one with a slender skin and one with a thicken skin. The round fruits are 4–5 cm in diameter and contain 4–6 seeds. They are green in shading, turning yellow when ripe. Their taste is sour, however not acidic, and not extremely sweet-smelling (https://www.fondazioneslowfood.com).

37.3 Distribution

Garcinia binucao Batuan trees are local toward the south eastern area of Asia these trees are found plentifully in woods that are situated in low elevations, generally in Vietnam and the Philippines.

37.4 Utilization

They can also be eaten raw. Batwan is used as a souring agent in soups typical of the area. The soured broth, called sinigang, is not common in other parts of the Philippines. In Ilonngo cooking, it is considered the star ingredient in a dish called

KBL, for kadyos (pigeon peas), baboy (pork) and langka (green jackfruit). Another common dish is paksiw, cooked with boiled batwan fruits. The fruits can also be preserved with salt and used as an appetizer. There is little documented history on the batwan tree, but it is known to have been used since before World War II. There is even a local riddle related to the species. It goes: "I went to the forest and saw many trees, but one (bat-wan) I like best. Which is it?" Despite this established connection to the local culture, fewer people in the younger generations are using the traditional fruit in home cooking, due to the commercial souring agents now available for purchase in supermarkets (https://www.fondazioneslowfood.com/en/ark-of-taste-slow-food/batwan-2/).

Binucao fruits are sour. So these cannot be eaten as such. So these are used as a souring agent while cooking fish and other foods. The young leaves of binucao trees also have a sour taste. So these are also used as a souring agent if the fruits are not available http://www.fruitipedia.com/2018/12/binucao_garcinia-binucao/.

Unripe batuan natural fruit can be cooked as a vegetable alongside fish. It can likewise be utilized to get ready chutney by bubbling batuan fruit. Individuals in Assam utilize the shriveled and safeguarded batuan cuts with dark green heartbeats to set up an extremely famous, yet to some degree acidic curry. The dried batuan organic fruits are a successful customary natural solution for diarrhea. In Ayurveda, the antiquated Indian drug framework, batuan fruits are utilized for treating various conditions, including dysentery and gastritis. Moreover, it is asserted that batuan has anti-inflammatory properties. At the point when the bark of batuan trees is removed it gives a yellowish tar known as gamboge, which has different employments. This gum is utilized in food, medication and paints. In addition, this resin can likewise be utilized as a rootstock for *Garcinia mangostana* (mangosteen). It is said that devouring new batuan fruits brings down pulse and, henceforth, they are valuable for individuals with hypertension. The sap or gamboge radiated from batuan is connected to the skin to mend disturbed skin (https://elmaskincare.com/herbs/herbs_batuan.htm).

37.5 Biochemistry of *Garcinia binucao*

Proximate analysis, physicochemical properties and supplement structure among parts and fruit development fluctuated to some dimension. The pulp, which comprised the greatest piece of the fruits and the immature ones contained high moisture and acidity that diminished as the fruit developed. Ash, protein, sugar, starch, absolute CHO, all out dissolvable solids, and sodium content were low in "batuan" fruits. The seeds contained high unrefined fat, crude protein, and tannin. The fruits were likewise discovered high in vitamin C, potassium, phosphorus, calcium, magnesium, iron, and follow dimensions of zinc, copper and manganese. Crude fiber and vitamin A were concentrated in the peel, pulp and ripe fruit (Quevedo et al. 2013).

Ragasa et al. (2014) isolated and identified β-sitosterol, stigmasterol, and monounsaturated and saturated triglycerides, from the fruits of *G. binucao*, which were reported to exhibit diverse biological activities. These authors reported oleic acid and stearic acid as the major fatty acid constituents of *G. binucao.*

In 1965, a pharmaceutically active ingredient (−)-hydroxy citric acid (-HCA) has been identified from the dried rind of *Garcinia* species. This has been shown to reduce obesity by accelerating fat burning and inhibit fatty acid synthesis. It is considered as one of the major non-wood forest produces in the Western Ghats (Hegde et al. 2010; Soni et al. 2004).

Bainto et al. (2018) directed an exploration concentrate to decide the best technique for hydroxycitric acid extraction from batuan fruit. They utilized three strategies to detach the acid including water, methanol and acetone as extraction solvents. Obtained hydroxycitric acid was measured utilizing spectrophotometric examination while thin layer chromatography was utilized to decide its purity. Their outcomes showed that *Garcinia binucao* is a potential source of hydroxycitric acid. The collected hydroxycitric acid from batuan added up to 4.81 + 0.12 g/100 g test utilizing water extraction strategy. Methanol and acetone extraction strategies gave 2.65 + 0.18 g/100 g sample and 2.76 + 0.08 g/100 g sample, separately. Thin layer chromatography showed that gathered isolate utilizing water extraction technique is pure. Water extraction strategy was discovered to be the best and proficient technique to separate hydroxycitric acid from batuan fruit (Bainto et al. 2018).

37.6 Medicinal and Therabeautic Uses

37.6.1 Bioactivities of G. binucao *Ethanolic Extract*

The ethanolic leaf, seed, and fruit concentrates of *G. binucao* were utilized in this examination to assess its antibacterial action against *Salmonella sp*, *Shigella sp* and *Vibrio cholerae*. Disk diffusion assay was utilized to decide the antibacterial action of *G. binucao* extracts on the chose gastrointestinal pathogens by estimating the zone of inhibition (ZOI) (100%) extracts for each demonstrated plant part was utilized as treatment. Accessible antibiotic disks (Tetracycline, Ampicillin, Ciprofloxacin) were utilized as control. Results demonstrated that *G. binucao* leaf and seed extracts did not show any inhibitory activity towards any of the chose gastrointestinal pathogens (ZOI = 6 mm). In any case, it was just the fruit extract that indicated inhibitory activity towards *Salmonella sp* (ZOI = 10.20 mm), *Shigella sp* (ZOI = 12.50 mm) and *Vibrio cholerae* (ZOI = 13.30 mm). Consequently, the inhibitory activity showed that the gastrointestinal pathogens are as yet resistant to the fruit extract. With the two-way ANOVA investigation, the inhibitory activity of the fruit extract demonstrates huge contrast ($p = 0.472557$) than the leaf and seed extracts, which implies that the fruit extricate is viewed as the most proficient part in repressing the development of the given enteric pathogens. Consequently, it is less productive than the given regular control antibiotics. Such inhibitory activity of the *G. binucao* fruit extract is demonstrative for the presence phytochemical constituents, for example, terpenes and phenols (Hipe and Laxamana 2015).

Exorbitant alcohol utilization is related with psychological and behavioral disabilities. Drosophila melanogaster is a generally utilized model for learning, memory, and motor function research. We used Garcinia binucao to study the capacity of its crude ethanolic leaf concentrate to lessen shortages in adapting, transient memory, and engine works because of alcoholic introduction. Chronic alcohol presentation essentially diminished motor function by 22% ($p < 0.05$), learning capacity by 37% ($p < 0.05$), and short-term memory by 33% ($p > 0.05$). Notwithstanding, in flies fed with *G. binucao* extract (GBE) incessantly presented to alcohol impacts of specific compounds, no huge decreases in engine, learning and in short-term memory capacities were watched; rather, these capacities increased ($p < 0.05$) in a dose-dependent way, recommending that the extract isn't just neuroprotective but at the same time is conceivably nootropic. GBE had low antioxidant activity regardless of the presence of triterpenes, flavonoids, and quinones, despite the fact that these may have contributed negligibly to the neuroprotective capacity of GBE. The neuroprotective capacity of GBE may then be clarified by different instruments not reliant on the immediate radical or oxidant scavenging properties of the phytochemicals in GBE, as in the regulation of signaling cascades which avoid neuronal apoptosis (Tantengco et al. 2018).

37.6.2 Antiobesity Property of Hydroxyl Citric Acid (HCA)

Hydroxycitric acid is a compound with anti-obesity and weight management property which is prevalent in *Garcinia binucao* in the Philippines. Fatty acid and cholesterol synthesis were blocked significantly by HCA and also that rats fed with HCA tended to eat less compared to the control animals, HCA lowered body fat levels with no loss of body protein in test animals (Sullivan et al. 1974). HCA exhibited antiobesity activity by inhibiting the ATP-citrate lyase, a catalyst for the conversion process of citrate to acetyl-coenzyme A, the building block for fatty acid and cholesterol synthesis (Tharachand et al. 2013).

Garcinia extracts and HCA have widely been used for obesity and weight control treatments and the long term continuous consumption demands systematic toxicity evaluation and a number of reports about the toxicity of *Garcinia* fruits and supplements are available in literature (Majeed et al. 1994). However, the potential contributions of HCA as a weight loss agent in humans were controversial, especially regarding the long term benefits and when the randomized, placebo-controlled clinical trials were counted. Also, some clinical studies reported various toxic effects such as toxicity towards spermatogenesis and hepatotoxicity. However, scientific evidence based on structure, mechanism of action and long history of the use of *Garcinia* had shown 'no observed adverse effect level' (NOAEL) at levels up to 2800 mg/day and suggests that HCA is safe for use (Chuah et al. 2013).

37.7 Nutritional Value and Food Uses

G. binucao fruit is regularly utilized for different purposes, for example, food, forage, and processing. A prior examination announced that *G. binucao* organic fruits have physicochemical properties and supplement substance which are equal to or stunningly better than certain fruits utilized for souring dishes. Another examination detailed that G. binucao seeds were great source of stearic and oleic acids. Moreover, the most elevated gelatin characteristics among the fruits studied were found in *Ficus nota* and *G. binucao* with an estimated gelatinizing level of 140–200 and 80–120, individually. Earlier studies reported the isolation and structure of α-mangostin, gartanin and isomangostin from the pericarp of Garcinia, α-Mangostin displayed high antibacterial activity against S. aureus. The confinement and recognizable proof of β-sitosterol (1), stigmasterol (2), and triglycerides (3) from the fruits of G. binucao was reported (Ragasa et al. 2014).

In light of the 9-point Hedonic scale for sensory assessment, the fresh and the dry, powdered "batuan" fruits were similar to each other regarding color, mouth feel, taste and general acceptability as a souring material for fish stew (Quevedo et al. 2013).

37.8 Different Uses

In Ayurvedic medicine, the yellow resin exuded by batuan plant is dried out and showcased under the mark Gamboge or Kokum. At the point when taken in portions of anything between 10 and 15 cgm, gamboge fills in as a purgative. Be that as it may, when this latex is utilized in higher dosages, for example, 30–50 cgm, it produces profuse evacuation accompanied by fierce colicky pains. Besides, when utilized in amazingly high dosages, gamboge fills in as a powerful toxic substance. Gamboge is regularly utilized as a hydragogue cathartic to empty out liquid constituents out of the blood, much similarly as in asthma, dropsy, just as cerebral and aspiratory clog. This pitch is likewise utilized as a vermifuge to remove different parasites and worms from the digestive organs. Gamboge is only here and there utilized freely, yet dependably controlled with some different herbs, for example, aloe vera, calomel, rhubarb, jalap and others.

In the Malabar, Canara, Sri Lanka (in the past known as Ceylon) and Singapore, individuals utilize the accompanying technique to extricate gamboge from batuan. At the beginning of the rainy season in the tropics and sub-tropical locales, individuals make a winding entry point in the bark enclosing generally 50% of the trees' trunk. At that point they embed a bamboo piece at where they need to gather the smooth juice, which oozes gradually from the cut over numerous months. After some time, the juice turns viscid and along these lines ends up strong when it is presented to the air. on average, one batuan tree yields resin that is adequate to fill no less than three inter-nodal areas of the bamboo – each fragment estimates about 50 cm long and anything between 3 and 5 cm over (https://elmaskincare.com/herbs/herbs_batuan.htm).

References

Bainto LC, Dizon EI, Laurena AC, Castillo-Israel KAT (2018) Isolation and quantification of hydroxycitric acid from batuan [Garcinia binucao (Blanco) Choisy] fruit. Int Food Res J 25(2):706–711

Chuah LO, Ho WY, Beh BK, Yeap SK (2013) Updates on antiobesity effect of Garcinia origin (−)-HCA. Evid Based Complement Alternat Med 2013:1–17

Glen FH (2004) Sappi what's in a name: the meanings of the botanical name of trees. Jacana Media, Johannesburg, 28p

Hegde L, Kuruvunshetty MS, Padnad M, Thimmaraju R (2010) Evaluation of morphological and chemical variation in two endangered medicinal tree species of garcinia. In: Vasudeva R, Janagoudar BS, Reddy BMC, Sthapit B, Singh HP (eds) Garcinia genetic resources: linking diversity, livelihood and management, proceedings of national symposium. College of Forestry, Sirsi, Karnataka, pp 74–78

Hemshekhar M, Sunitha K, Sebastin Santhosh M, Devaraja S, Kemparaju K, Vishwanath BS, Niranjana SR, Girish KS (2011) An overview on genus garcinia: phytochemical and therapeutical aspects. Phytochem Rev 10:325–351

Hipe AMP, Laxamana NPV (2015) Antibacterial activity of *Garcinia binucao* (Blanco) Choisy (batuan) extract on selected gastrointestinal pathogens. Bachelor of Science in Biology, De La Salle University – Dasmariñas City of Dasmariñas, Cavite, Philippines

http://www.fruitipedia.com/2018/12/binucao_garcinia-binucao/

https://elmaskincare.com/herbs/herbs_batuan.htm

https://www.fondazioneslowfood.com/en/ark-of-taste-slow-food/batwan-2/

Majeed M, Rosen R, McCarty M, Conte A, Patil D, Butrym E (1994) Citrin: a revolutionary, herbal approach to weight management. New Editions Publishing, Burlingame

Paull RE, Duarte O (2011) Tropical fruits volume 1, 2nd edn. CAB International, Wallingford, p 123

Quevedo ES, Laurena AC, Merca FE (2013) Physicochemical properties, nutritional and sensory quality of "Batuan" [(Blco.) Choisy] fruits *Garcinia binucao*. Ann Trop Res 35(2):1–21

Ragasa Y, Torres OB, Marasigan E, She C-C (2014) Sterols and triglyceride from the fruit of Garcinia binucao. Der Pharma Chem 6(6):229–232

Rameshkumar KB (2016) Diversity of Garcinia species in the Western Ghats. Phytochemical Perspective Jawaharlal Nehru Tropical Botanic Garden and Research Institute, Palode, Thiruvananthapuram, Kerala, India

Soni MG, Burdock GA, Preuss HG, Stohs SJ, Ohia SE, Bagchi D (2004) Safety assessment of (−)-hydroxycitric acid and Super CitriMax, a novel calcium/potassium salt. Food Chem Toxicol 42(9):1513–1529

Sullivan AC, Triscari J, Hamilton JG, Miller ON, Wheatley VR (1974) Effect of (−)-hydroxycitrate upon the accumulation of lipid in the rat: I. Lipogenesis. Lipids 9:121–128

Tantengco OAG, Tan JJE, Tan NRC, Sison MCC, Medina PMB (2018) *Garcinia binucao* crude ethanolic leaf extract prevents alcohol induced neurotoxic effects on learning, short-term memory, and motor functions in *Drosophila melanogaster*. J Appl Pharm Sci 8(10):106–112

Tharachand, Selvaraj I, Avadhani M (2013) Medicinal properties of Malabar tamarind, Garcinia Cambogia (Gaertn) DESR. Int J Pharm Sci Rev Res 19:101–107

Yapwattanaphum C, Subhadrabhandu S, Sugiura A, Yonemari K, Utsunomiya N (2002) Utilization of some Garcinia species in Thailand. Acta Hortic 575(2):563–570

Chapter 38
Some Wild *Elaeagnus* Species: Overview, Description, Biochemistry, and Utilization

Thamer Elamin Abdalla

38.1 Introduction

The Elaeagnaceae is a plant family, comprising small trees and shrubs, The family has about 60 species in three genera (Christenhusz and Byng, 2016). They are commonly thorny, with simple leaves often coated with tiny scales or hairs. Most of the species are xerophytes (found in dry habitats); several are also halophytes, tolerating high levels of soil salinity. The Elaeagnaceae often harbor nitrogen-fixing actinomycetes of the genus *Frankia* in root nodules, making them useful for soil reclamation (Winrock 1992). This characteristic, together with their production of plentiful seeds, often results in the Elaeagnaceae being regarded as weeds.

Elaeagnaceae are small trees, described by Bartish and Swenson (2004) as an anemophilous or entomophilous tree, shrub or rarely could be woody climbers; shoots sometimes reduced to spines; young branches and leaves, as well as calyx tube covered with peltate and stellate trichomes; It contains nitrogen fixing bacteria on the roots nodules.

38.2 Description and Vegetative Structures

Leaves alternate, opposite, or in pseudowhorls rarely, simple, petiolate, with no stipules; lamina entire, pinnately veined. Inflorescences axillary, short, fasciculate, either to be racemose or spicate, or rarely are flowers solitary. Flowers are apetalous, actinomorphic, often aromatic, perfect, or plants monoecious, mainly dioecious but sometimes are polygamous, there are no staminodes in female flowers, no pistil in male; sepals 2 or 4 (−6), linked with a hypanthium; hypanthium free,

T. E. Abdalla (✉)
Sudanese Standards and Metrology Oraganization, Khartoum, Sudan

A. A. Mariod (ed.), *Wild Fruits: Composition, Nutritional Value and Products*,
https://doi.org/10.1007/978-3-030-31885-7_38

narrow above the gynoecium, the color vary from white and cream to yellow; bisexual and female flowers are receptacle tubular, male flowers are mostly flat; Fruit drupe- or berry-like with thin, membranous pericarp, surrounded by the persistent calyx tube, which becomes fleshy. Seed solitary; testa hard; embryo straight and achlorophyllous; endosperm scanty or absent; cotyledons fleshy, plano-convex. Germination phanerocotylar. A family with three genera and 30–50 species, distributed mainly in the temperate regions of the Northern Hemisphere, some in tropical SE Asia and eastern Australia.

Elaeagnus are deciduous or evergreen bush or small trees. The alternate leaves and the tendrils are containing silvery to brownish dye making the plants with whitish to grey-brown color from a far. The flowers are tiny, with a four calyx and no petals, sometimes appeared fragrant. The fruit is a fleshy drupe with a single seed edible in many species of *Elaeagnus* (CABI 2018).

The stems and leaves are covered with peltate or scaly trichomes. The leaves are leathery, simple, entire and without stipules. Stomata are anomocytic. Vessels have simple perforations, and small to vestigial vestured pits are present, and non vestured pits are also common. Rays are heterogeneous to almost homogeneous in some species of *Elaeagnus*. Axial parenchyma is diffuse, sometimes very scanty. Phloem is usually tangentially stratified into hard and soft layers (Jansen et al. 2000).

According to Morphologically and molecular analyses, Eleaeagnaceae are placed in a squad of families in order Rosales sensu with Cecropiaceae, Moraceae, Barbeyaceae, Celtidaceae, Cannabaceae, Ulmaceae, Rosaceae, Rhamnaceae, Urticaceae and Dirachmaceae (Thulin et al. 1998; Richardson et al. 2000; Soltis 2000).

The genus *Elaeagnus* is comprised of about 70–80 species (Ge et al. 2013). It is known by various names as red berries, oleaster, silverberry, Persian olive, autumn olive, Russian olive, thorn olive and wild olive.

38.3 Habitat

The majority of the species are native to temperate regions of the Northern Hemisphere, south into subtropical regions of Asia. and Australia. *Elaeagnus triflora* extends to northeastern Australia from Asia south, while *E. commutata* is native to North America, *Elaeagnus philippinensis* native to Philippines. *E. angustifolia* is native in southeastern Europe many Asiatic species of *Elaeagnus* have become established as introduced species in North America, with some of these species being considered invasive, or even designated as noxious, in portions of the United States (FLEPPC 2012; and BONAP 2014).

38.4 Cultivation

Elaeagnus species majorly cultivated for their showy, sometimes variegated, vegetation, and numerous cultivars and hybrids have been driven. The fruit is acid and slightly shrivel and makes good tarts. Many species are cultivated to use their fruit, including *E. angustifolia*, *E. umbellata* and *E. multiflora* (gumi) (RHS 2013) (Table 38.1).

38.5 Ecology

The larvae of some Lepidoptera species such as *Coleophora elaeagnisell*a and the gothic moths are using *Elaeagnus* species as food plants. Also some bird species can use the thorny shrubs as a good nesting sites. Many species of *Elaeagnus* has ability to grow in low-nitrogen soil due to organisms content in their roots which have high activity of nitrogen fixing because of that they can invade many regions where they are established as exotic species (CABI 2018). The fleshy fruits of many of *Elaeagnus* are consumed by birds and may be distributed for wide distances. During winter, the dried *Hippophae* fruits are known to float and disperse by spring floods along rivers (Bartish and Swenson 2004). *Hippophae* and *Shepherdia* are dioecious, *Elaeagnus* is monoecious or polygamous.

38.6 Biochemistry and Photochemistry

The fruits of many species of *Elaeagnus* are consumed as such or processed as juice. It is also consumed as herbal tea, wine, soup, sauce, dessert, candy, pudding, ice cream topping, fruit leather, jam and jelly. Beside their edibility, the plant parts multiple uses in folk medicine as anti-inflammatory, muscle relaxant, antipyretic, analgesic, astringent and antiulcer agent (Saboonchian et al. 2014; Bendaikha et al. 2014). The fruits of *Elaeagnus* shown as rich in many bioactive components including flavonoids, phenolics, carotenoids, lipids and ascorbic acid (Wu et al. 2011). The pulp of the fruit has lycopene in profusion. The seeds have been assessed to be plentiful in proteins and fats. *E. angustifolia* seed had 8% unsaturated fatty acids mainly linoleic acid, oleic acid, and stearic acid. The protein content was 11%, mainly globulin and albumin (Kadir and Kuerban-jiang 2011). The lycopene content

Table 38.1 Notable species and hybrids in *Elaeagnus* cultivation (WC 2019)

Elaeagnus angustifolia	*Elaeagnus multiflora*
Elaeagnus commutata	*Elaeagnus pungens*
Elaeagnus submacrophylla	*Elaeagnus reflexa*
Elaeagnus macrophylla	*Elaeagnus umbellata*

of *E. umbellata* berries in fresh fruit, was multiple times higher than that of tomato. The comparative findings suggest the rationale of promoting elaeagnus berries as a rich source of lycopene (Fordham et al. 2001). The fruits of *E. angustifolia* have been previously characterized to have polysaccharides, flavonoids, coumarins, phenolcarboxylic acids, tannins, saponins and carotenoids (Abizov et al. 2008). Elaeagnaceae are known as strongly tanniniferous, as both of ellagitannins and condensed tannins are present as well as many other compounds including indol alkaloids, sinapinic acid, flavonols, pentacyclic triterpenes and L-quebrachit (Hegnauer 1966, 1989).

38.7 Economic Importance

Many of *Elaeagnus* species including *Elaeagnus angustifolia, E. pungens, E. macrophylla*, and *E. umbellata* are known to grow as ornamental shrubs and mainly are grown as deciduous or evergreen shrubs. In China, *Hippophae* plantations are also used to prevent soil erosion (Lu 1992). The fruits of wide range of the species are edible, such as *Elaeagnus angustifolia, Hippophae rhamnoides*, and *Shepherdia argentea*. Many fruits of *Elaeagnus* species have high content of vitamins specially A, C and E. in different regions mainly in Europe and Asia the fruits of the plant are used for the production of juices, beverages and jams. Sea buckthorn seed oil of act as anti-inflammatory, bactericidal, pain relief, and used for the promotion of tissue regeneration and have been clinically tested in Russia and China and listed in the Pharmacopoeia in 1977 (Xu 1994).

38.8 Major Uses for Selected Species of *Elaeagnaceae*

38.8.1 Elaeagnus triflora

Are hard shrubs that have the ability to adapt and tolerates moderate frost conditions. The stems are extremely strong with side-branchlets to provide supporting vegetation. Leaves appeared as silver-backed while the new growth. The shininess provided by the starry scales which are also present on the little whitish flowers. The ripe red fruits have a sweet flavor juicy flesh with the smell and consistency as well. In full sun and good soil, *Elaeagnus triflora* can easily spread to its neighbors. The arching canes can be clipped over and over again to encourage the softer flowering foliage (Ganguli and Kennedy 2013). *Elaeagnus triflora* retains green leaves throughout the year, sometimes spread into nearby trees for support and can grow to reach up to 10 meters tall. Older stems are usually supported with spikes that are derived from short shoots – and act to help and support the plant as it scrambles into the neighboring plants. The edible fruit is often collected from the wild to be eaten locally. The plant also has uses of local medicinal. It is often grown in gardens as a fruit plant. Sometimes all *Elaeagnus conferta* of *Elaeagnus latifolia* and *Elaeagnos triflora* addressed as three different species and sometimes as a same species (as *Elaeagnus latifolia*). There is also argument about the range of the species, with *Elaeagnus latifolia* in several different ways stated to be native to the Indian Continent or to continental Southeast Asia. In this treatment as per the Flora Malesiana, the three species are treated separately. Seed best to be sown as soon as it is mature in a cold mount. The germination period ranged within a few weeks and it may take 18 months. Stored seed can have very slow germination rate, sometimes it can take more than 18 months. Useful tool is warm stratification for about 4 weeks followed by about 12 weeks of cold stratification. The seed are usually quite well germinates (Van Steenis 1950).

38.8.1.1 Medicinal and Edible Uses

The flowers are astringent and cardiac. Raw or cooked fruits has a sweet flavor when ripe (Brown 1920). The ripe fruit is Juicy and pleasant to eat out of hand when fully ripe. The fruits have a sweet tangy flavor and make a good quality savory sauce or sweet jam and also used to make a highly colored jelly (Facciola 1998). Some forms are delicious (Nicholson and Nicholson 2007). Fruits are a drupe with a single stony seed (WINROCK 1992). Seed - raw or cooked. Peanut similar flavor, it is preferred to be eaten with the fruit, although the fibrous seed case would have to be spat out afterwards (Ken Fern 1997). The fruits of *Elaeagnus triflora* are also described to children sick by dysentery.

38.9 *Elaeagnus umbellata*

E. umbellata has been disseminated from its native Asia to Europe and North America for environmental benefits as well as revegetation purposes and also to provide wildlife cover and prevent soil erosion. It has been escaped from many and become naturalized especially in unstable sites and cultivation areas and determined as invasive in Italy, Belgium and France (Christenhusz and Uffelen 2001). It has high seed production and germination rates and has the potential to colonize new areas easily also it has ability of nitrogen fixation (CABI 2018). The plant is very ornamental, and it is a hardy plant has ability to grow well under dry conditions (Fig. 38.1). *E. umbellata* is bisexual, aromatic, and open-pollinated sometimes by insects. It grows well on a many types of soils such as sandy, loamy, and somewhat clayey soils rage between (pH 4.8–6.5), though prefers deep, relatively coarse-textured soils that are moderately to well-drained. It shows drought tolerance, though does not grow well on very wet or dry sites, nor shallow or poorly drained soils. Mature trees can produce about 14 kg of fruit annually (Munger 2003).

Fig. 38.1 *Elaeagnus umbellata* leaves and fruits (Source: https://commons.wikimedia.org)

38.9.1 Edible and Medicinal Uses

E. umbellata fruit is about 8 mm in diameter contains a single large seed can be consumed raw or cooked. The fruit is tasty, juicy with pleasantly acidity. Sometimes it can also be made into jams (Facciola 1998; Parmar and Kaushal 1982). The fruit must be fully ripe to be consumed raw, it will be quite astringent when slightly under-ripe. The fruit contains 8.3%, 4.5%, 1.0% of sugars, protein and ash respectively and about 12 mg per 100 g of vitamin C content (Parmar and Kaushal 1982). Flowers are used as astringent, stimulant and cardiac agents. The seeds are useful in the treatment of coughs as a stimulant. The extracted oil from the seeds described in the treatment of pulmonary affections (Chopra et al. 1986). The fruit are very rich source of vitamins and minerals, mainly in vitamins A, C and E as well as flavanoids and many other bioactive compounds and also known as a source of essential fatty acids. Because of its high content of antioxidant carotenoids, and lycopene, the plant associated to prevent many types of diseases and it is able to reduce and reverse the growth of cancers (Matthews 1994; Fordham et al. 2003).

38.10 *Elaeagnus pungens*

It is an evergreen shrub reach over 3–4 meters in height and 4 meters wide. It is in leaf during all the year, flowers appeared from November to February, the seeds matured from April to May. Flowers fragrant, silvery-white to brown, tubular and 1 cm long, few (1–3), in axillary clusters, pedicel 5–8 mm and scaly brown, calyx tube 6–7 mm long and abruptly narrowed at base, lobes ovate, apex rounded. Fruit oblong, 1.0–1.5 cm, brown scaly containing one nut, white changing to red with spotted brown scales when ripe (Figs. 38.2 and 38.3). The species has both male and female organs and fertilized mainly by bees. *E. pungens* is cultivated mainly in warm regions as a garden plant. It has ability to tolerate various environmental conditions. It is drought-tolerant and can grow in different soil types (Miller 2003; FCEC 2008).

38.10.1 Edible and Medicinal Uses

Fruit - raw or cooked (Kunkel 1984). A good sub-acidic flavor if fully ripped and astringent if eaten before full maturation. It can be converted into preserves and drinks. The oval fruit is seed - raw or cooked about 15 mm long can be eaten with the fruit though the seed case is rather fibrous, the taste similar of peanuts. The seed contains about 42% protein and 23% fat on a zero moisture basis (Duke and Ayensu 1985). The fruit of many *E. Pungens* members is a very rich source of vitamins and minerals, especially in vitamins A, C and E. Sometimes the leaves and stems are used in the incidents of asthma, cough, diarrhea and haemorrhoids. The seeds are used in watery diarrhea treatment. The roots are astringent and are used to sores and itchy skin and as many *Elagneacae* species, it has ability to reduce and surround cancer (Matthews 1994).

Fig. 38.2 *Elaeagnus pungens* Fruits

Fig. 38.3 *Elaeagnus pungens* leaves

38.11 *Elaeagnus angustifolia L.*

Deciduous tree or sometimes large shrub which is commonly known as Russian olive due to its appearance similar to an olive tree with small red-brownish, elliptic like fruits. Leaf appearance is generally described as lanceolate to oblong and often elliptic. Aromatic flowers are 3–12 mm long, in small axillary clusters on the twigs. Fruits are drupe- and sometimes berry-like, oval-shaped, about 1 to 2 cm long, and a single, relatively large, 6–13 mm, oblong achene is enclosed in the fleshy fruit. The optimum fruiting was observed in 40-year-old stands, with an average yield of 73.8 kg/ha (Baranov and Kositzyn, 2003). It is known to have a long lived tree (about 80–100 years) can grow fast to reach up to 10 meters in height and about 30 cm in diameter and can produce fruits after about 5–6 years (Kiseleva and Chindyaeva 2011) (Fig. 38.4). This tree can resist a wide range of unfavorable environmental conditions including flood, severe dryness, sandy and high salinity or soils with high alkalinity. Different parts of the *Elaeagnus angustifolia L.* have been used in a different of medicinal uses, in perfume industries, and musical instruments production.

38.11.1 Edible Uses and Nutritional Value

*Elaeagnus angustifolia L.*fruits known to have high nutritional values and rich content of proteins, sugar, minerals and vitamins (Fonia et al. 2009). The flowers are tiny yellowish-white, aromatic, and also have been used for honey bees as a source of nectars. The fruits of *E. angustifolia L.* have been traditionally used fresh or dried. The fruits have been known of it is rich content of various types of vitamins including tocopherol, carotene, thiamine, vitamin C, and minerals such as calcium, manganese, potassium, iron and magnesium (Boudraa et al. 2010).

Fig. 38.4 *Elaeagnus angustifolia L.* tree and fruits

The fruits also used in liqueur production as a flavoring agent. Fruit extraction indicate the availibility of many compunds such as flavonoid, polysaccharides, coumarines, phenol carboxylic acids, amino acids, saponins, sitosteroles, cardiac glycosides, terpenoids, carotenoids, vitamins, and tannins (Okmen and Turkcan 2013). The plant contains copper, lead, iron, cadmium, zinc, chromium, cobalt and nickel in different part including the root, root bark, stem bark branches and leaves. The major mineral found in *E. angustifolia L.* fruit is potassium (8504 mg/kg), Then sodium (1731 mg/kg) and phosphorus (635 mg/kg) (Cansev et al. 2011).

38.11.2 Medicinal Uses

The flowers of *Elaeagnus angustifolia L.* have been used in traditional medicine to treat tetanus (Hosseinzadeh et al. 2003). The fruit is consumed raw or boiled fruit for the treatment of many symptoms and diseases including flu, cough, sore throat, cold, fever, nausea, vomiting, jaundice, asthma, diarrhea and many other. In Asia, mainly in Iranian folk medicine. The fruits of the *E. angustifolia L.* used for the relief of pain and inflammation in rheumatoid arthritis patients to accelerate wound healing process in an injured region (Ahmadiani et al. 2000). Because of tannins content, *Elaeagnus angustifolia L* have been shown to have anti-cholesterol, anti-inflammatory, anti-cancer also shows cardio-protective and chemopreventive effects in human body. In addition it increase the angiogenesis activities in wounded skin which help to decresse the healing time (Zeng et al. 2009).

In some parts of Europe and Central Asia, the preparations of *E. angustifolia L.*fruits are used widely in the treatment of peptic ulcer (Gürbüz et al. 2003). The pshatin drug which is made of polyphenolic compounds concentration of the fruit of *E. angustifolia L.*, has been used in Armenia for a long time for colitis and other GI tract diseases treatment (Abizov et al. 2008). In Iran it has been used as an analgesic agent to relieve pain in rheumatoid arthritis patients. In recent study carried by (Hamidpour et al. 2017), signifies the presence of potintial antimicrobial agents in the fruit extracts of *E. angustifolia L.* and showed minimum inhibitory concentration values of the extraction of the fruit indicating antimicrobial activity against the targeted culture ranged between 0.1 mg/mL to 7.5 mg/mL. and strong antimicrobial activity of 1.62 mg/mL against E. coli . Also high effectiveness against the growth wide range of bacteria including *Bacillus subtilis, Staphylococcus aureus, Salmonella typhimurium* and *Yersinia enterocolitica* (Tamtaji et al. 2014; Okmen and Turkcan 2013).

38.12 *Elaeagnus x ebbingei*

It is an evergreen shrub can grow up to 5 metres high and about the same wide. The hedge of the plant could reach 1.5 metres tall and only about 45 cm wide. In the firsy year, the plant have slow establishment rate after that it have ability to grow about 75 cm each year. This plants are very tolerant of maritime exposure, growing well right by the coast (Ken Fern 1997). *E. x ebbingei* is a palnt have easily grown, it can gorw in most soils that are well-drained and prefers a soil that is only moderately fertile, can grow in both poor soils and in dry soils. The plant can tolerate unfavourable conditions and grow in very poor soils and can resist drought effectivly. It can grow near the sea and give good wind protection. Also it tolerant shade effevtivly and grows well under trees (Huxley 1992).

They have shown resistance to pests including the honey fungus, slugs, insects and diseases but sometimes whole branches die when grafted within the deciduous

E. multiflora and should be immediatly removed from the plant (Thomas 1992). Stored seed have very slow germination rate. It is recommended to be placed in a moist sand in plastic bag for four weeks at around 15 -20C then followed by 12 weeks of cold stratification at about 1C. *E. x ebbingei* will not breed well from seed. Seed can help the process of cultivars development and improvement. If fresh seed have sow in a cold greenhouse in the spring season, will usually germinate within a month or two. The seedlings should be planted into individual pots to be grown on in a cold greenhouse until the following spring before planting. This species has a symbiotic relationship with many types of soil bacteria which form nodules on the roots and fix atmospheric nitrogen and some of the fixed nitrogen utilized by the growing plant and can also be used by other plants growing nearby (Huxley 1992).

38.12.1 Edible and Medicinal Uses

The fruit is of a reasonable size and when fully ripe is very acceptable for dessert. It should be fairly easy to selectively breed for improved fruit size and flavour. The fruits are Fruit - raw or cooked, very soft, deep red-silvery color, ovale and about 2 cm or more long and 13 cm wide with larege seed (Kunkel 1984). Not all plants bear many fruits, though many have been seen that produce very heavy crops on a regular basis. Astringent when not fully but when fully ripe, it has a very rich flavor and makes pleasant and delecious tasting with a slight acidity. The fruits have acceptable flavour specially if the fruit stored for a day or two after being collected. Seed - raw or cooked, can be consumed with the fruit though the seed case is rather fibrous and have taste risemble for peanuts (Ken Fern 1997). The fruit of *E. x ebbingei* are very rich source of vitamins and minerals, mainly in vitamins A, C and E, flavanoids and many other bio-active compounds. It is also a good source of essential fatty acids. It is being investigated that is capable of reducing the incidence of cancer and also as a means of halting or reversing the growth of cancers (Matthews 1994).

38.13 *Elaeagnus multiflora*

E.multiflora is widly distributed in Australia, Britain, North Africa, Pacific, Philippines, Tibet and USA but it is native in China, Japan and Korea (Anna et al. 2017). It is deciduous Shrub estimated to grow up to 3 meters by 2 meters at average. This is a shrub growing up to 3 meters. Its shoots feature scarce thorns. The leaves are similar for *Elaeagnus* – the upper part of the leaf blade is green, and the bottom side appears silvery (Seneta and Dolatowski 2012). It's get flowered from April to May, the seeds completely ripen in July. The species pollinated by Bees and *Apidae* family insects. The self-fertility rate can vary from 0% to 70% (Radiuk and

Radiuk 1997). It has both male and female organs. It has ability to fix Nitrogen and can tolerate and grow in the atmospheric pollution. Easily to grow in most weel darained soils and can grow in light shade but prefer the sunny positions. Seed - best sown as soon as it get ripned. Germination period ranged between late winter or early spring. The germination rate of the stored seeds can be very slow and could take more than 18 months. A warm stratification for 4 weeks followed by cold stratification for 12 weeks can useful. The vegetation period of *E.multiflora* is long and it starts to drop leaves only after first ground frosts. Such a late leaf fall can lead to shoot freezing in hard winters. The shoots, even completely frozen, demonstrate good regeneration abilities and grow from the root neck in significant numbers (Kawecki et al. 2007).

38.13.1 Edible and Medicinal Uses

Fruit - raw or cooked with pleasant acidity when fully ripe with a taste similar that of red currant. It difficult to be collected without breaking the young shoots. It contains a single large seed and must be fully ripe before it can be consumed raw. The fruit is astringent when under-ripe. The fruits are ellipsoidal, drupe-like type, juicy, sour, colour is red up to 1 cm long and are set on stems. Fruits of *E.multiflora* can be used in home processing to prepare juice, compote, jam or jelly and other food products such as dessert, pies and preserves. Since silverberry flowers give off a strong aroma similar to the smell of cinnamon and vanilla, this plant can be used for flavouring cakes and other meals. Seed - raw or cooked. It can be consumed with the fruit.

The fruits is very rich of vitamins and minerals as many members of this genus it contains vitamins A, C and E as well as flavanoids and many bio-active compounds. The cherry silverberry fruit are rich in carotene, phenolic compounds, amino acids as well as macro and microelements (Szałkiewicz and Kawecki 2003). It is also known to has good content of essential fatty acids. Because of their rich chemical composition, reveal therapeutic and preventive values (Anna et al. 2017). The fruits of *E.multiflora* had antioxidant, anti-inflammatory and anti-proliferative effects and could be of healing importance in the treatment of colon cancer (Lee et al. 2010). The fruit of *E. multiflora* are also used in the treatment of watery diarrhoea and also demonstrate a toning effect and improve blood circulation. The fresh and processed fruit are recommended in abdominal pain The root decoction used in the treatment of foul sores and itch. The leaves of *E. multiflora* are used in the treatment of coughs and as per (Kawecki et al. 2007) it can be used as compresses for slow-healing wounds.

References

Abizov EA, Tolkachev ON, Mal'tsev SD (2008) Composition of biologically active substances isolated from the fruits of Russian olive (*Elaeagnus angustifolia*) introduced in the European part of Russia. Pharm Chem J 42:696–698

Ahmadiani A, Hosseiny J, Semnanian S, Javan M, Saeedi F, Kamalinejad M, Saremi S (2000) Antinociceptive and anti-inflammatory effects of Elaeagnus angustifolia fruit extract. J Ethnopharmacol 72(1–2):287–292

Anna B, Beata P, Marina S, Bogumil M, Ewa G (2017) Evaluation of yield, morphology and quality of fruits of cherry silverberry (*Elaeagnus Multiflora Thunb.*) biotypes under conditions of North-Eastern Poland. Pol J Nat Sci 32(1):61–70

Baranov AF, Kositzyn VN (2003) Productivity and stocks of fruits of *Elaeagnus angustifolia L.* in basin of the lower Volga. Rastitel'nye Resursy 39(4):54–60

Bartish IV, Swenson U (2004) Elaeagnaceae. In: Kubitzki K (ed) Flowering plants dicotyledons. The families and genera of vascular plants, vol 6. Springer, Berlin/Heidelberg

Bendaikha S, Gadaut M, Harakat D, Magid A (2014) Acylated flavonol glycosides from the flower of *Elaeagnus angustifolia L.* Phytochemistry 103(2014):129–136

BONAP (2014) "Elaeagnus" county-level distribution maps from the North American Plant Atlas (NAPA). Biota of North America Program. http://bonap.net/NAPA/TaxonMaps/Genus/County/*Elaeagnus*

Boudraa S, Hambaba L, Zidani S, Boudraa H (2010) Mineral and vitamin composition of fruits of five underexploited species in Algeria: *Celtis australis L., Crataegus azarolus L., Crataegus monogyna Jacq., Elaeagnus angustifolia L.* and *Zizyphus lotus L. Fruits* 65:75–84

Brown WH (1920) Minor products of Philippine forests. Bureau of Forestry, Manila

CABI (2018) *Elaeagnus umbellata* (autumn olive). Invasive Species Compendium. Centre for Agriculture and Biosciences International. https://www.cabi.org/isc/datasheet/20728

Cansev A, Sahan Y, Celik G, Taskesen S, Ozbey H (2011) Chemical properties and antioxidant capacity of *Elaeagnus angustifolia L.* fruits. *Asian J Chem* 23:2661–2665

Chopra RN, Nayar SL, Chopra IC (1986) Glossary of Indian medicinal plants (including the supplement). Council of Scientific and Industrial Research, New Delhi

Christenhusz MJM, Byng JW (2016) The number of known plants species in the world and its annual increase. Phytotaxa. Magnolia Press. 261(3):201–217. https://doi.org/10.11646/phytotaxa.261.3.1

Christenhusz MJM, Uffelen GA (2001) Naturalized Japanese plants in Netherlands. Gorteria 27(5):631–644

Duke JA, Ayensu ES (1985) Medicinal plants of China. Reference Publications, Michigan, USA. Inc. 1985 ISBN 0-917256-20-4

Facciola S (1998) Cornucopia II – *a source book of edible plants*. Kampong Publications, California

FCEC (2008) Flora of China Web. Flora of China Editorial Committee. Harvard University Herbaria, Cambridge, MA. http://flora.huh.harvard.edu/china/

Fern K (1997) Plants for a future: edible & useful plants for a healthier world. Permanent, Hampshire

FLEPPC (2012) Florida exotic pest plant council invasive plant lists. Florida Exotic Pest Plant Council. www.fleppc.org version 3.0

Fonia A, White IR, White JML (2009) Allergic contact dermatitis to *Elaeagnus* plant (Oleaster). Contact Derm 60:178–179

Fordham IM, Clevidence BA, Wiley ER, Zimmerman RH (2001) Fruit of autumn olive: a rich source of lycopene. HortScience 36:1136–1137

Fordham IM, Zimmerman RH, Black BL, Clevidence BM, Wiley ER (2003) Autumn olive: a potential alternative crop. Proceeding of the XXVI international horticultural congress 11–17 August 2002, Toronto

Ganguli NK, Kennedy IR (2013) Indigenous actinorhizal plants of *Australia*. J Biosci 38:749–759. https://doi.org/10.1007/s12038-013-9368-x

Ge Y, Dai Q, Wan D, Liu Q, Mei Z (2013) Relaxant effect of 1- butanol fraction from *Elaeagnus pungens* leaf through inhibiting L-type Ca2+ channel on Guinea pig tracheal smooth muscle. J Ethnopharmacol 150(1):196–201

Gürbüz I, Üstün O, Yesilada E, Sezik E, Kutsal O (2003) Anti-ulcerogenic activity of some plants used as folk remedy in Turkey. J Ethnopharmacol 88:93–97

Hamidpour R, Hamidpour S, Hamidpour M, Shahlari M, Sohraby M, Shahlari N, Hamidpour R (2017) Russian olive (*Elaeagnus angustifolia L.*): from a variety of traditional medicinal applications to its novel roles as active antioxidant, anti-inflammatory, anti-mutagenic and analgesic agent. J Tradit Complement Med 7:24–29

Hegnauer R (1966) Chemotaxonomie der pflanzen, vol 4. Birkhauser, Basel

Hegnauer R (1989) Chemotaxonomie der pflanzen. Eine Ubersicht uber die Verbreitung und die systematische Bedeutung der Pflanzenstoffe, vol 8. Birkhauser Verlag, Basel

Hosseinzadeh H, Ramezani M, Namjo N (2003) Muscle relaxant activity of Elaeagnus angustifolia L. fruit seeds in mice. J Ethnopharmacol 84(2003):275–278

Huxley A (1992) The new RHS dictionary of gardening. MacMillan Press, London, UK. ISBN 0-333-47494-5

Jansen S, Piesschaert F, Smets E (2000) Wood anatomy of Elaeagnaceae, with comments on vestured pits, helical thickenings, and systematic relationships. Am J Bot 87:20–28

Kadir M, Kuerban-jiang B (2011) Research on fat and protein compositions in seeds of wild *Elaeagnus angustifolia* from Xinjiang Uigur autonomous region. Med Plant 2:51–53

Kawecki Z, Łojko R, Pilarek B (2007) Mało znane rośliny sadownicze. Wyd. UWM, Olsztyn

Kiseleva TI, Chindyaeva LN (2011) Biology of oleaster (*Elaeagnus angustifolia L.)* at the north-eastern limit of its range. Contemp Probl Ecol 4:218–222

Kunkel G (1984) Plants for human consumption. Koeltz Scientific Books, Koenigstein, Germany. ISBN 3874292169

Lee MS, Lee YS, Park OJ (2010) Cherry silver berry (*Elaeagnus multiflora*) extracts exert anti inflammatory effects by inhibiting COX-2 and Akt signals in HT-29 colon cancer cells. Food Sci Biotechnol 19(6):1673–1677

Lu R (1992) Seabuckthorn: a multipurpose plant species for fragile mountains. ICIMOD Publication Unit, Katmandu

Matthews V (1994) Calochortus gunnisonii. *The New Plantsman*, vol 1. Royal Horticultural Society, London, UK

Miller JH (2003) Nonnative invasive plants of southern forests: a field guide for identification and control. General Technical Report – Southern Research Station, USDA Forest Service, No.SRS-62, 93 pp. Asheville, North Carolina, USA

Munger GT (2003) *Elaeagnus umbellata*. Fire effects information system [Online].USA: USDAFS. http://www.us/database/feis/plants/shrub/elaumb/all

Nicholson N, Nicholson H (2007) Australian rainforest plants volumes 1–6. Terania Rainforest Publishing, Lismore

Okmen G, Turkcan OA (2013) Study on antimicrobial, antioxidant and antimutagenic activities of *Elaeagnus angustifolia L.* leaves. *Afr J Tradit Complement Altern Med* 11:116–120

Parmar C, Kaushal MK (1982) Wild fruits of the sub-Himalayan region. Kalyani Publishers, New Delhi

Radiuk AF, Radiuk WA (1997) Płodowo jagodnyj sad. Urażaj, Mińsk, p 527

RHS (2013) *Eleagnus* × *ebbengei* 'Gilt Edge'. Royal Horticultural Society, London, UK

Richardson JE, Fay MF, Cronk QCB, Bowman D, Chase MW (2000) A phylogenetic analysis of *Rhamnaceae* using rbcL and trnL-F plastid DNA sequences. *Am J Bot* 87:1309–1324

Saboonchian F, Jamei R, Hosseini Sarghein S (2014) Phenolic and flavonoid content of *Elaeagnus angustifolia L.* (leaf and flower). Avicenna J Phytomedicine 4(2014):231–238

Seneta W, Dolatowski J (2012) Dendrologia. PWN, Warszawa

Soltis DE (2000) Morphological study of the genus *Hippophae L.*, including the new taxa recognized by A. Rousi. Ann Bot Fennici 8:228–236

Szałkiewicz M, Kawecki Z (2003) Oliwnik wielokwiatowy (*Elaeagnus multiflora Thunb.*) – nowa roślina sadownicza. Biul Nauk 22:285–290

Tamtaji OR, Taghizadeh M, Takhtfiroozeh SM, Talaei SAR (2014) The effect of *Elaeagnus angustifolia* water extract on scopolamine-induced memory impairment in rats. *Zanjan Univ Med Sci J ZUMS J* 22:101–111

Thomas GS (1992) Ornamental shrubs, and Bamboos. Murray, UK. ISBN 0-7195-5043-2

Thulin M, Bremer B, Richardson J, Niklasson J, Fay MF, Chase MW (1998) Family relationships of the enigmatic rosid genera Barbeya and Dirachma from the horn of Africa region. Pl Syst Evol 213:103–119

Van Steenis CGGJ (1950) Flora Malesiana, Series 1, vol 1. Centre for Research and Development in Biology/Rijksherbarium/Kluwer Academic Publishers, Bogor/Leiden/Dordrecht, Boston, London, pp 4–10

WC (2019) Elaeagnus. Wikipedia Contributors. In Wikipedia, The Free Encyclopedia. https://en.wikipedia.org/w/index.phptitle=Elaeagnus&oldid=904253256

Winrock (1992) Elaeagnus: a widely distributed temperate nitrogen fixer. Winrock International, Archived from the original on 18 May 2009. Retrieved 4 May 2009. https://www.winrock.org/factnet-a-lasting-impact/fact-sheets/elaeagnus-a-widely-distributed-temperate-nitrogen-fixer/

Wu M, Hu HT, Yang L, Yang L (2011) Proteomic analysis of up-accumulated proteins associated with fruit quality during autumn olive (*Elaeagnus umbellata*) fruit ripening. J Agric Food Chem 59(2011):577–583

Xu M (1994) The medical research and exploitation of sea buckthorn. Hip 7:32–34

Zeng F, Wang W, Zhan Y, Xin Y (2009) Establishment of the callus and cell suspension culture of *Elaeagnus angustifolia* for the production of condensed tannins. Afr J Biotechnol 2009:8

Chapter 39
Wild Species of *Vaccinium* Composition, Nutritional Value and Utilization

Mohamed E. S. Mirghani, Jamal I. Daoud, and Ahmed A. M. Elnour

Abbreviations

DPPH	2, 2-diphenyl-1-picrylhydrazyl
FRAP	Ferric Reducing Antioxidant Power
TPC	Total Phenolic Compounds

39.1 Introduction

Berries are small fleshy fruits with delicious flavour and taste, of wild or cultivated origin, which are commonly consumed in fresh and in processed forms. The term "berry" used to describe any kind of small edible fruit, usually juicy and round shaped, with seeds within the flesh, sweet to sour tasting and with a different intensity of colouration. Not all berries in the pomological sense are true berries in botanical terms of fruit descriptions. Beside the ovary, other flower parts often contribute to the development of the fruit and also more than one ovary can form a fruit (Aggregate and Multiple Fruit) (Zhao 2007). The continuously increasing demand for suitable, adequate, safe and nutritious food needs to be considered and addressed as a hot issue universally. The majority plants of the genus *Vaccinium* are terrestrial

M. E. S. Mirghani (✉)
Department of Biotechnology Engineering, Kulliyyah of Engineering, International Islamic University Malaysia (IIUM), Gombak, KL, Malaysia

International Institute for Halal Research and Training (INHART), International Islamic University Malaysia (IIUM), Gombak, KL, Malaysia
e-mail: elwathig@iium.edu.my

J. I. Daoud
Department of Science in Engineering, Kulliyyah of Engineering, International Islamic University Malaysia (IIUM), Gombak, KL, Malaysia

A. A. M. Elnour
Department of Biotechnology Engineering, Kulliyyah of Engineering, International Islamic University Malaysia (IIUM), Gombak, KL, Malaysia

A. A. Mariod (ed.), *Wild Fruits: Composition, Nutritional Value and Products*,
https://doi.org/10.1007/978-3-030-31885-7_39

shrubs, however, others of the same genus vary from epiphytes to trailing creepers or climbers, and some are trees (Luby et al. 1991). The popular species of *Vaccinium* inhabit the tropical, subtropical and temperate regions and most are grown on mountain slopes (Camp 1942). The tree of *Vaccinium myrtoides* (Bl.) Miq. that belongs to the plant family Ericaceae (Heath family) is more symbolic to a shrub (Fig. 39.1) growing from 0.3 to 2 m high in various landscapes ranging from bare steep rocky slopes to mossy forest undergrowth to summit vegetation, volcanic soil, open grassland and many more. *Vaccinium myrtoides* is well known in Southeast Asian countries such as the Philippines, Indonesia and neighboring. In the forests of the Philippines as elevation is gained at around 2000 meters the family of Vaccinium is found to grow as an invasive species, gathered by the locals for consumption of preserve or pies, the bluish-black fruit of *Vaccinium myrtoides* grows in the wild.

Racquel Tan Chua-Bacelo (2014) narrates that people of the Benguet believe in spirits that come from within the skies and underworld with power over mankind. Along with prayers, Benguetians have not only traditionally used *V. myrtoides* in spiritual occasions and religious rituals as a form of offering to please these spirits, but they also believed that planting these originally wild growing shrubs during ritualistic occasions such as during a full moon bestowed the power of being productive. Racquel, quantitatively measured the cultural significance of different ethnobotanical species within the Benguet based on the spread of use and the versatility of each species among the local Benguetians, concluding that *V. myrtoides* had the highest value of cultural significance. This may explained by the local belief in the supernatural power of these fruits. Crops planted after rituals were named "pack shell" and were believed to be fruitful and of high yield (Baucas 2003). Thanksgiving is another instance of a ritual during which offerings consisting of animals, food, rice wine, and other materials gifted to the spirits of the unseen worlds. Over time, belief in getting on the good side of the mentioned spirits became a necessity to ensure social and economic wellbeing as well as to guarantee health. Thus, the offerings became part of their folk medicine.

Fig. 39.1 Plant of *V. myrtoides*

39.2 Composition and Nutritional Value

As reported by Barcelo et al. (2015), the fruits of *V. myrtoides* (ayosep) (Fig. 39.2) can also be either consumed raw or as wine products for commercial purposes and it also has high polyphenol content of almost 60.00 mg GAE per 100 g fruit weight (**fw**). The same group of researchers also reported that the flavonoid content in *V. myrtoides* was higher than 1185 mg QE/100 g **fw**. They could also be a potent source of Vitamin E and antioxidants. These small fruits are rich sources of both anthocyanins and phenolic compounds (Özgen et al. 2010). Anthocyanins are coloured substances commonly extracted from plants and fruits. Blue-black berries were utilised in England in the First World War as a source of pigment when aniline dye was scarce. As a result, a small number of berries were left for making jams Agriculture and Agri-Food Canada, (Wolf and Bonanno 2013).

Phenolics refer to flavonoids, tannins and polyphenols. These are found in almost all higher plants, at various levels. The *Vaccinium* species are also a rich source of flavonoids, generally flavonols, in fresh and dry fruits. Ethanolic extracts containing flavonoids, phytosterols and phenolic compounds were found in the leaf of the plant. Tannins were present in the extract of *Vaccinium myrtoides*. However, the content of flavonoids is lower as compared to the level of polyphenols (Galvez 2016). These are commonly present in fruits, vegetables (Smirnoff 2005). Flavonoids and polyphenols possess antioxidant properties as proven in numerous studies

Fig. 39.2 Fresh fruit samples of *Vaccinium myrtoides*. (Source: Barcelo (2016))

(Barcelo et al. 2015; Hossain and Rahman 2011; Kumarappan et al. 2012; Matthes and Schmitz-Eiberger 2009). The ethanolic leaf extract of *Vaccinium myrtoides (Bl.)* showed strong antioxidant activity and potent inhibition of DPPH radical scavenging activity. As the concentration of the extract increased, the antioxidant activity also increased. These results revealed that the free radical scavenging (antioxidant) activities are due to the presence of flavonoids and phenolic compounds (Galvez et al. 2015).

There is, however, a lack of research regarding the phenolic and antioxidant activity of the local fruits (Belina-Aldemita et al. 2013). The data could prove to be beneficial to farmers and marketers in their products. *Vaccinium myrtoides* was found to have the highest polyphenol content (Pineli et al. 2011). Fruit samples had higher polyphenol content as compared to their wines which could be attributed to the fermentation process (Ivanova et al. 2012). Moreover, clarification as part of the winemaking procedure decreases their polyphenol content (Singh et al. 2015). Flavonoids are commonly found in the edible pulp of fruits and Vaccinium was one of the richest sources (Wang and Chen 2009). Testing the antioxidants activity using inhibition of DPPH radical scavenging activity; Galvez (2016) reported that the antioxidant activity of *Vaccinium myrtoides* was found to be the strongest compared to other *Vaccinium* spp (Fig. 39.3).

39.3 Uses

In general, the ripe fruit of *Vaccinium myrtoides* has a desirable bluish-blackish colour and nice admirable flavour that they can be consumed fresh or as an ingredient and condiment in other foods such as pies. Chua-Barcelo (2014) reported that *V. myrtoides* fruits are primarily used as a source of dye or ink in Benguet. In the

Fig. 39.3 Fresh fruit samples of *Vaccinium* spp

food industry, it is used as a food ingredient in jam jelly candies and wine. It is also reported to be used in traditional medicine for poor eyesight, diabetes, antioxidant, anticancer and flu (Chua-Barcelo 2014). It was reported as traditional uses in folk medicine in the Philippines that the decoction of stems and leaves taken orally to dismiss a cough and may use as a wash or antiseptic during fever. Also, the crushed leaves and stems are applied on burns (Galvez et al. 2015).

There is a great tendency towards the generation of inexpensive plant-based medications in the current age band. Secondary metabolites of the various parts of trees found to contribute to their antimicrobial potential by providing baseline information. Further investigations resulted in finding the antibacterial activity of the extracts of some parts of the plant. Knowledge is an effective way to enhance the effectivity of acquiring a new drug (Patwardhan 2005) and studying natural plant-based antimicrobial activity may provide a discovery of antimicrobials that are effective in the course of treating diseases (Saad et al. 2012). By utilizing the disc diffusion method to test antimicrobial activity, *Vaccinium myrtoides* showed a partially active mean region of 12 mm against *Pseudomonas aeruginosa*.

The antibacterial potential is attributed to the secondary metabolites present that could provide baseline information. Therefore, plant extract can developed further as plant-based antibacterial drugs. Interestingly, ethanolic leaf extraction and phytochemical analysis of the plant also showed that the leaf extract contained flavonoids, phytosterols, phenolic compounds, as well as tannins (Galvez 2016). These secondary metabolites thought to be responsible for the antimicrobial properties of the potentially antibacterial compounds extracted from this plant. Flavinoids antibacterial activity is attributed to multiple mechanisms of action that include disruption of cell wall synthesis, and inhibition of cell membrane synthesis as reported by Cushnie and Lamb (2011). Table 39.1 below shows the potential uses of fruits of *Vaccinium myrtoides* in various fields such as food, folk medicine, the industry as well as forage for birds.

Table 39.1 Potential uses of *Vaccinium myrtoides* fruits[a]

Areas of Potential utilisation of *Vaccinium myrtoides* Fruits				
	Food			
Forage	Fresh	Processed/preserved	Medicine	Industry
For Birds	Table Food	Jam	Poor eyesight	Ink (Colouring)
	Snak	Jelly	Diabetes	Condiment
	Desert	Candy	Antioxidant	Ingredient
	Ingredient	Juice	Flu & Cough	
		Condiment	Burns	
		Wine	Fever	
		Others (Colouring)	Anti-cancer	

[a]Source: Chua-Barcelo (2014)

39.4 Other Ericaceae Vaccinium

39.4.1 Vaccinium cylindraceum

Llorent-Martínez et al. (2017) report on other wild growing berries includes *Vaccinium cylindraceum* and *Vaccinium padifolium.* The shrubs of which grow wildly in Portugal. The berries of *V. cylindraceous* are blue-black and are cylindrically shaped. Llorent-Martínez et al. (2017) determined micronutrients contents in these berries using inductively coupled plasma-mass spectrometry. Looking into the micronutrient contents of these berries, the berries contain a sum of the essential minerals potassium, calcium, sodium, phosphorus and magnesium and found to be very low in toxic elements making them safe for human consumption.

Vaccinium cylindraceum is used in making edible jam and candy. Fruit and leaves extract from *V. cylindraceum* were reported to have phenolic compounds with bioactive antioxidant properties useful in the control of hyperglycaemia and obesity when tested in-vitro (Rivas et al. 2018). This may potentially contribute to innovative ways to treat or prevent diabetes and obesity.

With the rising global awareness of the benefits of antioxidants for human health, scientists have investigated different types of berries as potential rich sources of antioxidants. Wild grown berries have captivated a lot of researchers' interest for their folkloric usages and for the possibility of them to be utilized on a wider scale appealing to the rising number of health-aware consumers. Anthocyanins are bioactive pigment compounds with antioxidant activity found in fruits and berries as they ripen. Veberic R. et al., described the content of anthocyanins in different types of wild berries that included highbush blueberry (*Vaccinium corymbosum L.*), bilberry (*Vaccinium myrtillus L.*), American cranberry (*Vaccinium macrocarpon Aiton*), and lingonberry (*Vaccinium vitis-idea L.*). They concluded that the different profiles of anthocyanins give the berries its dark blue or scarlet colour as well as determines the richness of the colour. Highbush blueberry and bilberry were found to have more than five different types of anthocyanins.

39.4.2 Vaccinium padifolium

As mentioned above it was reported by Llorent-Martínez et al. (2017) on other wild growing berries includes *Vaccinium padifolium* shrubs that grow wildly in Portugal. Fruit and leaves extract from *V. padifolium* -mainly the young leaves were reported to have phenolic compounds with bioactive antioxidant properties useful in the control of hyperglycaemia and obesity when tested in-vitro (Spínola et al. 2018).

39.4.3 Uses

The berries of *Vaccinium padifolium* are used to make jam, and they are traditionally used for treating respiratory illnesses such as cough, colds, and bronchitis as well as dysentery (Llorent-Martínez et al. 2017).

39.4.4 **Vaccinium corymbosum** *L.*

Vaccinium corymbosum (VC) was found to be the primary sources of resveratrol at levels of 5884 ng/g DW. This resveratrol is naturally occurring as stilbenes, by which known to be potent antioxidants, and to serve as cancer chemopreventive activities, will add to the purported health benefits derived from the consumption of these small fruits (Rimando et al. 2004). Moreover, Moyer et al. (2002) have studied Fruits from 107 genotypes of Vaccinium. They found that certain Vaccinium fruits with pigmented flesh were lower in total anthocyanins (TACY), and ORAC comparison with those values in berries with nonpigmented flesh. In this chapter, they confirmed that the total anthocyanins (TACY) as ORAC values ranged from 19 to 131 μg TE/g (Moyer et al. 2002), and therefore, demonstrates the wide diversity of phytochemical levels and antioxidant capacities in Vaccinium fruits. Anthocyanins of all varieties increased during successive harvest stages; temporarily flavonols and hydroxycinnamic acids reduced from unripe green to ripe blue stage of berry ripening. (Castrejón et al. 2008). Thus, Blueberry antioxidant activity, as well as total phenolic content tended to decrease during ripening.

Recent advances in plant biotechnology have led to a reliable and reproductive method for genetic transformation of blueberry (Sanchez-Serrano and Salinas 2014). These efforts built on previous attempts at the transient and stable transformation of blueberry that demonstrated the potential of Agrobacterium tumefaciens-mediated transformation, and as well, the difficulties of selecting and regenerating transgenic plants (Sanchez-Serrano and Salinas 2014). Thus, the protocol has led to the successful regeneration of transgenic plants of four commercially crucial highbush blueberry cultivars. Uses of *Vaccinium corymbosum* in foodstuff has multi-functions, for examples, Vieira et al. (2016) have reported that evaluated VC extract is useful for coatings which extend blueberries' shelf-life for about 5 consecutive days when combining with chitosan. Hence, the liquid of VC fractions materials has prodigious potential in expanding the shelf life of VC fruits.

39.4.5 **Vaccinium myrtillus** *L.*

In this book, *Vaccinium myrtillus* L. is a valuable food resource for many mammals, including insects and birds in forests. It is reproduction influenced by canopy openness, but specific effects of flower availability, light conditions and pollinators on fruit production remain unclear (Eckerter et al. 2019), It was, therefore, flower and fruit abundance as well as quality traits for *V. myrtillus* fruits under shade. Canopy openness can trigger some positive additive effects on the number of flowers and hence potential fruits but also their quality due to an enhanced pollinator activity. (Eckerter et al. 2019). Tree species diversity can positively affect numerous forest functions in Central Europe since the eighteenth Century. This chapter discussed the impacts of Northern red oak introduction and forest continuity on bilberry (*Vaccinium myrtillus*).

However, it was suggested limiting admixture of invasive into pine monocultures in areas with abundant *V. myrtillus* cover (Woziwoda et al. 2019). Bilberry, *Vaccinium myrtillus*, have an essential role in nutrient cycling of boreal forests, metallic element (As, Ca, Cd, Co, Cu, Mn, Mo, Ni, Pb, Se, Zn) concentrations in leaves, berries and herbivorous larvae of *V. myrtillus* around a Finnish copper-nickel smelter and compare those with levels in relatively unpolluted reference sites, grown experimentally on *V. myrtillus*. In general, metal levels in leaves, berries and larvae were higher in the polluted area and comparable to those reported at other smelter sites in Europe. The concentration of the toxic metals (As, Ni, Cu, Cd, Pb) followed the general pattern: soil > bird faeces > leaves > larvae = berries, and levels in *V. myrtillus*, *E. autumnata* and *F. hypoleuca*. The lowest values were found to be in matrices that are the most important sources of food for birds and humans, i.e. leaf-eating larvae and berries, reducing risk of toxic effects. (Eeva et al. 2018). Moreover, Sodium (Na^+) stress occur in roadside forests soil. Its effect on anthocyanin concentrations and tissue water content decreased to 30% along in bilberry (*Vaccinium myrtillus* L.) plants were subjected to NaCl (0, 6, 30 and 60 g m^{-2}), where resulting water stress increases anthocyanin level for osmotic regulation. Also, anthocyanin concentrations decreased in aboveground stems. It is thus proposed that the direct effect of Na^+ ions, rather than Na^+-induced water stress, in bilberry (Tahkokorpi et al. 2012).

Bilberries and their products are popular worldwide and represent a very interesting source of dietary antioxidants. Vaccinium myrtillus L phenotypes have a higher polyphenolic content range, 220.06–3715.21 mg/100 g DW) and total monomeric anthocyanin (range, 206.18–867.52 mg/100 g DW) contents. Delphinidin (range, 5915.93–18108.39 µg / g DW) was the major anthocyanin moiety, while sinapic acid was the major phenolic acid in the free form (range, 0.01–6.06 µg / g DW), and p-coumaric acid in the ester (range, 26.39–110.78 µg / g DW), glycoside (range, 15.83–57.73 µg/ g DW) and ester-bound (range, 2.32–14.20 µg /g DW) forms. Antioxidant capacity was much higher in coloured (pink to blue/black) berry samples than in the white sample, and it was more related to the total phenolic concentration rather than to the anthocyanin concentration. (Colak et al. 2017).

Bilberry (*Vaccinium myrtillus* L.) is an important wild berry that has a long tradition as a food and medicinal plant in Europe. Nowadays, bilberries are picked commercially, especially in the northern and eastern parts of Europe. These berries are among the best sources of anthocyanin pigments, which have multiple health-beneficial properties. In addition to anthocyanins, bilberries also contain high yields other phenolic compounds and carotenoids and are a good-to-moderate source of vitamins and both macro-and micronutrients. The populations of this species still possess the natural variation developed as a result of long-term adaptation, under different growth conditions (Zoratti et al. 2016).

Total soluble polyphenols (TSP), total monomeric anthocyanins (TMA), radical scavenging activity (RSA), ferric reducing antioxidant power (FRAP), and some anthocyanins, phenolic acids, coumarins, flavanols, dihydrochalcones and flavonols were investigated in *Vaccinium myrtillus*and *Vaccinium uliginosum* subsp. *gaultherioides* Bigelow *V. myrtillus* berries showed much higher TSP, TMA, RSA and FRAP values than *V. uliginosum* subsp. *gaultherioides* fruits (Ancillotti et al. 2016).

*39.4.6 American cranberry (*Vaccinium macrocarpon*Aiton)*

American cranberry (*Vaccinium macrocarpon*) has a preventive effect against urinary tract infections (UTI) due to its phenolic constituents (Sun et al. 2015). The cranberry oligosaccharides, in addition to its phenolic constituents, may play a role in its preventive effects against urinary tract infections.

Vaccinium macrocarpon and *Vaccinium oxycoccos* species have an accumulation of high concentrations of a diverse set of biologically active substances. For these reasons Breeders tent to cranberry cultivars with high contents of biologically active substances in the Russia, Lithuania, Estonia and United States. Both cranberry species are sources of phenolic compounds, such as flavonols, anthocyanins, phenolic acids and flavonols. Flavonoids were known as the dominant compounds among over 150 individual components investigated in the cultivars of the large cranberry. (Česonienė and Daubaras 2016). Thus, a comparison of different cultivars demonstrated differences in biochemical composition as functions of the specific cultivars, berry ripeness stage, harvesting time, and environmental conditions.

In this book chapter, cranberry and lingonberry were Initially, varying concentrations of synthetic antifungal agents, such as sodium benzoate, potassium sorbate and butyl 4-hydroxybenzoate were tested against these fungi on wort agar containing 31% fructose at different pH values. The findings of this study demonstrate that these concentrates were able to inhibit the growth of visible colonies of xerophilic and non-xerophilic fungi (Ermis et al. 2015). Hence, cranberry and lingonberry concentrates are interesting candidates for natural preservation against fungal growth in sugar reduced fruit spreads.

According to Elberry et al. (2010), the doxorubicin (DOX) is a widely used cancer chemotherapeutic agent. However, it generates free oxygen radicals that result in serious dose-limiting cardiotoxicity. This can be attributed, at least in part, to CRAN's antioxidant compounds associated American cranberry as follows:

1. Supplementations with berries were proven effective in reducing oxidative stress associated with several ailments;
2. protective effect of cranberry extract (CRAN) against DOX-induced cardiotoxicity in rats;

3. CRAN was given orally to rats (100 mg/kg/day for 10 consecutive days) and DOX (15 mg/kg; i.p) for 7 days;
4. CRAN protected against DOX-induced increased mortality and ECG changes;
5. It significantly inhibited DOX-provoked glutathione (GSH) depletion and accumulation of oxidized glutathione (GSSG);
6. Malondialdehyde (MDA), and protein carbonyls in cardiac tissues;
7. The reductions of cardiac activities of catalase (CAT), superoxide dismutase (SOD), glutathione peroxidase (GSH-Px) and glutathione reductase (GR) were significantly mitigated;
8. Elevation of cardiac myeloperoxidase (MPO) activity in response to DOX treatment was significantly hampered.
9. Pre-treatment of CRAN significantly guarded against DOX-induced rise of serum lactate dehydrogenase (LDH), creatine phosphokinase (CK), creatine kinase-MB (CK-MB) as well as troponin I level.
10. CRAN alleviated histopathological changes in rats' hearts treated with DOX.
11. CRAN protects against DOX-induced cardiotoxicity in rats.

39.4.7 **(Vaccinium vitis-idea** ***(L)), Lingonberry***

A study conducted by Lee and Finn (2012) highlighted that the Lingonberries and their products are popular in Europe, though in the US. The on-going interest in potential health benefits from berry consumption has heightened interest in broadening the selection of berry/fruit crops in the US. Lingonberry contained the three anticipated anthocyanins cyanidin-3-galactoside, cyanidin-3-glucoside, and cyanidin-3-arabinoside respectively. Its total anthocyanin content found to be ranged from 27.4 to 52.6 mg/100 g fw. In addition, it contained 22 free amino acids (FFA) and total FAAs ranged from 28.92 to 70.38 mg/100 g fw. Asparagine (ASN) was the leading FAA (22–34% of the total FAAs) for all five cultivars. Anthocyanins, flavanols and benzoic acid derivatives were equally present in fruits. Stem and leaf are highly homologous with (+)-catechin, A- and B-type dimers/trimers, and two quercetin glycosides as major contributors (Bujor et al. 2018). Finally, the content in phenolic compounds is highly correlated with TPC and the DPPH radical scavenging activity although leaf and stem. The uses of Lingonberry fruit extract may reduce adipose tissue inflammation (Kowalska et al. 2019) and therefore, support immune cell homeostasis, and thus can be considered a natural tool for inflammation control for several mechanisms as follows:

1. For it is the ability to suppress the inflammatory response and mitigate oxidative stress in activated 3 T3-L1 adipocytes and RAW 264.7 macrophages;
2. Reduction of IL-6, MCP-1 and IL-1β production;
3. Decreasing adiponectin and IL-10 expression in TNF-α-induced adipocytes.
4. Exhibited a high anti-inflammatory potential in the macrophage cell culture;

5. Decreased intracellular ROS formation in inflamed adipocytes by enhancing expression of antioxidant defence enzymes (SOD, catalase, GPx) and inhibiting an oxidant enzyme (NADPH oxidase 4), which is one of the primary sources of ROS.

Lingonberry (*Vaccinium vitis-idea*), Leaves were analyzed for frost hardiness, tissue water content and osmotic concentrations, and photoinhibition (Fv/Fm). The effects of lingonberry (*Vaccinium vitis-idea* L.) berries ash as fertilizer burning on the P, K, Ca, Mg, Mn, S, Fe, Al, Cu, Zn, Cd, Cr, Ni and ^{137}Cs concentrations was discussed in this book chapter (Levula et al. 2000). *Vaccinium vitis-idea* L ash fertilization did not affect the heavy metal concentrations in the berries. As a result, the potassium was the only macronutrient whose concentration in the berries significantly increased after ash fertilization (5000 kg ha^{-1}). Prescribed burning increased the berry Cd concentrations, which, however, remained low even after prescribed burning. The berry ^{137}Cs concentrations decreased as a result of ash fertilization and prescribed burning (Levula et al. 2000). The reduction in ^{137}Cs concentrations caused by ash fertilization may be an important finding especially for areas where the picking and consumption of berries have to be restricted because of radioactive fallout.

39.4.8 Vaccinium barandanum *(Lusong)*

Both *V. barandanum* and *A. tuberosum* significantly differ ($P < 0.05$), together with the decreasing trend of absorbance as the concentration of extracts is increased, though not manifested by *A. tuberosum* against *P. aeruginosa.* A significant difference ($F < 0.05$) lie between positive control and the test extracts, disqualifying the extracts for quorum sensing inhibitors, but maybe antibiofilm. The mean of zones of inhibition measured using the extracts together with cefoxitin (17.33 mm) is lower than the inhibition of cefoxitin only (19 mm).

Significant inhibition of *Vaccinium barandanum* and *Allium tuberosum* on the biofilm formation of *P. aeruginosa* and *S. aureus,* though it is not suggestive of anti-quorum sensing activity. There is also a significant inhibition of the extracts to the growth of MRSA when combined with oxacillin, but none when combined with cefoxitin. (Buya 2016). Vaccinium cylindraceous (Azores blueberry) and V. padifolium (Madeira blueberry) are two endemic species, which polyphenolic content and beneficial activities necessary to be studied. Anthocyanins were predominant in fruits, while hydroxycinnamic acids (in particular 5-O-caffeoylquinic acid) and flavonoids were the main compounds in leaves. Inhibitions of glucosidases and glycation of proteins and mild activity towards α-amylase and pancreatic lipase (by comparison with reference compounds). The phenolic composition strongly correlated with reported bioactivities. In fact, 5-O-caffeoylquinic can be considered one of the main hypoglycemic and anti-glycation agents of analysed extracts (Spínola et al. 2018). Thus, consumption of Vaccinium species should be encouraged since berries could offer a dietary fibre in the prevention and control of diabetes and obesity, while leaves are good candidates for the development of nutraceuticals

1. In this chapter, the inhibition of advanced glycation end products (AGEs) by extracts of leaves, of *Vaccinium* L. spp. (Ferrier et al. 2012). In different ways which include:
2. Peoples have used *Vaccinium* species to treat symptoms of type I and II diabetes;
3. Sustained hyperglycaemia, often associated with diabetes, facilitates cross-linking of sugars with proteins;
4. Producing AGEs are a therapeutic target for diabetes atherosclerosis;
5. kidney, vascular, and neurological diseases;
6. Ethanolic extracts of leaves *Vaccinium* spp., potent inhibitors of AGEs.

39.5 Conclusion

Plants are key principal and primary sources of food. Exploring the safely edible, available and cultural significance of wild fruits can contribute to an increased utilization as food that may help in meeting the dietary needs of the people.

In conclusion, this review chapter portrayed the traditional value of these fruits combined with contemporary knowledge of their significance. Unfortunately, urbanization has vastly influenced not only the physical land causing transformations, but also on the lifestyle and practices of the people including dietary habits and perceptions. Efforts must made to protect and conserve wild fruits along with maintaining and improving their habitat. Plantations should be encouraged to fully utilize the land and make the plants settled and inhabited.

Therefore, this chapter proposes to increase the level of awareness regarding the edible wild fruits through information dissemination campaigns to boost the current value of these fruits, the additional declaration of protected areas, especially in the municipalities where the rare fruits are found and promoting the sustainable use of edible wild fruits in the society. Finally, information gathered from this review mainly contributes to the strategies of the biodiversity that is to expand and improve knowledge of the extent, characteristics, uses and values of natural resources. Strategic development of the sustainable natural resource, climate change adaptation plans and for improving agroforestry systems with the integration of wild species will aid in addressing food security issues, strengthening links between communities and the environment by promoting a sustained acceptance of wild fruits as important dietary components and their propagation as well.

References

Ancillotti C, Ciofi L, Pucci D, Sagona E, Giordani E, Biricolti S, Del Bubba M (2016) Polyphenolic profiles and antioxidant and antiradical activity of Italian berries from Vaccinium myrtillus L. and Vaccinium uliginosum L. subsp. gaultherioides (Bigelow) S.B. Young. Food Chem 204:176–184. https://doi.org/10.1016/j.foodchem.2016.02.106

Barcelo R (2016) Antioxidant property, total polyphenol and flavonoid content of selected fruits and fruit wines in Benguet. Cordillera Administrative Region, Philippines

Barcelo, R., Basilio, A., Calsiyao, I. D., Mabesa, C. B., Palconete, R. M., & Tobias, J. A. (2015). Antioxidant property and total polyphenol and flavonoid content of selected fruits and fruit wines

Baucas B L-l (2003) Traditional beliefs and cultural practices in Benguet. New Baguio Offset Press, La Trinidad

Belina-Aldemita MD, Sabularse VC, Dizon EI, Hurtada WA, Torio MAO (2013) Antioxidant properties of bignay [Antidesma bunius (L.) Spreng.] wine at different stages of processing. The Philippine Agricultural Scientist 96(3):308–313

Bujor O-C, Ginies C, Popa VI, Dufour C (2018) Phenolic compounds and antioxidant activity of lingonberry (Vaccinium vitis-idaea L.) leaf, stem and fruit at different harvest periods. Food Chem 252:356–365. https://doi.org/10.1016/j.foodchem.2018.01.052

Buya, J. T. (2016). Vaccinium barandanum and Allium tuberosum inhibit biofilm formation and antibiotic resistance of Bacteria

Camp W (1942) On the structure of populations in the genus Vaccinium. Brittonia 4(2):189–204

Castrejón ADR, Eichholz I, Rohn S, Kroh LW, Huyskens-Keil S (2008) Phenolic profile and antioxidant activity of highbush blueberry (Vaccinium corymbosum L.) during fruit maturation and ripening. Food Chem 109(3):564–572

Česonienė L, Daubaras R (2016) Chapter 8 – phytochemical composition of the large cranberry (Vaccinium macrocarpon) and the small cranberry (Vaccinium oxycoccos). In: Simmonds MSJ, Preedy VR (eds) Nutritional composition of fruit cultivars. Academic, San Diego, pp 173–194

Chua-Barcelo RT (2014) Ethno–botanical survey of edible wild fruits in Benguet, cordillera administrative region, the Philippines. Asian Pac J Trop Biomed 4:S525–S538

Colak, N., Primetta, A. K., Riihinen, K. R., Jaakola, L., Grúz, J., Strnad, M., . . . Ayaz, F. A. (2017). Phenolic compounds and antioxidant capacity in different-colored and non-pigmented berries of bilberry (Vaccinium myrtillus L.). Food Biosci, 20, 67–78. doi:https://doi.org/10.1016/j.fbio.2017.06.004

Cushnie TT, Lamb AJ (2011) Recent advances in understanding the antibacterial properties of flavonoids. Int J Antimicrob Agents 38(2):99–107

Eckerter T, Buse J, Förschler M, Pufal G (2019) Additive positive effects of canopy openness on European bilberry (Vaccinium myrtillus) fruit quantity and quality. For Ecol Manag 433:122–130. https://doi.org/10.1016/j.foreco.2018.10.059

Eeva T, Holmström H, Espín S, Sánchez-Virosta P, Klemola T (2018) Leaves, berries and herbivorous larvae of bilberry Vaccinium myrtillus as sources of metals in food chains at a Cu-Ni smelter site. Chemosphere 210:859–866. https://doi.org/10.1016/j.chemosphere.2018.07.099

Elberry AA, Abdel-Naim AB, Abdel-Sattar EA, Nagy AA, Mosli HA, Mohamadin AM, Ashour OM (2010) Cranberry (Vaccinium macrocarpon) protects against doxorubicin-induced cardiotoxicity in rats. Food Chem Toxicol 48(5):1178–1184. https://doi.org/10.1016/j.fct.2010.02.008

Ermis E, Hertel C, Schneider C, Carle R, Stintzing F, Schmidt H (2015) Characterization of in vitro antifungal activities of small and American cranberry (Vaccinium oxycoccos L. and V. macrocarpon Aiton) and lingonberry (Vaccinium vitis-idaea L.) concentrates in sugar reduced fruit spreads. Int J Food Microbiol 204:111–117. https://doi.org/10.1016/j.ijfoodmicro.2015.03.017

Ferrier, J., Djeffal, S., Morgan, H. P., Vander Kloet, S. P., Redžić, S., Cuerrier, A., . . . Arnason, J. T. (2012). Antiglycation activity of Vaccinium spp.(Ericaceae) from the Sam Vander Kloet collection for the treatment of type II diabetes. Botany, 90(5), 401–406

Galvez M (2016) Antibacterial activity and phytochemical screening of selected folkloric medicinal plants of Maggok, Hungduan, Ifugao, Cordillera administrative region, Philippines. Int J Sci Res Publ 6(1):460

Galvez ME, Jacot R, Scheffe J, Cooper T, Patzke G, Steinfeld A (2015) Physico-chemical changes in Ca, Sr and Al-doped La–Mn–O perovskites upon thermochemical splitting of CO 2 via redox cycling. Phys Chem Chem Phys 17(9):6629–6634

Hossain MA, Rahman SM (2011) Total phenolics, flavonoids and antioxidant activity of tropical fruit pineapple. Food Res Int 44(3):672–676

Ivanova V, Vojnoski B, Stefova M (2012) Effect of winemaking treatment and wine aging on phenolic content in Vranec wines. J Food Sci Technol 49(2):161–172

Kowalska K, Olejnik A, Zielińska-Wasielica J, Olkowicz M (2019) Inhibitory effects of lingonberry (Vaccinium vitis-idaea L.) fruit extract on obesity-induced inflammation in 3T3-L1 adipocytes and RAW 264.7 macrophages. J Funct Foods 54:371–380. https://doi.org/10.1016/j.jff.2019.01.040

Kumarappan C, Thilagam E, Mandal SC (2012) Antioxidant activity of polyphenolic extracts of Ichnocarpus frutescens. Saudi J Biol Sci 19(3):349–355

Lee J, Finn CE (2012) Lingonberry (Vaccinium vitis-idaea L.) grown in the Pacific Northwest of North America: Anthocyanin and free amino acid composition. J Funct Foods 4(1):213–218. https://doi.org/10.1016/j.jff.2011.10.007

Levula T, Saarsalmi A, Rantavaara A (2000) Effects of ash fertilization and prescribed burning on macronutrient, heavy metal, Sulphur and 137Cs concentrations in lingonberries (Vaccinium vitis-idaea). For Ecol Manag 126(3):269–279. https://doi.org/10.1016/S0378-1127(99)00110-3

Llorent-Martínez EJ, Zengin G, Fernández-de Córdova ML, Bender O, Atalay A, Ceylan R, Guler GO (2017) Traditionally used Lathyrus species: phytochemical composition, antioxidant activity, enzyme inhibitory properties, cytotoxic effects, and in silico studies of L. czeczottianus and L. nissolia. Front Pharmacol 8:83

Luby JJ, Ballington JR, Draper AD, Pliszka K, Austin ME (1991) Blueberries and cranberries (Vaccinium). Genet Resour Temp Fruit Nut Crops 290:393–458

Matthes A, Schmitz-Eiberger M (2009) Polyphenol content and antioxidant capacity of apple fruit: effect of cultivar and storage conditions. J Appl Bot Food Qual 82(2):152–157

Moyer RA, Hummer KE, Finn CE, Frei B, Wrolstad RE (2002) Anthocyanins, phenolics, and antioxidant capacity in diverse small fruits: Vaccinium, Rubus, and Ribes. J Agric Food Chem 50(3):519–525

Özgen M, Scheerens JC, Reese RN, Miller RA (2010) Total phenolic, anthocyanin contents and antioxidant capacity of selected elderberry (Sambucus canadensis L.) accessions. Pharmacogn Mag 6(23):198–203

Patwardhan B (2005) Ethnopharmacology and drug discovery. J Ethnopharmacol, 100(1–2):50–52

Pineli L d L d O, Moretti CL, dos Santos MS, Campos AB, Brasileiro AV, Córdova AC, Chiarello MD (2011) Antioxidants and other chemical and physical characteristics of two strawberry cultivars at different ripeness stages. J Food Compos Anal 24(1):11–16

Rimando AM, Kalt W, Magee JB, Dewey J, Ballington JR (2004) Resveratrol, pterostilbene, and piceatannol in vaccinium berries. J Agric Food Chem 52(15):4713–4719

Rivas M, Spínola M, Arrieta H, Faife-Cabrera M (2018) Effect of extreme climatic events resulting in prolonged precipitation on the reproductive output of sea turtles. Anim Conserv 21(5):387–395

Saad MA, Bovik AC, Charrier C (2012) Blind image quality assessment: a natural scene statistics approach in the DCT domain. IEEE Trans Image Process 21(8):3339–3352

Sanchez-Serrano JJ, Salinas J (2014) Arabidopsis protocols. Springer, New York

Singh P, Sahota PP, Bhadra F, Singh RK (2015) Optimization, production and scale up of debittered kinnow beverage by [alpha]-L-rhamnosidase producing yeast. Emirates J Food Agric 27(7):548–556

Smirnoff, N (2005) Antioxidants and reactive oxygen species in plants: Wiley Online Library

Spínola V, Pinto J, Castilho PC (2018) Hypoglycemic, anti-glycation and antioxidant in vitro properties of two Vaccinium species from Macaronesia: a relation to their phenolic composition. J Funct Foods 40:595–605
Sun, J., Marais, J. P. J., Khoo, C., LaPlante, K., Vejborg, R. M., Givskov, M., . . . Rowley, D. C. (2015). Cranberry (Vaccinium macrocarpon) oligosaccharides decrease biofilm formation by uropathogenic Escherichia coli. J Funct Foods, 17, 235–242. doi:https://doi.org/10.1016/j.jff.2015.05.016
Tahkokorpi M, Taulavuori E, Laine K, Taulavuori K (2012) Severe salt stress in Vaccinium myrtillus (L.) in response to Na+ ion toxicity. Environ Exp Bot 76:49–53. https://doi.org/10.1016/j.envexpbot.2011.10.008
Vieira JM, Flores-López ML, de Rodríguez DJ, Sousa MC, Vicente AA, Martins JT (2016) Effect of chitosan–Aloe vera coating on postharvest quality of blueberry (Vaccinium corymbosum) fruit. Postharvest Biol Technol 116:88–97
Wang J, Chen C (2009) Biosorbents for heavy metals removal and their future. Biotechnol Adv 27(2):195–226
Wolf SA, Bonanno A (2013) The neoliberal regime in the Agri-food sector: crisis, resilience, and restructuring. Routledge, London
Woziwoda B, Dyderski MK, Jagodziński AM (2019) Effects of land use change and Quercus rubra introduction on Vaccinium myrtillus performance in Pinus sylvestris forests. For Ecol Manag 440:1–11. https://doi.org/10.1016/j.foreco.2019.03.010
Zhao Y (2007) Berry fruit: value-added products for health promotion. CRC press, Boca Raton
Zoratti L, Klemettilä H, Jaakola L (2016) Chapter 4 – Bilberry (Vaccinium myrtillus L.) ecotypes. In: Simmonds MSJ, Preedy VR (eds) Nutritional composition of fruit cultivars. Academic Press, San Diego, pp 83–99

Chapter 40
Flacourtia rukam: Biochemistry, Functions and Utilization

Khozirah Shaari and Soo Yee Lee

40.1 Introduction

Flacourtia rukam Zoll. & Moritzi, is a plant of the family Flacourtiaceae. This fruit tree is native to China, India, and a number of Southeast Asian countries, including the Philippines, Malaysia and through New Guinea to the Solomon Islands (Walter and Sam 2002). The plant is locally known in Malaysia as rokam or rukam. In some localities, it is also known as rukam gajah or rukam manis which literally translate to elephant rukam and sweet rukam, respectively. The English vernacular names are governor's plum, Indian plum, and Indian prune. The taxonomy and the vernacular names given to this plant by the local people from different parts of the world are listed in Tables 40.1 and 40.2, respectively.

Flacourtia rukam is an evergreen, perennial tree that grows and thrives in tropical rainforests, often along rivers. It is usually considered a wild fruit tree, but it can be cultivated in home gardens. The tree can grow up to 15–20 m tall, forking into several branches (Fig. 40.1). *Flacourtia rukam* trees that grow wild are rather spiny (Fig. 40.2) compared to cultivated ones.

The young leaves of *F. rukam* are red or brown, turning dark green on the upper surface and pale green underneath. There are several forms of the species, differing in leaf shape and size. The narrow-leafed forms are known as *F. peninsula* Elmer., while the big-leafed forms are known as *F. megaphylla* Ridley and *F. euphlebia*

K. Shaari (✉)
Laboratory of Natural Products, Institute of Bioscience, Universiti Putra Malaysia, Serdang, Selangor, Malaysia

Department of Chemistry, Faculty of Science, Universiti Putra Malaysia, Serdang, Selangor, Malaysia
e-mail: khozirah@upm.edu.my

S.Y. Lee
Laboratory of Natural Products, Institute of Bioscience, Universiti Putra Malaysia, Serdang, Selangor, Malaysia

A. A. Mariod (ed.), *Wild Fruits: Composition, Nutritional Value and Products*,
https://doi.org/10.1007/978-3-030-31885-7_40

Table 40.1 Taxonomical classification of *Flacourtia rukam*

Kingdom	Plantae
Division	Magnoliophyta
Class	Magnoliopsida
Order	Violales
Family	Flacourtiaceae
Genus	*Flacourtia*
Species	*Flacourtia rukam* Zoll. & Moritzi

Table 40.2 Vernacular names of *Flacourtia rukam*

Region	Vernacular names	Language[1]/Ethnic group[2]/ Region[3]
Malaysia	Rokam, Rukam, Rukam Gajah, Rukam Manis	Malay[1]
China	Da Ye Ci Li Mu	Chinese[1]
France	Prunier Café, Prunier De Chine, Prunier Malgache	
Indonesia	Tenggolan	Batak[2], Sumatra[3]
	Ganda rukem, rukam	Java[3]
	Klang tatah kutang	Borneo[3]
Laos	Kén	
Philippines	Amaiit, Bitoñgol	Tagalog[1]
	Aganas	Bisaya[1,2]
	Kalominga	Igorot[1,2]
Thailand	Takhop-Thai	Central[3]
	Khrop-Dong	Pattani[3]
Vietnam	Mung Guan Ru'ng	
Samoa	Filimoto	

Source: Subhadrabandhu (2001)

Merr. (Lim 2013). However, all these forms are now categorized within the natural variations of *F. rukam*. The fruit tree bears small clusters of greenish-yellow flowers at the leaf axils. The fruits are depressed-globose or obovate in shape, turning from green to purple-red upon ripening (Fig. 40.3). *Flacourtia rukam* can be propagated by seeds, and asexually, by the root suckers produced from the tree (Kennard and Winters 1960). The fruits of *F. rukam* are very similar to those of *F. indica, F. inermis* and *F. jangomas* found in India and Sri Lanka. From observations of trees growing in several areas in Malaysia, fruiting is discontinuous, but not seasonal. Although it can be cultivated, it is still quite an uncommon plant, mainly found self-sown in home gardens and small fruit orchards, as well as fruit arboreta and botanical parks for conservation and educational purposes. In Malaysia, it is not known to be sold in local markets, Despite its wide distribution and similar uses in several Asian countries, the plant does not seem to be planted on a commercial scale and considered to be underutilized.

40.2 Uses as Food and Ethnomedicinal Significance

Like other fruits of its allied species, the fruits of *Flacourtia rukam* are edible. In some cultivars, the cherry-like drupes can be quite sweet but in general the ripe rukam fruits are rather tart and sour in flavor. To reduce the astringent taste, it is a usual practice to rub them between the hands before consuming the bruised berries. The ripe berries may also be used as an ingredient of a local spicy fruit salad called 'rojak'. In addition, the ripe fruits are used to make jam, jelly, juice and wine. A survey on the traditional uses and cultural importance of 36 edible wild fruits in Benguet, Philippines conducted by Chua-Barcelo (2014) reported that the cultural importance index of the species to be ranked 8th for jam-making, 12th for jelly-making, 6th for juicing and 7th for wine-making. Medicinally, the boiled water prepared from the unripe fruits is drunk to treat diarrhoea and to regulate menstrual cycle (Burkill 1966). The shoots and young leaves of *F. rukam* are also edible, eaten

Fig. 40.1 *Flacourtia rukam* tree.(Photo was taken in Taman Pertanian Universiti, Universiti Putra Malaysia)

Fig. 40.2 Forked and woody spines on the trunk and old branches of *Flacourtia rukam*. (Photo was taken in Taman Pertanian Universiti, Universiti Putra Malaysia)

raw as a local vegetable salad or 'ulam' (Lim 2013). The wood of *F. rukam* is hard and strong, offering a useful source of building material for making household appliances and for simple house construction in rural areas (FAO 1984; Lim 2013). Moreover, the wood exhibits moderate resistance against the subterranean termite (*Coptotermes curvignathus*) and has been classified under natural durability class 3, according to Indonesian standard SNI 01.7207.2006 (Febrianto et al. 2015).

Apart from being a source of food and wood material, *F. rukam* is also a traditional medicine plant. The leaf-juice of the species may be applied on eyelids to relieve inflammation, while the sap from the roasted leaves is administered externally to heal boils and wounds (Christophe 2006; Quattrocchi 2016). The use of *F. rukam* leaves as a cure for gonorrhea appeared in records of local practices in the district of Poncokusumo Malang, Province East Java, Indonesia (Batoro and Siswanto 2017). Meanwhile, the juice obtained from unripe fruits is used as medicine to treat diarrhea and dysentery (Subhadrabandhu 2001). The fruits are also consumed to relieve pain during menstruation (Lim 2013). The roots of *F. rukam* are

Fig. 40.3 Unripe and ripe fruits of *Flacourtia rukam.* (Photo was taken in Taman Pertanian Universiti, Universiti Putra Malaysia)

utilized for several therapeutic purposes as well. The Higaonon tribe in Philippines prescribed a thrice a day consumption of the root decoction for relieving muscle pain and fatigue, for post-partum care, stomach ulcer, lung infection and anemia (Olowa et al. 2012). Similarly, the Kadazandusun tribe of Sabah, Malaysia, used the roots to treat colic (Kulip 1997). Meanwhile, the bark of the tree is used to treat inflammation and bone fracture by the Samoans (Cox 1993; Uhe 1974).

40.3 Phytochemical Constituents

Although *F. rukam* is widely distributed across many tropical countries and a common ethnomedicinally used plant species, it has received little attention from the perspective of both phytochemical and pharmacological research. Only a few studies on the plant have been carried out based on a perusal of the relevant literature. Phytochemical screening of wild fruits in Benguet, Philippines, revealed the presence of steroids, flavonoids, saponins, tannins and polyphenols in an 80% methanolic fruit extract of *F. rukam* (Barcelo 2015). The same study also showed that the total phenolic content of the fruit extract was higher than its total flavonoid content. Another study reported the isolation of monogalactosyl diacylglycerol (**1**), β-sitosteryl-3β-glucopyranoside-6'-*O*-fatty acid esters (**2**), β-sitosterol (**3**), triacylglycerols (**4**), and chlorophyll a (**5**) from the dichloromethane extract of different parts (pulp, peel and seed) of the fruit (Ragasa et al. 2016). Since the fruits of *F. ruk*am are very similar to those of its allied species in the Indian continent, it is highly probable that they share very similar chemistry. Extensive phytochemical studies carried out on *F. indica* and *F. inermis* showed that the fruits of these species are notable source of phenolic compounds such as chlorogenic acids (CGAs), flavonoids, anthocyanins, benzoyl glucosides and hydroxybenzyl alcohol glycosides, besides containing terpenoids, alkaloids, anthraquinones and saponin (Alakolanga et al. 2014). It is highly likely that these classes of phytochemicals could also be present in the fruits of *F. rukam*.

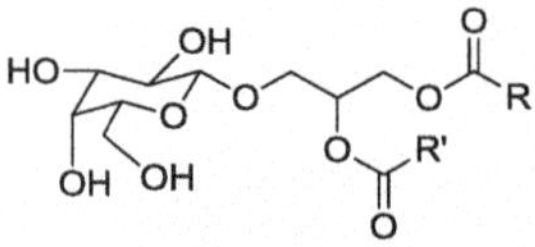

R, R' = long chain fatty acid alkyls

1

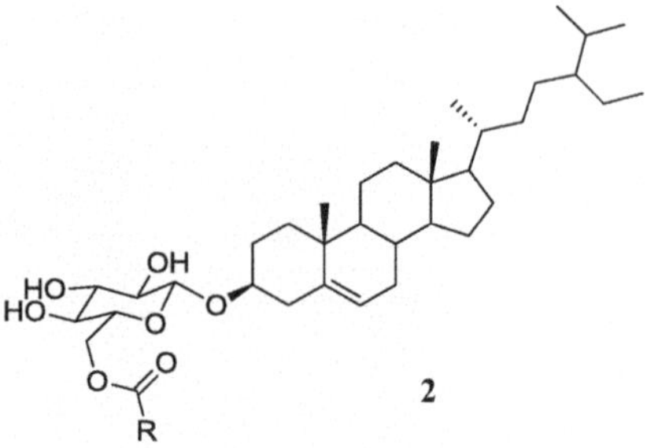

2

R = long chain fatty acid

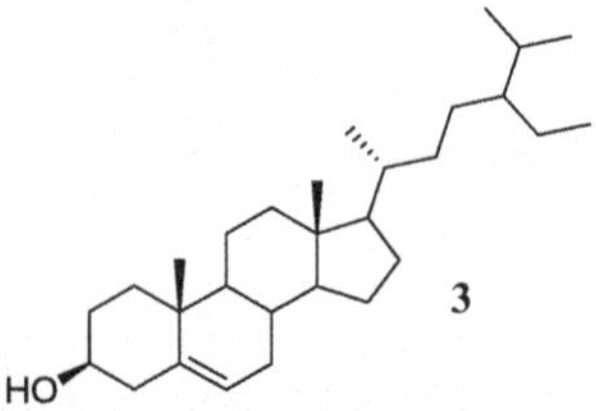

3

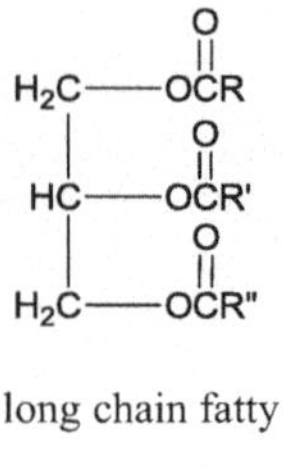

4

R, R', R''' = long chain fatty acid alkyls

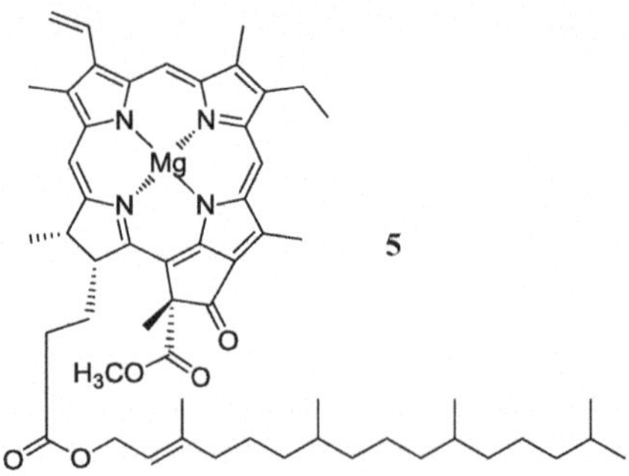

5

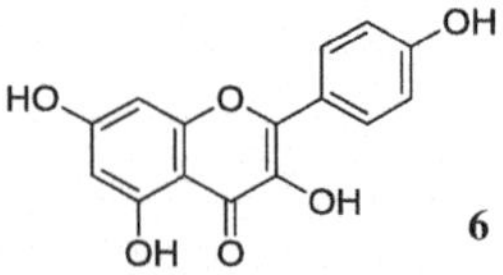

6

40.4 Pharmacological Studies

The antioxidant capacity of *F. rukam* fruit was evaluated by Ikram et al. (2009) using ferric reducing antioxidant potential (FRAP) and 2,2-diphenyl-1-picrylhydrazyl (DPPH) assays. The authours reported high antioxidant capacity of the fruit, which they associated with the high total phenolic content (Ikram et al. 2009). The authors opined that the high antioxidant property could be contributed by anthocyanins, phenolic acids and flavonoids present in the fruit. However, there was no further analysis on this. *Flacourtia rukam* fruit was also screened for *in vitro* photo-cytotoxic activity, but there were no observable activity for both cytotoxicity and light-activated cytotoxicity against HL60 promyelocytic leukemia cells (Ong et al. 2009).

Flacourtia rukam collected from Western Samoa was evaluated for antiviral, antitumour, antihypertensive and antibacterial properties in a preliminary biological activity study (Norton et al. 1973). While the 60% ethanolic extracts of different parts (bark, fruit, leaf, and stem) of the plant did not significantly inhibit Columbia SK virus and Ehrlich ascites tumor cells *in vivo,* the 60% ethanolic leaf extract showed appreciable antihypertensive effect *in vivo*, in which the arterial blood pressure of the rat model was reduced by more than 20%, 30 min after receiving the treatment. The extracts were also evaluated for antibacterial activity against mycobacteria and nocardia, in vitro. However, all the extracts were inactive.

In another study on muscarinic receptor activity of the plant, the methanolic leaf extract showed 21–40% binding inhibition of the radioligand, [^{3}H] *N*-methylscopolamine, to muscarinic receptors (Chung et al. 2005b). The fruit and leaf methanolic extracts of *Flacourtia rukam* also showed less than 50% radioligand binding activity towards several central nervous system receptors which included 5-hydroxy tryptamine (5HT), γ-aminobutyric acid (GABA) and dopamine (Chung et al. 2005a).

40.5 Conclusions

In this review chapter, effort has been taken to gather and compile the information regarding the ethnomedicinal, phytochemical and pharmacological aspects of *F. rukam*, an underutilized tropical fruit tree. The ethnobotanical literature indicated that the plant has been widely used by indigenous people for medicinal purposes, particularly for inflammation-related diseases or conditions. However, very little is known about the plant and it seemed to have received very little attention compared to other fruit tree species. In terms of its phytochemistry, only six compounds have been reported from the fruits of *F. rukam*. Although, there were some biological studies done, the studies were very preliminary and the information was still limited and incomprehensive. Currently, the only value of this plant appears to be for self-consumption. Very little effort has been made to develop the species for economic use, which presumably is due to the lack of information on its economic potential. Although it may be possible to sell the fruit in some urban markets, its potential for commercial production seems limited unless its economic potential as a health giving food crop is properly evaluated. More research is therefore needed to ascertain the true value of the species, in particular the nutritional and medicinal properties of the fruits.

References

Alakolanga AGAW, Siriwardane AMDA, Savitri Kumar N, Lalith Jayasinghe L, Jaiswal R, Kuhnert N (2014) LC-MS^n identification and characterization of the phenolic compounds from the fruits of *Flacourtia indica* (Burm. F.) Merr. and *Flacourtia inermis* Roxb. Food Res Int 62:388–396

Barcelo R (2015) Phytochemical screening and antioxidant activity of edible wild fruits in Benguet, Cordillera administrative region, Philippines. Elect J Bio 11(3):80–89

Batoro J, Siswanto D (2017) Ethnomedicinal survey of plants used by local society in Poncokusumo district, Malang, East Java Province, Indonesia. Asian J Med Biol Res 3(2):158–167

Burkill IH (1966) A dictionary of the economic products of the Malay Peninsula. Revised reprint 2 volumes. Ministry of Agriculture and Co-operatives, Kuala Lumpur. Vol 1 (A-H) pp 1–1240, vol 2 (I-Z). pp 1241–2444

Christophe W (2006) Medicinal plants of the Asia-Pacific: drugs for the future? World Scientific, Singapore

Chua-Barcelo RT (2014) Ethno–botanical survey of edible wild fruits in Benguet, Cordillera administrative region, the Philippines. Asian Pac J Trop Biomed 4(Suppl 1):S525–S538

Chung LY, Goh SH, Imiyabir Z (2005a) Central nervous system receptor activities of some Malaysian plant species. Pharm Biol 43(3):280–288

Chung LY, Yap KF, Mustafa MR, Goh SH, Imiyabir Z (2005b) Muscarinic receptor activity of some Malaysian plant species. Pharm Biol 43(8):672–682

Cox PA (1993) Saving the ethnopharmacological heritage of Samoa. J Ethnopharmacol 38(2–3):181–188

FAO (Roma) (1984) Food and fruit-bearing forest species 2: examples from Southeastern Asia. Food and Agriculture Organization of the United Nations, Rome, Italy

Febrianto F, Pranata AZ, Septiana D, Arinana A, Gumilang A, Hidayat W et al (2015) Termite resistance of the less known tropical woods species grown in West Java, Indonesia. J Korean Wood Sci Technol 43(2):248–257

Ikram EHK, Eng KH, Jalil AMM, Ismail A, Idris S, Azlan A et al (2009) Antioxidant capacity and total phenolic content of Malaysian underutilized fruits. J Food Compos Anal 22(5):388–393

Kennard WC, Winters HF (1960) Some fruits and nuts for the tropics (No. 801). US Department of Agriculture, Washington, D.C

Kulip J (1997) A preliminary survey of traditional medicinal plants in the west coast and interior of Sabah. J Trop For Sci 10(2):271–274

Lim TK (2013) Flacourtia rukam. In: Lim TK (ed) Edible medicinal and non-medicinal plants. Springer Netherlands, Dordrecht, pp 776–779

Norton TR, Bristol ML, Read GW, Bushnell OA, Kashiwagi M, Okinaga CM, Oda CS (1973) Pharmacological evaluation of medicinal plants from Western Samoa. J Pharm Sci 62(7):1077–1082

Olowa LF, Torres MAJ, Aranico EC, Demayo CG (2012) Medicinal plants used by the Higaonon tribe of Rogongon, Iligan City, Mindanao, Philippines. Adv Environ Biol 6(4):1442–1449

Ong CY, Ling SK, Ali RM, Chee CF, Samah ZA, Ho ASH et al (2009) Systematic analysis of in vitro photo-cytotoxic activity in extracts from terrestrial plants in Peninsula Malaysia for photodynamic therapy. J Photochem Photobiol B Biol 96(3):216–222

Quattrocchi U (2016) CRC world dictionary of medicinal and poisonous plants: common names, scientific names, eponyms, synonyms, and etymology. CRC Press, Boca Raton, Florida

Ragasa CY, Reyes JMA, Tabin TJ, Tan MCS, Chiong ID, Brkljača R, Urban S (2016) Chemical constituents of *Flacourtia rukam* Zoli. & Moritzi fruit. Int J Pharmaceut Clin Res 8(12):1625–1628

Subhadrabandhu S (2001) Under-utilized tropical fruits of Thailand. FAO Regional Office for Asia and the Pacific, Bangkok

Uhe G (1974) Medicinal plants of Samoa – a preliminary survey of the use of plants for medicinal purposes in the Samoan Islands. Econ Bot 28(1):1–30

Walter A, Sam C (2002) Fruits of Oceania. ACIAR Monograph No 85. Australian Centre for International Agricultural Research, Canberra

Chapter 41
Muntingia calabura: Chemical Composition, Bioactive Component and Traditional Uses

Siti Nurulhuda Mastuki, Siti Munirah Mohd Faudzi, Norsharina Ismail, and Norazalina Saad

Abbreviations

$AEMC_{fr}$	Aqueous extract *M. calabura*
$CECMC_L$	Chloroform extract *M. calabura* leaves
$EEMC_L$	Ethanol extract *M. calabura* leaves
GCMS	Gas chromatography–mass spectrometry
HPLC	High Performance Liquid Chromatography
LCMS	Liquid chromatography–mass spectrometry
LD_{50}	Lethal dose at 50%
$MEMC_{fr}$	Methanol extract *M. calabura* fruit
$MEMC_L$	Methanol extract *M. calabura* leaves

S. N. Mastuki (✉)
Laboratory of Natural Products, Institute of Bioscience, Universiti Putra Malaysia, Serdang, Selangor, Malaysia
e-mail: ctnurulhuda@upm.edu.my

S. M. M. Faudzi
Laboratory of Natural Products, Institute of Bioscience, Universiti Putra Malaysia, Serdang, Selangor, Malaysia

Department of Chemistry, Faculty of Science, Universiti Putra Malaysia, Serdang, Selangor, Malaysia
e-mail: sitimunirah@upm.edu.my

N. Ismail
Laboratory of Molecular Biomedicine, Institute of Bioscience, Universiti Putra Malaysia, Serdang, Selangor, Malaysia
e-mail: norsharina@upm.edu.my

N. Saad
Laboratory of UPM – MAKNA Cancer Research, Institute of Bioscience, Universiti Putra Malaysia, Serdang, Selangor, Malaysia
e-mail: norazalina@upm.edu.my

A. A. Mariod (ed.), *Wild Fruits: Composition, Nutritional Value and Products*,
https://doi.org/10.1007/978-3-030-31885-7_41

MIC	Minimum Inhibitory Concentration
NMR	Nuclear magnetic resonance
$PEEMC_L$	Petroleum ether extract *M.calabura* Leaves

41.1 Introduction

Muntingia calabura or commonly known as Jamaican cherry, is originates from the Muntingiaceae family. *M. calabura* was formerly classified in the family of Elaeocarpaceae or Tiliaceae, before further investigations led to the discovery of a new family, namely Muntingiaceae, based on its morphological characterizations and molecular data (Bayer et al. 1998). To date, *M. calabura* is the only species in Muntingia genus and is a synonym of *M. rosea* H. Karst. This flowering plant is natively from Southern Mexico and America continents, and is widely cultivated in a warm area of Asian regions such as India, Malaysia, Philippines, Indonesia, Thailand, Vietnam and Singapore (Morton 1987; Sani et al. 2012; Yusof et al. 2011; Zakaria et al. 2006a, 2007a, 2008, 2010, 2011). The Portuguese were previously reported as the main Jamaican cherry fruit distributor and was firstly transported to Thailand or Vietnam, before subsequently spread to Malaya (Malaysia) (Blench 2008) (Fig. 41.1).

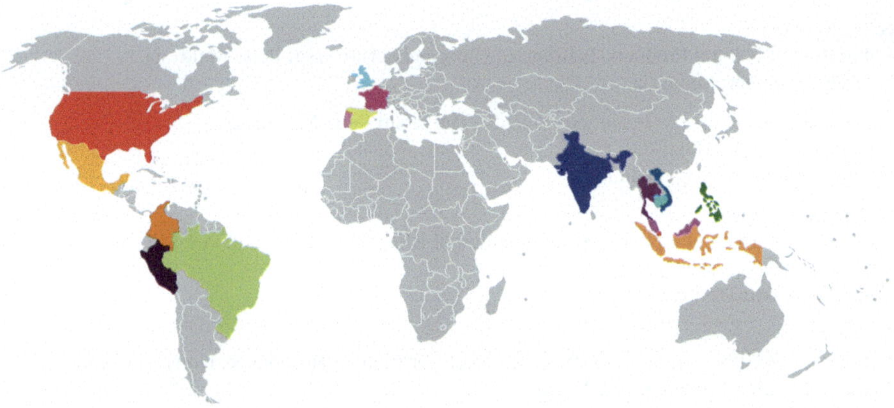

Fig. 41.1 The distribution of *Muntingia calabura*. (Source: https://commons.wikimedia.org)

41.2 Ethnobotany

41.2.1 Vernacular Names

Generally known as the Jamaican Cherry, *M. calabura* has various vernacular names from region to region, including bolaina or capulı'n blanc (Spain), pau de seda and calabura (Brazil), pu'an and capulin rojo (Mexico) and bois ramier (France). Meanwhile, in Asian countries, *M. calabura* is known as gasagase hanninamara (India), kerukup siam and buah cheri (Malaysia), kersen and talok (Indonesia), alatris, sarisa and cereza (Philippines), takhop farang (Thailand), cay trung ca (Vietnam) and krakhob barang (Cambodia) (Mahmood et al. 2014).

41.2.2 Botanical Description

Jamaican cherry grows between 7.5 and 12.0 meters tall, with unique horizontal spreading branches. It composes of the either oblong ovate or lanceolate (5.0–12.5 cm long) shape leaves, with sticky, hairy green upper surface and brown underside, with saw-toothed margins. *M. calabura* flowers are approximately 2 cm in wide, consist of 5 white petals and green sepals each that are surrounded by the yellow stamens. Its pollens are pollinated by insects and bats. Meanwhile, cherry fruits are in round or globular shape, approximately 1–1.5 cm wide, with a smooth surface, thin, tender, and the pale green colour fruits turning red upon ripening, where it takes 6–8 weeks to ripen. The ripe cherries have a sweet juicy pulp with musky, along with a fig-like flavor containing elliptic greyish yellow or dark brown seed (Morton 1987) (Fig. 41.2).

Fig. 41.2 *Muntingia calabura* tree (**a**), flower (**b**) and fruits (**c** – white arrow). (Source: https://commons.wikimedia.org)

41.3 *Muntingia calabura* Chemical Compositions

The *M. calabura* entails of the fruits, peel, leaf, stem/bark and root. Many researchers have discovered numbers of beneficial phytochemicals in the bark, leaves, fruits and peel of this plant since 1987 to date, through a various chromatography and spectroscopy approaches.

41.3.1 Muntingia calabura *Fruits Chemical Compositions*

The fresh cherry fruits are widely eaten and sold in Mexican markets, as well as used as an ingredient for food recipes. According to Morton (1987), every 100 g of fresh edible portion consist of 778.0 g moisture, 0.324 g protein, 1.56 g fat, 4.6 g fiber, 1.14 g ash, 124.6 mg calcium, 84.0 mg phosphorus, 1.18 mg iron, 0.019 mg carotene, 0.065 mg thiamine, 0.037 mg riboflavin, 0.554 mg niacin and 80.5 mg ascorbic acid. This fruit is rich in protein (6.44 ± 0.15 mg/g protein) and lower carbohydrate content (75.33 ± 4.61 mg/g), as compared to *M. calabura* leaves (Krishnaveni and Dhanalakshmi 2014). Recently, Pereira et al. (2018) reported that the fruits consist of a high level of total content of sugars (7.52 g/100 g fresh weight (fw)), in comparisons to cherry (7.20 g/100 g fw), blackberry (6.8 g/100 g fw) and raspberry (6.58 g/100 g fw). However, the fermentable oligosaccharide content of *M. calabura* fruits is lower (0.01 g/100 g fw) than the fruits as mentioned above (Jovanovic-Malinovska et al. 2013). This lower content of fermentable oligosaccharide value may provide several health benefits including for the treatment of gastrointestinal-related illnesses such as blocking abdominal discomfort or pain (Varney et al. 2017).

Ragasa et al. (2015) indicated that dichloromethane extracts of dried cherries, harvested from San Pedro, Laguna, Philippines yielded a squalene, triglycerides, linolenic acid, α-linolenic acid, β-sitosterol and stigmasterol. Whilst, GCMS analyses revealed the presence of 25 constituents in polyphenolic extract of ripen Jamaican cherries (Gomathi et al. 2013), with the most abundant phytochemical was phytol (26.26%), *n*-hexadecanoic acid (11.97%), cyclopropaneoctanoic acid (10.26%), isoprenoids including γ-sitosterol (11.15%), stigmasterol (7.20%) and campesterol (4.48%). This study too discovered minor components namely ethyl linolenate (4.48%), ethyl linoleate (3.14%), ethyl hexadecanoate (3.65%), neophytadiene (3.03%), isoamyl acetate (2.32%). 1-Deoxy-d-mannitol (1.86%), α-tocopherol (1.27%), β-cholest-5-en-3-ol (0.97%), ethyl stearate (0.88%), γ-tocopherol (0.82%), (2*E*)-3,7,11,15-tetramethyl-2-hexadecen-1-ol (1.81%), 3-cyclopentylpropionic acid (0.75%), 2,3-dihydro-3,5-dihydroxy-6-methyl-4*H*-pyran (0.78%), 2-dimethylaminoethyl ester (0.66%), 1,3,5-triazine-2,4,6- triamine (0.56%), octanoic acid (0.54%), 2,3-dihydrobenzofuran (0.41%), 1,2,3-propanetriol, monoacetate (0.40%) and *n*-nonanoic acid (0.33%). Meanwhile, Pereira et al. (2018) stated that the sesquiterpene β-farnesene (28.7%) and dendrolasin (15.4%) were the major volatile components in fruits, followed by β-himachalene (3.9%), α-curmane (3.7%), limonene (3.1%), 3-hexen-1ol (2.8%) and methyl salicylate (2.7%).

On the other note, *M. calabura* cherries are known to be enriched with phenolic compounds that are associated with their antioxidant capacity. These valued phenolic compounds are reported as the gallocatechin, epigallocatechin, catechin, flavanols, naringenin, quercetin, gallic acid, vanillic acid, chlorogenic acid, caffeic acid, ρ-coumaric acid, ferulic acid, Ɛ-hydroxycinnamic acid and myricetin (Pereira et al. 2018; Rotta et al. 2017).

41.3.2 **Muntingia calabura** *Leaves Chemical Compositions*

Numerous works have been conducted to analyze the phyto-constituents of *M. calabura* leaves. Su et al. (2003) mentioned that 25 compounds were isolated and identified from ethyl acetate partition of methanol leaves extract (collected in Purus, Peru). The reported phytochemicals comprising of a new (2*R*,3*R*)-7-methoxy-3,5,8-trihydroxyflavanone and 24 known compounds of several classes including (2*S*)-7-hydroxyflavanone, (2*S*)-5,7-dihydroxyflavanone, (2*R*,3*R*)-3,5,7-trihydroxyflavanone, (2*S*)-5-hydroxy-7-methoxyflavanone, 7-hydroxyflavone, 5,7-dihydroxyflavone, 3-methoxy-5,7,4′-trihydroxyflavone, 3,3′-dimethoxy-5,7,4′-trihydroxyflavone, 3,8-dimethoxy-5,7,4′-trihydroxyflavone, 3,5-dihydroxy-7,4′-dimethoxyflavone, 3,5-dihydroxy-7,8-dimethoxyflavone, 5-hydroxy-3,7,8-trimethoxyflavone, 5,4′-dihydroxy-3,7,8-dimethoxyflavone, 5-hydroxy-3,7,8,4′-tetramethoxyflavone, 2′,4′-dihydroxychalcone, 4,2′,4′-trihydroxychalcone, 7-hydroxyisoflavone, 7,3′,4′-trimethoxyisoflavone, (2*S*)-5′-hydroxy-7,8,3′,4′-tetramethoxyflavan, 2′,4′-dihydroxydihydrochalcone, 3,4,5-trihydroxybenzoic acid, lupenone and (2α,3β)-2,3-dihydroxyolean-12-en-28-oic acid.

Investigation by Chen et al. (2005) revealed the characterization of 4 new constituents of dihydroxychalcones, flavanone and flavonol, respectively, namely the 2′,4′-dihydroxy-3′-methoxydihydrochalcone, (-)-3′-methoxy-2′,4′,β-trihydroxydihydrochalcone, (2*S*)-(-)-5′-hydroxy-7,3′,4′-trimethoxyflavanone, and muntingone in chloroform and butanol soluble leaves fractions. Along with that, 16 known compounds were also identified comprising of 10 flavones (5-hydroxy-7-methoxyflavone, 3,7-dimethoxy-5-hydroxyflavone, 5-hydroxy-3,6,7-trimethoxyflavone, 6,7-dimethoxy-5-hydroxyflavone, 3,5-dihydroxy-7-methoxyflavone, 3,5-dihydroxy-6,7-dimethoxyflavone, 8-methoxy-3,5,7-trihydroxyflavone, 5,7-dihydroxy-3,8-dimethoxyflavone, galangin, and chrysin), 3 flavanones (7-hydroxyflavanone, 7-hydroxy-8-methoxyflavanone, and 4′-hydroxy-7-methoxyflavanone), 2 chalcones (2′,4′-dihydroxychalcone and 2′,4′-dihydroxy-3′-methoxychalcone) and a dihydrochalcone, 2′,4′-dihydroxydihydrochalcone. Chen et al. (2007) also described the identification of 22 compounds through the same soluble fractions of methanol leaves extract collected from Kaohsing city, Taiwan in 2001. Among these isolates, 2 new dihydrochalcones (2,3-dihydroxy-4,3′,4′,5′-tetramethoxydihydrochalcone and 4,2′,4′-trihydroxy-3′-methoxydihydrochalcone), and a new flavanone, (2*R*,3*R*)-(-)-3,5-dihydroxy-6,7-dimethoxyflavanone were established. Whereas, the other 19 known compounds were including 7-methoxyflavone, 5,7-dihydroxy-3-methoxyflavone, 5,7-dihydroxy-6-methoxyflavone, 5,4′-dihydroxy-3,7-dimethoxyflavone, (2*S*)-7,8,3′,4′,5′-pentamethoxyflavan, (2*S*)-5′-hydroxy-7,8,3′,4′-tetramethoxyflavan, methyl 4-hydroxybenzoate, isovanillic acid, *p*-nitrophenol, methyl gallate, *trans* isomer of methyl *p*-coumarate, β-sitostenone, β-sitosterol, stigmasterol, β-amyrenone, α-tocopherylquinone, δ-tocopherol, α-tocospiro and α-tocospiro B.

A recent study by Mohamad Yusof et al. (2013) has led to the discovery of a mixture of new and known metabolites responsible for anti-nociceptive through the bioassay-guided fraction of petroleum ether leaves extract. The most active fraction was recorded to consist a new 8-hydroxy-6-methoxyflavone, and other known compounds; 5-hydroxy-3,7,8-trimethoxyflavone, 3,7-dimethoxy-5- hydroflavone, and 2′,4′-dihydroxy-3′-methoxychalcone, respectively.

41.3.3 Muntingia calabura *Root Chemical Compositions*

Despite a growing number of researches perform on Jamaican cherry and its leaves, there are limited information available on chemical constituents of *M. calabura* root. Nonetheless, the first isolation of bioactive compounds from roots were described by Kaneda et al. (1991), where 12 flavonoids of methanolic extract have been isolated and identified as (2*S*)-5′-hydroxy-7,3′,4′-trimethoxyflavan, (2*S*)-7,8,3′,4′,5′-pentamethoxyflavan, (2*S*)-2′-hydroxy-7,8,3′, 4′,5′-pentamethoxyflavan, (2*S*)-5′-hydroxy-7,8,3′,4′-tetramethoxyflavan, (2*S*)-8-hydroxy-7,3′,4′,5′-tetramethoxyflavan, (2*S*)-8,2′-dihydroxy-7,3′,4′,5′-tetramethoxyflavan, (2*S*)-8,5′-dihydroxy-7,3′,4′-trimethoxyflavan, 7,8,3′,4′,5′-pentamethoxyflavone, (*M*),(2*S*),(2”*S*)-,(*P*),(2*S*),(2”*S*)-8,8″-5′-trihydroxy-7,7′-3′, 3″’-4′,4″’-5″’-heptamethoxy-5,5″-biflavan, 5′-hydroxy-7,8,3′4’-tetramethoxyflavone, (*M*),(2*S*),(2”*S*)-,(*P*),(2*S*),(2”*S*)-8,8″-5′-5″’-tetrahydroxy-7′,7″-3′,3″’-4′,4″’-hexamethoxy-5′,5″’-biflavan and 8,5′-dihydroxy-7,3′,4′-trimethoxyflavone. In addition, findings from Rajesh et al. (2014) confirmed the presence of stigmasterol in *M. calabura* roots using the GCMS, IR and NMR spectroscopic characterizations in supporting the Kaneda et al. (1991) results.

41.3.4 **Muntingia calabura** *Stem Chemical Compositions*

Stem bark of *M. calabura* was previously reported to contain high concentrations of flavonoids, triterpenes, saponins, glycosides and tannins (Buhian et al. 2016). Two new flavones, 8-hydroxy-7,3′,4′,5′-tetramethoxyflavone and 8,4′-dihydroxy-7,3′,5′-trimethoxyflavone, and 13 known compounds including 6,7-dimethoxy-5-hydroxyflavone, 5,7-dimethoxyflavone, 3,5-dihydroxy-6,7-dimethoxyflavone, (2*S*)-5′-hydroxy-7,8,3′,4′-tetramethoxyflavan, β-sitostenone, 6β-hydroxystigmast-4-en-3-one, β-sitosterol, syringic acid, vanillic acid, 3-hydroxy-1-(3,5-dimethoxy-4-hydroxyphenyl)propan-1-one, tetracosyl ferulate, 1-tetracosanol and 1-hexacosanol were amongst isolated compounds from stem extracts by Chen et al. (2004). In addition to this, Kuo et al. (2014) has reported to successfully characterized a new biflavan (*M*),(2*S*),(2"*S*)-,(*P*),(2*S*),(2"*S*)-7,8,3′,4′,5′,7″,8″,3″',4″',5″'-decamethoxy-5,5″ biflavan, a new flavone, 4′-hydroxy-7,8,3′,5′-tetramethoxyflavone, and a new dihydrochalcone, (*R*)-2′,β-dihydroxy-3′,4′-dimethoxydihydrochalcone with other known isolated compounds.

41.4 Pharmacological Properties of *Muntingia calabura*

Various parts of *M. calabura* has been claimed to possess diverse medicinal values, in which supported the traditional uses of the plants in treating numbers of ailments. The respective parts of *M. calabura* plant were prepared as an extract using various types of solvents and extraction methods before further undergone the *in vitro* and *in vivo* experiments, in proving their potential scientifically. All the previous findings were summarized in the following sub-chapters.

41.4.1 Acute Toxicity Study

Sridhar et al. (2011) was the first research group that reported on the *M. calabura* acute toxicity. They have determined the oral acute toxicity LD_{50} (lethal dose at 50%) value of methanol extract leaves ($MEMC_L$) collected in Gharpur Station, Andhra Pradesh, India in the rat's *in vivo* system. The $MEMC_L$, in dosage of 300, 500 and 200 mg/kg was administrated orally, and the results showed neither mortality nor abnormal visible clinical signs in animal model even in a higher dose, in further supported from histopatological, hematological and serum biochemical tests data. Another acute toxicity study by Balan et al. (2013) also recorded a similar trend, in which the higher dose of 2000 mg/kg of $MEMC_L$ did not caused toxicity and death in the treated animal.

41.4.2 Cytotoxicity Activity

Determination of the cytotoxicity of $MEMC_L$ was firstly made by Kaneda et al. (1991). The tested plant was collected in Sarabut Province, Thailand and subjected for the cytotoxicity test on human breast cancer (BC1), human fibrosarcoma (HT-1080), human lung cancer (Lu1), human melanoma (Me12), human colon cancer (Co12), human nasopharyngeal carcinoma (KB), vincristine-resistant KB (KB-V), and murine lymphocytic leukemia (P-388) cell lines. All the isolated compounds (7 falvans, 1 biflavan and 3 flavones) demonstrated a significant cytotoxicity activity against the cultured P-388 cells, where the flavans were observed to exhibit a stronger activity than flavones. Likewise, Chen et al. (2005) too have investigated the cytotoxicity properties of several compounds isolated from $MEMC_L$ against P-388 and HT-29 cell lines using the 3-(4,5-dimethythiazol-2-yl)-2,5-diphenyltetrazolium bromide (MTT) assay. It was reported that the (2*S*)-5′-hydroxy-3′-methoxyflavanone, 4′-hydroxy-7-methoxyflavanone, 2′4’-dihydroxychalcone and 2′,4′-dihydroxy-3-methoxychalcone exhibited cytotoxicity with the half maximal inhibitory concentration (IC_{50}) values less than 4 μg/ml against both cell lines. Besides that, Sufian et al. (2013) has performed the bioassay-guided fractionation on cytotoxicity of $MEMC_L$ against several cancer cell lines. Two compounds, 2′4’-dihydroxychalcone and 5-hydroxy-3,8-flavonone were found to exerted the most remarkable cytotoxicity against HL-60 (human acute lymphoblastic leukemia) and MCF-7 (human breast adenocarcinoma) with the IC_{50} values of 3.43 and 3.34 μg/ml; and 11.78 and 18.85 μg/ml, respectively.

41.4.3 Anti-platelet Aggregation Activity

The anti-platelet aggregation activity of *M. calabura* was assessed by Chen et al. (2007) using the rabbit platelet induced by thrombin, arachidonic acid, collagen and platelet-activating factor with the aspirin was used as the standard anti-platelet agent. The research discovered a mixture of flavone analogues and benzenoids including 5,7-dihydroxyflavone, 5,7-dihydroxy-6-methoxyflavone, 5,4′-dihydroxy-3,7-dimethoxyflavone and methyl gallate from the cold $MEMC_L$ exhibited a positive anti-platelet aggregation effect. Among these bioactive compounds, methyl gallate was served as the most potent agent against the arachidonic acid- and collagen-induced platelet aggregation in rabbit, with IC_{50} values of 2 and 5 μg/ml respectively.

41.4.4 Quinone Reductase Activity

Quinone reductase is an enzyme responsible for an induction of phase 2 drug-metabolizing system, which might be serves as a potential and an effective strategy in providing a protection against the toxic and neoplastic effects from carcinogens

(Talalay 2000). The potency of M. calabura as a quinone reductase inhibitor has been evaluated in the Hepa1c1c7 murine hepatoma cell line, where the enzyme activity was expressed as CD with the determination of IC_{50} and CI (chemoprevention index, IC_{50}/CD). It was recorded that (2*S*)-5′-hydroxy-3′-methoxyflavanone is having the CD value of <0.56 μM as significant as the reference drug, sulpfuraphene (0.43 μM). (Su et al. 2003). Furthermore, 2′4'-dihydroxychalcone from *M. calabura* leaves exhibited a potent quinone reductase-inducing activity, in which in-line with the previous work of Nshimo et al. (1993). This chalcone was found to be cytotoxic against human and murine tumor cell lines.

41.4.5 Anti-inflammatory Activity

The anti-inflammatory activity of chloroform extract of *M. calabura* leaves ($CEMC_L$) was initiated by Zakaria et al. (2007a), where all concentrations (10, 50 and 100%) of the tested samples demonstrated an inconsistency in the carrageenan-induced paw edema assay. Through the same method and concentrations, the aqueous extract ($AEMC_L$) shown a concentration-independent activity, in which 10 and 50% $AEMC_L$ was observed to last up until 7 hours of its administration in compared to 6 hours in 100% group (Zakaria et al. 2007e). Meanwhile, the *M. calabura* fruit extracts ($MEMC_{Fr}$ and $AEMC_{Fr}$) exhibited a dose-dependent inhibition in the carrageenan-induced localized edema test, with both doses (200 and 400 mg/kg) gave 24.5% and 44.2% ($MEMC_{fr}$) and 20.4% and 46.2% ($AEMC_{Fr}$) inhibition, respectively (Preethi et al. 2010). A similar trend was recognized in Preethi et al. (2012), with the 100, 200 and 300 mg/kg $MEMC_{fr}$ were subjected for the carrageenan-induced paw edema experiment. Recently, Kuo et al. (2014) has published a new way to combat inflammatory-related diseases by targeting the suppression of an extensive activation of neutrophils using drugs. Based on this study, a group of flavonoids including the 5-hydroxy-7-methoxyflavone, quercetin and (2S)-7-hydroxyflavanone demonstrated a potent inhibition against the formyl-*L*-methionyl-*L*-leucyl-*L*-phenylalanine-induced superoxide anion generated by the human neutrophils, with the IC_{50} values of 1.77 ± 0.7, 3.82 ± 0.46 and 4.92 ± 1.71 μM, respectively.

41.4.6 Anti-pyrectic Activity

Zakaria et al. (2007a) in his study revealed that the effects of chloroform extract of *M. calabura* leaves ($CEMC_L$) at 10, 50 and 100% concentration in the Brewer's yeast-induced pyrexia assay was less active in comparison to the standard drug, acetyl salicylic acid (ASA). However, he reported again by showing a different extract, the $AEMC_L$, to exhibited a stronger concentration-dependent anti-pyretic activity, in which is more effective than 100 mg of ASA. It is interesting to note that the activity was last longer up until 480 min (Zakaria et al. 2007e).

41.4.7 Cardioprotective Activity

Nivethetha et al. (2009) indicated the capability of *M. calabura* as cardioprotective agents by treated the isoprotarenol-induced myocardial infarction in rats with $AEMC_L$. The results demonstrated the 200 and 300 mg/kg $AEMC_L$ were significantly reduced the activity of the marker enzymes in the serum and heart tissues of the treated rats including the aspartate transaminase (AST), alanine transaminase (ALT), lactase dehydrogenase (LDH) and creatinine phosphokinase (CK).

41.4.8 Anti-ulcer Activity

It was observed that ethanol extract of *M. calabura* leaves ($EEMC_L$) exhibited a great anti-ulcer effect in a dose-dependent manner in 250 and 500 mg/kg ethanol-induced gastric ulcer in Sprague-Dawley rats. The gastric ulcer injuries area were reduced at 112.5 ± 2.11 and 95.08 ± 2.18 mm^2 in relative to the omeprazole-reference drug group (90.33 ± 2.02 mm^2) (Ibrahim et al. 2012). This noteworthy potency also were seen in Balan et al. (2013) work, which concluded a lower dosage of 50 mg/kg $MEMC_L$ was greatly reduced the gastric ulcer formation at 95%, a slightly higher than the 100 mg/kg ranitidine reference group (70% reduction). The probable cause of the protection effects of $MEMC_L$ is expected to be reversing the toxicity of ethanol and indomethacin induction, in which is comparable towards ranitidine protective effects.

41.4.9 Anti-diabetic Activity

In first experiment, Sridhar et al. (2011) reported the hypoglycemic effects of $MEMC_L$ with 24.8% reduction of blood glucose at 500 mg/kg dose after 6 hours of treatment in normal fasted rats. Due to this potential result, the active dose was further investigated in oral glucose tolerance test (OGTT). It was interestingly to highlight that pre-treatment of 500 mg/kg $MEMC_L$ successfully exhibited 27% reduction of blood glucose in comparison to the 37% reduction in 5 mg/kg glimepiride (reference drug) in OGTT test.

41.4.10 Anti-microbial Activity

Yasunaka et al. (2005) was the first group to discover the anti-bacterial property of *M. calabura* fruits and leaves. The dried samples (collected in State of Puebla and State of Veracruz, Mexico) were subjected for methanolic-extraction. The results from anti-bacterial assays disclosed a lower minimum inhibitory concentration (MIC) values in the fruits-treated samples in comparison to the leaves part against *Staphylococcus aureus* and *Escherichia coli,* with MIC values of 128 and 256 μg/ml; and 512 and 1024 μg/ml, respectively. Meanwhile, Zakaria et al. (2006b) demonstrated the potency of $MEMC_L$ towards *Shigella flexneri, Bacillus cereus, Proteus vulgaris, Aeromonas hydrophila, Kocuria rhizophila* and *S. aureus.* Furthermore, the $AEMC_L$ was reported to remarkably inhibit the growth of *Corynebacterium diphtheria, Staphylococcus epidermidis, P. vulgaris, and A. hydrophila* as compared to the $CEMC_L$ that can inhibited the *S. aureus* growth only. The following studies mentioned that a new batch of *M. calabura* sample collection (June, 2006) exerted the *Staphylococcus sp.* growth inhibition by the $AEMC_L$, $CEMC_L$ and $MEMC_L$ (Zakaria et al. 2007b), with the MIC values ranged between 1.25–5.00 μg/ml. Apart from the anti-bacteria efficacy determination, all parts of *M. calabura* were also tested for their anti-fungal potential. However, the $AEMC_L$, $CEMC_L$ and $MEMC_L$ extracts were recorded for a lower activity against *Candida albicans.* In contrast, Sibi et al. (2012) analyzed that the $MEMC_L$, $MEMC_B$ and $MEMC_{Fr}$ were significantly inhibited the fungal growth of *Aspergillus oryzae, Fusarium sp.* and *Penicillum sp.*

Moreover, Sufian et al. (2013) have isolated three known compounds from fraction F5 of ethyl acetate partition $MEMC_L$. These compounds including 2′,4′-dihydroxychalcone, galangin and 5-hydroxy-3,8-dimethoxyflavone were selectively inhibited the methicillin sensitive *S. aureus* (MSSA) and methicillin-resistant *S. aureus* (MRSA) growth at 200, 50, and 200 mg/ml, and 400, 100, and 400 mg/ml, respectively. Recently, Rajesh et al. (2014) extracted the $MEMC_L$, $CEMC_L$ and petroleum ether extract ($PEEMC_L$) of dried powdered roots and tested them against selected fungal pathogens. The resultant active $MEMC_L$ extract was furthered fractionated by petroleum ether and ethyl acetate to yield the fraction 21, in which contain a stigmasterol as the bioactive compound, with MIC value of 1 mg/ml against *Alternaria solani.* The molecular docking studies revealed the strong binding of stigmasterol to the cryptogein elicitors secreted by the fungal, hence preventing the pathogen attack towards the host, and thus explain the strong anti-fungal activity. In 2016, Buhian et al. examined the anti-microbial activity of 95% ethanol *M. calabura* extracts of stem and leaves collected in Paniqui, Philippines (2005), on several strains. Data recorded both leaves and stem extracts were potentially inhibited the *S. aureus* growth at 37.7 and 24.7 mm in diameter, respectively. This potent activity was the first report published for *M. calabura* stem.

41.4.11 Anti-nociceptive Activity

The anti-nociceptive properties of *M. calabura* leaves were determined by Zakaria et al. (2007a). The assay was set using the abdominal construction test with either the L-arganine, nitric oxide, or cyclic guanosine monophosphate pathway as the targeted parameter. All of the concentrations (27, 135 and 270 mg/kg $AEMC_L$)

tested have produced a significant anti-nociceptive activity for both early and late phases, except for 135 mg/kg. The capability to inhibit both phases is important as this mimicking the centrally acting drug, morphine. More evidences were collected by Zakaria et al. (2007b); interestingly the analgesics percentage of 10% $AEMC_L$ was equivalent to 0.8 mg/kg morphine, while 50% $AEMC_L$ was producing 68.4% of analgesia effects as equivalent to 100 mg/kg ASA effects. Further studies by Zakaria et al. (2008) recorded 5% and 50% concentrations of $AEMC_L$ were able to deliberate a potent anti-nociceptive activity as reference drugs (morphine and ASA). In addition, Yusof et al. (2011) investigated that fraction D from petroleum ether extract ($PEEMC_L$) (300 mg/kg) exerted the highest activity of early (66.2%) and late (81.4%) phases in formalin test, while Sani et al. (2012) reported the possible mechanism of this anti-nociceptive activity. Both studies concluded the potential mechanism for the specified bioactivity involving both the vanilloid receptors and glutamatergic systems. Moreover, Zakaria et al. (2014) reported the anti-nociceptive activity of $MEMC_L$ is related to involvement of the activation of μ-, δ-, ϒ-opioid and non-opioid including the adenosinergic, α-nor and β-adenergic receptors, regulation of ATP-sensitive K+ channel and inhibition of protein kinase C and bradykinin. Based on these scientific records, the traditional usage of this plant for the treatment of various ailments related to pain can be justified.

41.4.12 Anti-hypertensive Activity

Apart from cardioprotective action, this plant was claimed to possess the anti-hypertensive property. Shih et al. (2006) was reported to assess the anti-hypertensive potency of *M. calabura* water-soluble fraction from $MEMC_L$ extract by injecting the fraction into the femoral vein of animal model. Findings recorded that the water-soluble fraction successfully decreased the initial phase of the mean systemic arterial pressure (MSAP) and returned back to the pre-injection baseline. This action was observed to be happened in 10 minutes without any heart rate changes. In 2009, Shih et al. reported the butanol-soluble fraction (10–100 mg/kg) has induced a hypotensive and bradycardiac responses in a spontaneous-hypertension rat in comparison to the normo-tensive model. This mode of action was mediated by the activation of nitric oxide-dependent cGC/cGMP/PKG signaling pathways.

41.4.13 Antioxidant Activity

The earliest antioxidant study was performed by Eindond et al. (2004), using the Jamaican cherries collected from the Fruit and Spice Park, Florida, United State of America. The yielded cherries were subjected to the extensive methanol extraction, prior the partitioning step in resulting the water:methanol fractions before tested with the 2,2-diphenyl-1-picrylhydrazyl (DPPH)-radical scavenging assay. From this report,

M. calabura fruits gave a low IC_{50} value of 6.5 ± 0.6 μg/ml, which is comparable to the standard ascorbic acid (7.9 ± 0.5 μg/ml). For the *M. calabura* leaves fraction, Zakaria (2007) reported that $AEMC_L$ was having an antioxidant capacity of 94.8 ± 1.14% (DPPH) and 83.40 ± 2.05% (superoxide anion radical scavenging). A series of antioxidant assays also recorded ripened fruits of *M. calabura* exerted the highest total phenolic content (TPC) in methanol (148.6 ± 0.028 mg GAE/100 fw) than the other extracts. The higher content of TPC was translated into an antioxidant IC_{50} values at 90.0 ± 0.04 μg/ml (DPPH), 79.2 ± 0.04 μg/ml (superoxide anion scavenging assay), 49.98 ± 0.20 μg/ml (hydroxyl radical scavenging; HRS assay) and 187.0 ± 0.60 μg/ml (nitric oxide; NO radical scavenging assay). In addition, Vijayanand and Thomas (2016) have published positive correlations between TPC and antioxidant activity of *M. calabura* fruits with the value of 3.8 mg (TPC), 70% (FRAP assay) and 33.3% (CUPRAC assay). For the stem part, Buhian et al. (2017) indicated that 4 mg/ml of 95% ethanol extract exhibited 90% of DPPH inhibition in relative to the standard gallic acid. Meanwhile, Rotta et al. (2017) determined the positive antioxidant property in *M. calabura* peel extract, which is proportional to its high TPC content. Through the depth studies evidence, Pereira et al. (2018) stated that the high antioxidant effects were contributed by the high flavonoids content in the fruits, including catachin, gallocatechin, epigallacatechin, naringenin and quercetin. In conclusion, based on these remarkable scientific findings, this plant has confirmed to possess various beneficiary effects to the human body.

41.5 Traditional Uses

Generally, various parts (leaves, fruits, roots and barks) of *M. calabura* were used by the traditional folklore in treating various ailments and diseases. However, there is limited information available on *M. calabura* traditional uses as in Malaysia as this plant is undervalued or neglected (Zakaria et al. 2006a). Previously, the leaves were taken as infusion or decoction and were believed to be medicinally useful to reduced prostate gland swelling and gastric ulcer among Peruvian folks. This infused water was also used for the treatment of fever and headache as reported in Morton (1987). In the same community, the flowers and barks of *M. calabura* were used to relieve the lower extremities swelling and as antiseptic (Zakaria et al. 2006a, 2007c). Meanwhile, the Mexican applied these plants onto measles scars and mouth pimples in aiding the healing process as well as to relieve the stomachache (Yasunaka et al. 2005). In Philippines, the flowers have been widely used to lessen the headache and fever. On the other notes, the roots part has been applied as an abortifacient and emmenogogue, particularly in Malay and Vietnamese culture, respectively (Su et al. 2003). Despite claimed as a traditional medicinal plant, its cherry fruits are widely sold in Mexican market and freshly eaten. The fruits are also used as the main ingredient in making tarts and jam. While the flower infusion is drink as hot or cold beverage, or as tranquillizer and tonic (Zakaria et al. 2007d; Kaneda et al. 1991; Perez-Arbealaez 1975).

41.6 Conclusion

Even though numbers of phytochemicals have been isolated and characterized from all parts of *M. calabura*, their medicinal effectiveness over the traditional claims are not scientifically prove and studied in-depth just yet. The specific pharmacological activities and mechanism underlying the event remains unsolved. Thus, more researches are in needed to explore and investigate new candidates from *M. calabura* towards the discovery and drug development in the future.

References

Balan T, Mohd Sani MH, Suppaiah V, Mohtarrudin N, Suhaili Z, Ahmad Z, Zakaria ZA (2013) Antiulcer activity of Muntingia calabura leaves involves the modulation of endogenous nitric oxide and nonprotein sulfhydryl compounds. Pharm Biol 52:410–418

Bayer C, Chase MW, Fay MF (1998) Muntingiaceae, a new family of dicotyledons with Malvalean affinities. Taxon 47(1):37–42

Blench R (2008) A history of fruits on the southeast Asian mainland. Ed: Toshiki Osada and Akinori Uesugi. In: Occasional paper 4: linguistics, archeology and the human past. Indus Project Research Institute for Humanity and Nature, Kyoto. ISBN 978-4-902325-33-1

Buhian WPC, Rubio RO, Valle DLJ, Martin-Puzon JJ (2016) Bioactive metabolite profiles and antimicrobial activity of ethanolic extracts from *Muntingia calabura* L. leaves and stems. Asian Pac J Trop Biomed 6(8):682–685

Buhian WPC, Rubio RO, Puzon JJM (2017) Chromatographic fingerprinting and free-radical scavenging activity of ethanol extracts of Muntingia calabura L. leaves and stems. Asian Pacific Journal of Tropical Biomedicine 7(2): 139–143

Chen JJ, Lin RW, Duh CY, Huang HY, Chen IS (2004) Flavones and cytotoxic constituents from the stem bark of *Muntingia calabura*. J Chin Chem Soc 51:665–670

Chen JJ, Lee HH, Duh CY, Chen IS (2005) Cytotoxic chalcones and flavonoids from the leaves of *Muntingia calabura*. Planta Med 71:970–973

Chen JJ, Lee HH, Shih CD, Liao CH, Chen IS, Chou TH (2007) New dihydrochalcones and anti-platelet aggregation constituents from the leaves of *Muntingia calabura*. Planta Med 73:572–577

Einbond LS, Reynertson KA, Luo XD, Basile MJ, Kennelly EJ (2004) Anthocyanin antioxidants from edible fruits. Food Chem 84:23–28

Gomathi R, Anusuya N, Manian S (2013) A dietary antioxidant supplementation of Jamaican cherries (*Muntingia calabura* L.) attenuates inflammatory related disorders. Food Sci Biotechnol 22:787–794

https://commons.wikimedia.org/w/index.php?curid=2446203

https://commons.wikimedia.org/w/index.php?curid=2446210

https://commons.wikimedia.org/w/index.php?curid=8267519

Ibrahim IAA, Abdulla MA, Abdelwahab SI, Al-Bayaty F, Abdul Majid N (2012). Leaves extract of *Muntingia calabura* protects against gastric ulcer induced by ethanol in Sprague–Dawley rats. J Clin Exp Pharmacol S5:004. doi:https://doi.org/10.4172/2161-1459.S5-004. [Epub ahead of print]

Jovanovic-Malinovska R, Kuzmanova S, Winkelhausen E (2013) Oligosaccharide profile in fruits and vegetables as sources of prebiotics and functional foods. Int J Food Prop 17:949–965

Kaneda N, Pezzuto JM, Soejarto DD, Kinghorn AD, Farnsworth NR, Santisuk T, Tuchinda P, Udchachon J, Reutrakul V (1991) Plant anticancer agents, XLVIII. New cytotoxic flavanoids from *Muntingia calabura* roots. J Nat Prod 54:196–206

Krishnaveni M, Dhanalakshmi R (2014) Qualitative and quantitative study of phytochemicals in *Muntingia Calabura* L. leaf and fruit. World J Pharm Res 3(6):1687–1696

Kuo WL, Liao HR, Chen JJ (2014) Biflavans, flavonoids, and a dihydrochalcone from the stem wood of *Muntingia calabura* and their inhibitory activities on neutrophil pro-inflammatory responses. Molecules 19:20521–20535

Mahmood ND, Nasir NLM, Rofiee MS, Tohid SFM, Ching SM, Teh LK, Salleh MZ, Zakaria ZA (2014) *Muntingia calabura*: a review of its traditional uses, chemical properties, and pharmacological observations. Pharm Biol 52(12):1598–1623

Mohamad Yusof MI, Salleh MZ, Teh LK, Ahmat N, Nik Azmin NF, Zakaria ZA (2013) Activity-guided isolation of bioactive constituents with antinociceptive activity from *Muntingia calabura* L. leaves using the formalin test. Evid Based Complement Alternat Med 715074:1–9

Morton JF (1987) Jamaica cherry. In: Morton JF (ed) Fruits of warm climates, vol 2. Timber Press, Miami, pp 65–69

Nivethetha M, Jayasari J, Brindha P (2009) Effects of *Muntingia calabura* L. on isoproterenol-induced myocardial infarction. Singap Med J 50:300–302

Nshimo CM, Pezzuto JM, Kinghorn AD, Farnsworth NR (1993) Cytotoxic constituents of Muntingia calabura leaves and stems collected in Thailand. Int J Pharmacogn 31(1):77–81

Pereira GA, Arruda HS, de Morais DR, Eberlin MN, Pastore GM (2018) Carbohydrates, volatile and phenolic compounds composition and antioxidant activity of calabura (*Muntingia calabura* L.) fruit. Food Res Int 108:264–273

Perez-Arbealaez E (1975) In: Salazar H (ed) Plants Medicinales y Venenosas de Colombia. Medellin, Colombia, p 192

Preethi K, Vijayalakshmi N, Shamna R, Sasikumar JM (2010) In vitro antioxidant activity of extracts from fruits of *Muntingia calabura* Linn. from India. Pharm J 2:11–18

Preethi K, Premasudha P, Keerthana K (2012) Anti-inflammatory activity of *Muntingia calabura* fruits. Pharm J 4:51–56

Ragasa CY, Tan MC, Chiong ID, Shen CC (2015) Chemical constituents of *Muntingia calabura* L. Der Pharma Chemica 7(5):136–141

Rajesh R, Jaivel N, Marimuthu P (2014) Antifungal metabolite from *Muntingia calabura* root against early leaf blight of tomato. J Med Plant Res 8(17):646–656

Rotta EM, Haminiuk CWI, Maldaner L, Visentainer JV (2017) Determination of antioxidant activity and phenolic compounds of *Muntingia calabura* Linn. peel by HPLC-DAD and UPLC-ESI-MS/MS. Int J Food Sci Technol 52:954–963

Sani MH, Zakaria ZA, Balan T, Teh LK, Salleh MZ (2012) Antinociceptive activity of methanol extract of *Muntingia calabura* leaves and the mechanisms of action involved. Evid Based Complement Alternat Med 890361(1):1–10

Shih CD (2009) Activation of nitric oxide/cGMP/PKG signalling cascade mediates antihypertensive effects of *Muntingia calabura* in anesthetized spontaneously hypertensive rats. Am J Chin Med 37:1045–1058

Shih CD, Chen JJ, Lee HH (2006) Activation of nitric oxide signalling pathway mediates hypotensive effect of *Muntingia calabura* L. (Tiliaceae) leaf extract. Am J Chin Med 34:857–872

Sibi G, Naveen R, Dhananjaya K, Ravikumar KR, Mallesha H (2012) Potential use of *Muntingia calabura* L. extracts against human and plant pathogens. Pharm J 4:44–447

Sridhar M, Thirupathi K, Chaitanya G, Kumar BR, Mohan GK (2011) Antidiabetic effect of leaves of *Muntingia calabura* L., in normal and alloxan-induced diabetic rats. J Pharmacol 2:626–632

Su BN, Jung PE, Vigo JS, Graham JG, Cabieses F, Fong HH, Pezzuto JM, Kinghorn AD (2003) Activity-guided isolation of the chemical constituents of *Muntingia calabura* using a quinine reductase induction assay. Phytochemistry 63:335–341

Sufian AS, Ramasamya K, Ahmat N, Zakaria Z (2013) Isolation and identification of antibacterial and cytotoxic compounds from the leaves of *Muntingia calabura* L. J Ethnopharmacol 146:198–204

Talalay P (2000) Chemoprotection against cancer by induction of phase 2 enzymes. Biofactors 12:5–11

Varney J, Barrett J, Scarlata K, Catsos P, Gibson PR, Muir JG (2017) FODMAPs: food composition, defining cutoff values and international application. J Gastroenterol Hepatol 32(1):53–61

Vijayanand S, Thomas AS (2016) Screening of *Michelia champacca and Muntingia calabura* extracts for potential bioactives. Int J Pharma Sci Res 7:266–273

Yasunaka K, Abe F, Nagayama A, Okabe H, Lozada-Pérez L, Lopez-Villafranco E, Muñiz EE, Aguilar A, Reyes-Chilpa R (2005) Antibacterial activity of crude extracts from Mexican medicinal plants and purified coumarins and xanthones. J Ethnopharmacol 97:293–299

Yusof MM, Teh LK, Zakaria ZA, Ahmat N (2011) Antinociceptive activity of the fractionated extracts of *Muntingia calabura*. Planta Med 77:PF21

Zakaria ZA (2007) Free radical scavenging activity of some plants available in Malaysia. Iran J Pharmacol Ther 6:87–91

Zakaria ZA, Sulaiman MR, Jais AMM, Somchit MN, Jayaraman KV, Balakhrisnan G, Abdullah FC (2006a) The antinociceptive activity of *Muntingia calabura* aqueous extract and the involvement of L-arginine/nitric oxide/cyclic guanosine monophosphate pathway in its observed activity in mice. Fundam Clin Pharmacol 20:365–372

Zakaria ZA, Fatimah CA, Jais AMM, Zaiton HH, Henie EFP, Sulaiman MR, Somchit MN, Thenamutha M, Kasthuri D (2006b) The in vitro antibacterial activity of *Muntingia calabura* extract. J Pharmacol 2:290–293

Zakaria ZA, Kumar GH, Zaid SNH, Ghani MA, Hassan MH, Hazalin NAMN, Khamis MM, Devi GR, Sulaiman MR (2007a) Analgesic and antipyretic actions of *Muntingia calabura* leaves chloroform extract in animal models. Orient Pharm Exp Med 7:34–40

Zakaria ZA, Jais AMM, Mastura M, Mat Jusoh SH, Mohamed AM, Mohd Jamil NS, Rofiee MS, Sulaiman MR (2007b) In vitro antistaphylococcal activity of the extracts of several neglected plants in Malaysia. Int J Pharmacol 3:428–431

Zakaria ZA, Hassan MH, Aqmar NMNH, Abd Ghani M, Mohd Zaid SN, Sulaiman MR, Hanan Kumar G, Fatimah CA (2007c) Effects of various nonopioid receptor antagonists on the antinociceptive activity of *Muntingia calabura* extracts in mice. Methods Find Exp Clin Pharmacol 29:515–520

Zakaria ZA, Mustapha S, Sulaiman MR, Mat Jais AM, Somchit MN, Abdullah FC (2007d) The antinociceptive action of aqueous extract from *Muntingia calabura* leaves: the role of opioid receptors. Med Princ Pract 16:130–136

Zakaria ZA, Hazalin NAMN, Zaid SNHM, Abdul Ghani M, Hassan MH, Gopalan HK, Sulaiman MR (2007e) Antinociceptive, anti-inflammatory and antipyretic effects of *Muntingia calabura* aqueous extract in animal models. J Nat Med 61:443–448

Zakaria ZA, Somchit MN, Sulaiman MR, Jais AMM, Fatimah CA (2008) Effects of various receptor antagonists, pH and enzymes on *Muntingia calabura* antinociception in mice. Res J Pharmacol 2:31–37

Zakaria ZZ, Sufian AS, Ramasamy K, Ahmat N, Sulaiman MR, Abdul KA, Ahmad Z, Abdullah MNH (2010) In vitro antimicrobial activity of Muntingia calabura extracts and fractions. Afr J Microbiol Res 4(4):304–308

Zakaria ZA, Mohamed AM, Jamil NS, Rofiee MS, Hussain MK, Sulaiman MR, Teh LK, Salleh MZ (2011) In vitro antiproliferative and antioxidant activities of the extracts of *Muntingia calabura* leaves. Am J Chin Med 39:1–18

Zakaria ZA, Mohd Sani MH, Cheema MS, Abdul Kader A, Teh LK, Salleh MZ (2014) Antinociceptive activity of methanolic extract of *Muntingia calabura* leaves: further elucidation of the possible mechanisms. J Int Soc Complement Med Res 14:63–75

Correction to: *Irvingia gabonensis*: Phytochemical Constituents, Bioactive Compounds, Traditional and Medicinal Uses

Gustav Komla Mahunu, Lydia Quansah, Haroon Elrasheid Tahir, and Abdalbasit Adam Mariod

Correction to: Chapter 21 in: A. A. Mariod (ed.), *Wild Fruits: Composition, Nutritional Value and Products*, https://doi.org/10.1007/978-3-030-31885-7_21

In the original version of this book, Chapter 21 was inadvertently published with the corresponding author's name as "Lydiah Quansa". The author's name has now been corrected to "Lydia Quansah".

The updated online version of this chapter can be found at
https://doi.org/10.1007/978-3-030-31885-7_21

A. A. Mariod (ed.), *Wild Fruits: Composition, Nutritional Value and Products*,
https://doi.org/10.1007/978-3-030-31885-7_42

Index

A. A. Mariod (ed.), *Wild Fruits: Composition, Nutritional Value and Products*,
https://doi.org/10.1007/978-3-030-31885-7

H

The manufacturer's authorised representative in the EU is Springer Nature Customer Service Centre GmbH, Europaplatz 3, 69115 Heidelberg, Germany. If you have any concerns regarding our products, please contact ProductSafety@springernature.com

Printed and bound by CPI Group (UK) Ltd, Croydon, CR0 4YY
30/09/2026
02225651-0018